AF595477

Mining Safety and Sustainability II

Editors

Longjun Dong
Yanlin Zhao
Wenxue Chen

MDPI • Basel • Beijing • Wuhan • Barcelona • Belgrade • Manchester • Tokyo • Cluj • Tianjin

Editors
Longjun Dong
Central South University
China

Yanlin Zhao
Hunan University of Science and Technology
China

Wenxue Chen
University of Sherbrooke
Canada

Editorial Office
MDPI
St. Alban-Anlage 66
4052 Basel, Switzerland

This is a reprint of articles from the Topic published online in the open access journals *Energies* (ISSN 1996-1073), *Minerals* (ISSN 2075-163X), *Safety* (ISSN 2313-576X), *Sensors* (ISSN 1424-8220), and *Sustainability* (ISSN 2071-1050) (available at: https://www.mdpi.com/topics/Mining_Safety).

For citation purposes, cite each article independently as indicated on the article page online and as indicated below:

LastName, A.A.; LastName, B.B.; LastName, C.C. Article Title. *Journal Name* **Year**, *Volume Number*, Page Range.

ISBN 978-3-0365-6398-5 (Hbk)
ISBN 978-3-0365-6399-2 (PDF)

Contents

About the Editors

Longjun Dong

Longjun Dong, Ph.D, is a Professor and the Director of Department of Safety Science and Engineering, Central South University, China. His research interests include rock mechanics, rock acoustics, engineering seismicity, and applied geophysics in mining engineering. He obtained Ph.D., M.Sc., and B.Sc. in 2013, 2009, and 2007, respectively. He was appointed as Professor, Associate Professor, and Lecturer at Central South University in 2017, 2015, and 2013, respectively. He was invited as a research assistant at Australian Center for Geomechanics (ACG) from 2012 to 2013. He presided over 20 projects of the National Key Research and Development Program, International Cooperation and Exchange of the National Natural Science Foundation of China (NSFC), etc. He is the recipient of the NSFC for Excellent Young Scholars (2018), the Young Elite Scientist sponsored by the China Association for Science and Technology (CAST) (2016), the leading talent of Science and Technology Innovation of Hunan Province (2021), and the Distinguished Young Scholars Fund of Hunan Province (2018). He has 166 publications in peer-reviewed journals, with 4998 citations and 42 h-index in Google Scholar, and has 51 Chinese patents. He has obtained 10 First Prizes (and above) of provincial and ministerial level, as well as industrial science and technology awards. He is an individual ISRM Member and an IEEE Senior Member. He was invited to present over 10 Keynote Presentations and Invited Presentations at international conferences. He is a Topical Editor in Chief of Arabian Journal of Geosciences (Springer Nature); an Associate Editor of Journal of the Acoustical Society of America Express Letter, and Shock and Vibration; an Editorial Board Member of Safety Science (Elsevier), Soils and Foundations (Elsevier, the official journal of the Japanese Geotechnical Society), Scientific Reports (Springer Nature), and International Journal of Distributed Sensor Networks; an Editorial Advisory Board Member of Archives of Mining Sciences (established by the Polish Academy of Sciences since 1956), and a guest editor of Induced Seismicity in Scientific Reports. In 2020 and 2021, he was selected into the list of "Highly Cited Chinese Researchers (Elsevier)" and the "Top 2% of Scientists in the World (Stanford University)".

Yanlin Zhao

Yanlin Zhao, professor and doctoral supervisor, is the Deputy Dean of the School of Resources Environment and Safety Engineering, Hunan University of Science and Technology. He mainly studies rock mechanics and rock ground control, multi-field coupling theory and engineering response of deep mining rock mass. He presided over and completed 2 general programs of National Natural Science Foundation of China, more than 20 scientific research projects such as provincial and ministerial level and enterprise cooperation, and participated in the basic theoretical research on the mechanism and prevention of coal mine water inrush (973 Program) and the key program of the National Natural Science Foundation of China. He is a visiting scholar at the University of Arizona, a young backbone teacher in ordinary colleges and universities in Hunan Province, an outstanding expert in Xiangtan City, an outstanding doctoral dissertation winner in Hunan Province, a Xiangjiang scholar at Hunan University of Science and Technology, a director of the Hunan Rock Mechanics and Engineering Society, and a national technology forecasting expert of the Ministry of Science and Technology, an excellent master thesis advisor in Hunan Province, and a guest editor-in-chief of the international SCI journals "Frontiers in Earth Science" and "Geofluids". He was selected in the 2020 Elsevier Highly Cited Chinese Scholars and Top 2% Global Top Scientists in the field of "Mining Engineering". He has published more than 100 academic papers, 3 academic monographs, of which more than 80 papers were indexed by SCI and EI, 6 papers in international Top journals, 9 papers in the world's top 1% ESI highly cited papers, 4 hot papers, 1 paper in "Frontrunner 5000" top articles in outstanding S&T journals of China (F5000), and 1 paper won the 2019 Excellent Paper of the Chinese Society of Rock Mechanics and Engineering—Chen Zongji Award.

Wenxue Chen

Wenxue Chen, Adjunct Professor of Sherbrook University, is Industrial Research Chair of Canadian Natural Science and Engineering Research Fund (NSERC) "Innovative Composite Fiber Materials Concrete Engineering Application Project" and the Principle Scientist of SFTec. His main research fields include safety technology and engineering, underground engineering, structural engineering and engineering application of new composite materials etc. He is a committee member of the Infrastructure Products Committee of the American Precast Association (NPCA-TIPC), a committee member of the American Composites Manufacturing Industry Council (ACMA), and an associate editor of the International Journal of Safety Science (IJSS). He has published 65 academic papers, of which more than 20 were indexed by SCI. And there are 25 authorized patents and won a number of provincial and ministerial awards.

Preface to "Mining Safety and Sustainability II"

The mining industry provides important support for development of the global economy and society. With further economic development and the continuous improvement in people's living standards, the demand for mineral resources is gradually increasing. However, with the increasing depth of mining, the difficulty of mining also increases, thus placing higher requirements for mining equipment and safety. Detection of the mineral exploration environment, improving the safety of mining operations, developing intelligent tunneling equipment, and ensuring the coordination of human–machine–environment in the mining production system have become necessary conditions for promoting the transformation of mining methods and achieving the goal of "double carbon". Exploring safe, efficient, and sustainable methods for mining mineral resources has become urgent. In the past, scholars around the world have carried out extensive research work on safe and sustainable mining, but because mining operations are resource-consuming and characterized by productive behaviors, applying the principles of safety and sustainability to mining itself is challenging.

It is important to reduce the risk of mining accidents and disasters, enhance the safety of mining operations, and improve the efficiency and sustainability of mineral resources development. Therefore, it is necessary to extract useful knowledge from the process of mineral resources development and resource integration by means of experimental techniques, simulation methods, data mining, theoretical innovation, technological development, and other methods. This will provide further theoretical and technical support for guiding the normative, green, safe, and sustainable development of the mining industry. This Reprint aims to present new research and recent advances in the safety and sustainability of mining.

Last but not least, thanks to all authors for their excellent and meaningful contributions to this topic. We welcome more experts and scholars to present your new ideas regarding safety mining, sustainable mining, mineral resource management, technology of intelligent mining, research and development of intelligent mining equipment, sustainable development, new mining methods, and related topics.

Longjun Dong, Yanlin Zhao, and Wenxue Chen
Editors

Article

Numerical Study on Characteristics of Bedrock and Surface Failure in Mining of Shallow-Buried MCS

Guangchun Liu [1,2], Wenzhi Zhang [1,*], Youfeng Zou [1], Huabin Chai [1,*] and Yongan Xue [1]

1 School of Surveying and Land Information, Henan Polytechnic University, Jiaozuo 454000, China; liugch168@home.hpu.edu.cn (G.L.); zouyf@hpu.edu.cn (Y.Z.); xueyongan@tyut.edu.cn (Y.X.)
2 School of Resouce and Civil Engineering, Liaoning Institute of Science and Technology, Benxi 117004, China
* Correspondence: zhangwenzhi@hpu.edu.cn (W.Z.); chaihb@hpu.edu.cn (H.C.)

Abstract: Coal is one of the important energy sources for industry. When it is mined, it will cause the destruction of bedrock and surface. However, it is more severe in mining shallow-buried multi coal seams (SBMCS). To better reveal the characteristics of the bedrock and surface damage, we have carried out a theoretical analysis, as well as used numerical simulations and field monitoring methods to study the surface and bedrock damage caused by the mining of SBMCS. The characteristics of bedrock and surface failure structure, settlement, and stress distribution were studied and analyzed. The findings show that the collapsed block, formed by the rupture of the overlying stratum, interacts with the surrounding rock to form large cavities and gaps, and the stress concentration occurs between them. The maximum downward vertical concentration stress is about 9.79 MPa. The mining of the lower coal seam can lead to repeated failure of the upper bedrock and goaf. The settlement of bedrock presents gradient change, and the settlement of upper bedrock is large, about 8.0 m, and the maximum settlement is 8.183 m, while that of lower bedrock is small and about 3.5–4.0 m. The weak rock stratum in the bedrock is crushed by the change stress of repeated mining, and formed a broken rock stratum. The cracks in the bedrock develop directly to the ground. On the ground, tensile cracks, compression uplift, stepped cracks, and even collapse pits are easy to cause in mining SBMCS. Affected by repeated mining, the variation of surface vertical stress is complex and disorderly in the middle of the basin, and the variation of horizontal stress is mainly concentrated on the edge of the basin. The maximum stress reaches 100 KPa, and the minimum stress is about 78 KPa. Through theoretical analysis and discussion, the size of the key blocks is directly related to the thickness and strength of the rock stratum.

Keywords: coal mining; multi coal seam; CDEM; numerical simulation; bedrock failure; surface damage

Citation: Liu, G.; Zhang, W.; Zou, Y.; Chai, H.; Xue, Y. Numerical Study on Characteristics of Bedrock and Surface Failure in Mining of Shallow-Buried MCS. *Energies* **2022**, *15*, 3446. https://doi.org/10.3390/en15093446

Academic Editors: Longjun Dong, Yanlin Zhao, Wenxue Chen and Vamegh Rasouli

Received: 9 April 2022
Accepted: 6 May 2022
Published: 9 May 2022

1. Introduction

Coal is a non-renewable fuel energy, which is gradually replaced by renewable energy. However, in today's world, it is still one of the most important industrial energy sources. Coal resource exploitation still plays a vital role in the energy planning and management of all countries in the world. The safe and efficient mining of coal is related to the national industrial and economic development. After coal mining, the bedrock layer and surface are damaged, affecting and threatening the safety of working face and surface. Coal fields are widely distributed in China, and the occurrence conditions of coal seams are different [1,2]. Qinshui coalfield, in Shanxi Province, is the basin coalfield with the most abundant coal in North China. It belongs to the coal-bearing strata of the Benxi Formation, Taiyuan Formation, and Shanxi formation developed in the late Paleozoic [3,4]. It is a typical geological condition for the occurrence of multiple coal seams. The mining thickness of coal seams in Qinshui coalfield varies greatly. According to [1], coal seam in Benxi

Formation is thick in the Northeast and West, and thin in the central and South; coal seam in Taiyuan formation is thick in the East and thin in the West; coal seam in Shanxi formation is the thickest coal bearing group, which is thick in the North, thin in the South, thick in the East, and thin in the West. The geological conditions of shallow burial multi coal seams (SBMCS) refer to the geological conditions in which the coal seams are buried in a shallow depth, the composition and structure of the bedrock are relatively simple, and there are three or more coal seam occurrence geological conditions. The mining of SBMCS is affected by factors such as coal seam depth, coal seam thickness, bedrock and interlayer lithology, geological structure, and fault [5]. Many scholars have conducted in-depth research on multi coal seam (MCS) mining, under different geological mining conditions, from the perspectives of bedrock failure structure characteristics, roof structure characteristics, rock movement law and stability, and fracture development characteristics, and they achieved fruitful research results. Zhang [6], Lin [7] and others studied the characteristics of overburden movement and fracture development, affected by repeated mining in MCS, by using a similar simulation method, and they determined the relationship between fracture development height, overburden lithology, and thickness. The fracture development, rock fracture, and water loss of MCS mining are studied. The stress field distribution of bedrock and surface needs to be further discussed. Qin [8], Dong [9], Yang [10], and others used a combination of various methods to study the structural characteristics of mining MCS and analyzed roof instability and working face pressure, the calculation method of coal mine support resistance and the breaking impact effect of roof structure, and obtained the laws of MCS bedrock failure and roof structure characteristics. Based on the study of coal seam group, the stability and rock failure mechanical characteristics are analyzed. The bedrock and surface damage caused by SBMCS mining need to be further studied. Chen [11], Yan [12], Dong [13], and others studied the surface movement and deformation characteristics, surface dynamic prediction, bedrock damage, and migration law of SBMCS mining by means of numerical simulation, similarity test, and field monitoring. They found the relationship between the surface movement, bedrock damage, and bedrock fracture development of SBMCS and the bedrock buried depth and lithology. Their research on SBMCS is more comprehensive, through on–site monitoring, numerical simulation and theoretical analysis. However, they are aimed at the geological conditions with thick loose layers, and most of the research areas are located in the desert area of Northwest China. According to the geological conditions of Qinshui coalfield, further research is needed. Xie [14] et al. studied the basic roof structure characteristics, movement, and deformation law of MCS mining, providing the basis for the mechanical calculation of bedrock movement and deformation. The equivalent basic roof structure and activity law of overburden between MCS are studied. The mechanical analysis of surface failure and the fracture structure of bedrock in MCS are not studied in detail. Zhao [15] and others used the discrete element UDEC numerical simulation software and mining subsidence prediction software to simulate and study the surface residual settlement after mining, which provided a theoretical feasibility demonstration for coal mining under a high-speed railway. Wu [16] et al. Used FLAC3D to simulate the conditions of different bedrock heights and determined the impact of repeated mining on the height of "two zones". UDEC is a discrete element method, which can better simulate the characteristics of an overburden collapse zone caused by mining, and it has certain limitations on the simulation of initial stress before mining and the interaction between broken rock and soil. FLAC belongs to the finite element method and adopts the algorithm based on the Lagrange continuity equation. It also has certain limitations in the simulation of collapse and cracking. The above research shows that the mining of SBMCS is a complex problem related to geological conditions. The bedrock and surface failure characteristics and stress distribution law of SBMCS mining, under the geological conditions of Qinshui coalfield, need to be studied further. In numerical simulation, the discreteness of bedrock failure and the continuity of initial stress after mining cannot be well treated by using the finite element or discrete element method, respectively. Discrete element and finite element methods are widely used in geotechnical engineering and underground engineering. Then, before mining, the rock stratum is a continuous medium, and it becomes a discrete block structure after mining. Therefore, a new method of coupling a finite element and discrete element, which is Continuous Discontinuous Element Method (CDEM), can solve this problem [17,18].

This paper attempts to use the CDEM and Generalized Digital Emulation Manufacturer(GDEM) software numerical simulation platform to study the bedrock failure characteristics, structural form, and bedrock stress distribution characteristics of SBMCS [19]. We take Majiazhuang coal mine in the southeast of Qinshui coalfield as the engineering background. It provides a numerical analysis basis for SBMCS mining in the southeast of Qinshui coalfield, and it provides a theoretical basis for safe mining and green mining of mines with MCS geological conditions. The remainder of this manuscript is organized as follows. In Section 2, Methodology, we mainly expound the mechanical analysis of key blocks and CDEM algorithm. In Section 3, Materials and Methods, we mainly describe the engineering background of the engineering research area, modeling, and simulation process. In Section 4, Results, we analyzed the failure characteristics of bedrock, surface stress distribution, and settlement, according to the results of numerical simulation and field monitoring. In Section 5, Discussion, we obtain bedrock failure characteristics and processes, and we discuss the impact of water on mining and the CDEM numerical simulation method. In Section 6, Conclusions, we describe the research results for the research content, as well as the conclusions obtained.

2. Methodology

2.1. The Calculation of Key Stratum Failure

After the coal seam is mined, the key layer in the rock stratum can support the overlying rock. However, the key layer will collapse when it reaches the ultimate bearing capacity [20]. The key block structure is formed after collapse, which interacts with the rock stratum and floor to form a stable structure. According to relevant literature research [21], it is found that, after the coal seam is mined, the key block plays a very important role. It forms a relatively stable structure to support the rock mass on it. We need to establish a mechanical model to deeply study and discuss the process.

Mechanical analysis shall be carried out from the initial stage of key block collapse. With the continuous advancement of the working face, the direct roof of the upper coal seam reaches the limit span, resulting in separation and collapse [22]. The key stratum bears the upper bedrock load. When the key stratum reaches the ultimate bearing capacity, cracking and disconnection will occur at both ends, as well as midspan of the rock beam, in the key stratum [23]. Under the action of friction and extrusion force between key blocks, the blocks produce rotary motion and finally reach a static equilibrium state, as shown in Figure 1.

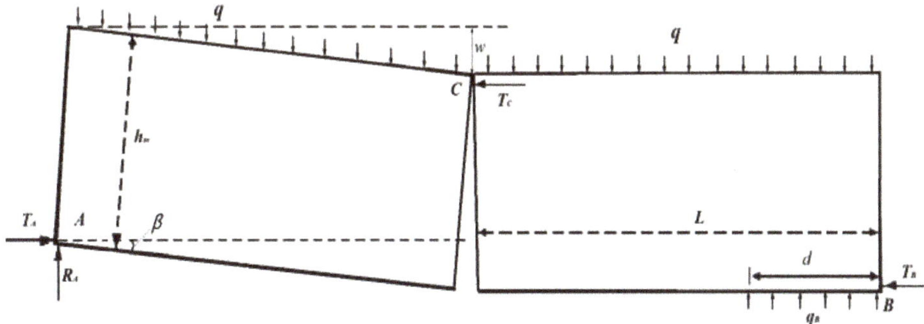

Figure 1. Force analysis of the basic structure of the key stratum failure.

After bedrock collapse, the key blocks interact with the coal seam floor and surrounding rock to form a relatively stable structure. According to the "masonry beam" theory [24,25], there is a height difference between the horizontal thrusts T_A and T_B on both sides, and the compressive stress passes through the overhanging fault block to form the rear arch. The compression force T_C in the fault block increases with the increase in the overhanging area of the block, where the two blocks are squeezed and broken, and the friction force of the block decreases. When the friction force between the two blocks is less than the shear force on the contact surface, the broken block slides down and sinks along the contact surface, resulting in rotational deformation and instability. The key stratum fractured for the first time, and the formed key block constituted an asymmetric three–hinge arch structure [26,27].

As for the geometry in Figure 1, assuming that the rotation angle of the key block is β, the subsidence amount is w, the length of the key block is equal to L, the upper part of the block is uniformly distributed with load q, and the generated horizontal thrusts are T_A and T_B, it functions with points A and B. The thickness of the block is h_m, point A is subjected to frictional force R_A, and point B is subjected to a supporting force q_B of upward distance d. Point C is the hinge point of the fractured block, which transmits the internal force T_C between the two block systems. It is a force per unit length (1 m).

According to the overall equilibrium state of the two key block structures, the equilibrium equation can be obtained:

$$\left.\begin{array}{l}\sum F=0\\ \sum M=0\end{array}\right\} \tag{1}$$

where F is the force in different directions, N;

M is the moment generated by each force, N·m;

According to the force analysis of the key blocks, as shown in Figure 1 and Formula (1), T_C can be calculated that:

$$T_C=\frac{qL(L\cos\beta+h_m\sin\beta)}{2(h_m\cos\beta-w)} \tag{2}$$

where T_C is the internal force between the two blocks, N/m;

q is the load above the block, MPa;

β is the tilting rotation angle of the block, °;

h_m is the height of the block rock beam, m;

W is the subsidence caused by block rotation, m.

According to the geometric relationship, as shown in Figure 1, we can be known that

$$\left\{\begin{array}{l}\sin\beta=\frac{w}{L}\\ \cos\beta=\frac{\sqrt{L^2-w^2}}{L}\end{array}\right. \tag{3}$$

Substituting Formula (3) into Formula (2), the relationship between thrust T_C and settlement w can be calculated:

$$T_C=\frac{qL}{2}\frac{\sqrt{L^2-w^2}+\frac{h_m}{L}w}{\frac{h_m}{L}\sqrt{L^2-w^2}-w} \tag{4}$$

In the initial stage of key stratum failure, when periodic failure occurs, the rotation angle β is usually small, and the ratio of rock layer thickness to failure distance (h_m/L) is usually small and can be ignored in the calculation process. We can get:

$$\left.\begin{array}{l}\sqrt{L^2-w^2}\approx L\\ \frac{h_m}{L}\approx\left(\frac{1}{50}\sim\frac{1}{20}\right)\end{array}\right\} \tag{5}$$

Therefore, Formula (4) can be approximately simplified to

$$T_C = \frac{qL}{2}\frac{L}{h_m - w} \tag{6}$$

According to the relevant knowledge of mechanics, rock beam collapse is usually caused by tensile failure in the middle, so the relationship between the initial collapse distance L_{ini} and the tensile strength of rock stratum $\sigma_{\max}$ can be calculated as follows [11,24].

$$L_{ini} = h_m\sqrt{\frac{\sigma_{\max}}{q}} \tag{7}$$

where L_{ini} is the initial collapse distance of bedrock, m;

q is the load applied on the collapse block, Mpa;

$\sigma_{\max}$ is the maximum tensile strength of broken block, Mpa.

From the above analysis, it can be concluded that the size of the key blocks is directly related to the thickness and strength of the rock stratum, as well as the load on the bedrock.

The failure of key stratum has a direct relationship with the properties of the rock formation, and the failure distance to reach the initial failure can be calculated, as shown in Formula (7). Therefore, the horizontal thrust T_C can be expressed by the tensile strength of the rock stratum $\sigma_{\max}$, and the relationship between h_m and L can be obtained.

$$T_C = \frac{1}{2}\frac{h_m^2}{h_m - w}\sigma_{\max} \tag{8}$$

The large tensile strength and thickness of bedrock make the thrust T_C between blocks larger. Therefore, the friction between collapses is large, and it is not easy to slip. When the horizontal thrust reaches the limit, the indirect contact surface of the blocks is broken, resulting in slip and collapse. This also shows that the length and thickness of the collapsed block, as well as the tensile strength of the block itself, affect the size of the space under it and even the duration of the cavity.

2.2. CDEM Algorithm

Before coal mining, each rock mass layer and coal seam are considered to be rock layers with continuous properties. When the coal seam is mined, it first causes the separation and collapse of the direct roof, and then, with the advance of the working face, the basic roof separation and collapse form a discrete block structure. After the coal seam is mined out, the bedrock fractures are developed, which shows the discreteness of cracking. The collapsed blocks fall into the goaf to form a discrete block structure. Therefore, the CDEM, based on the Lagrange equation, should be selected as the numerical simulation method of coal mining. This method can not only calculate the deformation state of coal and rock mass but also simulate the plasticity, damage, and the fracture process of coal and rock mass after applying the Mohr Coulomb criterion of brittle fracture on the virtual contact surface.

CDEM algorithm is an algorithm for solving the explicit solution of the dynamic equation of fractured elements, according to the theory of the generalized Lagrange system [17,18]. The algorithm is based on the strict generalized Lagrange equation, as shown in Formula (9), combined with the global dynamic relaxation algorithm, to iteratively solve the explicit solution of the equation.

$$\frac{d}{dt}(\frac{\partial L}{\partial x_i}) + (\frac{\partial L}{\partial y_i}) = Q_i \tag{9}$$

where x_i, y_i are generalized coordinates;

L is the energy of the Lagrangian system;

Q_i is the work done by non-conservative force.

The basic structure of the CDEM algorithm is composed of two parts: blocks and interface. The block element is usually composed of quadrangles. It has only elastic deformation and no plastic deformation. The plastic deformation is concentrated on the virtual crack. For the direction of crack development and actual cracking, the theoretical direction is perpendicular to the direction of the maximum principal stress of the surrounding element, while the actual direction is the closest element boundary or along a diagonal direction [23]. Interface element is a constitutive model describing the relationship between adjacent blocks in the block discrete element, also known as contact surface. Three attributes, connecting two block elements and the normal vector of the contact surface, are used to represent the force between block elements. CDEM algorithm is the same grid shared by the Lagrangian element algorithm and the discrete element algorithm.

The core calculation formula of CDEM is shown in Formula (10) [28]:

$$M\{\ddot{u}(t)\} + C\{\dot{u}(t)\} + K\{u(t)\} = \{Q(t)\} \tag{10}$$

where M is quality matrix;

$\{\ddot{u}(t)\}$ is acceleration matrix;

C is damping matrix;

$\{\dot{u}(t)\}$ is velocity matrix;

K *is* stiffness matrix;

$\{u(t)\}$ is displacement matrix;

When the cell is not embedded before the medium cracks, the node only contains unbalanced force (only elastic force and external force); when the medium is cracked or embedded, the node force changes, and the corresponding parameters of node velocity and acceleration motion state are generated. CDEM algorithm couples continuous and discontinuous media algorithms, and it realizes the perfect unity of the finite element, block discrete element, and particle discrete element [29]. GDEM numerical simulation platform, developed based on CDEM algorithm, is independently developed by the Institute of mechanics, Chinese Academy of Sciences. It is a constitutive model based on CDEM method. By means of CPU/GPU and behavior acceleration, it can not only simulate the elastic, plastic, damage, and fracture processes of geological bodies and artificial materials under multi field coupling but also develop unit constitutive and contact constitutive by using C++ or JavaScript, according to actual needs. The platform software contains 14 built-in block constitutive models and 12 built-in interface constitutive models [19]. In this study, the elastic damage fracture constitutive model is adopted, and the mechanical parameters of rock stratum include density, elastic modulus, Poisson's ratio, tensile strength, and internal friction angle. The brittle damage fracture constitutive model is applied on the virtual interface, and the mechanical parameters of the interface include normal stiffness, tangential stiffness, cohesion, internal friction angle, and tensile strength [30].

3. Materials and Methods

3.1. Geological and Mining Conditions

The study area is located in the southeast of Qinshui coalfield and the east of Jincheng mining area, belonging to the edge of Qinshui Basin. Qinshui coalfield has simple geological structure conditions, a flat coal seam dip angle, multi coal seam occurrence, and strong regularity, and most of the hydrogeological conditions are not complex. The geological conditions of the mine are the most superior mining conditions in China. Geographical location of the study area is shown in Figure 2.

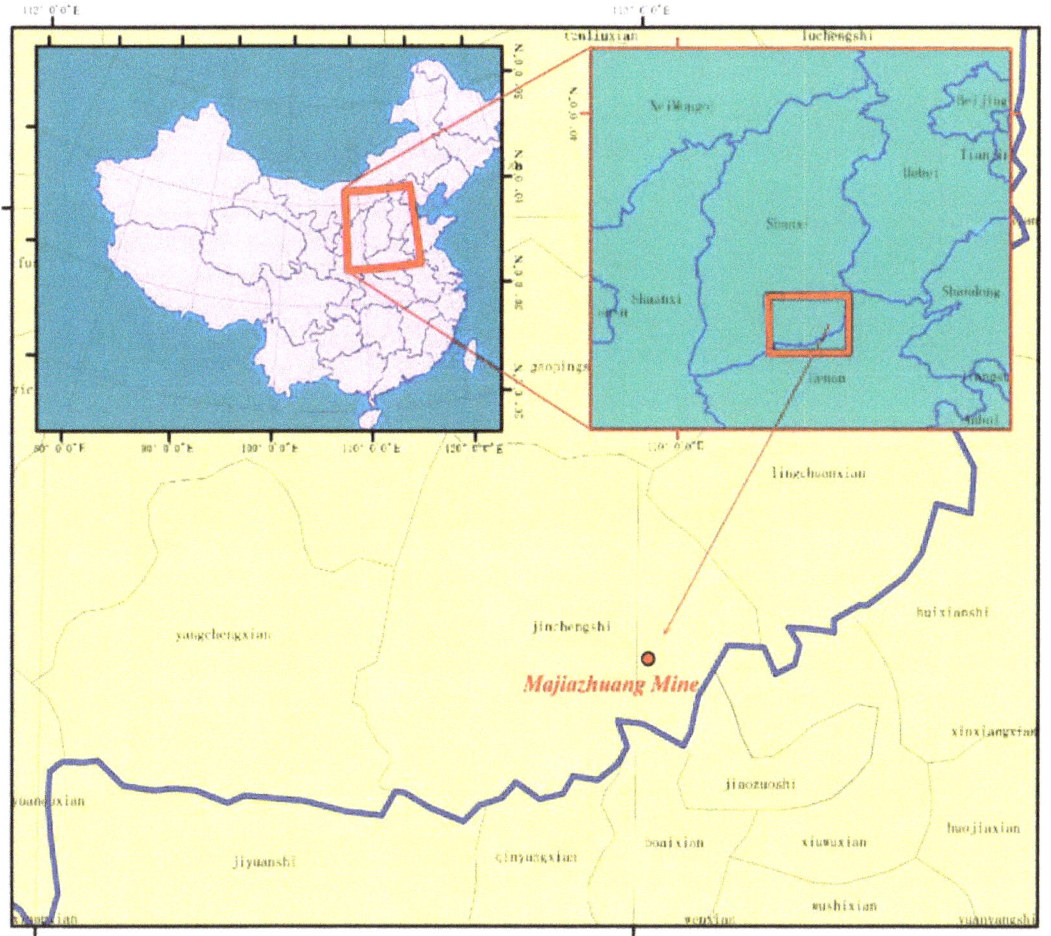

Figure 2. Geographical location of the study area.

The coal bearing strata of coal mines in this area are the Permian Shanxi Formation and Carboniferous Taiyuan formation, with a total of 13–15 layers of coal, with a total thickness of nearly 13 m. According to the analysis of the geological data of Majiazhuang coal mine, the stratigraphic structure is mainly sandstone and mudstone, including limestone and siltstone. There are three stable minable coal seams, of which the buried depth of 3# coal seam is 31.4 m, and the average coal thickness is 2.9 m; the buried depth of 9# coal seam is 84.6 m, and the average coal thickness is 1.0 m; the burial depth of 15# coal seam is 120.4 m, and the average coal thickness is 4.3 m. Other coal seams are unstable coal seams or coal lines, which are not minable or are partially minable. The coal seam dip angle of the mine is less than 3°, which is a nearly horizontal coal seam, belonging to the typical SBMCS occurrence condition on the southeast boundary of Qinshui coalfield, as shown in Figure 3.

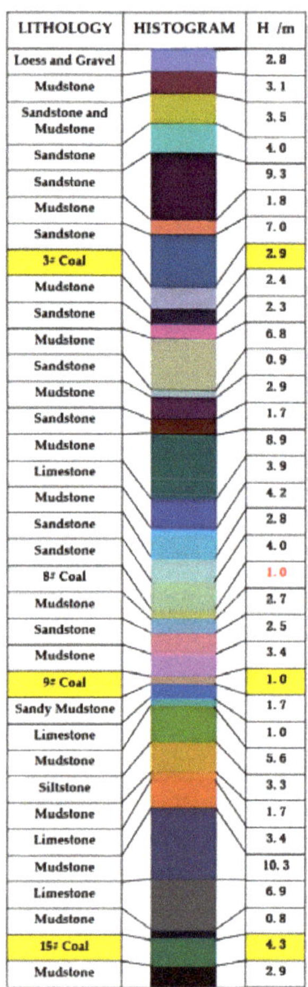

Figure 3. Comprehensive geological histogram of Majiazhuang mine.

3.2. Model and Mechanical Parameters

This simulation takes the geological conditions in the southeast of Qinshui coalfield as the background, and it establishes a 1:1 numerical simulation model. A 2D model of 35 coal and rock strata is designed, including seven groups of rock strata with different properties and three coal seams. The design dimension of the model is 127.5 m high and 600 m wide. The mining length of the simulated working face is 300 m. The mining thickness is designed according to the actual average mining thickness of each layer of coal. According to the boundary conditions of the model, the left and right boundaries are fixed by means of zero velocity in the X-axis direction; in the Y-axis direction, the bottom boundary is completely fixed, and the upper boundary is free, as shown in Figure 4.

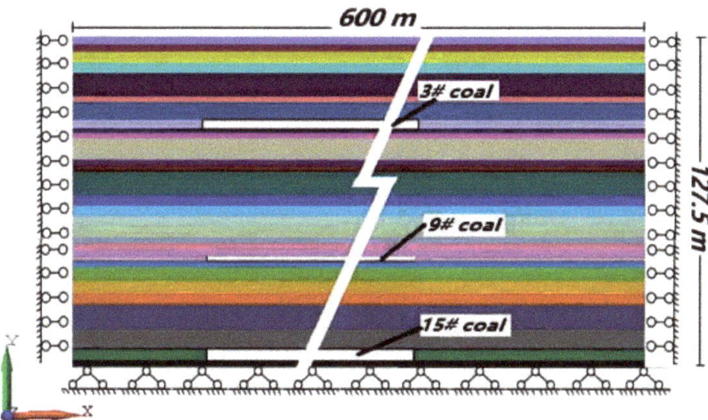

Figure 4. Schematic of model size and boundary conditions.

The initial in-situ stress of this simulation is calculated by using the linear elastic constitutive model, and the unbalance rate is set to 1×10^{-5}. The mechanical parameters required for calculation are shown in Table 1.

Table 1. Calculation parameters of rock mechanics.

LITHOLOGY	ρ (/kg/m^3)	E (/GPa)	μ	c (/MPa)	σ_t (/MPa)	φ (/°)
Loess and Sand Gravel	1900	0.015	0.30	0.02	0.60	17.0
Mudstone	2200	1.31	0.24	2.16	3.73	25.0
Sandstone and Mudstone	2440	2.23	0.25	3.80	4.25	30.0
Sandstone	2640	3.54	0.25	5.50	4.78	35.0
Limestone	2800	2.66	0.23	6.20	7.98	35.0
Sandy Mudstone	2613	0.90	0.26	5.50	4.78	30.0
Siltstone	2540	3.16	0.35	2.65	1.36	31.0
Coal	1400	1.00	0.29	1.25	1.60	25.0

3.3. Simulation Scheme and Steps

This time simulates the mining process of three stable and minable coal layers in the southeast of Qinshui coalfield, studies the characteristics of bedrock failure and surface movement, and analyzes the characteristics of bedrock stress distribution. The numerical calculation is divided into two stages and three times of coal seam excavation. In the initial geostress calculation stage, the virtual mass calculation method is adopted. During the simulation, the coal and rock strata are discretized as a whole. The strength and stiffness of the contact surface are inherited from the cell strength, and the strength of the contact surface is set as 10% of the cell strength to obtain the initial geostress state under the action of gravity. The initial strain of coal is converted into the constitutive model of brittle rock after the initial strain of the coal-softening element is converted into the constitutive model of brittle rock. In the stage of the coal seam excavation solution, in order to better simulate the bedrock breaking process, the displacement and change gradient within the calculation range are zeroed. Then, calculate the coal seam excavation stage, excavate in three times, and set it as "*None*" model, according to the designed coordinate range of coal seam goaf, to realize coal seam excavation. After each coal seam excavation, the solution time step is set to 10,000. In order to master the variation law of surface settlement, monitoring points with an interval of 10 m are set on the surface. When simulating the excavation of different coal seams, the characteristics of bedrock fracture and surface failure are recorded.

The numerical model is corrected and adjusted by using surface monitoring data. During 3# coal mining, surface safety monitoring was conducted, and monitoring points were arranged along the advancing direction of the working face to monitor surface settlement.

4. Results

4.1. Analysis of Bedrock Failure Characteristics

This simulation adopts a 2D plane model. After the excavation calculation of three layers of coal, the failure characteristics of bedrock are obtained, as shown in Figure 5. Figure 5a shows that, when mining the upper coal seam, the coal seam floor near the coal pillar at the edge of the goaf is uplifted, and the floor is cracked and damaged. The key bedrock layer above the edge of the goaf breaks down to form a key block structure. The inclined broken block contacts the coal seam floor and the upper rock mass to form a triangular suspended space. The key blocks in the middle of the goaf are unstable and broken, and the "masonry beam" structure is formed by stacking in the goaf [19]. Affected by the failure of bedrock, there are the tensile zone, compression zone, and compressive uplift zone on the surface.

With the mining of the middle coal seam, the floor uplift of the upper coal seam gradually disappears. The mining thickness of the middle coal seam is small, and the change of bedrock failure characteristics is not obvious. Figure 5 shows the bedrock failure characteristics after mining of the lower coal seam. When the lower coal seam is mined, the coal seam floor is broken and lifted. The failure of the lower direct roof rock layer develops upward and affects the goaf and bedrock of the middle and upper coal seams, resulting in its instability and fracture again, and the floor rock layer of the middle coal seam breaks down, which makes the key rock layer above the middle coal seam unstable and collapse. A key block structure is formed at the edge of the goaf to support the overlying strata. The rock stratum breaks and collapses, and the cracks of the bedrock develop directly to the surface, resulting in great damage to the surface, forming huge staggered platforms, cracks, or collapse pits, with large widths and depths of damage. The development of fissures connects the goaf and the surface of the three layers of coal, resulting in the discharge of harmful gases. At the same time, it also forms a water diversion channel, and the water and sediment are fed into the working face and goaf. This not only pollutes the air and water near the mining area but also threatens the safety of the working face and surface water.

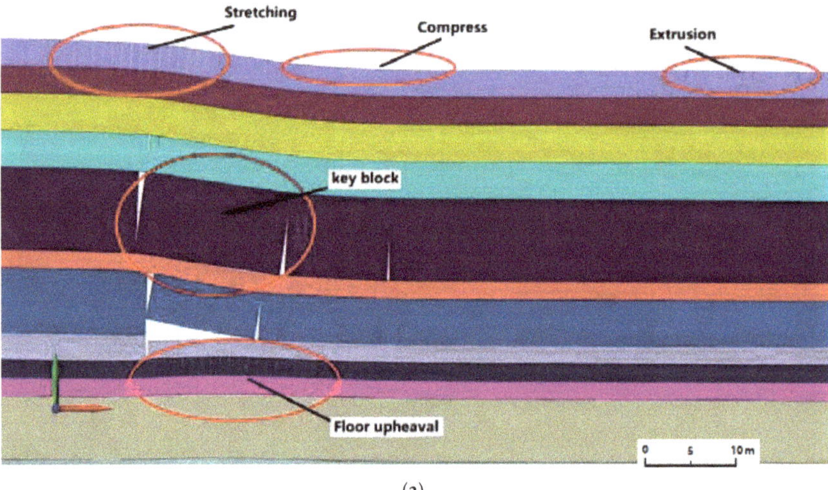

(a)

Figure 5. *Cont.*

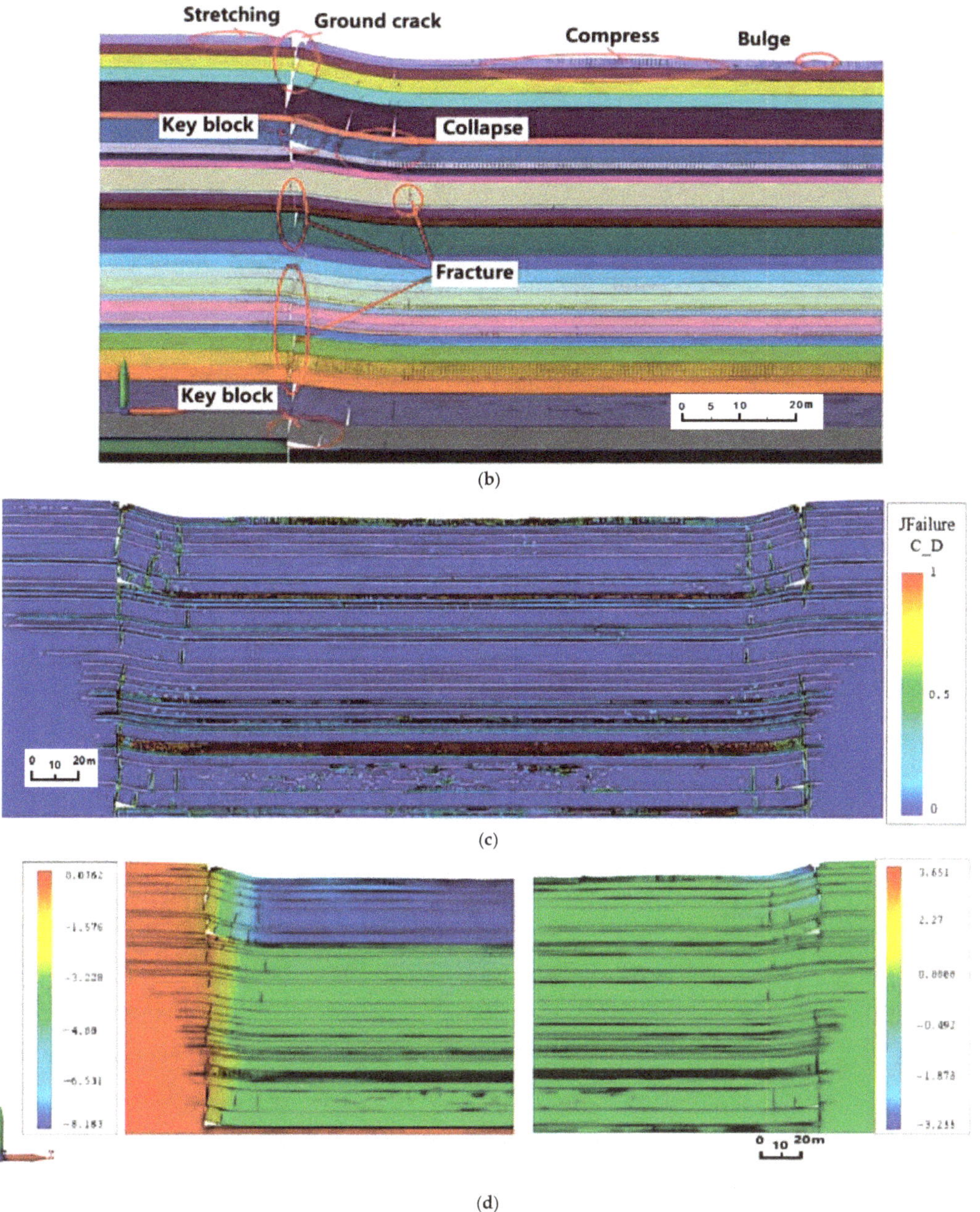

(b)

(c)

(d)

Figure 5. Characteristics of bedrock and surface damage. (**a**) Bedrock failure in upper coal seam mining; (**b**) Bedrock failure of all coal mining; (**c**) Cloud map of stress distribution of bedrock; (**d**) Cloud map of bedrock settlement (left is vertical, right is horizontal).

According to the analysis of Figure 5c, the stress change of bedrock has obvious characteristics, and the failure of rock stratum starts from the surface of rock stratum. The bedrock above the edge of the goaf is mainly broken, cracked, and with tensile damage, while the middle part is mainly compressive and shear damage. The weak rock stratum in the bedrock is damaged by shear and tension, and the broken rock stratum belt is formed under the action of repeated mining. The damage boundary in bedrock decreases suddenly with the increase in mining depth. The damage boundary line presents a parabola shape and is symmetrically distributed on both sides of the goaf. The content shown in Figure 5d is the cloud map of bedrock settlement, and the bedrock settlement shows three gradients with the change of depth. The settlement of the lower bedrock is only affected by the mining of 15# coal seam, and the settlement is small, about 3.5–4.0 m; the subsidence of the middle bedrock is larger than that of the lower bedrock. Affected by the mining of the middle and lower coal seams, the subsidence caused by the bedrock has a superposition effect, and the subsidence is about 4.88–5.5 m; the upper bedrock has the largest settlement, which is affected by the mining of all coal seams, and the cumulative effect is more obvious. The maximum settlement is 8.183 m. From the gradient settlement caused by bedrock failure, the bedrock settlement caused by MCS mining is related to the mining thickness of the coal seam, and the influence is cumulative. The mining of the lower coal seam will lead to the repeated destruction of the rock mass of the upper bedrock and goaf, and the subsidence of the bedrock and surface bedrock, caused by the mining of the lower coal seam, is the most severe.

4.2. Bedrock Failure Stress Analysis

The stress balance of bedrock is broken during coal mining, resulting in bedrock instability, fracture, and fragmentation. After all the coal is mined, the stress distribution nephogram of bedrock failure shows left-right symmetry. The horizontal stress and vertical stress are selected for comparative analysis. It can be seen from Figure 6 that the left part is the cloud diagram of vertical stress distribution, and the right part is the cloud diagram of horizontal stress distribution. Before the instability and collapse of the key bedrock layer, it exists in the form of "cantilever beam" and suddenly breaks after reaching the ultimate bearing capacity. The key block is subjected to the pressure of the upper rock mass, and the vertical stress concentration occurs in the rock stratum above the coal pillar. The maximum downward vertical stress is about 9.79 MPa, and the maximum upward vertical stress is about 2.73 MPa. At the same time, the coal pillar is also affected by vertical stress, resulting in bending deformation and even coal explosion and rock explosion accidents. Before the bedrock is broken, it is stretched and bent, and there is a horizontal stress concentration area in the upper part of each rock stratum. After the rock stratum is broken, the stress on the upper and lower sides of the key block changes: the upper part is subjected to compressive stress, the lower part is subjected to tensile stress, and then, there is a horizontal stress concentration area. In the broken rock stratum, the interaction between rock blocks also produces stress concentration points. After the bedrock is broken, the maximum horizontal stress value is 4.9–6.1 MPa, which is symmetrically distributed left and right, in the opposite direction, and points to the middle of the basin.

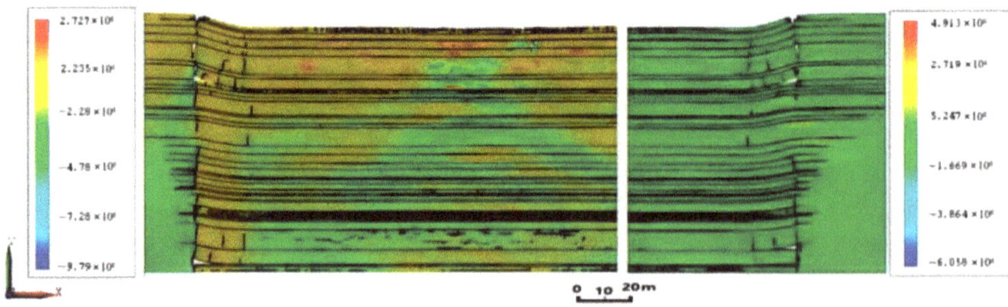

Figure 6. Cloud map bedrock stress distribution (**left** is vertical, **right** is horizontal).

4.3. Surface Deformation and Stress Characteristics

There is a tensile zone, a compression zone, and a compressive fracture zone in the surface basin. The tensile zone is located at the edge of the surface subsidence basin, and the surface fractures are relatively developed, resulting in step fractures and tensile fractures [31]. The compression zone is located at the bottom of the edge of the subsidence basin, the cracks are closed, and the topsoil is compressed without crushing damage. The compression and fracture area is in the middle of the surface basin [32]. The topsoil is in the state of compression and fracture, resulting in compression and shear failure, and compression uplift occurs in some areas. These surface damage phenomena are consistent with the characteristics of surface damage caused by SBMCS [31,33].

In order to better analyze the characteristics of surface movement and deformation, in the process of numerical simulation, a monitoring point is set every 10 m on the surface, and a total of 61 surface monitoring points are arranged to monitor the settlement value and stress distribution of the surface during the simulation process, as shown in Figure 7. The characteristics of surface subsidence accord with the law of surface subsidence in the actual mining process. The surface presents a flat bottom moving basin, and the characteristics of surface subsidence caused by MCS mining are also obvious. According to Figure 7a, the three dense areas of the curve in the middle of the basin show the maximum subsidence curves of the three coal seam excavation stages, respectively. According to the 3# coal surface monitoring data collected, the maximum settlement is 2.622 m, and the maximum settlement obtained by numerical simulation is 3.0 m. The simulation state is ideal, and the difference of settlement is 0.378 m, which is a reasonable range. According to Figure 7b, the distribution law of surface horizontal stress is that the stress concentration fluctuates up and down at 10 KPa (downward), the stress in the middle of the basin (within 200~400 m) changes frequently, complexly, disorderly, and greatly, the stress value changes greatly, and the difference between the maximum value and the minimum value is greater than 26 KPa; however, the stress changes at the edge and outside of the basin are small [34].

According to Figure 7c, the overall variation characteristics of surface horizontal stress are obvious. Near the edge of the mobile basin (within 100 m), the horizontal stress changes significantly, and the horizontal stress changes slightly in the middle and outside of the mobile basin. The maximum stress reaches 100 KPa, and the minimum stress is about 77 KPa. Therefore, within a certain range of the surface, at the edge of the goaf, the horizontal stress changes actively, and the phenomena of tension and compression are obvious. The opening and closing movement of surface cracks in this area is obvious. Combined with the change of vertical stress, it is easy to produce staggered cracks and surface uplift. The stress change has obvious stratification phenomenon. The mining activity of the lower coal seam has an impact on the upper goaf and its bedrock. The stress increases due to the influence of lower coal seam mining.

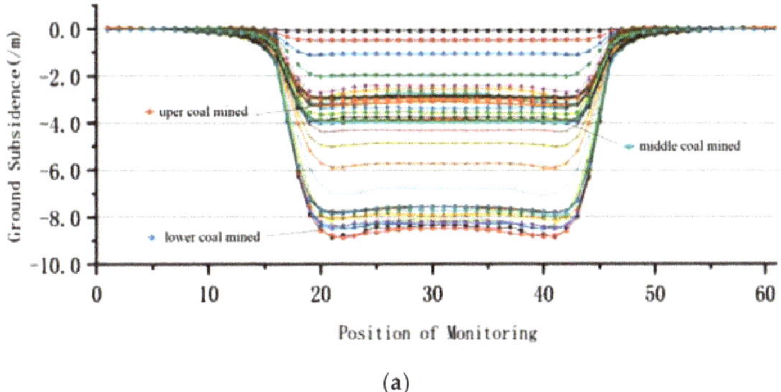

(a)

Figure 7. *Cont.*

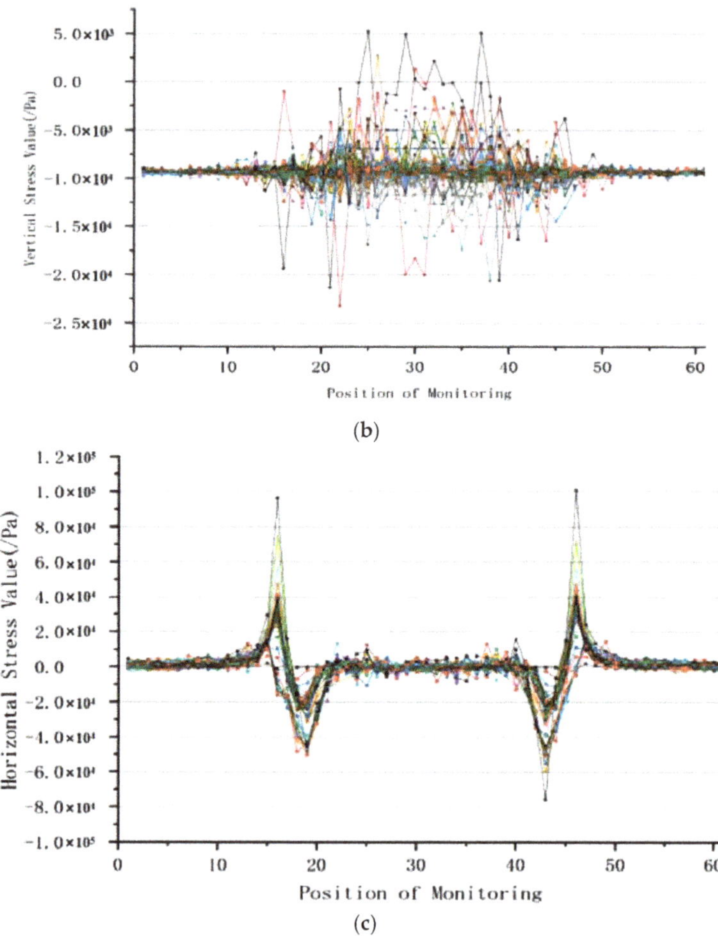

Figure 7. Characteristic analysis curve of surface movement and deformation. (**a**) Surface subsidence characteristic curve; (**b**) Characteristic of surface vertical stress distribution curve; (**c**) Characteristic of surface horizontal stress distribution curve.

5. Discussion

5.1. The Bedrock Failure Process and Characteristics of MCS

After each layer of coal is mined, a caving zone is formed. Then, the roof collapses directly, forming a collapse zone. The rock blocks falling into the goaf are relatively broken, and the particle size is small, which is mainly affected by the impact force of the mining. The volume of the broken rock and soil mass of the roof becomes larger, due to the influence of rock formation fragmentation. In the bedrock between different coal seams, when the key stratum reaches the failure limit, it presents a cantilever beam structure to support the broken rock layer above [20,35]. As the working face advances, a sudden brittle fracture occurs when the cantilever beam exceeds the ultimate bearing capacity. As the height increases, the filling space becomes narrow, and after the key layer above breaks down, an ordered block structure is formed as shown in Figure 8. The blocks are discrete and stacked on each other to form a masonry beam structure [12,36]. Therefore, the settlement of the broken bedrock layer, composed of the collapse zone and "masonry beam", needs many years to be gradually compacted, and the surface of the goaf can reach a relatively stable state. The bedrock damage caused by MCS mining is more serious than that caused by single seam mining, and the damage recovery time may be longer.

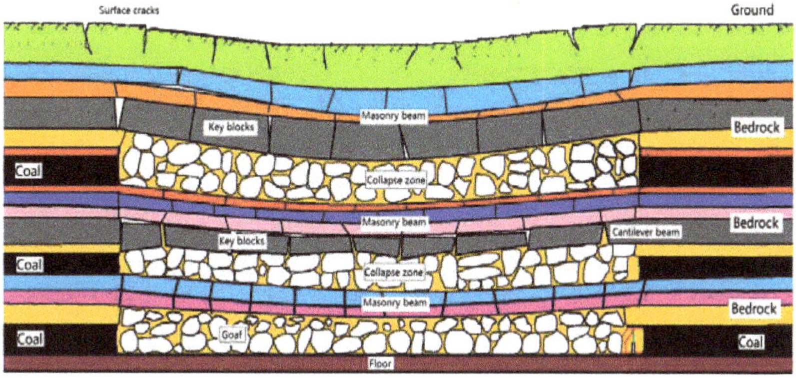

Figure 8. Schematic of bedrock failure in mining of MCS.

The mining of SBMCS causes the damage state of bedrock and surface, which is directly related to the mining geological conditions. In Northwest China, the coal seam is thicker, and the bedrock is thinner. The surface of many coal mines is covered with thick loose aeolian sand layer, and the bedrock is thin [11,30]. The destruction of bedrock and surface caused by mining is more serious. However, the loose layer can repair the surface damage automatically and slowly. There are few buildings on the surface, which is suitable for mechanized high-intensity mining [37]. There are many differences between the multi coal seam mining in the Qinshui coalfield and the MCS mining in the Shendong mining area. Qinshui Coalfield does not have a thick loose layer on the surface, but it has thinner Loess and Sand Gravel. Therefore, the self-healing ability of the surface is poor. As a result, the failure characteristics and stress field distribution of the bedrock are different.

5.2. Impact of Water on Mining Safety

Underground mining engineering is a very dangerous thing if it is under the city or underwater. The water body has a great impact on the mining of coal mine. The mining causes the destruction of bedrock, and the crack is a good channel for the water body, which poses a great threat to the mining face. The mining of shallowly buried multi coal seams leads to bedrock collapse and relatively developed fractures, which leads to the direct connection between the surface and the working face and forms a water diversion channel. Surface water, groundwater, underground phreatic water, and other water bodies are affected by bedrock fissures and feed into the mining face, resulting in a flood in the working face.

The interaction between water body and coal rock mass is a complex physical and chemical process. Chemical actions such as dissolution, hydration, dissolution, and oxidation-reduction may occur between the water body and coal rock mass, which will not only change the composition and structure of coal rock mass but also affect the mechanical properties of coal rock mass [38] The infiltration of water into the rock and soil will also change its physical properties. The water in the fractured rock mass will produce lubrication at the fracture contact surface, reduce the friction resistance at the fracture surface, enhance the shear stress effect, and induce the shear movement along the fracture surface. Under the action of water softening and argillization, the filler in the fracture surface of rock mass begins to develop from solid state to plastic state to liquid state with the increase in water content, which reduces the mechanical properties of fractured rock mass, cohesion, and friction angle [39]. At the same time, the surface water is fed into the goaf, so that a large number of water bodies are stored in the goaf and overburden fractures, and the harmful substances in the goaf will also pollute the water bodies. For the research on the influence of overburden on water, it is necessary to establish water monitoring (water level monitoring, water quality monitoring, pore water pressure monitoring) and relevant tests of various rock masses under different saturated water conditions. The research on the coupling effect of water body and rock mass needs further research.

5.3. Numerical Simulation Method

In terms of numerical simulation, the CDEM algorithm has many advantages in the process. The GDEM platform can organically combine the finite element and discrete element, and it has a good effect on the numerical simulation of bedrock and surface damage in coal mining. UEC software, based on the discrete element, and FLAC software, based on the finite element, can simulate and analyze underground mining. However, compared with the GDEM platform, based on the CDEM method, the GDEM platform is closer to engineering practice, has more constitutive models, and the mining process can be redeveloped. If there is a water-resisting layer in the goaf geological conditions, corresponding measures shall be taken to protect the water resisting layer. In mining MCS, attention should be paid to controlling the development height of bedrock fractures and keeping the water resisting layer from being damaged. We can regard the aquifer as an elastic template, build a balanced differential equation according to the theory of elastic mechanics and elastic foundation, and solve the differential equation, according to the relevant boundary conditions in the mining process. Furthermore, the mechanical analysis and calculation of multiple water barriers can be realized.

6. Conclusions

Through the research of this paper, it is found that the bedrock and surface damage caused by coal mining have strong regional characteristics. Different geological conditions have different characteristics. Bedrock damage and fracture development, caused by SBMCS mining, have regional characteristics. Different lithology, different water conditions, different burial depths, mining thickness conditions, and method of mining are all important factors affecting the failure characteristics of overlying rocks. The conclusions of this paper are as follows:

(1) Affected by the repeated mining SBMCS, the bedrock is seriously damaged, and the bedrock is broken above the edge of the goaf, forming collapse pits, step cracks, and other failure forms. The weak rock stratum in the bedrock is affected by repeated mining to form a broken rock stratum belt. Repeated mining leads to a gradient change of bedrock subsidence. The settlement of mining 15# coal seam, and the settlement of bedrock between 15# coal and 9# coal is small, about 3.5–4.0 m, while the maximum settlement is 8.183 m on the ground.

(2) After the bedrock breaks, it forms a broken block, which overlaps with the coal seam floor and surrounding rock to form a large suspended space. The size of the key blocks is directly related to the thickness and strength of the rock stratum, as well as the load on it. After discussion and analysis, the horizontal thrust between the collapsed blocks of the overlying rock is related to the tensile strength and thickness of the overlying rock. The cracks in the bedrock are relatively developed. The mining of the lower coal seam leads to a large number of pores and cracks in the upper bedrock and the collapsed rock mass in the goaf. The crack is connected with the surface, which makes it easy to form sand and water burst accidents.

(3) The surface damage caused by SBMCS mining is more serious and complex. The amount of surface subsidence has the superposition effect of surface subsidence caused by repeated mining of MCS. The horizontal stress changes significantly along the edge of the surface subsidence basin, which is prone to tensile and compressive failure. The central part of the mobile basin is affected by repeated mining and presents complex vertical stress changes. The maximum stress reaches 100 KPa, and the minimum stress is about 77 KPa.

(4) The CDEM algorithm has many advantages in the process of numerical simulation. The GDEM platform can organically combine a finite element and a discrete element, and it has a good effect on the numerical simulation of bedrock damage in coal mining. In this study, it is necessary to further study and analyze the coupling effect between water body and bedrock. The coupling mechanical model of water body and broken rock mass is established to study the influence of the coupling model on the stress and settlement of bedrock and surface.

Author Contributions: Conceptualization, Y.Z. and H.C.; methodology, G.L.; software, G.L. and W.Z.; validation, Y.X., H.C. and G.L.; formal analysis, Y.Z.; investigation, W.Z.; resources, G.L.; data curation, G.L.; writing—original draft preparation, G.L.; writing—review and editing, Y.Z.; visualization, G.L.; supervision, W.Z.; project administration, W.Z.; funding acquisition, Y.Z. and W.Z. All authors have read and agreed to the published version of the manuscript.

Funding: This research was funded by the National Natural Science Foundation of China, grant number U1810203 and U21A20108, and This research was funded by Henan province science and technology tackling key project, China, grant number 212102310404. And This research was supported by the Fundamental Research Funds for the Universities of Henan Province, China, grant number NSFRF220432.

Institutional Review Board Statement: Not applicable.

Informed Consent Statement: Not applicable.

Data Availability Statement: The data used to support the findings of this study are included within the article.

Conflicts of Interest: The authors declare no conflict of interest.

References

1. Li, B.; Wang, X.; Liu, Z.; Li, T. Study on multi-field catastrophe evolution laws of water inrush from concealed karst cave in roadway excavation: A case of Jiyuan coal mine. *Geomat. Nat. Hazards Risk* **2021**, *12*, 222–243. [CrossRef]
2. Dong, L.; Tong, X.; Ma, J. Quantitative Investigation of Tomographic Effects in Abnormal Regions of Complex Structures. *Engineering* **2020**, *7*, 1011–1022. [CrossRef]
3. Liu, G.; Zou, Y.; Zhang, W.; Chen, J. Characteristics of Overburden and Ground Failure in Mining of Shallow Buried Thick Coal Seams under Thick Aeolian Sand. *Sustainability* **2022**, *14*, 4028. [CrossRef]
4. Xu, Z.; Liu, Q.; Zheng, Q.; Cheng, H.; Wu, Y. Isotopic composition and content of coalbed methane production gases and waters in karstic collapse column area, Qinshui Coalfield, China. *J. Geochem. Explor.* **2016**, *165*, 94–101. [CrossRef]
5. Zhang, Z.X.; Zhang, Y.B.; Zhao, Z.H.; Zhang, L.M. Similar simulation experimental study on overburden movement and surface deformation law in multi coal seam mining. *Hydro Eng. Geo* **2011**, *38*, 130–134.
6. Lin, Z.Y. *Study on Coalfield Structure and Structural Coal Control in Coal Bearing Area of North China*; China University of Mining and Technology: Beijing, China, 2021.
7. Ma, L.; Jin, Z.; Liang, J.; Sun, H.; Zhang, D.; Li, P. Simulation of water resource loss in short–distance coal seams disturbed by repeated mining. *Environ. Earth Sci.* **2015**, *74*, 5653–5662. [CrossRef]
8. Qin, Y. *Research on Rock Movement Law in Multi Coal Seam Mining*; China University of Geosciences: Beijing, China, 2021. [CrossRef]
9. Dong, C. *Research on Overlying Rock and Surface Movement and Deformation under Repeated Mining*; Taiyuan University of Technology: Taiyuan, China, 2020. [CrossRef]
10. Li, Z.C.; Li, L.C.; Wang, S.R.; Ma, S.; Zhang, Z.L.; Li, A.S.; Huang, B.; Zhang, L.Y.; Wang, Z.L.; Zhang, Q.S. Formation of X-shaped hydraulic fractures in deep thick glutenite reservoirs: A case study in Bohai Bay Basin, East China. *J. Cent. South Univ.* **2021**, *28*, 2814–2829. [CrossRef]
11. Chen, J.J.; Nan, H.; Yan, W.T.; Guo, W.B.; Zou, Y.F. Dynamic surface movement and deformation characteristics of shallow buried deep high–intensity mining. *Coal Sci. Technol.* **2016**, *44*, 158–162.
12. Yan, W.T.; Chen, J.J.; Chai, H.B.; Yan, S.G. Dynamic prediction model of surface damage in high–intensity mining area. *J. Agric. Eng.* **2019**, *35*, 267–273.
13. Dong, L.; Chen, Y.; Sun, D.; Zhang, Y. Implications for rock instability precursors and principal stress direction from rock acoustic experiments. *Int. J. Min. Sci. Technol.* **2021**, *31*, 789–798. [CrossRef]
14. Xie, S.R.; Pan, H.; Chen, D.D.; Gao, M.M.; He, S.S.; Song, B.H. Equivalent basic roof structure between layers and its activity law in multi coal seam mining. *J. China Univ. Min. Technol.* **2017**, *46*, 1218–1225.
15. Zhao, X.L.; Wang, Y.Q.; Gao, G.Q.; Wu, Y. Safety analysis of coal mining pavement under expressway based on UDEC and settlement prediction. *Coal Technol.* **2020**, *39*, 123–127.

16. Wu, Y.L.; Yu, Z.R.; Yan, Z.H.; Zhu, S.Y. Numerical simulation of development characteristics of "two zones" of repeated mining in the first mining face of shallow mining area. *Coal Mine Saf.* **2019**, *50*, 199–203.
17. Feng, C.; Li, S.H.; Liu, X.H.; Zhang, Y.N. A semi–spring and semi–edge combined contact model in CDEM and its application to analysis of Jiweishan landslide. *J. Rock Mech. Geotech. Eng.* **2014**, *6*, 26–35. [CrossRef]
18. Gao, H.; Feng, C.; Zhu, X.; Zhang, Y.; Wang, X.; Xin, X.; Shen, Y.; Sun, Z. Three dimensional analysis of coal seam gas fracturing mining based on continuous discontinuous element. *J. Shandong Univ.* **2021**, *6*, 119–128. Available online: https://kns.cnki.net/kcms/detail/37.1391.T.20210906.0941.010.html (accessed on 6 September 2021).
19. Beijing Jidao Chengran Technology Co. *Operation Manual of Block Dynamics Analysis Software GDEM Blockdyna (v1.0)*; Beijing Jidao Chengran Technology Co., Ltd.: Beijing, China, 2021.
20. Qian, M.; Miao, X.; Xu, J. Theoretical study on key stratum in strata control. *J. China Coal Soc.* **1996**, *21*, 12–23.
21. Xu, J.; Qian, M. Method for identifying the location of key overburden layers. *J. China Univ. Min. Technol.* **2000**, *29*, 463–467.
22. Li, S.H.; Wang, J.G.; Liu, B.S.; Dong, D.P. Analysis of criticalexcavation depth for a jointed rock slope using a face-to-face discrete element method. *Rock Mech. Rock Eng.* **2007**, *40*, 331–348. [CrossRef]
23. Ma, C.C.; Chen, K.Z.; Li, T.B.; Zeng, J.; Ma, J.J. Numerical simulation of stress structure rockburst based on GDEM. *Tunn. Undergr. Eng. Disaster Prev.* **2020**, *2*, 85–94.
24. Hou, Z.J. Discussion on "short masonry beam", "stepped rock beam" structure and masonry beam theory in shallow coal seam. *J. Coal* **2008**, *11*, 1201–1204.
25. Zhang, J.P. *Study on Development Law and Prediction Method of Overburden Fractures in Coal Seam Mining in Shendong Mining Area*; Henan Polytechnic University: Henan, China, 2017.
26. Zhao, Y.; Zhou, C.; Zhang, X.; Yuan, C. The pressure arching effect and distribution characteristics offractured strata of single key layer under shallow buried condition. *J. China Coal Soc.* **2020**, *45*, 1–11. [CrossRef]
27. Sawamura, Y.; Ishihara, H.; Otani, Y.; Kishida, K.; Kimura, M. Deformation behavior and acting earth pressure of three-hinge precast arch culvert in construction process. *Undergr. Space* **2019**, *4*, 251–260. [CrossRef]
28. Wang, D.; Guo, F.; Jia, L.; Xu, X. Three–dimensional stability analysis of soft rock slope along the slope of CDEM open–pit coal mine. *J. Saf. Environ.* **2021**, *21*, 195–200. [CrossRef]
29. Wang, X.B. *Study on the Coupling Continuous Discontinuous Method of Lagrangian Element and Discrete Element*; Science Press: Beijing, China, 2021.
30. Ju, J.; Xu, J.; Zhu, W. Longwall chock sudden closure incident below coal pillar of adjacent upper mined coal seam under shallow cover in the Shendong coalfield. *Int. J. Rock Mech. Min. Sci.* **2015**, *77*, 192–201. [CrossRef]
31. Deguchi, T.; Kato, M.; Akcin, H.; Kutoglu, H.S. Monitoring of Mining Induced Land Subsidence Using L– and C–band SAR Interferometry. In Proceedings of the IEEE International Geoscience and Remote Sensing Symposium 2007, Barcelona, Spain, 23–28 July 2007; pp. 2122–2125. [CrossRef]
32. Li, G.; Peng, S.; He, B.; Peng, X.; Yuan, C.; Hu, C. Detection of Coalbed Fractures with P–wave Azimuthal AVO in 3–D Seismic Exploration. *Chin. Sci. Bull.* **2005**, *50*, 146–150. [CrossRef]
33. Cross, T.A. Stratigraphic controls on reservoir attributes in continental strata. *Earth Sci. Front.* **2000**, *7*, 322–350.
34. Ahmad, A.; Kourosh, S. Prediction of surface subsidence due to inclined coal seam mining using profile function (Prediction of surface subsidence due to inclined coal seam mining using profile function). *J. Sci. Technol.* **2005**, *16*, 9–18.
35. He, M.; Zhu, G.; Guo, Z. Longwall mining "cutting cantilever beam theory" and 110 mining method in China—The third mining science innovation. *Chin. J. Rock Mech. Geotech. Eng. Engl. Ed.* **2015**, *5*, 10. [CrossRef]
36. Wang, A.; Wang, J. Study on Underground Pressure Law of Large Mining Height Fully Mechanized Face in Soft Rock Shallow Coal Seam. *Sci. Technol. Datong Coal Min. Adm.* **2016**, *4*, 20–24.
37. Zou, Y. *Evolution Mechanism and Regulation of Surface Ecological Environment under High Intensity Mining*; Science Press: Beijing, China, 2019.
38. Guo, H.; Ji, M.; Chen, K.; Zhang, Z.; Zhang, Y.; Zhang, M. The Feasibility of Mining Under a Water Body Based on a Fuzzy Neural Network. *Mine Water Environ.* **2018**, *37*, 703–712. [CrossRef]
39. Li, Y.; Zhou, H.; Zhu, W.; Li, S.; Liu, J. Numerical Study on Crack Propagation in Brittle Jointed Rock Mass Influenced by Fracture Water Pressure. *Materials* **2015**, *8*, 3364–3376. [CrossRef]

MDPI

Article

Effect of Loading Rate and Confining Pressure on Strength and Energy Characteristics of Mudstone under Pre-Cracking Damage

Hanghang Zheng, Zhenqian Ma *, Lang Zhou, Dongyue Zhang and Xuchao Liang

School of Mining, Guizhou University, Guiyang 550025, China; zhh1878696@163.com (H.Z.); zhoulang20161941@163.com (L.Z.); 17885906234@163.com (D.Z.); liangxcgzu@163.com (X.L.)

* Correspondence: zqma@gzu.edu.cn

Abstract: In order to explore the deformation and failure law of deep surrounding rock roadway disturbed by strong dynamic pressure, the triaxial mechanical properties of mudstone samples under pre-cracking damage conditions were tested to study the deformation and failure characteristics and energy evolution mechanism in the damage process, under different loading rates and confining pressures. In the mechanical experiment, the specimen is pre-cracked to simulate the damage and failure of surrounding rock during roadway excavation, and the damage degree model of rock specimen is established. The results show that the loading rate and confining pressure have significant effects on the peak strength and energy characteristics of mudstone at the average damage degree of 0.12, and the peak strength increases with the increase in confining pressure and loading rate. Under the same confining pressure, the energy increases first, and then decreases with the increase in loading rate, and the loading rate at the turning point is called the critical loading rate. Under the same confining pressure, the closed stress of mudstone gradually increases with the increase in loading rate, and the closed stress and loading rate show a good linear relationship. Through the fitting relationship, it is found that the fitting correlation coefficient between the closed stress of mudstone and the loading rate is as high as 0.99. The elastic strain energy ratio presents a composite function of exponential function with natural constant e, which is a nonlinear process.

Keywords: loading rate; confining pressure; peak strength; presplitting; critical loading rate

Citation: Zheng, H.; Ma, Z.; Zhou, L.; Zhang, D.; Liang, X. Effect of Loading Rate and Confining Pressure on Strength and Energy Characteristics of Mudstone under Pre-Cracking Damage. *Energies* **2022**, *15*, 3545. https://doi.org/10.3390/en15103545

Academic Editors: Longjun Dong, Yanlin Zhao and Wenxue Chen

Received: 18 April 2022
Accepted: 10 May 2022
Published: 12 May 2022

1. Introduction

In recent years, with the depletion of shallow coal resources, mining has gradually turned to deep mining. However, due to the high ground stress environment of deep mining, deep surrounding rock is often disturbed by roadway blasting excavation and high strength mining of the working face, which leads to the aggravation of rock damage and a decrease in strength. Accidents of roof caving and two-side shrinkage occur frequently, which will lead to the overall instability and failure of the roadway. In underground mining engineering, the influence of high ground stress surrounding the rock properties of the roadway, and artificial mining [1–4] has become the main reason for roadway failure.

Mudstone is the most widely distributed rock in the world, and mudstone is also a common rock type in coal mines. Mud soft rock roadway deformation is a process of the slow release of deformation energy induced by coal mining or other dynamic loads. At present, many scholars have carried out a large number of mechanical test studies on the failure disasters and support of coal and rock roadways under static and dynamic loads [5–9]. Bai et al. [10], by theoretical analysis and numerical methods, studied the temporal and spatial distribution law of mining stress and dynamic disturbance in the process of dynamic pressure roadway excavation, and proposed the sectional dynamic support technology. Zang et al. [11] studied the variation law of surrounding rock under a static load and different disturbances of intensity by numerical simulation, and proposed the combined support of shotcrete anchor net and reinforced anchor cable. Du et al. [12] studied a new test system coupled with true triaxial static load and local dynamic disturbance with regard to the strength characteristics and crushing law of rock under different dynamic and static combined loads. Su and Gao [13], through field investigation, numerical simulation and field testing, studied the stress and deformation evolution characteristics of surrounding rock under the influence of multiple excavation and mining. The mating support technology of high pre-stressed bolt and short anchor cable was proposed. From the above research contents, it can be seen that a large number of scholars have carried out in-depth research on the influence of the dynamic load strain rate on roadways, but the study of coal mine dynamic disturbance (10^{-5}/s < strain rate < 10^{-3}/s) on the mechanical properties and failure law of argillaceous rock mass is lacking. In particular, there is a lack of research on the mechanical behavior and engineering dynamic response characteristics of an argillaceous rock mass under static load and then under different disturbance loads. Therefore, it is of great significance to study and analyze the deformation and instability mechanism of roadways under dynamic disturbance.

In addition, from the perspective of energy, the material failure process is essentially a state instability phenomenon driven by energy. Many scholars at home and abroad have done a lot of research on rock failure processes. Zhang et al. [14] carried out failure tests on marble under different unloading confining pressure, to study the variation law between deformation and failure of marble and energy. Wen et al. [15] redefined the damage variable from the perspective of energy dissipation, and analyzed the damage evolution based on a triaxial test. Wang et al. [16] analyzed the energy, energy dissipation and energy conversion mode in the process of rock failure by cyclic loading and unloading. Based on the theory of damage mechanics, rock mechanics and energy conservation, the quantification of the rock failure degree was discussed. The study from Zhao et al. [17], combined with the experimental results, theoretical analysis and numerical simulation, demonstrated that the energy conversion characteristics in the process of rock deformation and failure are summarized according to the energy storage, release and dissipation mechanism of the rock. Hou et al. [18] studied the influence of the loading rate on the energy dissipation characteristics of shale, and discussed the variation in the strain energy conversion rate with the loading rate.

In summary, this paper adopts the method of axial loading pre-compression to produce certain micro-fracture damage inside the rock sample, to simulate the mining influence process of deep rock excavation and artificial mining in the roadway, which is called pre-cracking damage. The triaxial mechanical properties under different confining pressures and loading rates are tested, and the failure characteristics, strength characteristics and deformation characteristics, under different loading rates and confining pressures, are studied. The internal relationship between deformation and failure, pre-cracking damage and energy dissipation of specimens is analyzed, which has important research significance for roadway support and disturbance dynamic disaster analysis.

2. Test Materials and Test Schemes

2.1. Sample Preparation

The test rock samples were taken from the roof rock block of 1200 transport roadway in the Shanjiaoshu Coal Mine. The Z1Z-GT-230 core drilling rig was used to drill the core of the rock block. At the same time, the STJ2500-1-4 track trimming machine and SHM-200 double face leveling machine were used to process the core specimens. The core is shown in Figure 1. In order to better meet the test requirements, the standard specimen developed by the International Society of Rock Mechanics was adopted. The specimen was a cylinder with a diameter of 50 mm and a height of 100 mm. At the same time, the flatness of the two ends of the rock was controlled within 0.02 mm, and the non-parallelism of the two ends was less than 0.05 mm.

Figure 1. Core standard specimens.

2.2. Test Scheme

The test loading equipment adopts a DSZ-1000 stress–strain controlled triaxial shear test system, as shown in Figure 2. The maximum static main pressure of the test machine is 1000 kN, the force resolution is 10 N, the measurement accuracy is ±0.5% FS, the maximum vertical main pressure moving step distance is 300 mm, and the maximum static confining pressure is less than 60 MPa. The system control mode has three kinds: force control, displacement control, axial strain and transverse strain control, which can ensure the reliability of the experimental loading process data and the authenticity of the simulation of the roadway dynamic disturbance.

Figure 2. DSZ-1000 mechanical experimental system.

In order to study the influence of the loading rate and confining pressure on the strength and energy characteristics of rock samples after pre-cracking damage, three groups of different confining pressures were set in this experiment, representing the different stress environments of rock. There were three different confining pressures W (1 MPa, 2 MPa, 3 MPa). Under each group of confining pressure, there are three different loading rates S (0.3 mm/min, 0.6 mm/min, 1.8 mm/min). Due to too many specimens, this paper only selected nine representative specimens for research and analysis.

The surrounding rock on the surface of the deep mine roadway is broken due to mine pressure and artificial mining. The surrounding rock is in a low confining pressure state or zero confining pressure state (uniaxial compressive state). The maximum principal stress of the surrounding rock is axial stress, and the minimum principal stress is circumferential stress. In the process of coal mining, the stress will transfer, the surrounding rock has been damaged or failed and the rock will be further in three-dimensional stress or a uniaxial loading state. In order to study the whole influence process of deep rock excavation and artificial mining, the rock mass is divided into three main stages. The first stage: high-stress triaxial loading stage, which simulates the original rock stress state and original rock damage process of deep rock mass. The second stage: stress unloading process, which simulates the unloading process of deep rock mass after excavation under the stress state of the original rock. The third stage: loading stage of low confining pressure and different rate, this stage simulates the peak stress transfer under dynamic disturbance, and then loading a low confining pressure and high vertical stress on pre-cracking damaged rock specimens in the unloading stage. The confining pressure and the loading rate in the vertical direction after the unloading stage affect the failure factors of the deep rock mass.

The main process of pre-cracking damage is as follows: (1) The testing machine σ_3 is set to 20 MPa by the stress control mode, and the stress point moves from O point to A point, which simulates the real stress state of the roadway before excavation. (2) When the stress point reaches A, stress control is used to unload σ_3 at A certain rate, to a confining pressure of 15 MPa. In this process, σ_1 is continuously increased, and the increase in σ_1 is 50–70% of the peak strength. This process simulates the stress adjustment after roadway excavation (sections A–B). (3) When the stress point value reaches point B, the stress control reduces σ_1 to the same value as σ_3 (to prevent specimen failure), and then the specimen is unloaded to 0 MPa at the same time. In this process, the specimen forms cracks (B–O segment). The loading path was shown in Figure 3.

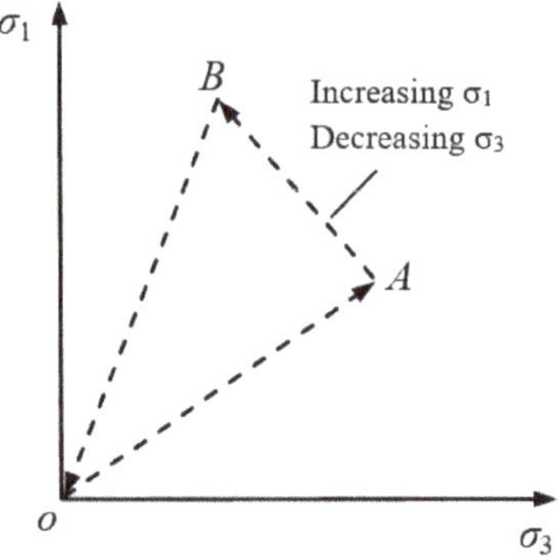

Figure 3. Specimen loading path.

3. Pre-Cracking Damage Analysis

3.1. Principle of Energy Dissipation

In the process of the rock experiment, the test machine inputs energy to the specimen through mechanical work, and the energy will exist in the form of accumulation and dissipation in the rock. The energy accumulated by the specimen is mainly divided into potential energy and thermal energy. In general, the energy conversion into heat energy is relatively small, which has little effect on the strength of the specimen, and can be attributed to the dissipation energy. The potential energy is mainly converted into elastic potential energy and also becomes elastic strain energy. The dissipation of energy is mainly manifested in the friction between the structural planes inside the rock and the extrusion between microscopic particles, and finally the thermal energy forms in the process of forming new microscopic cracks. Assuming that the specimen does not have other forms of energy exchange, the work done by the testing machine on the specimen is denoted as U, U^e as the elastic potential energy of the specimen during the test, and U^d as the dissipation energy of the internal damage of the rock, from the literature (Wen et al. [15]):

$$U = U^d + U^e \tag{1}$$

For elastomers, the total energy provided by the testing machine will be converted into elastic potential energy. For an elastic–plastic body under triaxial compression, its energy input can be expressed by the following formula:

$$U = \int_0^{\varepsilon_1} \sigma_1 d\varepsilon_1 + \int_0^{\varepsilon_2} \sigma_2 d\varepsilon_2 + \int_0^{\varepsilon_3} \sigma_3 d\varepsilon_3 \quad (2)$$

Among them: σ_1 indicates that the specimen is subjected to axial stress, MPa. σ_2 and σ_3 represent the confining pressure of the specimen, MPa. ε_1 denotes the axial strain of the rock specimen. ε_2 and ε_3 represent the circumferential strain of the rock specimen.

For pseudo-triaxial compression, when confining pressure $\sigma_2 = \sigma_3$, the energy absorbed by the rock specimen is:

$$U = \int \sigma_1 d\varepsilon_1 + 2\int \sigma_3 d\varepsilon_3 = \sum_{i=1}^{n} \frac{1}{2}(\sigma_{1i} + \sigma_{1i-1})(\varepsilon_{1i} - \varepsilon_{1i-1}) - 2\sum_{i=1}^{n} \sigma_3(\varepsilon_{3i} - \varepsilon_{3i-1}) \quad (3)$$

For the calculation, it can release elastic strain energy U^e of the unit rock mass, the unloading elastic modulus Eu can be replaced by the elastic modulus E of the elastic section before the peak. The released elastic strain energy at any point on the stress–strain curve can be calculated according to the formula:

$$U^e = \frac{\sigma_{1i}^2}{2E} \quad (4)$$

For the triaxial stress state, the released elastic energy is composed of axial and circumferential parts. By consulting the literature [19,20], the energy released by the circumferential elastic strain energy is small, and the elastic strain energy released by the axial is negligible, so only the axial elastic strain energy is considered.

3.2. Establishment of Damage Degree Model Based on Energy Dissipation

The deformation and failure of the rock is essentially the process of generation, expansion, connection, penetration and slippage of micro-cracks inside the rock. When new cracks are generated, the rock needs to absorb energy, and the slippage friction between cracks will consume energy. Therefore, the deformation and failure process of rock is the process of energy accumulation and dissipation. According to the law of thermodynamics, energy dissipation is the essential attribute of rock deformation and failure. Therefore, the dissipation and release of energy eventually lead to the deformation and failure of the rock, and the dissipation of energy is the main reason for the damage of the internal structure of the specimen. Considering the damage of mudstone in different degrees during deformation and failure, the damage degree D is defined, which can be expressed as:

$$D = \frac{U^d}{U} \quad (5)$$

Formula: U is the total energy absorbed by the specimen, MJ/m^3. U^d is the energy consumed in the simulated unloading process, MJ/m^3.

Because this experiment is carried out by the pseudo-triaxial test, therefore, using the pseudo-triaxial formula analysis, the synthesis of Formulas (1), (3), (4), and (5), obtained in the process of deformation and failure of mudstone damage, with regard to mathematical expression is:

$$D = 1 - \frac{\sigma_{1i}^2}{2(E\int \sigma_1 d\varepsilon_1 + 2\int \sigma_3 d\varepsilon_3)} \quad (6)$$

Figure 4 shows the stress–strain curves of some rock specimens under pre-cracking damage. Since there are many test rocks, some rocks are selected to analyze the damage degree of the pre-cracking stage. As can be seen from the figure, when the stress is loaded to the set load and then unloaded, the unloading curve does not return along the path of the original loading curve, but is lower than the loading curve. The area below the loading curve is the work done by the external load, while the area below the unloading curve is the elastic energy released by the rock, that is, the elastic deformation energy of the rock under the current load. In addition to the increase in elastic deformation energy of the rock sample, another part of the total work is dissipated, and the dissipated energy will not be released from the rock sample with unloading, so the unloading curve is lower than the loading curve. The total work minus the elastic deformation energy of the rock sample is the dissipated energy, namely the area between the loading and unloading curves. The damage degree D of each rock sample can be calculated by calculating the area, which proves that the greater the dissipation area, the greater the damage degree. According to Equation (7) and the test results, the damage degrees under pre-cracking conditions are 0.13, 0.11 and 0.12, respectively. It can be seen that this process is mainly to simulate the fracture specimens formed by roadway excavation, so it has little effect on the damage degree of the rock specimens at this stage. There is an inflection point in the curve in Figure 4c, indicating that there are cracks in the specimen before the pre-cracking test. During the test, the cracks need to absorb energy when they close slowly, and the energy absorbed in the pre-cracking stage after closure will be less than that before.

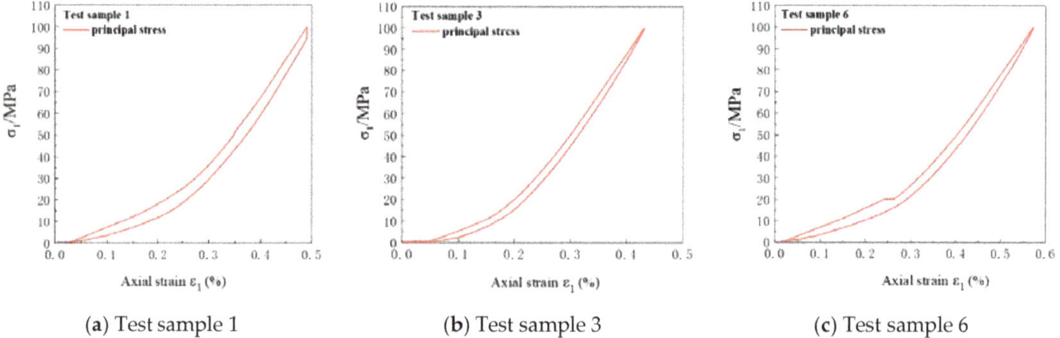

(**a**) Test sample 1 (**b**) Test sample 3 (**c**) Test sample 6

Figure 4. Stress–strain curve under pre-cracking damage.

4. Strength Characteristics under Pre-Cracking Damage Condition

4.1. Strength Characteristics Analysis

The stress–strain curves of mudstone rock samples under different loading rates and confining pressures are shown in Figure 5, by sorting out and analyzing the triaxial compression test results under pre-cracking damage conditions, and the main mechanical index parameters of the test results are given in Table 1. It can be seen from Figure 5 that, with the increase in confining pressure, the stress–strain curves of mudstone samples show different degrees of strain characteristics. The deformation and failure process of rock samples can be roughly divided into four stages, namely, the compaction stage, elastic deformation stage, elastic–plastic deformation stage and post-peak failure stage. In the initial compaction stage, the cracks in the rock specimens are continuously compacted with the increase in stress. In the elastic deformation stage, the stress–strain curve is linear, and the elastic modulus E remains constant. In the elastic–plastic deformation stage, the internal cracks of the rock samples begins to expand and is accompanied by a large number of cracks. In the post-peak failure stage, with the increase in strain, the stress gradually drops and the ability to continue bearing is lost. At the same loading rate, the peak strength of the rock samples increases with the increase in confining pressure. When the loading rate S is 0.3 mm/min, the confining pressure increases W from 1 MPa to 3 MPa, and the peak strength increases from 110.96 MPa to 138.7 MPa, with an increase of 25%. When the loading rate S was 0.6, the confining pressure increased W from 1 MPa to 3 MPa, and the peak strength increased from 169.18 MPa to 229.15 MPa, with an increase of 35.5%. When the loading rate was S 1.8, the confining pressure increased W from 1 MPa to 3 MPa, and the peak strength increased from 138.7 MPa to 265.86 MPa, with an increase of 91.7%.

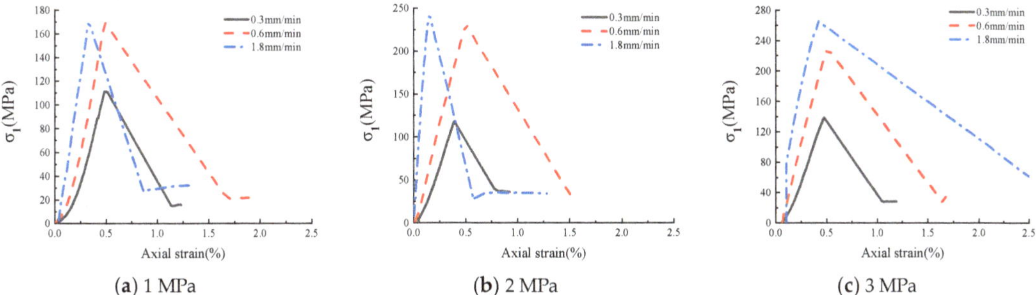

(a) 1 MPa (b) 2 MPa (c) 3 MPa

Figure 5. Stress–strain curves of specimens under different confining pressures.

Table 1. Triaxial compression test results.

Sample No	Confining Pressure /MPa	Loading Rate /mm min^{-1}	Peak Strength /MPa	Elastic Modulus /GPa
1	1.0	0.3	110.96	34.2
2	1.0	0.6	169.18	49.17
3	1.0	1.8	171.83	58.74
4	2.0	0.3	117.67	38.2
5	2.0	0.6	226.88	46.65
6	2.0	1.8	245.45	53.95
7	3.0	0.3	138.7	37.96
8	3.0	0.6	229.15	55.84
9	3.0	1.8	265.86	62.08

It can be seen from Table 1 that under the same confining pressure, the faster the loading rate is, the greater the strength of the specimen will be. When the confining pressure is 1 MPa, the loading rate increases from 0.3 mm/min to 1.8 mm/min, and the strength increases from 110.96 MPa to 171.83 MPa, with an increase of 54.8%. When the confining pressure was 2 MPa, the loading rate increased from 0.3 mm/min to 1.8 mm/min, and the strength increased from 117.67 MPa to 245.45 MPa, with an increase of 108.5%. When the confining pressure is 3 MPa, the loading rate increases from 0.3 mm/min to 1.8 mm/min, and the strength increases from 138.7 MPa to 265.86 MPa, with an increase of 91.6%.

Figure 6 shows the deformation and failure characteristics of the specimens and the binary treatment. It can be seen from the figure that the failure cracks of the specimen are mainly longitudinal in development, showing shear failure. The surface cracks of the specimen were relatively small at a small loading rate until after the failure. With the increase in the loading rate, the number of cracks in the specimen gradually increased, and cones were generated at both ends. Rock blocks with uneven sizes fell off, on to the surface, and most of them were detritus, indicating that, with the increase in the loading rate, a large amount of elastic energy accumulated in the specimen after the failure was transformed into the plastic dissipation energy that broke the rock.

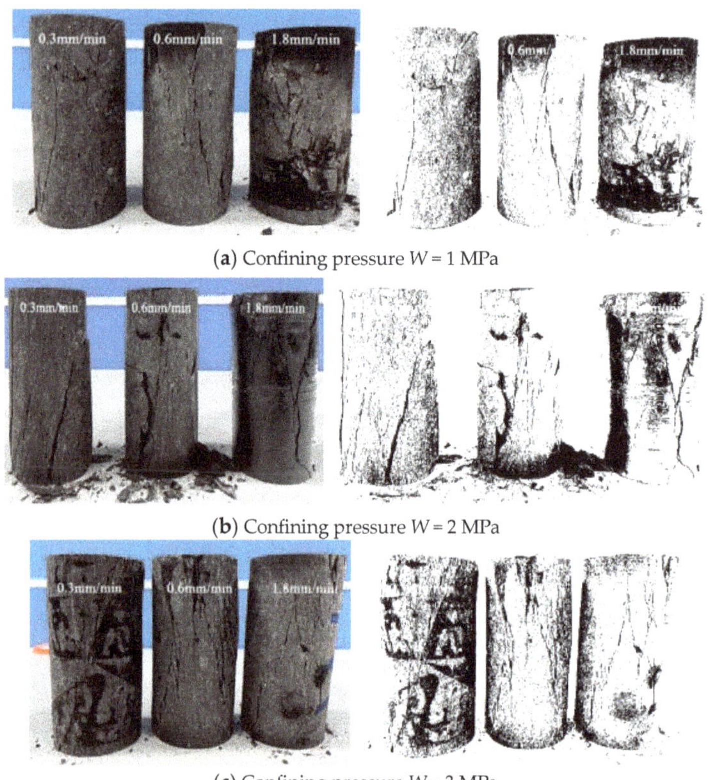

(**a**) Confining pressure $W = 1$ MPa

(**b**) Confining pressure $W = 2$ MPa

(**c**) Confining pressure $W = 3$ MPa

Figure 6. Failure characteristics and binary treatment of specimens.

4.2. Characteristic Stress Analysis

Martin [21,22] and others conducted a more in-depth study of rock mechanics experiments, which found that the fracture development of rock in the failure process will be divided into four stages, namely, the closure stage, the initiation stage, the expansion stage and the interactive penetration stage. Cracks in the four stages of the development process will have a great correlation with the overall strength and mechanical properties of the specimen. In the process of specimen crack closure compaction, due to the rock specimen, it has its own defects and the test machine loading first crack closure compaction, will produce a stress threshold denoted as crack closure stress σ_{cc}. When the testing machine is loaded to the stress threshold, it indicates that the specimen has been completely closed and will enter the elastic deformation stage. The change in this stage is mainly controlled by the mechanical parameters such as the elastic modulus E and Poisson' s ratio V of rock. With the increase in the stress of the specimen, there will be stable new cracks in the specimen. This process is a stable crack propagation stage. The stress of the specimen at the beginning of the crack is recorded as the initiation stress σ_{ci}. According to the research results of a large number of scholars, the initiation stress is roughly 30–60% of the peak strength of the specimen. When the stress of the specimen continues to increase, unstable crack propagation will occur inside the rock. At this time, the corresponding stress is denoted as the damage stress σ_{cd}. After the specimen reaches the damage stress, the internal crack will always be in an unstable state. Academia also regards the damage stress σ_{cd} as the long-term strength of the rock. After the damage stress, the internal cracks of the specimen continuously interact with each other, and gradually develop from a large number of surrounding cracks to macroscopic cracks and shear bands. Finally, the macroscopic failure of the specimen occurs, and the specimen reaches the peak strength σ_p.

In summary, the specimen has four main characteristic stresses in the compression process, including closure stress σ_{cc}, initiation stress σ_{ci}, damage stress σ_{cd} and peak strength σ_p. The stress threshold of each stage corresponds to different stages in the failure process of the test specimen. Therefore, the study of the stress eigenvalue has important guiding significance for the strength damage of the specimen under different confining pressures and different loading rates.

There is the volumetric strain method and acoustic emission method for determining characteristic stress values. The conventional mechanical experiments can be approximately calculated by the volume strain ε_v curve:

$$\varepsilon_v \approx 2\varepsilon_1 + \varepsilon_2 \tag{7}$$

In the formula, ε_1 represents the circumferential strain measured during the test and ε_2 represents the axial strain. When the specimen is subjected to external force, the internal cracks will be closed, and crack initiation, expansion stage and interactive penetration occur. The volumetric strain curve will occur at an offset inflection point, damage stress σ_{cd} and peak strength σ_p. The stress points corresponding to the circumferential strain curve are closed stress σ_{cc} and crack initiation stress σ_{ci}, as shown in Figure 7.

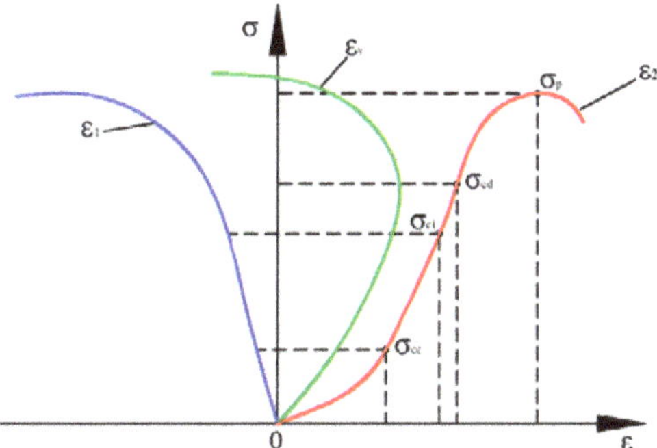

Figure 7. Diagram of characteristic stress corresponding value (Cai et al. [23]).

In this paper, the closed stress σ_{cc}, crack initiation stress σ_{ci}, damage stress σ_{cd} and peak strength σ_p of rock specimens under different loading rates and confining pressures in triaxial tests, are obtained by using volumetric strain and circumferential strain. Figure 7 is the characteristic stress curve of the specimen. It can be seen from Figure 7 that, with the increase in confining pressure, the characteristic stress increases in varying degrees, indicating that the expansion of rock specimens will require higher stress after the confining pressure increases. From another aspect, confining pressure has an inhibitory effect on the expansion of specimens. At high confining pressure, the inflection point of the volumetric strain will be delayed, so the crack initiation stress and damage stress will be increased.

It can be seen from the closed stress σ_{cc} in Figure 8a that, under the same confining pressure, the closed stress of mudstone gradually increases with the increase in loading rate, and the closed stress has a good linear relationship with the loading rate. It is found that the closed stress and loading rate of mudstone can be described by a linear relationship:

When the confining pressure W is 1 MPa, the loading rate S increases from 0.3 mm/min to 1.8 mm/min, and the closure stress increases from 12.5 MPa to 42.5 MPa, with an increase of 240%. The fitting relationship is $y = 20.925x + 6.715$, and the fitting correlation coefficient between the two is as high as 0.998.

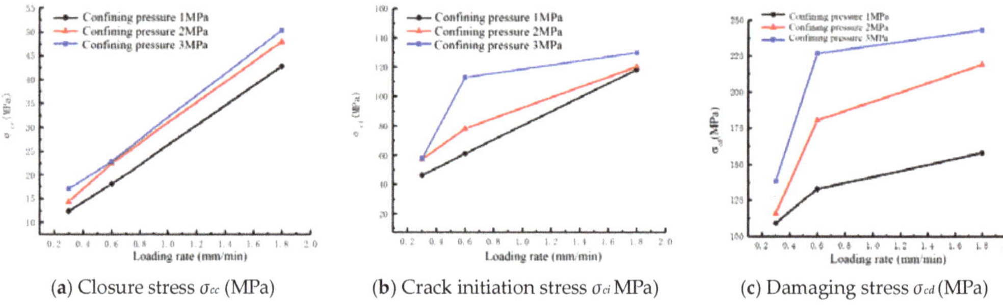

(a) Closure stress σ_{cc} (MPa) (b) Crack initiation stress σ_{ci} MPa) (c) Damaging stress σ_{cd} (MPa)

Figure 8. Characteristic stress curves of specimens.

When the confining pressure W is 2 MPa, the loading rate S increased from 0.3 mm/min to 1.8 mm/min, and the closure stress increases from 14.2 MPa to 48.2 MPa, with an increase of 239%. The fitting relationship is $y = 23.475x + 9.758$, and the fitting correlation coefficient between the two is as high as 0.996.

When the confining pressure W is 3 MPa, the loading rate S increases from 0.3 mm/min to 1.8 mm/min, and the closure stress increases from 16.8 MPa to 50.5 MPa, with an increase of 201%. The fitting relationship is $y = 22.78x + 10.133$, and the fitting correlation coefficient between the two is as high as 0.999.

When the loading rate is low, the reaction time for the deformation of mudstone mineral particles is sufficient, and the crack propagation can be coordinated with the increase in load. At a high loading rate, the crack propagation in the material lags behind the increase in the load, and the absorbed energy is accumulated inside the material, which shows that the closure stress increases with the increase in the loading rate.

From Figure 8b, the initiation stress will increase with the increase in loading rate. The initiation stress of confining pressure 1 MPa and 2 MPa can still maintain a significant linear relationship. When the specimen is 3 MPa, the increase rate of initiation stress will decrease. It can be seen from Figure 8c that under three confining pressures, the damage stress of the specimen increases with the increase in the loading rate, but the increase rate slows down.

5. Energy Characteristic under Pre-Cracking Damage Condition

5.1. Energy Evolution Characteristics under Different Loading Rates and Confining Pressures

Figure 9 shows the energy evolution curves of rock specimens under different loading rates and confining pressures after pre-cracking damage. It can be seen from the figure that the energy change characteristic curves have similar characteristics. Before the peak strength of the rock specimen was reached, most of the energy absorbed by the specimen at this stage was converted into elastic energy E, accounting for about 75.5–93.4% of the total energy, and only a small part of the energy was converted into plastic dissipation energy. With the increase in confining pressure, the elastic energy increases, indicating that confining pressure can inhibit the circumferential deformation of the specimen, forcing the axial specimen to accumulate large elastic strain energy, As shown in Figure 9, when the confining pressure W was 1 MPa, the accumulated elastic energy was 18.78 kJ$\cdot$m^{-3}. When the confining pressure W increased to 3 MPa, the accumulated elastic energy was 47.06 kJ$\cdot$m^{-3}, and the elastic energy increased by 150.58%. When the specimen was damaged, the plastic dissipation strain energy accounted for more than 99% of the total energy, and almost all the energy was consumed in the form of plastic dissipation energy, such as the energy consumed by friction between cracks.

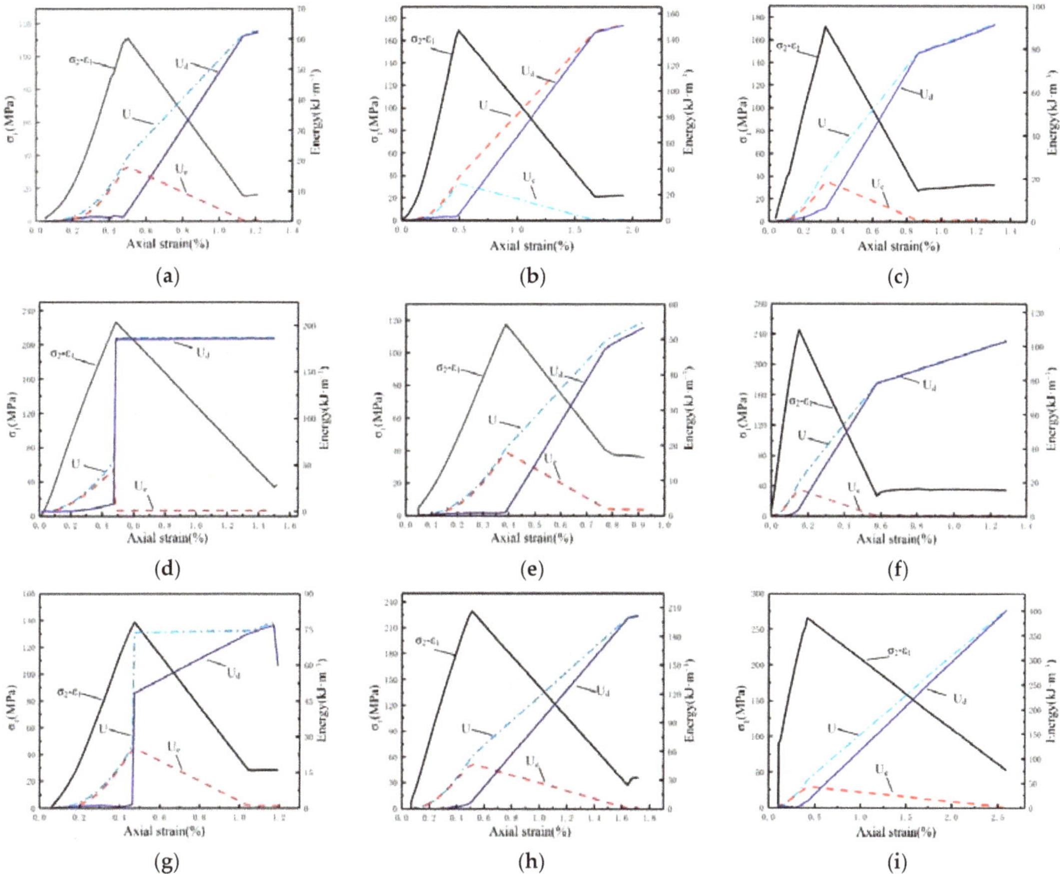

Figure 9. Energy evolution curve of rock specimens. (**a**) Test sample 1 (W = 1 MPa, S = 0.3 mm/min), (**b**) Test sample 2 (W = 1 MPa, S = 0.6 mm/min), (**c**) Test sample 3 (W = 1 MPa, S = 1.8 mm/min), (**d**) Test sample 4 (W = 2 MPa, S = 0.3 mm/min), (**e**) Test sample 5, (W = 2 MPa, S = 0.6 mm/min), (**f**) Test sample 6 (W = 2 MPa, S = 1.8 mm/min), (**g**) Test sample 7 (W = 3 MPa, S = 0.3 mm/min) (**h**) Test sample 8 (W = 3 MPa, S = 0.6 mm/min), (**i**) Test sample 9 (W = 3 MPa, S = 1.8 mm/min).

It can be seen from Figure 10 that, under the same confining pressure, when the loading rate increases to a certain value, the absorbed energy no longer increases with the increase in the loading rate, on the contrary, it decreases. For this rock sample, the loading rate S at a turning point is 0.6 mm/min, and the total strain energy absorbed by the specimen at the loading rate of 0.6 mm/min is the largest. This loading rate is called the critical loading rate; on the one hand, the faster loading rate S limits the development of cracks, which is conducive to bearing, but on the other hand, it also makes the load-bearing part of the micro-element store more deformation energy so that it is closer to failure, which is not conducive to bearing. Therefore, there is an optimal loading rate S that makes the energy absorbed by the specimen reach the maximum value. The confining pressure W was 1 MPa, and the absorbed total strain energy was 33.46 kJ·m^{-3}. When the confining pressure W increased by 2 Mpa, the total strain energy absorbed was 54.21 kJ·m^{-3}, which increased by 62.01%. When the confining pressure W increased by 3 Mpa, the total strain energy absorbed was 54.71 kJ·m^{-3}, which increased by 63.51%. With the increase in confining pressure W, the total strain energy absorbed by the rock is increasing.

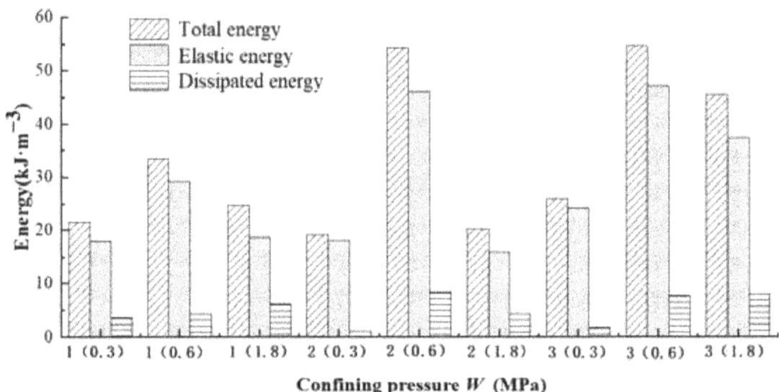

Figure 10. Energy distribution chart of rock sample.

From the elastic strain energy conversion, it can be seen that when the confining pressure *W* is 1 MPa and the loading rate *S* is 0.3 mm/min, the elastic strain energy conversion is 83.5%. When the loading rate *S* increases to 0.6 mm/min, the elastic strain energy conversion rate reaches 87.1%. When the loading rate *S* increases to 1.8 mm/min, the elastic strain energy conversion rate decreases to 75.5%. When the confining pressure *W* is 2 MPa and the loading rate *S* is 0.3 mm/min, 0.6 mm/min and 1.8 mm/min, the elastic strain energy is transformed by 94.57%, 81.61% and 78.42%, respectively. When the confining pressure *W* is 3 MPa and the loading rate *S* is 0.3 mm/min, 0.6 mm/min and 1.8 mm/min, the elastic strain energy is transformed by 93.45%, 86.01% and 82.39%, respectively. With the increase in the axial loading rate of the specimen, the conversion rate of the elastic strain energy of the specimen decreases continuously, which indicates that the elastic deformation of the rock cannot occur within the specimen at a high loading rate, thus the accumulation of elastic strain energy is small. The parenthesis in Figure 10 is the loading rate under the current confining pressure.

5.2. Proportion of Elastic Strain Energy

The elastic strain energy ratio represents the energy ratio absorbed by the elastic deformation of the specimen during triaxial compression, and this process can analyze the dynamic process of the energy accumulation state of the specimen. Figure 11 shows the relationship between the proportion of elastic strain energy and the strain before the loading peak of the rock specimen. It can be seen from Figure 11 that the curves of the proportion of elastic strain energy have similar characteristics. When the specimen is loaded to 30–40% of the peak strain, the proportion of elastic strain energy increases sharply, showing a linear upward trend. When the specimen is loaded to about 70% of the peak strain, the proportion of elastic strain energy of the specimen begins to be slow, and the growth process gradually tends to a stable maximum and remains for a period of time. When loading to the peak strain, the elastic strain energy ratio decreases gradually. In the initial stage of loading, the proportion of elastic strain energy in specimens 4, 6 and 8 decreased. In this stage, when the specimen was initially made of fractured specimens, some micro-cracks were generated inside the specimen. When the specimen was loaded, the micro-cracks inside the rock and other micro-structural surfaces were closed and friction slipped, thus consuming elastic strain energy. When the specimen reaches the peak stage, there are new cracks in the specimen. The strain energy consumed by unit strain is greater than the accumulated elastic strain energy, and the elastic energy decreases. It can be seen from the overall fitting curve that the elastic strain energy accumulation of the specimen presents a composite function of the exponential function of the natural constant e, and its change is obviously nonlinear.

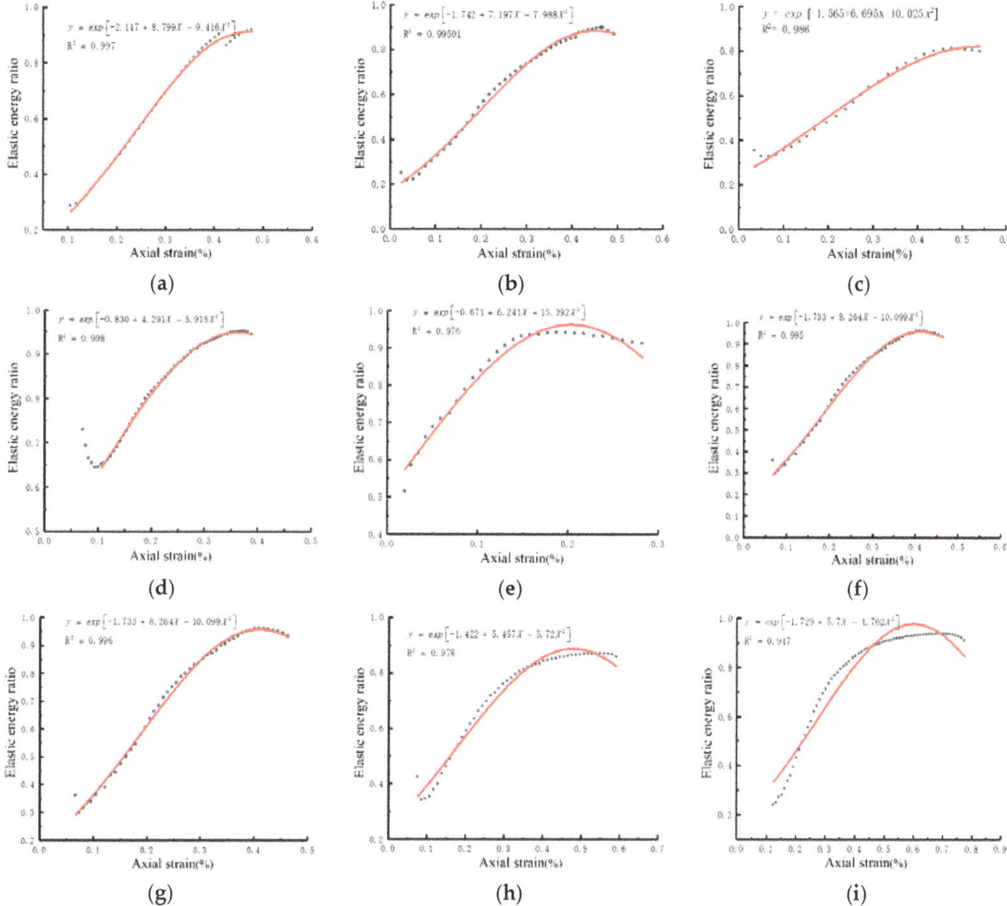

Figure 11. Elastic strain energy ratio, strain curve diagram. (**a**) Test sample 1 (W = 1 MPa, S = 0.3 mm/min), (**b**) Test sample 2(W = 1 MPa, S = 0.6 mm/min), (**c**) Test sample 3 (W = 1 MPa, S = 1.8 mm/min), (**d**) Test sample 4 (W = 2 MPa–0.3 mm/min), (**e**) Test sample 5 (W = 2 MPa, S = 0.6 mm/min), (**f**) Test sample 6 (W = 2 MPa, S = 1.8 mm/min), (**g**) Test sample 7 (W = 3 MPa, S = 0.3 mm/min) (**h**) Test sample 8 (W = 3 MPa, S = 0.6 mm/min), (**i**) Test sample 9 (W = 3 MPa, S = 1.8 mm/min).

6. Conclusions

(1) The damage constitutive model of the mudstone specimen is established and verified by pre-cracking. The results show that the average damage degree D of the specimen is 0.12, and the model is reasonable.

(2) Through triaxial mechanical experiments under different confining pressures W and loading rates S, the results show that under the same confining pressure W, the closed stress of mudstone increases gradually with the increase in loading rate S, and the closed stress has a good linear relationship with the loading rate S. Through the fitting relationship, it is found that the fitting correlation coefficient R, between the closed stress of mudstone and the loading rate S, is as high as 0.998.

(3) Under the same confining pressure W, the energy absorbed by the rock sample is the largest when the loading rate S is 0.6 mm/min, and the energy absorbed by the rock sample decreases with the increase in the loading rate. The loading rate S at

the turning point is 0.6 mm/min, indicating that the energy increases first and then decreases with the increase in the loading rate. The loading rate at the turning point is called the critical loading rate.

(4) The proportion of elastic strain energy is used to analyze the dynamic process of the energy accumulation state of the specimen. It can be seen from the overall fitting curve that the elastic strain energy accumulation of the specimen presents a composite function of the exponential function of the natural constant *e*, and its change is obviously a nonlinear process.

Author Contributions: Conceptualization, Z.M.; methodology, H.Z. and X.L.; validation, L.Z. and D.Z.; data curation, X.L. and L.Z.; writing—original draft preparation, H.Z.; writing—review and editing, Z.M. All authors have read and agreed to the published version of the manuscript.

Funding: This paper is financially supported by the National Natural Science Foundation of China (No. 51904080 and No. 52164003) and the Science and Technology Project of Guizhou Province (No. 2021 general 352 and No. [2021]5610) during the research.

Institutional Review Board Statement: The study did not require ethical approval.

Informed Consent Statement: Informed consent was obtained from all subjects involved in the study.

Data Availability Statement: This study did not report any data.

Acknowledgments: The authors of this paper would like to express their gratitude for the National Natural Science Foundation of China (No. 51904080 and No. 52164003) and the Science and Technology Project of Guizhou Province (No. 2021 general 352 and No. [2021]5610) during the research.

Conflicts of Interest: The authors declare that there is no conflict of interest regarding the publication of this paper.

References

1. Xu, J.H.; Kang, Y.; Wang, X.C.; Feng, G.; Wang, Z.F. Dynamic characteristics and safety criterion of deep rock mine opening under blast loading. *J. Int. J. Rock Mech. Min.* **2019**, *119*, 156–167. [CrossRef]
2. Li, X.B.; Weng, L. Numerical investigation on fracturing behaviors of deep-buried opening under dynamic disturbance. *J. Tunn. Undergr. Sp. Technol.* **2016**, *54*, 61–72. [CrossRef]
3. Zhang, Q.; Wang, J.Q.; Guo, Y.M.; Chen, Y.; Sun, Q. Study on deformation and stress evolution law of surrounding rock under repeated mining in close coal seam. *J. Energ. Source Part A* **2020**. [CrossRef]
4. Jing, H.D.; Li, Y.H.; Li, K.M. Study on the Deformation Mechanism of Soft Rock Roadway under Blasting Disturbance in Baoguo Iron Mine. *J. Shock Vib.* **2018**, *2018*, 4349810. [CrossRef]
5. Zhao, Q.C.; Fu, B.J.; Yin, J.D.; Wen, Z.J. Deformation Mechanism and Control Technology of the Surrounding Rock of the Floor Roadway under the Influence of Mining. *J. Adv. Civ. Eng.* **2020**, *2020*, 6613039. [CrossRef]
6. Kan, J.G.; Wang, P.; Wang, P. Influencing Factors of Disturbance Effects of Blasting and Driving of Deep Mine Roadway Groups. *J. Shock Vib.* **2021**, *2021*, 8873826. [CrossRef]
7. Tian, M.L.; Han, L.J.; Meng, Q.B.; Ma, C.; Zong, Y.J.; Mao, P.Q. Physical Model Experiment of Surrounding Rock Failure Mechanism for the Roadway under Deviatoric Pressure form Mining Disturbance. *J. KSCE J. Civ. Eng.* **2020**, *24*, 1103–1115. [CrossRef]
8. Lai, X.P.; Xu, H.C.; Shan, P.F.; Kang, Y.L.; Wang, Z.Y.; Wu, X. Research on Mechanism and Control of Floor Heave of Mining-Influenced Roadway in Top Coal Caving Working Face. *J. Energ.* **2020**, *13*, 381. [CrossRef]
9. Liu, X.S.; Song, S.L.; Tan, Y.L.; Fan, D.Y.; Ning, J.G.; Li, X.B.; Yin, Y.C. Similar simulation study on the deformation and failure of surrounding rock of a large section chamber group under dynamic loading. *J. Int J. Min. Sci. Technol.* **2021**, *31*, 495–505. [CrossRef]
10. Bai, J.B.; Shen, W.L.; Guo, G.L.; Wang, X.Y.; Yu, Y. Roof Deformation, Failure Characteristics, and Preventive Techniques of Gob-Side Entry Driving Heading Adjacent to the Advancing Working Face. *J. Rock Mech. Rock Eng.* **2015**, *48*, 2447–2458. [CrossRef]
11. Zang, C.W.; Chen, Y.; Chen, M.; Zhu, H.M.; Qu, C.M.; Zhou, J. Research on Deformation Characteristics and Control Technology of Soft Rock Roadway under Dynamic Disturbance. *J. Shock Vib.* **2021**, *2021*, 6625233. [CrossRef]
12. Du, K.; Tao, M.; Li, X.B.; Zhou, J. Experimental Study of Slabbing and Rockburst Induced by True-Triaxial Unloading and Local Dynamic Disturbance. *J. Rock Mech. Rock Eng.* **2016**, *49*, 3437–3453. [CrossRef]
13. Su, M.; Gao, X.H. Research of the Surrounding Rock Deformation Control Technology in Roadway under Multiple Excavations and Mining. *J. Shock Vib.* **2021**, *2021*, 6681184. [CrossRef]
14. Zhang, L.M.; Gao, S.; Wang, Z.Q.; Ren, M.Y. Analysis on Deformation Characteristics and Energy Dissipation of Marble under Different Unloading Rates. *Tehnički Vjesnik* **2014**, *21*, 987–993.

15. Wen, T.; Tang, H.M.; Ma, J.W.; Liu, Y.R. Energy Analysis of the Deformation and Failure Process of Sandstone and Damage Constitutive Model. *J. KSCE J. Civ. Eng.* **2019**, *23*, 513–524. [CrossRef]
16. Wang, C.L.; He, B.B.; Hou, X.L.; Li, J.Y.; Liu, L. Stress-Energy Mechanism for Rock Failure Evolution Based on Damage Mechanics in Hard Rock. *J. Rock Mech. Rock Eng.* **2020**, *53*, 1021–1037. [CrossRef]
17. Zhao, Z.H.; Ma, W.P.; Fu, X.Y.; Yuan, J.H. Energy theory and application of rocks. *J. Arab. J. Geosci.* **2019**, *12*, 474. [CrossRef]
18. Hou, Z.K.; Gutierrez, M.; Ma, S.Q.; Almrabat, A. Mechanical Behavior of Shale at Different Strain Rates. *J. Rock Mech. Rock Eng.* **2019**, *52*, 3531–3544. [CrossRef]
19. Huang, D.; Li, Y.R. Conversion of strain energy in Triaxial Unloading Tests on Marble. *J. Int. J. Rock Mech. Min.* **2014**, *66*, 160–3544. [CrossRef]
20. Zhang, Y.B.; Zhao, T.B.; Taheri, A.; Tan, Y.L.; Fang, K. Damage characteristics of sandstone subjected to pre-peak and post-peak cyclic loading. *J. Acta Geodyn. Geomater.* **2019**, *16*, 143–150. [CrossRef]
21. Martin, C.D. The Strength of Massive Lac du Bonnet Granite around Underground Openings. Ph.D. Thesis, University of Manitoba, Winnipeg, MB, Canada, 1993.
22. Eberhardt, E. Brittle Rock Fracture and Progressive Damage in Uniaxial Compression. Ph.D. Thesis, University of Saskatchewan, Saskatoon, SK, Canada, 1998.
23. Cai, M.F.; He, M.C.; Liu, D.Y. *Rock Mechanics and Engineering Science Press*, 2nd ed.; Sciences Publishing House: Beijing, China, 2013; pp. 63–64.

 MDPI

Article

A Field Study on the Law of Spatiotemporal Development of Rock Movement of Under-Sea Mining, Shandong, China

Jia Liu [1,2,3], Fengshan Ma [1,2,*], Jie Guo [1,2], Guang Li [1,2], Yewei Song [1,2,3] and Yang Wan [1,2,3]

1 Key Laboratory of Shale Gas and Geoengineering, Institute of Geology and Geophysics, Chinese Academy of Sciences, Beijing 100029, China; liujia18mails@mail.iggcas.ac.cn (J.L.); guojie@mail.iggcas.ac.cn (J.G.); liguang@mail.iggcas.ac.cn (G.L.); songyewei20@mails.ucas.ac.cn (Y.S.); wanyang19@mails.ucas.ac.cn (Y.W.)
2 Innovation Academy for Earth Science, Chinese Academy of Sciences, Beijing 100029, China
3 University of Chinese Academy of Sciences, Beijing 100049, China
* Correspondence: fsma@mail.iggcas.ac.cn

Abstract: The phenomenon of rock movement in mining areas has always been a difficult problem in mining engineering, especially under complicated geological conditions. Although the backfilling method mitigates the destruction of the surrounding rock, deformation can still exist in the mining area. In order to ensure the safety of under-sea mining, it is necessary to study the rock movement laws and the mechanisms. This paper focuses on a settlement analysis of the monitoring data of the No. 55 prospecting profile. By analyzing the shape of the settlement curves, the spatial distribution characteristics of settlements of different mining sublevels are summarized. Additionally, the fractal characteristics of the settlement rate under different space–time conditions are studied. We also discuss the relationship between the fractal phenomenon and the self-organized criticality (SOC) theory. The findings are of great theoretical value for the further study of mining settlements to better understand the physical mechanisms of internal movement and rock mass failures through the fractal law of the settlement. Furthermore, elucidating the rock movement law is an urgent task for the safety of seabed mining.

Keywords: rock movement; high-dip metal deposits; fractal character; the self-organized criticality (SOC) theory

Citation: Liu, J.; Ma, F.; Guo, J.; Li, G.; Song, Y.; Wan, Y. A Field Study on the Law of Spatiotemporal Development of Rock Movement of Under-Sea Mining, Shandong, China. *Sustainability* **2022**, *14*, 5864. https://doi.org/10.3390/su14105864

Academic Editors: Longjun Dong, Yanlin Zhao and Wenxue Chen

Received: 17 March 2022
Accepted: 11 May 2022
Published: 12 May 2022

1. Introduction

Rock movement of different degrees often occurs in mining areas due to the disturbance of mining activities [1–4]. The movement, deformation, and destruction of rock masses caused by underground mining activities may further lead to ground subsidence which has caused growing social stability and ecological problems worldwide [5–7]. Therefore, many scholars have been attaching great importance to the study of rock movement phenomena; they have proposed relevant theories and numerical models [8–10]. They have conducted much research on rock movement law and relevant profound theories from different perspectives [8,11–13]: geometric theory, mechanical theory, and mathematical–statistical theory, for instance, the geometric method [14], the circular integral grid method [15], key strata theory [16,17], random medium theory [18], numerical methods [19,20], and fractal theory [21,22]. In addition, a large number of numerical models have emerged, including the Knothe function model [23], the visco-elastic model [24], the anisotropic material model [25], FDM predictive methodology of subsidence [26], the improved Knothe function [27–29], the "four-zone" model [30], and the improved key stratum theory [31]. Regarding fractal theory, the fractal concept was created in 1967 [32]. For the study object, if there is a power-law (or scaling-law) relationship between some of its variables within a certain range, it is considered to have a fractal phenomenon [33,34]. The power-law and fractal theory are closely linked and complement each other. The fractal theory has been widely applied in social sciences [35], agriculture [36], geography [32], geology [37,38], and other fields in recent years. Although the application of fractal theory in geological science started relatively late, it shows strong vitality and broad application prospects due to its powerful expression and concise form. Most geological phenomena are fractal in the statistical sense. In addition, the research and exploration of rock mechanics have also benefited from the development of fractal theory [22,35,36], for instance, in the study of the fractal characteristics of land subsidence [16] and the acoustic emissions of rock mass [37–39]. Fractal theory is widely used to guide production practices.

In summary, rock movement theory continues to develop and mature. However, most mining-induced rock movement studies focus on coal mines; the problem of rock movement caused by the mining of high-dip metal deposits on the seabed is little studied, especially in the presence of faults [40–44]. As one of the popular mining methods, backfill mining is increasingly applied in China and other mining countries. The filled materials consolidate and compress together with the surrounding rock, and the stress of the surrounding rock gradually shifts to the filling body over time. Finally, the mine reaches a state of equilibrium. However, the movement of the surrounding rock is inevitable, which is affected by the mining operation, the low-stiffness filling body, the filling method, the existence of unfilled void spaces, and repeated mining activities [10,45,46]. Severe rock movement or even surrounding rock damage can pose a great threat to the safety of mining personnel and mining production activities. Nonlinear scientific theories are increasingly widely used in geological hazard evaluation and prediction [47–51]. Some scholars have used the self-organized criticality (SOC) theory to analyze the progressive failure process of rocks under stress. We know that many natural phenomena exhibit fractal features, and geological phenomena are no exception. As for the physical mechanisms underlying the fractal phenomenon, the self-organized criticality (SOC) theory offers a better explanation of the characteristics of nonlinear complex systems [47,48,51]. Simply put, SOC means that a system will spontaneously evolve to a critical state. The stratum is in a dynamic equilibrium state before excavation. The excavation process breaks the equilibrium and leads to changes in the energy, so the stratum instinctively adjusts itself and organizes itself to reach a new equilibrium through the adjustment of stress and deformation.

In this study, we used the following methods to explore the law of rock movement produced by mining high-dip seabed deposits. We took the No. 55 prospecting line of the Xinli mining area as the study object, summarized the deformation characteristics of various sublevels (−200 m sublevel, −240 m sublevel, −320 m sublevel, −400 m sublevel, −480 m sublevel, and −600 m sublevel), analyzed the vertical settlement rate using fractal theory, and studied the possible deformation mechanism using the self-organized criticality (SOC) theory. Hereafter, further studies on the mechanisms of rock movement phenomena caused by mining high-dip deposits will be conducted, including physical simulation and numerical simulation research.

2. Study Area

2.1. Geographical Environment and Geology

The Xinli mining area of the Sanshandao gold deposit is the first hard rock mine that belongs to the under-sea deposit in Shandong province, China. It is located in Laizhou bay, Laizhou city, bordering the Bohai Sea in the northwest and connected to the land in the southeast (Figure 1). Many scholars have conducted much research on it [52–55]. The study area belongs to the continental semi-humid climate of the temperate monsoon region. In addition, it has both oceanic and continental climate characteristics. The annual average temperature is 12.5 °C, and the precipitation is 612.1 mm [56].

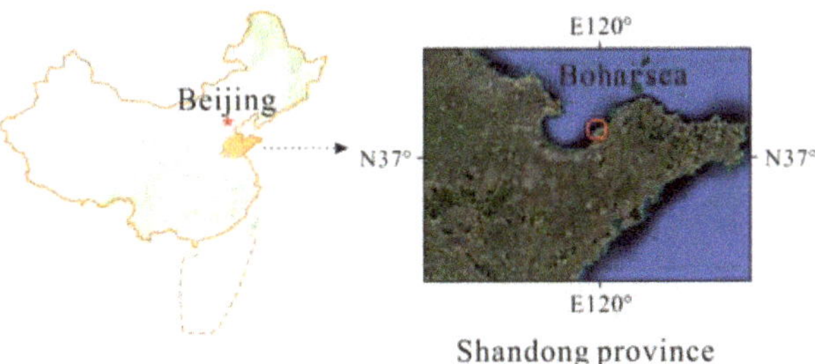

Figure 1. The study area (indicated in the red circle).

The study area is located to the east of the Tan–Lu fault, the north-limb of the Qixia anticlinorium, and the northeastern part of the Sanshandao–Cangshang fault zone. As for the Xinli mining area, the gold-bearing alteration zone is 1300 km long in the strike, and its width varies from 70 to 185 m. It is produced in a relatively simple large vein shape, with a stable occurrence and good mineralization continuity. The altered rocks are zoned along the strike and inclination. From top to bottom, the altered rocks are pyrite sericite granite, pyrite sericite granitic cataclysm, fault gouge (existing below the main fracture), pyrite sericite cataclysm, pyrite sericite granitic cataclysm, and pyrite sericite granite. The pyrite-sericite cataclysm zone in the range of 0–35 m below the main fracture has the strongest alteration and gold mineralization, and it is also the location of the main ore body. The quaternary sediments are distributed in the shallow part of the surface, exposed in the southeast, with a thickness of 16–46 m, with sandy clay, fine sand, medium-coarse sand, and coarse gravel. The sedimentary sequence is diverse, and the regularity is poor [57]. The direction of the maximum principal stress is roughly the horizontal direction (NWW), and its value is greater than the self-weight stress of the rock mass [7]. There are three main faults (F1, F2, and F3) and some typical secondary joints in the study area (Figure 2). The gold ore body is an altered rock type gold deposit controlled by the F1 fault (a reverse fault), whose strike is 40° and the dip angle is about 38°. The F2 fault (a normal fault) lies west of the F1 fault, with a strike of 15° and a dip angle of 80°. The F3 fault (a strike-slip fault) stretches northwest to the sea and southeast to the continent, with a strike of 300° and a dip angle of 90°.

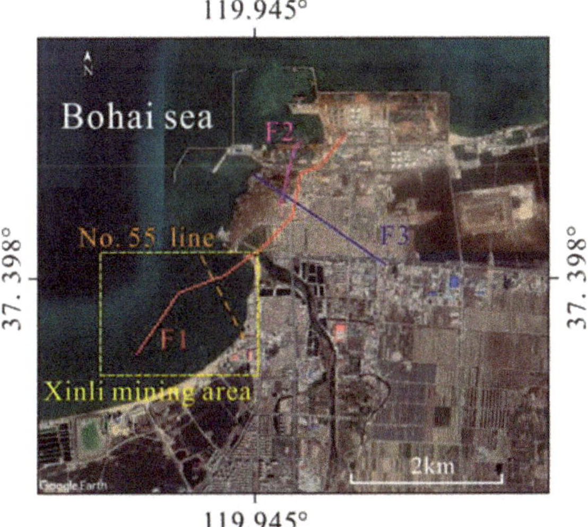

Figure 2. Three main faults and the Xinli mining area.

2.2. Mining Situation

The Xinli mining area of the Sanshandao gold deposit began its basic construction in 2005, with a designed production capacity of 1500 t/d. The Xinli mining area adopted the mechanical upward horizontal slice stopping-filling method with pointed pillars. In order to improve production efficiency, simultaneous upward mining in several sublevels has been designed. The mining activities are conducted from bottom to top by slicing and filling at a slice height of 2.5 m. After the excavation, the mined-out area is immediately backfilled with hardened solids mixed with mineral tailings and cement slurry (binder). The stope length is 100 m, and its width is the thickness of the ore body. Three vertical shafts and one return air shaft have been established in the mining area, and several sublevels (−200 m, −240 m, −320 m, −400 m, −480 m, and −600 m.) have been developed (Figure 3). In addition, a transportation roadway was developed at −440 m; a water silo, pump room, and power distribution room at −600 m; and a fine ore recovery roadway at −667 m.

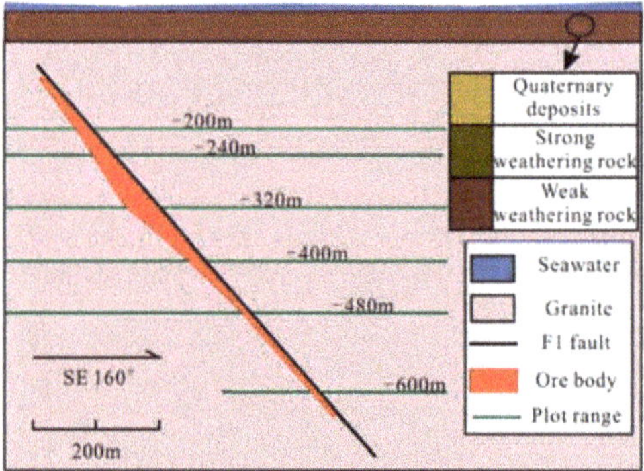

Figure 3. The simplified geological condition of the No. 55 prospecting profile (the green lines show the plot range of the following settlement curves).

3. Methodology

3.1. Displacement Monitoring of the Roadway

In August 2015, a settlement monitoring system was established in the Xinli mining area. In the whole mine, only the No. 55 prospecting profile line passes through the ore body vertically from the hanging wall, that is, perpendicular to the mining direction, forming the most favorable monitoring space conditions (Figure 2). Compared with other profiles, the monitoring network of the No. 55 prospecting profile is relatively complete, covering several sublevels of different depths. The monitoring network is mainly deployed in several sublevels of the No. 55 prospecting profile and other roadways, and the number of monitoring points exceeds 250 (Figure 4); the monitoring intervals are usually one or two months. All of the monitoring uses the fourth-level leveling method. A leveling survey is a method used to determine the height difference between different points on the ground by using instruments (such as level and leveling staff) that provide level realization. It is a vital technique of classical geodetic surveying [58–60]. This method played a positive role in China's height measurement campaign of Mount Qomolangma from 2019 to 2020 [61]. The total length of the leveling route is 7100 m. In the early stage, the monitoring interval was one month. Later, due to the increased monitoring workload and relatively stable settlement data in the study region, the monitoring interval was adjusted to two months after 2017. In order to ensure the efficiency of the monitoring work and to save costs, the distance between the monitoring points was nearly 50 m, and it became smaller in the area close to the ore body, where the deformation is intense. The density of the monitoring points depends on the deformation degree of surrounding rock; areas of intense deformation often require more detailed monitoring data. It should be noted that the deformation of the roadway caused by the mining disturbance is composed of the convergence displacement of the roadway itself and the displacement of the rock mass. When considering the time effect, it can be assumed that the convergence displacement of the roadway itself is close to zero. Therefore, the monitoring data of the roadway were taken as the results of the rock movement.

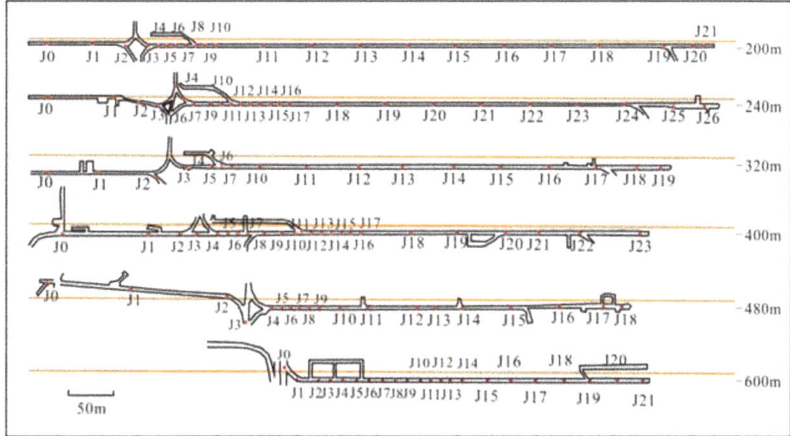

Figure 4. The monitoring network of the No. 55 prospecting profile.

3.2. Fractal Theory

Regarding fractal theory, the capacity dimension is also called the box-counting dimension, and its definition is based on coverage [62]. The functional model is expressed as $N(r) = C \times r^{-D}$. In the equation, r represents characteristic scale; C represents the proportional factor; D represents the fractal dimension; and N (r) represents the number or sum of the scale greater than or equal to r. According to the different meanings of N (r), it can be divided into two function models of the capacity dimension: the content–number method and the content-sum method [63]. In order to obtain the fractal dimension value D, the value of r is constantly changed, and a series of corresponding N (r) values are obtained. The two series of data obtained are projected onto a logarithmic coordinate graph. If there is a straight line in a log–log plot, it indicates that the data conform to the fractal distribution. In order to analyze the fractal characteristics of settlement rate evolution in the Xinli mining area, we adopted long-term monitoring data of the −200 m, −320 m, −400 m, and −480 m sublevels from May 2016 to November 2021.

4. Results

4.1. Settlement Monitoring Curve

Although it is well known that the backfilling mining method is an effective method used to control the severe deformation of surrounding rock [64–67], the rock movement phenomenon still exists in the mining region. Furthermore, the strength of the surrounding rock has decreased due to mining disturbance. We reorganized the monitoring results of several sublevels of the No. 55 prospecting profile from May 2016 to November 2021. The monitoring curves are displayed in Figures 5–7. It should be noted that different colors imply different monitoring dates, and we only present some of the monitoring dates in order to keep the figures more concise and readable. Due to the arrangement of the administrative staff, the monitoring of the −240 m sublevel was interrupted in April 2019, and the monitoring of the −600 m sublevel was started in December 2017. In addition, the settlement value increased by 1000 times in the following figures, and the horizontal distance is the real value. From the final form of the settlement monitoring curves, the backfilling method cannot completely constrict rock movement and deformation of the surrounding rock. According to the years of the monitoring results, we found that the vertical settlement curves can be divided into two categories based on the shape of the curves: (1) a "pan"-type curve. The upper sublevels (−200 m and −240 m) of the No. 55 prospecting profile tend to show "pan" type curves (Figure 5). Most monitoring points have similar settlement values, and the curves are smooth. (2) a "funnel"-type curve. The lower sublevels (−320 m, −400 m, −480 m, −600 m) of the No. 55 prospecting profile tends to show "funnel"-type curves (Figures 6 and 7). The settlement values of most of the monitoring points have obvious differences, and the curves are somewhat rough. Although there are also some differences between the two curves, there are some similarities, regardless of the sublevels of the No. 55 prospecting profile; the monitoring points on the left side of the monitoring line have a slight upward trend. After mining, the surrounding rock moves and deforms as a result of disturbance, and the closer the monitoring point is to the mining area, the larger the displacement is. Moreover, the maximum settlements of the six monitoring sublevels from top to bottom are 296 mm, 146 mm, 308.5 mm, 332.5 mm, 268.5 mm, and 82.5 mm, respectively. In the −200 m, −320 m, −400 m, and −480 m sublevels, the horizontal distances between the monitoring points reaching a 150 mm settlement are 341.6 m, 251.3 m, 152.3 m, and 97.9 m, respectively.

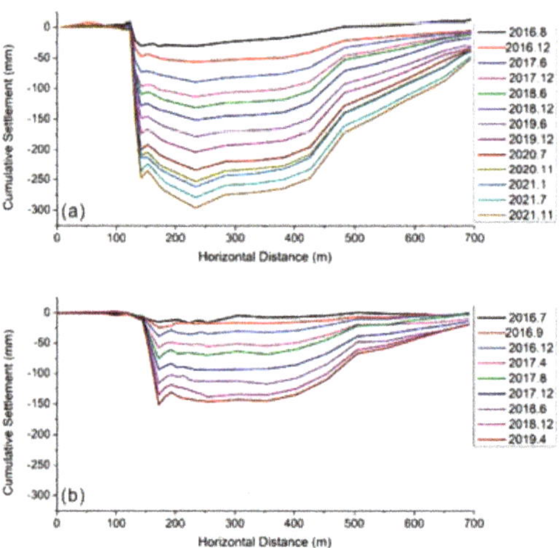

Figure 5. The "pan"-type curve (**a**): −200 m sublevel (**b**): −240 m sublevel.

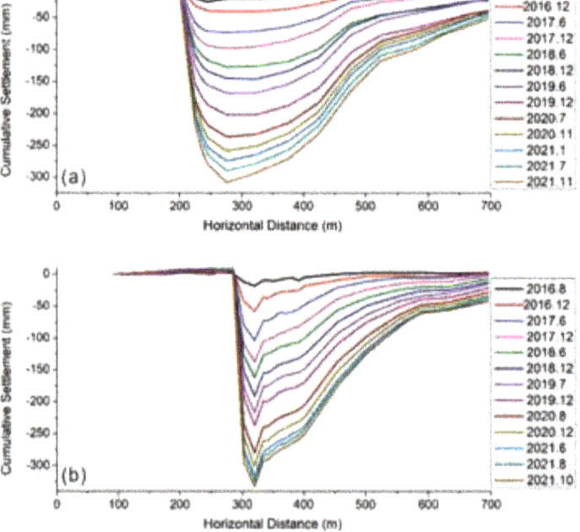

Figure 6. The "funnel" type curve (**a**): −320 m sublevel (**b**): −400 m sublevel.

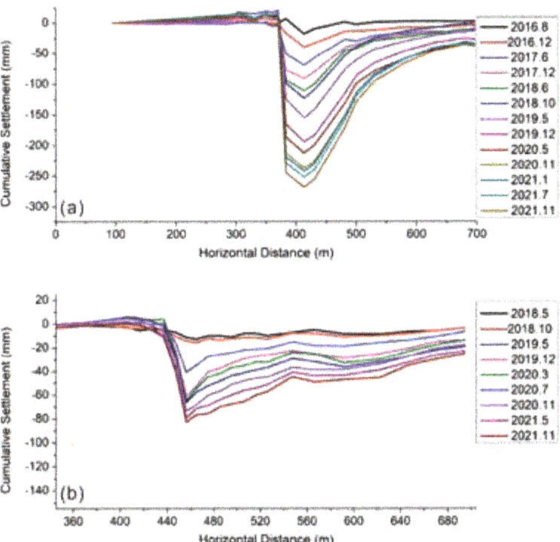

Figure 7. The "funnel" type curve (**a**): −480 m sublevel (**b**): −600 m sublevel.

4.2. Fractal Character of the Settlement Rate

Studies have shown that the dynamic settlement curve of the ground surface points caused by mining has statistical fractal characteristics [22,68].

We used two function models of the capacity dimension to analyze the monitoring data (Tables 1 and 2). The value of r (settlement rate) was constantly changed, and a series of corresponding N (r) values were obtained. It should be noted that the absolute values of the settlement are used here for convenience.

Table 1. Content-number method.

r/(mm/month)	3	6	9	12	15	18
N_1 (r)	894	175	45	17	4	1

Table 2. Content-sum method.

r/(mm/month)	3	6	9	12	15	18
N_2 (r)	4473.7	1441.7	517.0	236.8	68.1	19.7

The two series of data obtained were projected onto a logarithmic coordinate graph, respectively (Figure 8). For the content-number method, there is a straight line that was fitted by the least square method in a log–log plot, with an R^2 value of 0.95. For the content-sum method, there is a straight line that was fitted by the least square method in a log–log plot, with an R^2 value of 0.92. Whichever method we used, they both indicated that the data conformed to the fractal distribution.

The settlement rate data of any provided monitoring sublevel from May 2016 to November 2021 also showed a fractal character. When taking the −400 m sublevel as an example, the settlement rate data also have an obvious fractal character (Table 3 and Figure 9). There is a straight line that was fitted by the least square method in a log–log plot, with an R^2 value of 0.97. In addition, the settlement rate data of the −400 m sublevel from June 2016 to September 2019 also showed a fractal character (Table 4 and Figure 10). The settlement events over different space–time scales both show fractal characteristics. The settlement events reflect the scale invariance of the settlement system.

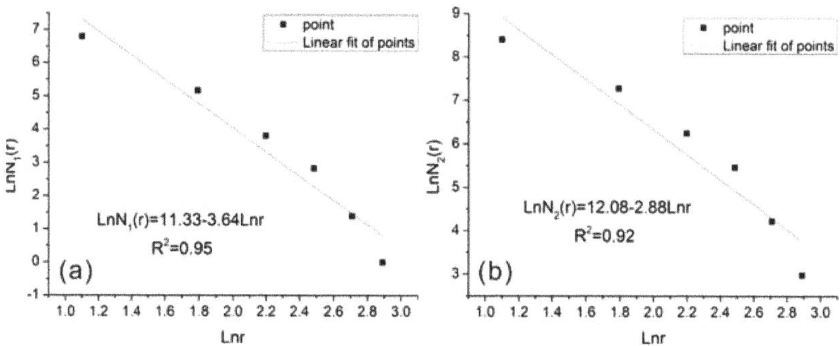

Figure 8. The settlement rate distribution of the whole No. 55 prospecting profile from May 2016 to November 2021. (**a**) Content-number method; (**b**) content-sum method.

Table 3. The settlement rate distribution of the −400 m sublevel from May 2016 to November 2021 (content-number method).

r/(mm/month)	3	6	9	12	15
N_3 (r)	287	59	12	3	1

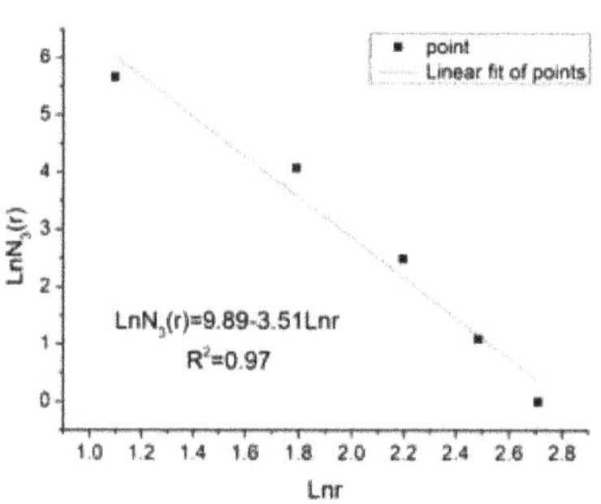

Figure 9. The log–log plot of the settlement rate distribution of the −400 m sublevel from May 2016 to November 2021.

Table 4. The settlement rate distribution of the −400 m sublevel from June 2016 to September 2019 (content-number method).

r/(mm/month)	3	6	9	12	15
N_4 (r)	208	46	12	3	1

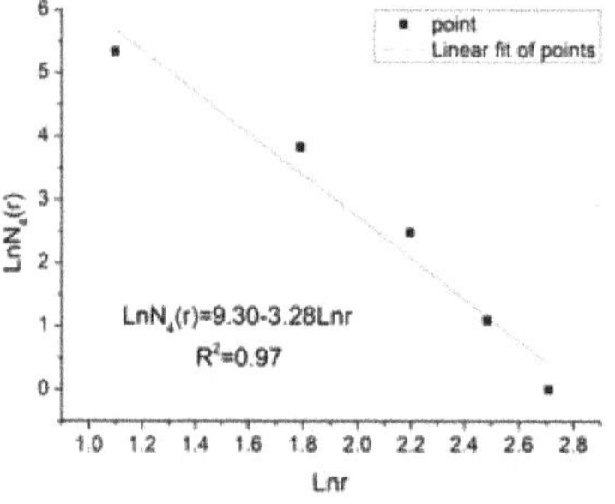

Figure 10. The log–log plot of the settlement rate distribution of the −400 m sublevel from June 2016 to September 2019.

5. Discussion

The subsidence monitoring curves at different depths have distinct characteristics. As the depth increases, the difference in settlement values between the monitoring points becomes larger. As mentioned earlier, the horizontal distance between the monitoring points reaching 150 mm subsidence is decreasing. There are many factors that could contribute to this result: the mining depth, the thickness of the ore body, the mining history, the mining method, backfilling effect, and tectonic geological conditions. Furthermore, the spatial accumulative effect may be another factor. Tectonic stress, hydrogeological conditions, and temperature vary at different mining depths; the thickness of the ore body determines the excavation scale; the mining history determines the original disturbance and deformation of the surrounding rock; the mining method and backfilling method also affect the settlement of the surrounding rock; and ore-controlling fractures have a significant influence on the deformation of mining areas [41]. All of these factors could lead to the result that the upper part may suffer more severe disturbances. Dynamic disturbances often cause local instability of the surrounding rock and goaf, which further lead to different degrees of settlement. When the lower surrounding rock is disturbed by mining, deformation occurs, and the filling is not sufficient to form goaf. The upper part tends to deform by the influence of the excavation of the lower part. This is a natural occurrence that happens under the influence of gravity. Moreover, the upper part can deform when mining activities are implemented in the upper rock mass. This means that the uppermost sublevel is the most affected by the spatial accumulative effect.

Based on the results of settlement monitoring curves, we also found apparent asymmetric characteristics, similar to other scholars who have conducted much research on subsidence [69–72]. The steep geometry of the orebody and the existence of the fault both have potential effects on rock movement phenomena. The steep geometry of an orebody and the existence of a fault can affect rock movement by forming different stress conditions. In this paper, the monitoring points on the right side of the monitoring line have a downward trend; the monitoring points on the left side of the monitoring line have a slight upward trend. We hold that the flexural cantilever beam model [73,74] is suitable to explain this phenomenon. Figure 11 is the schematic diagram. At first, the inclined orebody and the surrounding rock are under prominent horizontal tectonic stress [75]. Due to mining disturbance and tectonic stress, the deep rock mass produces a trend of upward bending, and the shallow rock is pulled; then, the fault activation of the shallow surrounding rock under tensile stress leads to the formation of a sedimentary basin. The hanging wall and footwall fault blocks can be regarded as two bending cantilevers interacting with each other. The equilibrium force generated by extension causes cantilever uplift of the footwall fault block and settlement of the footwall fault block.

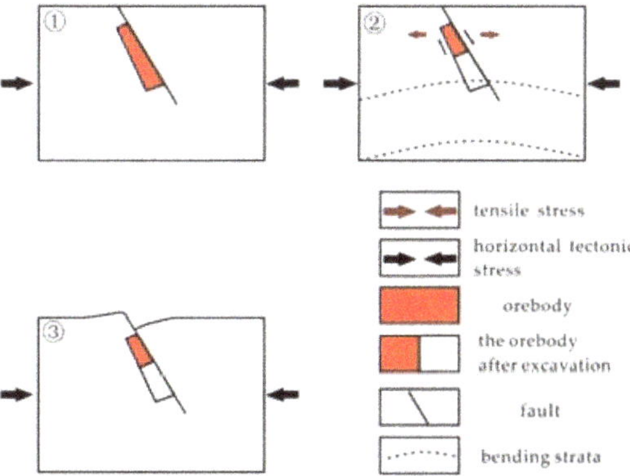

Figure 11. Schematic diagram of settlement development.

The rock mass is heterogeneous and consists of rock blocks and structural planes. Because of the difference in the physical and mechanical properties of rock block, as well as the complexity of the structural plane, mining-induced rock mass failures are strongly nonlinear, and the dynamic settlement curve of the surrounding rock is fluctuant. Taking the "J12" monitoring point in the −400 m sublevel as an example, the settlement rates show obvious fluctuation characteristics. (Figure 12). Mining-induced rock mass failures show a strong nonlinearity, which makes the dynamic settlement curves of the No. 55 prospecting profile show a non-smooth shape (Figures 5–7). The evolution process of rock movement is essentially a self-organizing evolution process, from disorder to order and from low levels to high levels under the action of certain external conditions and internal nonlinear mechanisms [49]. Underground mining activities may cause movement of the surrounding rock due to the advance of the excavation fronts and the progressive closure or collapse of the mineral extraction galleries. During this excavation process, the stress balance of the rock mass was disturbed, and stress redistribution occurred. The rock movement system of the mining area is an open system because the exchange of material and energy is always occurring between the rock movement system and the outside. The surrounding rock stress field, rock mass structure, and other factors all change with respect to two scales: time and space; and the nonlinear interaction between them makes this system very complicated. Thus, the rock movement system remains far from being an equilibrium, although it may have stable statistical properties [75]. The fractal character of subsidence is related to the SOC of the rock movement system. Thus, the internal movement and failure mechanisms of the rock mass are still unexplained problems. This paper reveals the fractal characteristics of rock movement, similar to other natural phenomena. According to the fractal characteristics and the SOC theory, we may find rock movement anomalies in time. In addition, it is necessary to use fractal theory to conduct in-depth studies on rock movement mechanisms.

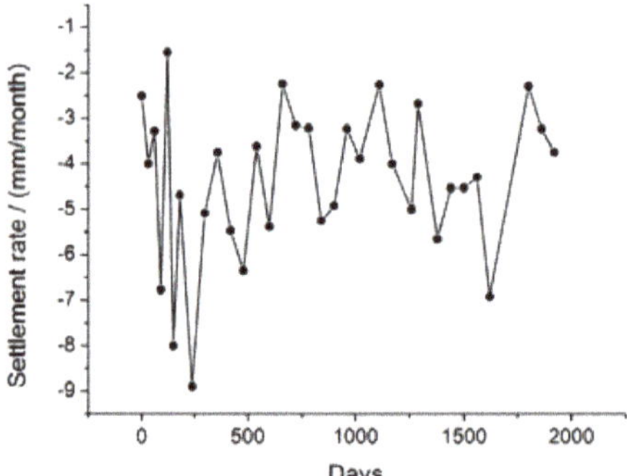

Figure 12. The settlement rates of the "J12" monitoring point in the −400 m sublevel.

6. Conclusions

We used several years' worth of monitoring data of the No. 55 prospecting profile of the Xinli mining area to study the deformation characteristics. The results show that the dynamic settlement curves of the monitoring sublevels have interesting differences. The vertical settlement curves can be divided into two categories based on the shape of the curves: (1) a "pan"-type curve (−200 m and −240 m sublevels) and (2) a "funnel"-type curve (−320 m, −400 m, −480 m, and −600 m sublevels). This phenomenon is not only related to the basic geological and mining conditions but also to the cumulative effect. The upper part may deform as a result of the mining activities in this part as well as due to the deformation of lower sublevels. This is a natural occurrence that happens under the influence of gravity.

Mining-induced rock mass failures are strongly nonlinear due to the randomness of rock mass structure, which is reflected in the fluctuating settlement curves. In addition, the settlement rates of the monitoring points of the No. 55 prospecting profile have good statistical fractal characteristics. The settlement events at different spatial and temporal scales all have fractal characteristics, which reflects the scale invariance of the settlement system. Based on the above information, the self-organized criticality (SOC) theory offers a better explanation of the fractal characteristics of nonlinear complex systems. In this paper, the physical mechanisms of rock movement have been preliminarily investigated through the collection and analysis of the monitoring data, and the findings have theoretical value for the further study of mining settlements.

Author Contributions: Conceptualization, J.L. and F.M.; methodology, J.L. and J.G.; formal analysis, J.L. and F.M.; investigation, J.G. and G.L.; resources, Y.W.; data curation, Y.S.; writing—original draft preparation, J.L.; writing—review and editing, F.M.; supervision, J.G.; project administration, G.L. and Y.W.; funding acquisition, F.M. All authors have read and agreed to the published version of the manuscript.

Funding: This research was funded by The National Natural Science Foundation of China (Grants number 41831293, 42072305).

Institutional Review Board Statement: Not applicable.

Informed Consent Statement: Not applicable.

Data Availability Statement: Not applicable.

Acknowledgments: The authors would like to express their sincere gratitude to the Sanshandao Gold Mine for their data support. In addition, the authors are grateful to the assigned editor and the anonymous reviewers for their enthusiastic help and valuable comments, which have greatly improved this paper.

Conflicts of Interest: The authors declare no conflict of interest.

References

1. Li, N.; Nguyen, H.; Rostami, J.; Zhang, W.G.; Bui, X.N.; Pradhan, B. Predicting rock displacement in underground mines using improved machine learning-based models. *Measurement* **2022**, *188*, 110552. [CrossRef]
2. Zhao, H.J.; Ma, F.S.; Xu, J.M.; Guo, J. In situ stress field inversion and its application in mining-induced rock mass movement. *Int. J. Rock Mech. Min. Sci.* **2012**, *53*, 120–128. [CrossRef]
3. Li, G.; Ma, F.S.; Guo, J.; Zhao, H.J.; Liu, G. Study on deformation failure mechanism and support technology of deep soft rock roadway. *Eng. Geol.* **2020**, *264*, 105262. [CrossRef]
4. Zeng, C.L.; Zhou, Y.J.; Zhang, L.M.; Mao, D.G.; Bai, K.X. Study on overburden failure law and surrounding rock deformation control technology of mining through fault. *PLoS ONE* **2022**, *17*, e0262243. [CrossRef] [PubMed]
5. Guillaume, M.; Cecile, D.; Frederic, M. Time evolution of mining-related residual subsidence monitored over a 24-year period using InSAR in southern Alsace, France. *Int. J. Appl. Earth Obs. Geoinf.* **2021**, *102*, 102392.
6. Helm, P.R.; Davie, C.T.; Glendinning, S. Numerical modelling of shallow abandoned mine working subsidence affecting transport infrastructure. *Eng. Geol.* **2013**, *154*, 6–19. [CrossRef]
7. Liu, J.; Ma, F.S.; Li, G.; Guo, J.; Wan, Y.; Song, Y.W. Evolution assessment of mining subsidence characteristics using SBAS and PS interferometry in Sanshandao gold mine, China. *Remote Sens.* **2022**, *14*, 290. [CrossRef]
8. Brady, B.H.G.; Brown, E.T. *Rock Mechanics for Underground Mining*; Kluwer: Dordrecht, The Netherlands, 2004.
9. Szurgacz, D.; Brodny, J. Analysis of the influence of dynamic load on the work parameters of a powered roof support's hydraulic leg. *Sustainability* **2019**, *11*, 2570. [CrossRef]
10. Li, X.; Wang, S.J.; Liu, T.Y.; Ma, F.S. Engineering geology, ground surface movement and fissures induced by underground mining in the Jinchuan Nickel Mine. *Eng. Geol.* **2004**, *76*, 93–107. [CrossRef]
11. Indraratna, B.; Nemcik, J.A.; Gale, W.J. Review and interpretation of primary floor failure mechanism at a longwall coal mining face based on numerical analysis. *Geotechnique* **2000**, *50*, 547–557. [CrossRef]
12. Wang, Y.H.; Deng, K.Z.; Wu, K.; Guo, G.L. On the dynamic mechanics model of mining subsidence. *Chin. J. Rock Mech. Eng.* **2003**, *22*, 352–357.
13. Coulthard, M.A. Applications of numerical modeling in underground mining and construction. *Geotech. Geol. Eng.* **1999**, *17*, 373–385. [CrossRef]
14. Institute of Scientific and Technical Information of China. *Ground Building in Polish Mined-Out Area*; China Science Press: Beijing, China, 1978.
15. Brauner. *Subsidence due to Underground Mining*; Bureau of Mines: Raleigh, NC, USA, 1973.
16. Qian, M.G.; Liao, X.X.; Xu, J.L. Research on key strata theory in rock strata control. *J. China Coal Soc.* **1996**, *3*, 6.

17. Xu, J.L.; Qian, M.G. Study on the influence of key strata movement on subsidence. *J. China Coal Soc.* **2000**, *2*, 122–126.
18. Guo, Z.Z.; Yin, Z.R.; Wang, J.Z. Random medium shiver movement probability and surface subsidence. *J. China Coal Soc.* **2000**, *3*, 264–267.
19. Wu, K.; Wang, Y.H.; Deng, K.Z. Application of dynamic mechanics model of overlying strata movement and damage above goaf. *J. China Univ. Min. Technol.* **2000**, *29*, 34–36.
20. Gong, Y.Q.; Guo, G.L.; Wang, L.P.; Zhang, G.J.; Zhang, G.X.; Fang, Z. Numerical study on the surface movement regularity of deep mining underlying the super thick and weak cementation overburden, a case study in Western China. *Sustainability* **2022**, *14*, 1855. [CrossRef]
21. Bagde, M.N.; Raina, A.K.; Chakraborty, A.K.; Jethwa, J.L. Rock mass characterization by fractal dimension. *Eng. Geol.* **2002**, *63*, 141–155. [CrossRef]
22. Yu, G.M.; Sun, H.Q.; Zhao, J.F. The fractal increment of dynamic subsidence of the ground surface point induced by mining. *Chin. J. Rock Mech. Eng.* **2001**, *20*, 34–37.
23. Knothe, S. *Time Influence on Formation of a Subsidence Surface*; Archiwum Gornictwa I Hutnictwa: Krakow, Polish, 1952; Volume 1, p. 1.
24. Singh, R.P.; Yadav, R.N. Prediction of subsidence due to coal mining in Raniganj coalfield, West Bengal, India. *Eng. Geol.* **1995**, *39*, 103–111. [CrossRef]
25. McNabb, K.B. Three dimensional numerical modelling of surface subsidence induced by underground mining Div. *Geomech. Tech. Rep.* **1987**, *146*, 20.
26. Alejano, L.; Ramrez-Oyanguren, P.; Taboada, J. FDM predictive methodology for subsidence due to flat and inclined coal seam mining. *Int. J. Rock Mech. Min. Sci.* **1999**, *36*, 475–491. [CrossRef]
27. Chang, Z.Q.; Wang, J.Z. Study on time function of surface subsidence: The improved Knothe time function. *Chin. J. Rock Mech. Eng.* **2003**, *22*, 1496–1499.
28. Wu, L.X.; Wang, J.Z. Study of deformation model of a controlling holding-plate when large area is extracted continuously. *J. China Coal Soc.* **1994**, *19*, 233–242.
29. Tomaž, A.; Goran, T. Prediction of subsidence due to underground mining by artificial neural networks. *Comput. Geosci.* **2003**, *29*, 627–637.
30. Gao, Y.F. "Four-zone" model of rock movement and inverse analysis of dynamic displacement. *J. China Coal Soc.* **1996**, *1*, 6.
31. Sun, Y.J.; Zuo, J.P.; Karakus, M.; Wang, J.T. Investigation of movement and damage of integral overburden during shallow coal seam mining. *Rock Mech. Min. Sci.* **2019**, *119*, 63–75. [CrossRef]
32. Mandelbrot, B.B. *The Fractal Geometry of Nature*; WH Freeman and Company: New York, NY, USA, 1982; pp. 25–50.
33. Delsanto, P.P.; Gliozzi, S.A.; Bruno, L.E.C.; Pugno, N.; Carpinteri, A. Scaling laws and fractality in the framework of a phenomenological approach. *Chaos Solitons Fractals* **2009**, *41*, 2782–2786. [CrossRef]
34. Wang, R.D.; Yang, M.G. Some problems in fractal statistical research on geological phenomenon. *Geoscience* **1998**, *12*, 6.
35. Shen, G.Q. Fractal dimension and fractal growth of urbanized areas. *Int. J. Geogr. Inf. Sci.* **2002**, *16*, 419–437. [CrossRef]
36. Ioelovich, M. Fractal dimensions of cell wall in growing cotton fibers. *Fractal Fract.* **2020**, *4*, 6. [CrossRef]
37. Xie, H.P.; Pariseau, W.G. Fractal character and mechanism of rock bursts. *Chin. J. Rock Mech. Eng.* **1993**, *30*, 28–37. [CrossRef]
38. Biancolini, M.E.; Brutti, C.; Paparo, G.; Zanini, A. Fatigue cracks nucleation on steel, acoustic emission and fractal analysis. *Int. J. Fatigue* **2006**, *28*, 1820–1825. [CrossRef]
39. Wu, X.Z.; Liu, X.X.; Liang, Z.Z.; You, X.; Yu, M. Experimental study of fractal dimension of AE serials of different rocks under uniaxial compression. *Rock Soil Mech.* **2012**, *33*, 3561–3569.
40. Xie, H.P.; Yu, G.M.; Yang, L.; Zhou, H.W. The influence of proximate fault morphology on ground subsidence due to extraction. *Int. J. Rock Mech. Min. Sci.* **1998**, *35*, 1107–1111. [CrossRef]
41. Bruneau, G.; Hudyma, M.R.; Hadjigeorgiou, J.; Potvin, Y. Influence of faulting on a mine shaft a case study part II—Numerical modelling. *Int. J. Rock Mech. Min. Sci.* **2003**, *40*, 113–125. [CrossRef]
42. Zhao, H.J.; Ma, F.S.; Li, G.Q.; Ding, D.M.; Wen, Y.D. Fault effect due to underground excavation in hangingwalls and footwalls of faults. *Chin. J. Geotech. Eng.* **2008**, *30*, 1372–1375.
43. Yan, S.; Bai, J.B.; Li, W.F.; Chen, J.G.; Li, L. Deformation mechanism and stability control of roadway along a fault subjected to mining. *Int. J. Min. Sci. Technol.* **2012**, *22*, 559–565. [CrossRef]
44. Sun, Q.H.; Ma, F.S.; Guo, J.; Li, G.; Feng, X.L. Deformation failure mechanism of deep vertical shaft in Jinchuan mining area. *Sustainability* **2020**, *12*, 2226. [CrossRef]
45. Cao, W.H.; Wang, X.F.; Li, P.; Zhang, D.S.; Sun, C.D.; Qin, D.D. Wide strip backfill mining for surface subsidence control and its application in critical mining conditions of a coal mine. *Sustainability* **2018**, *10*, 700. [CrossRef]
46. Cai, W.Y.; Chang, Z.C.; Zhang, D.S.; Wang, X.F.; Cao, W.H.; Zhou, Y.Z. Roof filling control technology and application to mine roadway damage in small pit goaf. *Int. J. Min. Sci. Technol.* **2019**, *29*, 477–482. [CrossRef]
47. Bak, P.; Tang, C.; Wiesenfeld, K. Self-organized criticality an explanation of the 1/f noise. *Phys. Rev. Lett.* **1987**, *59*, 381–394. [CrossRef] [PubMed]
48. Hergarten, S.; Neugebauer, H.J. Self-organized criticality in a landslide model. *Geophys. Res. Lett.* **1998**, *25*, 801–804. [CrossRef]
49. Qin, S.Q. Catastrophe model and chaos mechanism of ramp instability. *Chin. J. Rock Mech. Eng.* **2000**, *19*, 486–492.
50. Xu, Q.; Huang, R.Q. Systematic analysis principles of geological hazards. *Mt. Res.* **2000**, *3*, 272–277.

51. Iwahashi, J.; Watanabe, S.; Furuya, T. Mean slope-angle frequency distribution and size frequency distribution of landslide masses in Higashikubiki area, Japan. *Geomorphology* **2003**, *50*, 349–364. [CrossRef]
52. Li, Z.S.; Miao, S.J.; Ren, F.H. Research on comprehensive prediction of rock burst in Sanshandao Gold Mine. *Appl. Mech. Mater.* **2012**, *170*, 1612–1617.
53. Liu, G.; Ma, F.S.; Zhao, H.J.; Li, G.; Cao, J.Y.; Guo, J. Study on the fracture distribution law and the influence of discrete fractures on the stability of roadway surrounding rock in the Sanshandao coastal Gold Mine, China. *Sustainability* **2019**, *11*, 2758. [CrossRef]
54. Chen, Y.; Zhao, G.Y.; Wang, S.F.; Wu, H.; Wang, S.W. A case study on the height of a water-flow fracture zone above undersea mining: Sanshandao Gold Mine, China. *Environ. Earth Sci.* **2019**, *78*, 122. [CrossRef]
55. Song, M.C.; Ding, Z.J.; Zhang, J.J.; Song, Y.X.; Bo, J.W.; Wang, Y.Q.; Liu, H.B.; Li, S.Y.; Li, J.; Li, R.X.; et al. Geology and mineralization of the Sanshandao supergiant gold deposit (1200 t) in the Jiaodong Peninsula, China: A review. *China Geol.* **2021**, *4*, 686–719. [CrossRef]
56. Liu, G.W.; Ma, F.S.; Liu, G.; Zhao, H.J.; Guo, J.; Cao, J.Y. Application of multivariate statistical analysis to identify water sources in A coastal Gold Mine, Shandong, China. *Sustainability* **2019**, *11*, 3345. [CrossRef]
57. Ma, F.S.; Zhao, H.J.; Guo, J. Investigating the characteristics of mine water in a subsea mine using groundwater geochemistry and stable isotopes. *Environ. Earth Sci.* **2015**, *74*, 6703–6715. [CrossRef]
58. Kersten, T.; Kobe, M.; Gabriel, G.; Timmen, L.; Schön, S.; Vogel, D. Geodetic monitoring of subrosion-induced subsidence processes in urban areas. *J. Appl. Geod.* **2017**, *11*, 21–29. [CrossRef]
59. Li, S.M.; Wang, Z.M.; Yuan, L.W.; Li, X.X.; Huang, Y.Z.; Guo, R. Mechanism of land subsidence of plateau lakeside Kunming city cluster (China) by MT-InSAR and leveling survey. *J. Coast. Res.* **2020**, *115*, 666–675. [CrossRef]
60. Nojo, M.; Waki, F.; Akaishi, M.; Muramoto, Y. The investigation of a new monitoring system using leveling and GPS. *Proc. Int. Assoc. Hydrol. Sci.* **2015**, *372*, 539–542. [CrossRef]
61. Dang, Y.M.; Guo, C.X.; Jiang, T.; Zhang, Q.T.; Chen, B.; Jiang, G.W. 2020 height measurement and determination of Mount Oomolangma. *Acta Geod. Et Cartogr. Sin.* **2021**, *50*, 556–561.
62. Xie, Y.S.; Tan, K.X. Fractal research on fracture structures and application in geology. *Earth Environ.* **2002**, *1*, 71–77.
63. Guo, C.Y.; Gao, B.F.; Xing, X.W. Comparison of lower limit value determined by two fractal methods. *Gold* **2008**, *3*, 13–17.
64. Nooshin, F.; Michel, A.; Li, L. Numerical investigation of the geomechanical response of adjacent backfilled stopes. *Can. Geotech. J.* **2015**, *52*, 1507–1525.
65. Zhang, J.X.; Zhou, N.; Huang, Y.L.; Zhang, Q. Impact law of the bulk ratio of backfilling body to over-lying strata movement in fully mechanized backfilling mining. *Miner. Min. Technol.* **2011**, *47*, 73–84.
66. Liu, Z.X.; Lan, M.; Xiao, S.Y.; Guo, H.Q. Damage failure of cemented backfill and its reasonable match with rock mass. *Trans. Nonferrous Met. Soc. China* **2015**, *25*, 954–959. [CrossRef]
67. Deveci, H.; Ercikdi, B.; Kesimal, A. Cemented paste backfill of sulphide-rich tailings: Importance of binder type and dosage. *Cem. Concr. Compos.* **2009**, *31*, 268–274.
68. Zhang, D.M.; Yin, G.Z.; Wei, Z.A.; Zhang, W.Z. Fractal characteristics and prediction of rock layer movement in coal mining. *J. Min. Saf. Eng.* **2003**, *2*, 98–100.
69. Donnelly, L.J.; Culshaw, M.G.; Bell, F.G. Longwall mining-induced fault reactivation and delayed subsidence ground movement in British coalfields. *Q. J. Eng. Geol. Hydrogeol.* **2008**, *41*, 301–314. [CrossRef]
70. Jiang, J.P.; Zhang, Y.S.; Yan, C.H.; Luo, G.Y. Discussion on fault effect of rock movement in underground engineering. *Chin. J. Rock Mech. Eng.* **2002**, *8*, 1257–1262.
71. Wang, H.; Qin, Y.; Wang, H.B.; Chen, Y.; Liu, X.C. Process of overburden failure in steeply inclined multi-seam mining: Insights from physical modeling. *R. Soc. Open Sci.* **2021**, *2*, 75. [CrossRef]
72. Das, A.J.; Mandal, P.K.; Bhattacharjee, R.; Tiwari, S.; Kushwaha, A.; Roy, L.B. Evaluation of stability of underground workings for exploitation of an inclined coal seam by the ubiquitous joint model. *Int. J. Rock Mech. Min. Sci.* **2017**, *93*, 101–114. [CrossRef]
73. Kusznir, N.J.; Marsden, G.; Egan, S.S. *A Flexural-Cantilever Simple-Shear/Pure-Shear Model of Continental Lithosphere Extension: Applications to the Jeanne d'Arc Basin, Grand Banks and Viking Graben, North Sea*; Special Publications; Geological Society: London, UK, 1991.
74. Ren, J.Y.; Lei, C.; Yang, H.Z.; Yin, X.Y. Lithosphere stretching model of deep water in Qiongdongnan basin, northern continental margin of south China sea, and controlling of the post-rift subsidence. *Earth Sci.* **2009**, *34*, 963–974.
75. Hui, X.; Ma, F.S.; Zhao, H.J.; Xu, J.M. Monitoring and statistical analysis of mine subsidence at three metal mines in China. *Bull. Eng. Geol. Environ.* **2018**, *78*, 3983–4001. [CrossRef]

MDPI

Article

Research on Factors Affecting Mine Wall Stability in Isolated Pillar Mining in Deep Mines

Jiang Guo [1], Xin Cheng [1], Junji Lu [2], Yan Zhao [1,*] and Xuebin Xie [1]

[1] School of Resources and Safety Engineering, Central South University, Changsha 410083, China; guojiang@csu.edu.cn (J.G.); chengxin129@163.com (X.C.); xbxie@csu.edu.cn (X.X.)
[2] Hunan Lianshao Construction Engineering (Group) Co., Ltd., Changsha 410001, China; lujjlujj780902@126.com
* Correspondence: zyzhaoyan@csu.edu.cn

Abstract: This study takes the Dongguashan Copper Mine as its engineering background. Based on the mechanical model of the mine wall under the trapezoidal load of the backfill, a comprehensive evaluation index is proposed, and its calculation equation is derived. On this basis, an orthogonal test is designed to explore the influence of mining design parameters on mine wall stability. The results show that the width of the mine wall is the main factor affecting its stability, and increasing the width of the mine wall can significantly improve its stability. When the width of the mine wall is kept above 4 m, its stability is better. When the mechanical parameters of the backfill are poor, the mine wall is prone to overturning failure. The width of the mine room has an influence on the multi-directional loading of the mine wall, but the influence on the stability of the mine wall is low. According to the regression equation calculation, the mine wall safety factor is about 1.46 under the design of G5 mining of Dongguashan Line 52, the stability of the mine wall is good after actual mining and the engineering application effect is ideal, which can provide a theoretical basis for the design of isolation pillar mining in deep mines.

Keywords: mine wall stability; Platts arch; safety factor; orthogonal experiment; regression analysis

Citation: Guo, J.; Cheng, X.; Lu, J.; Zhao, Y.; Xie, X. Research on Factors Affecting Mine Wall Stability in Isolated Pillar Mining in Deep Mines. *Minerals* **2022**, *12*, 623. https://doi.org/10.3390/min12050623

Academic Editor: Yosoon Choi

Received: 15 April 2022
Accepted: 12 May 2022
Published: 13 May 2022

1. Introduction

With the continuous development and utilization of resources, the mining of mineral resources has developed towards a deeper level and a larger span, and deep mining has become the mainstream trend in mining [1–6]. As a mining method, fill mining can make full use of tailings resources and control the deformation of surrounding rock effectively, and it is widely used in deep mining [7,8]. After the goaf is filled, the backfill body can support the rock formation and control the stope pressure activities. In order to improve the utilization rate of mineral resources, the ore pillars in the stope often need to be mined after the goaf is filled. During the mining of the ore pillar, in order to ensure the stability of the backfill and the goaf, mine walls are often left on both sides of the pillar. The wall can improve the stability of the backfill and play a temporary supporting role to avoid the collapse of the backfill and the falling of the roof during the mining of the pillar, which increase dilution losses from mining techniques. Therefore, it is of great significance to carry out research on the stability of the mine wall and optimize the structural parameters of the mine wall, which is of great significance to reduce the loss of mine wall ore volume and ensure the safety of mining pillar.

At present, many scholars have achieved a lot of research results through theoretical calculation, reliability analysis and numerical simulation [9–13]. Elasticity theory, thin plate theory and catastrophe theory have been widely used in the study of mine wall stability. Wei [14] improved the limit equilibrium method by applying the stress area superposition method, in which the stress distribution of the elastic area and the inelastic area of the pillar was obtained and the stability of the pillar in the strip-mining process was analyzed.

Huang [15] established a mechanical model of mine wall bending failure; the critical bending stress of the mine wall was calculated by the energy method and the stability changes in its size were obtained and the correctness of the theoretical analysis was verified by numerical simulation. Xu [16] proposed the concept of stripping degree and established a corresponding secondary stripping model to analyze the stability of hyperbolic coal pillars. He considered the critical condition of pillar instability under the influence of coal pillars and high temperatures. Based on the instability theory and cusp catastrophe model, Wang [17] deduced the roof deflection curve under the condition of the unequal span of adjacent stopes, and then the pillar instability condition under the condition of asymmetric mining was determined. Based on reliability analysis, Idris [18] introduced the method of using an artificial neural network in pillar stability research. The pillar reliability index and failure probability were calculated, and he analyzed the influence of variation coefficient on pillar stability and determined it according to the minimum acceptable risk of pillar failure. The optimal pillar size was determined. Liu [19] studied the failure mechanism of the pillar group by using the method of the renormalization group, and the safety factor of the pillar system was obtained. The safety of the pillar group scheme was analyzed with the fuzzy comprehensive evaluation theory, and the pillar structure design in seabed mining was optimized. Ding [20] used the Stochastic Gradient Boosting (SGB) model to analyze pillar stability and established an evaluation index system based on five factors: pillar width, pillar height, pillar aspect ratio, rock uniaxial compressive strength and pillar stress. The parameter sensitivity was studied based on the relative variable importance, and the main variables affecting the pillar stability were obtained. With the development of computers, numerical simulation has also become an important means of research on pillar stability. Zhang [21] studied the relationship between various factors in multi-coal strip mining and the vertical displacement of the pillar based on FLAC numerical simulation software. Zhang [22] used the layered cover model for pillar design and stability analysis, eliminating the influence of boundary effects on the analysis results, and determined the main factors affecting pillar stability. Yang [23] combined the experimental data and the research method of numerical simulation, studied the pillar size of the Zhaozhuang Coal Mine in Shanxi Province, and proposed differentiated support technology, which improved the bearing strength of the pillar and effectively reduced the deformation of the surrounding rock.

In general, there are many research methods for mine wall stability analysis, and research results have also been obtained. Among them, reliability analysis can comprehensively consider the influence of various factors on the stability of the mine wall, and it is widely used in engineering. However, the current stability and reliability analysis of the mine wall is mostly related to unidirectional vertical load; the main consideration is the yield failure form of the mine wall, ignoring the influence of backfill on mine wall stability [24,25]. Furthermore, the backfill method is mostly used for mining in deep mines. As well as the yield failure of the overlying surrounding rock, overturning failure may also occur under the lateral load of the backfill, so existing research results are not suitable for the mine wall design of the backfill stope. Therefore, taking the Dongguashan Copper Mine as an example, this study analyzes the backfill-wall bearing mechanism after the isolated pillar is mined. The expression of the safety factor of the mine wall under the action of overlying strata and backfill is deduced, and orthogonal experiments are designed to study the influence of different parameters on the stability of the mine wall. The equation for calculating the comprehensive safety factor of the mine wall is obtained. The research results are expected to provide a reference for the design of isolated pillar mining in deep mines.

2. Project Overview

The Dongguashan Copper Mine is located in Tongling, Anhui Province. It is a typical deep deposit with a burial depth of −690 to −1000 m, a main ore body of 1810 m in length, 500 m in width and an average thickness of 20 to 50 m. The main ore body of the Dongguashan Copper Mine strikes NE35−40°, the average dip angle is 20°, and the maximum dip angle is located on the two sides of the ore body, about 30–40°. The middle part of the ore body is relatively gentle, and the lateral angle is less than 15°. The ore-bearing rock mass is mainly copper-bearing skarn and copper-bearing magnetite. The roof-surrounding rock has good properties; the surrounding rock is mainly composed of marble, and the floor surrounding rock is mainly composed of siltstone and diorite.

Due to the deep burial of the ore deposit and the high stress state of the surrounding rock, the mining period of the ore rock is long and the mining is difficult. In order to achieve the goals of efficient, safe, and low-cost mining, the main ore body adopts the "temporary isolation of the pillar stage and the subsequent filling mining method". The method is characterized by dividing the panel along the strike direction of the ore body; the panel span is 100 m, and the panel size is length (ore body width) × width (100 m) × height (ore body thickness), and there are temporary isolation pillars in each panel interval to achieve independent mining of each panel. The stopes are divided every 18 m along the vertical ore body strike (the width direction of the ore body) inside each panel. The mining process is carried out in three steps. First, the one-step stopes are mined according to the method of "interval mining". After the one-step stope is mined, full tailings are used for cementing and filling. After the first-step stopes are filled, the stopes are recovered and filled in a two-step process. At this stage, the one-step and two-step stopes' mining work has been completed. In order to fully recover the mineral resources, it is necessary to carry out three-step mining for the isolation pillar between the panel. After the isolation pillar is mined, mine walls are left on both sides to prevent a large number of backfill in the stope from collapsing. Figure 1a is a top view of the stope layout and Figure 1b is a vertical cross-section view of the stope layout.

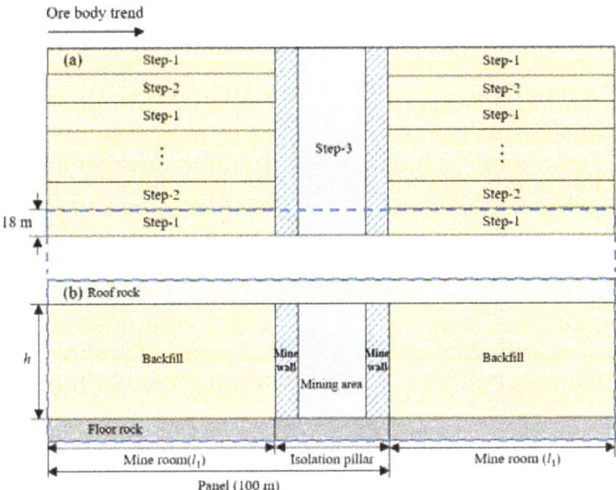

Figure 1. (**a**) Top view of Dongguashan stope layout. (**b**) Vertical cross section view of Dongguashan stope layout.

3. Analysis of Mine Wall Safety Factor

Mine wall instability includes various failure forms, among which are mainly brittle failure, ductile failure and weak plane shear failure [26,27]. The mine wall is simultaneously subjected to the load of the overlying surrounding rock and the backfill, which may cause brittle compression failure under the action of the overlying surrounding rock or shear failure under the lateral load of the backfill [28–31]. In addition to the above-mentioned strength failure, the mine wall may also suffer from overturning failure. Combined with the actual situation of Dongguashan, this study takes the comprehensive safety factor as an evaluation index of mine wall safety. Among them, when the safety factor value is greater than 1, it means that the mine wall is in a stable state. When the safety factor is equal to

1, it means that the mine wall is in a critical state. When the safety factor is less than 1, it means that the mine wall is in an unstable state.

3.1. Mine Wall Mechanics Model

According to Platts' ground pressure theory, when excavation is carried out, the original initial stress state of the surrounding rock will be destroyed, and the stress will be redistributed [29]. When the stress field becomes stable again, a pressure arch containing the plastic zone will be formed, namely the Platts arch [30]. After the mining room is mined, the self-weight of the surrounding rock in the upper arched plastic area is borne by the backfill, which is denoted as G_1. Taking the isolated ore pillar as the research object, when the isolated ore pillar is mined, the rock body above the goaf will cave in, and a temporary stable caving arch will gradually form. The self-weight of the surrounding rock of the caving arch is G_2. Figure 2a shows the bearing mechanism diagram of the backfill–mine wall system. Considering the backfill as a homogeneous medium, the effect of the backfill body on the mine wall increases linearly along the depth direction. When the backfill bears the direct upper load, the initial value of the top load on the backfill is not 0, and there is an initial lateral effect on the top of the mine wall. This study considers the force form of the mine wall under the trapezoidal lateral load. The mechanical model of the mine wall is shown in Figure 2b.

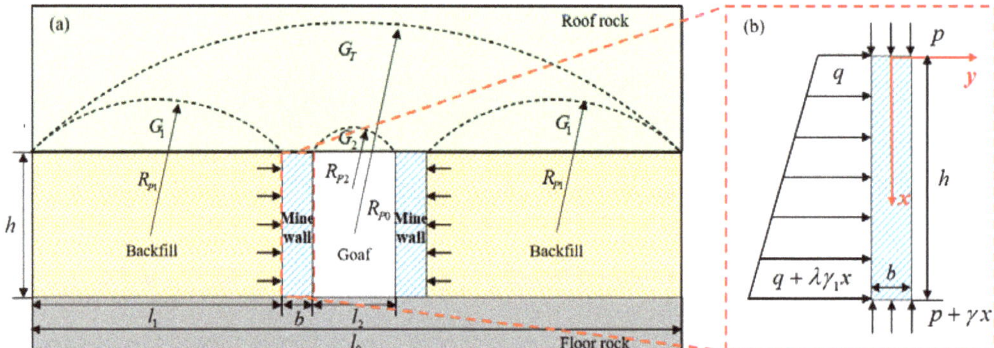

Figure 2. (**a**) Bearing mechanism diagram of the backfill–mine wall system. (**b**) Mine wall mechanic model.

In Figure 2a: G_T is the total weight of the overlying surrounding rock of the backfill–mine wall system; G_1 is the self-weight of the upper surrounding rock for the backfill; G_2 is the self-weight of the caving surrounding rock in the isolated pillar goaf; R_{P0} is the equivalent radius of the plastic zone of the backfill–mine wall system; R_{P1} is the equivalent radius of the upper plastic zone of the backfill; R_{P2} is the equivalent radius of the plastic zone in the upper part of the goaf; l_0 is the span of the backfill–mine wall system; l_1 is the width of the mine room; l_2 is the span of the goaf.

In Figure 2b: p is the initial upper load; q is the initial lateral load; λ is the lateral pressure coefficient; γ_1 is the bulk density of the backfill; γ is the bulk density of the mine wall; h is the height of the mine wall; b is the width of the mine wall, x is the vertical distance from the origin of the coordinates, y is the lateral distance to the origin of the coordinates.

According to our previous research [31–33], the stress expression of the mine wall under trapezoidal load was obtained by the semi-inverse solution method of elastic mechanics:

$$\begin{cases} \sigma_x = -\frac{2\lambda\gamma_1 y}{b^3}x^3 - \frac{6qy}{b^3}x^2 + \left(\frac{4\lambda\gamma_1 y^3}{b^3} + \frac{6qhy}{b^3} - \frac{3\lambda\gamma_1 y}{5b} + \frac{6\gamma y}{b}\right)x \\ \quad + \frac{4qy^3}{b^3} + \frac{6qhy}{b^3} - \frac{3qhy}{5b} - p - \gamma x \\ \sigma_y = \left(-\frac{2q}{b^3} - \frac{2\lambda\gamma_1 x}{b^3}\right)y^3 + \left(\frac{3q}{2b} + \frac{3\lambda\gamma_1 x}{2b}\right)y - \frac{\lambda\gamma_1 x}{2} - \frac{q}{2} \\ \tau_{xy} = \left(\frac{3\lambda\gamma_1}{4b} - \frac{3\lambda\gamma_1 y^2}{b^3}\right)x^2 + \left(-\frac{3q}{2b} - \frac{6qy^2}{b^3}\right)x + \lambda\gamma_1\left(\frac{b}{80} - \frac{3y^2}{10b} + \frac{y^4}{b^4}\right) \\ \quad -\gamma_1\left(\frac{b}{4} + y - \frac{3y^2}{b}\right) - \frac{3qh}{4b} - \frac{3qhy^2}{b^3} \end{cases} \tag{1}$$

In Equation (1): σ_x is the vertical stress of the mine wall, σ_y is the lateral stress of the mine wall, τ_{xy} is the shear stress of the mine wall.

According to Platts' theory, the plastic zone radius correction coefficient ξ is [26,27]:

$$\xi = \left[\frac{(\gamma_2 H_0 + C_t \cos\varphi_1)(1 + \sin\varphi_1)}{C_t \cos\varphi_1}\right]^{\frac{1-\sin\varphi_1}{2\sin\varphi_1}} \tag{2}$$

The dead weights of the backfill–mine wall system, the backfill, and the caving surrounding rock in the goaf are, respectively,

$$G_T = \left[\sqrt{\left(\frac{h}{2}\right)^2 + \left(\frac{l_0}{2}\right)^2}\xi - \frac{h}{2}\right] l_0\gamma_2 \tag{3}$$

$$G_1 = \left[\sqrt{\left(\frac{h}{2}\right)^2 + \left(\frac{l_1}{2}\right)^2}\xi - \frac{h}{2}\right] l_1\gamma_2 \tag{4}$$

$$G_2 = \left[\sqrt{\left(\frac{h}{2}\right)^2 + \left(\frac{l_2}{2}\right)^2}\xi - \frac{h}{2}\right] l_2\gamma_2 \tag{5}$$

Then, the initial vertical load of the mine wall p is:

$$p = \frac{G_T - 2G_1 - G_2}{Nb} \tag{6}$$

The initial lateral load q is calculated by the Rankine earth pressure equation:

$$q = \frac{\lambda G_1}{l_1} = \tan^2\left(45 - \frac{\theta}{2}\right)\left(\sqrt{\left(\frac{h}{2}\right)^2 + \left(\frac{l_1}{2}\right)^2}\xi - \frac{h}{2}\right)\gamma_2 \tag{7}$$

where γ_2 is the bulk density of the roof rock; H_0 is the mining depth; C_t is the cohesion of the roof rock; φ_1 is the friction angle of the roof rock; N is the number of pressure-bearing mine walls; θ is the friction angle of the backfill.

3.2. Compression Safety Factor K_σ

The compressive safety factor represents the compressive capacity of the mine wall, and its calculation equation is the ratio of the compressive strength of the mine wall to the maximum compressive stress on the mine wall. The maximum compressive stress on the mine wall is the maximum value of σ in Equation (1). For the value of the compressive strength of the mine wall, some scholars have deduced a variety of calculation equations for the bearing strength of the mine wall. Among them, the empirical equation proposed by Bieniawski [34] is more accurate and widely used. The equation for calculating the compressive strength of the mine wall is:

$$\sigma_p = \sigma_c \left[0.64 + 0.36\left(\frac{b}{h}\right)\right]^{\alpha} \tag{8}$$

In Equation (8): σ_p represents the compressive strength of the mine wall; σ_c represents the uniaxial compressive strength of the complete rock mass; b represents the width of the mine wall; h represents the height of the mine wall; α is a constant, and its value depends on the size of the mine wall. When the ratio of the width to height dimension of the mine wall is greater than 5, the value of α is 1.4, and when the aspect ratio is less than 5, the value of α is 1.0.

Therefore, the calculation equation of the mine wall compression safety factor is:

$$K_\sigma = \frac{\sigma_p}{\max(\sigma)} = \frac{\sigma_c\left[0.64 + 0.36\left(\frac{b}{h}\right)\right]^{\alpha}}{\max\left\{\begin{array}{l} -\frac{2\lambda\gamma_1 y}{b^3}x^3 - \frac{6qy}{b^3}x^2 + \left(\frac{4\lambda\gamma_1 y^3}{b^3} + \frac{6qhy}{b^3} - \frac{3\lambda\gamma_1 y}{5b} + \frac{6\gamma y}{b}\right)x \\ +\frac{4qy^3}{b^3} + \frac{6qhy}{b^3} - \frac{3qhy}{5b} - p - \gamma x, \\ \left(-\frac{2q}{b^3} - \frac{2\lambda\gamma_1 x}{b^3}\right)y^3 + \left(\frac{3q}{2b} + \frac{3\lambda\gamma_1 x}{2b}\right)y - \frac{\lambda\gamma_1 x}{2} - \frac{q}{2} \end{array}\right\}} \tag{9}$$

3.3. *Shear Safety Factor* K_τ

The mine wall shear safety factor is an index reflecting the size of the mine wall shear resistance, and its calculation equation is the ratio of the shear strength of the mine wall shear plane to the maximum shear stress on the mine wall. The maximum shear stress on the mine wall is the maximum value of τ_{xy} in Equation (1). The shear strength of the mine wall can be calculated by the Mohr–Coulomb strength criterion, and its calculation expression is:

$$\tau_p = \sigma_0 \tan\varphi + c \tag{10}$$

where σ_0 is the normal stress in the normal direction of the shear plane of the mine wall; φ is the friction angle of the mine wall; c is the cohesion of the mine wall.

Therefore, the calculation equation of the mine wall shear safety factor is:

$$K_\tau = \frac{\tau_p}{\max(\tau_{xy})} = \frac{\sigma_0 \tan\varphi + c}{\max\left\{\begin{array}{l} \left(\frac{3\lambda\gamma_1}{4b} - \frac{3\lambda\gamma_1 y^2}{b^3}\right)x^2 + \left(-\frac{3q}{2b} - \frac{6qy^2}{b^3}\right)x + \lambda\gamma_1\left(\frac{b}{80} - \frac{3y^2}{10b} + \frac{y^4}{b^4}\right) \\ -\gamma_1\left(\frac{b}{4} + y - \frac{3y^2}{b}\right) - \frac{3qh}{4b} - \frac{3qhy^2}{b^3} \end{array}\right\}} \tag{11}$$

3.4. *Overturning Safety Factor* K_c

The overturning safety factor is the anti-overturning ability of the mine wall, and its calculation expression is the ratio of the anti-overturning moment of the mine wall to the force and bending moment of the backfill on the mine wall. According to the force of the mine wall, the expression of the anti-overturning moment of the mine wall is:

$$M_K = \frac{b}{2}(G_p + G) \tag{12}$$

In Equation (12): M_K is the anti-overturning moment of the mine wall; G_p is the load of the overlying surrounding rock on the mine wall and its calculation equation is: $G_p = pb$; G is the gravity of the mine wall per unit thickness, and its calculation equation is: $G = \gamma hb$.

The force of the backfill on the mine wall can be calculated by Rankine earth pressure Equation (13):

$$F_y = qh + \frac{1}{2}\lambda\gamma_1 h^2 \tag{13}$$

The pressure of the backfill on the side of the mine wall is linearly distributed in a trapezoid shape. According to the moment calculation, the action point can be obtained $\frac{3qh+\lambda\gamma_1 h^2}{3(2q+\lambda\gamma h)}$ from the bottom surface, so the overturning safety factor of the mine wall is:

$$K_c = \frac{\frac{b}{2}(G_p + G)}{\left(qh + \frac{1}{2}\lambda\gamma_1 h^2\right)\frac{3qh+\lambda\gamma_1 h^2}{3(2q+\lambda\gamma h)}} \tag{14}$$

3.5. Comprehensive Safety Factor K

Considering the various failure forms of the mine wall, the comprehensive safety factor is used as an index to evaluate the stability of the mine wall, and its calculation equation is as follows:

$$K = \min\{K_\sigma, K_\tau, K_c\} \tag{15}$$

In Equation (15): K is the comprehensive safety factor, when $K = K_\sigma$, the main failure form of the mine wall is compression failure; when $K = K_\tau$, the main failure form of the mine wall is shear failure; when $K = K_c$, the main failure form of the mine wall is overturning failure.

4. Analysis of Factors Affecting Mine Wall Stability

According to Equation (15), the factors affecting the stability of the mine wall mainly include the width of the mine wall, the height of the mine wall, the friction angle of the backfill, the bulk density of the backfill, the width of the mine room, the mining depth, the uniaxial compressive strength of the mine wall and the mechanical properties of the surrounding rock. Combined with the current mining situation of the Dongguashan Copper Mine. Since the mining depth, mine wall height, mine wall uniaxial compressive strength and mechanical properties of the overlying surrounding rock are determined by the mine geological conditions, it is difficult to artificially design and adjust, so the main consideration of mine wall stability is the width of the mine room, the width of the mine wall, the friction angle of the backfill and the bulk density of the backfill.

According to the measured data of the mine, the mining depth $H_0 = 700$ m, the uniaxial compressive strength of the mine wall $\sigma_p = 104.5$ Mpa, the thickness of the ore body is about 20–50 m. and the weighted average thickness is about 40 m. For the convenience of calculation, we equivocate it with a model with a height of 40 m, therefore $h = 40$ m. Furthermore, the mechanical parameters of the mine wall and the overlying surrounding rock are shown in Table 1.

Table 1. Mechanical parameters of mine wall and overlying surrounding rock.

Name	Bulk Density/KN·m^{-3}	Poisson	Cohesion/Mpa	Friction/°
Roof rock	27.1	0.2087	36.50	35.3
Mine wall	32.2	0.3124	21.43	50.21

The above four factors are analyzed by the orthogonal range analysis method, combined with the actual situation of the Dongguashan project. Each factor is assigned within an appropriate range, and the calculation results of the mine wall safety factor under the conditions of each mine wall mining design scheme are shown in Table 2. The stability coefficient range results are shown in Table 3.

From the orthogonal test results in Table 2, it can be seen that the comprehensive safety factor of the mine wall is mostly the compression safety factor or the overturning safety factor. The main form of wall instability is brittle compression failure or overturning failure. According to the extreme difference of each factor in Table 3, the important factors affecting the stability of the mine wall are the width of the mine wall, the bulk density of the backfill, the friction angle of the backfill, and the width of the mine room. In order to further study the variation law between the factors affecting mine wall stability and the mine wall safety factor, the research factors are regarded as variables and other factors are regarded as quantitative, and the quantitative relationship between each factor and the mine wall comprehensive safety factor is studied by the control variable method.

Table 2. Orthogonal test table of mine wall safety factor.

Test Number	l_1/m	b/m	φ/°	γ_1/KN·m^{-3}	K_σ	K_τ	K_c	K
1	60	2	20	20	0.3870	1.9867	0.7080	0.6863
2	60	3	28	32	0.7225	2.5390	0.8276	0.7225
3	60	4	36	24	1.9760	4.9425	2.2466	1.9760
4	60	5	24	36	1.5123	2.7669	1.0465	1.0465
5	60	6	32	28	3.2470	4.9980	2.4223	2.4223
6	65	2	36	32	0.4456	2.6540	0.6855	0.4456
7	65	3	24	24	0.8124	2.3587	0.9161	0.8124
8	65	4	32	36	1.2796	3.0240	1.0286	1.0286
9	65	5	20	28	1.6343	2.5561	1.1373	1.1373
10	65	6	28	20	3.6705	4.7018	2.7559	2.7559
11	70	2	32	24	0.4952	2.3765	0.7222	0.4952
12	70	3	20	36	0.5020	1.4942	0.4386	0.4386
13	70	4	28	28	1.3959	2.7750	1.0948	1.0948
14	70	5	36	20	3.3561	5.2323	2.7465	2.7465
15	70	6	24	32	2.3529	2.9248	1.2510	1.2510
16	75	2	28	36	0.3050	1.5236	0.3536	0.3050
17	75	3	36	28	1.0779	2.9706	0.9955	0.9955
18	75	4	24	20	1.6238	2.5372	1.1667	1.1667
19	75	5	32	32	2.1605	3.2314	1.2491	1.2491
20	75	6	20	24	2.6039	2.6640	1.3099	1.3099
21	80	2	24	28	0.3370	1.3522	0.3552	0.3370
22	80	3	32	20	1.2449	2.6173	1.0063	1.0063
23	80	4	20	32	0.9804	1.6180	0.5517	0.5517
24	80	5	28	24	2.4243	2.9227	1.2762	1.2762
25	80	6	36	36	3.2303	3.8513	1.4194	1.4194

Table 3. Stability coefficient range analysis results.

The Mean of K	l_1/m	b/m	θ/°	γ_1/KN·m^{-3}
K_1	1.3707	0.4538	0.8248	1.6723
K_2	1.2360	0.7951	0.9227	1.1739
K_3	1.2052	1.1636	1.2309	1.1974
K_4	1.0052	1.4911	1.2403	0.8440
K_5	0.9181	1.8317	1.5166	0.8476
R	0.4526	1.3779	0.6918	0.8284

4.1. Quantitative Relationship between Mine Wall Safety Factor and the Width of the Mine Wall

Other factors affecting the safety factor of the mine wall are regarded as fixed values, and study the relationship between the safety factor of the mine wall and the width of the mine wall, the width of the mine room $l_1 = 70$, the friction angle of the backfill $\theta = 28°$, and the bulk density of the backfill $\gamma_1 = 24$ KN·m^{-3}. Figure 3 is a diagram showing the relationship between the safety factor and the width of the mine wall.

As shown in Figure 3, with the increase of the width of the mine wall, the bearing strength and the anti-overturning capacity of the mine wall will increase. Therefore, the mine wall compression safety factor, shear safety factor and overturn safety factor all increase with the increase of the width of the mine wall. Analysis of the change rate of the safety factor shows that with the increase of the width of the mine wall, the increase rate of the shear safety factor and the overturning safety factor is basically stable, and the increase rate of the compression safety factor increases, which indicates that the width of the mine wall has the most significant effect on the compression safety factor. Under the value level of the above factors, when the width of the mine wall is less than 3 m, the mine wall comprehensive safety factor is less than 1, the mine wall is in an unstable state, and the main failure forms are compression failure and overturning failure. With the increase in the width of the mine wall, the mine wall compression safety factor increases sharply. When the width of the mine wall is greater than 4 m, the mine wall safety factor is greater than 1, the mine wall compression safety factor is greater than the overturning safety factor, and the backfill has a more significant effect on the mine wall bending.

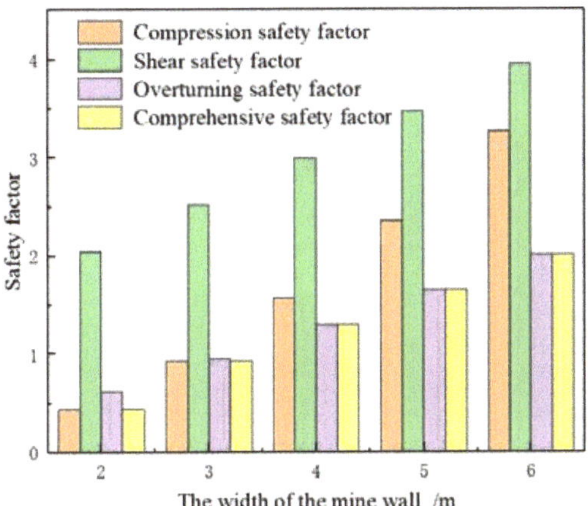

Figure 3. Relationship between safety factor and the width of the mine wall.

4.2. Quantitative Relationship between Safety Factor of Mine Wall and Bulk Density of the Backfill

The width of the mine wall is taken as 4 m, the width of the mine room is taken as 70 m, and the friction angle of the backfill is taken as 28°. The quantitative relationship between the safety factor of the mine wall and the bulk density of the backfill is studied. The results are shown in Figure 4.

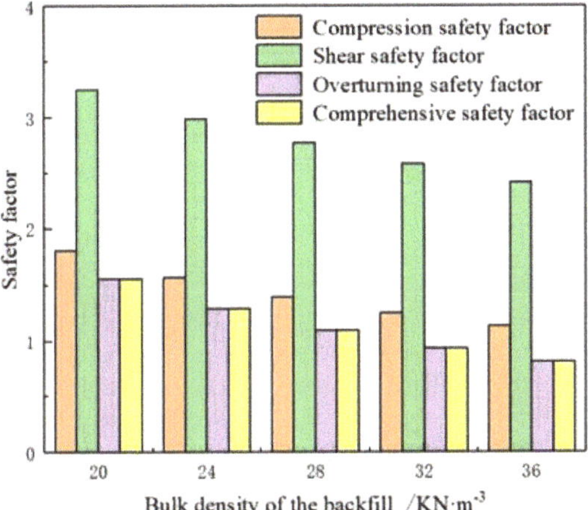

Figure 4. Relationship between safety factor and bulk density of the backfill.

It can be seen from Figure 4 that the mine wall compression safety factor, shear safety factor and overturning safety factor all decrease with the increase of the bulk density of the backfill, but the decreasing rate of all three decreases gradually. Comparing the mine wall compression safety factor and the change rate of the overturning safety factor, we can see that the overturning safety factor of the mine wall changes faster with the bulk density of the backfill. Therefore, when considering the influence of the bulk density of the backfill on the stability of the mine wall, the overturning failure form of the mine wall should be considered.

4.3. Quantitative Relationship between the Mine Wall Safety Factor and the Friction of the Backfill

Controlling other variables, the variation law of the mine wall safety factor with the friction angle of the backfill is analyzed. The width of the mine wall and the width of the mine room are 4 m and 70 m, respectively, and the bulk density of the backfill is 24 KN·m^{-3}. Figure 5 is a diagram showing the relationship between the mine wall safety factor and the friction angle of the backfill.

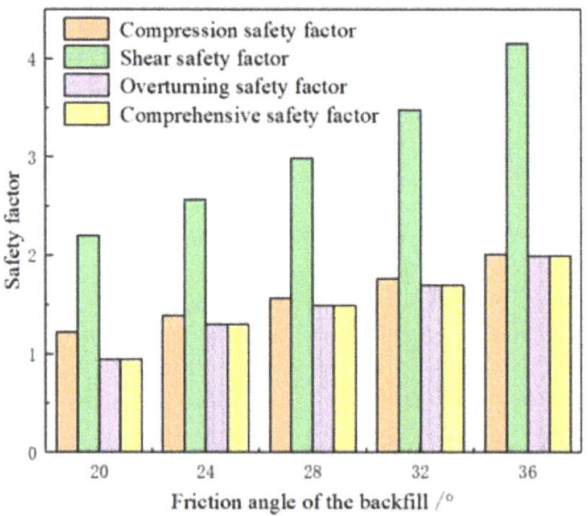

Figure 5. Relationship between safety factor and friction angle of the backfill.

It can be seen from Figure 5 that with the increase of the friction angle of the backfill, the safety factor of the mine wall increases, and the increase rate of the compression safety factor remains basically unchanged, while the increase rate of the shear safety factor and the overturning safety factor gradually increases. The analysis shows that when the friction angle of the backfill increases, the lateral pressure coefficient decreases, the effect of the backfill on the mine wall decreases, and the safety factor of the mine wall increases accordingly. Because the friction angle in the backfill mainly affects the side of the backfill on the mine wall. Therefore, the shear safety factor and the overturning safety factor are more affected by the friction angle of the backfill.

4.4. Quantitative Relationship between Mine Wall Safety Factor and the Width of the Mine Room

The width of the mine wall is 4 m, the bulk density and friction angle of the backfill are 24 KN·m^{-3} and 28°, respectively, and the variation law of the safety factor of the mine wall with the width of the mine room is studied. The results are shown in Figure 6.

With the increase of the width of the mine room, the load of the overlying surrounding rock on the backfill increases, the load on the overlying surrounding rock on the mine wall decreases, and the lateral load of the backfill increases. Therefore, with the increase in the width of the mine room, the compression safety factor of the mine wall increases gradually and the shear safety factor and overturning safety factor of the mine wall both decrease, which is consistent with the trend shown in Figure 6. At the level of the above factors, when the width of the mine room exceeds 80 m, the mine wall comprehensive safety factor is taken as the overturning safety factor and the mine wall is more prone to overturning failure.

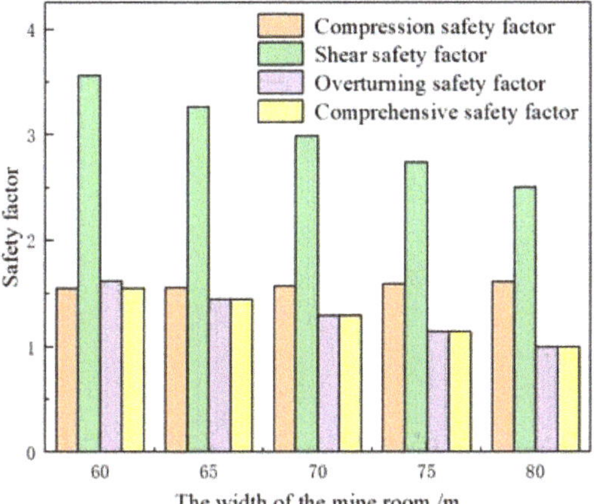

Figure 6. Relationship between safety factor and the width of the mine room.

4.5. Multi-Factor Analysis of Mine Wall Safety Factor

In order to comprehensively analyze the variation law of the comprehensive safety factor of the mine wall with the width of the mine wall, the width of the mine room, the bulk density of the backfill and the friction angle of the backfill, a regression equation of the safety factor including four factors was established, and the quantitative relationship is used to quantitatively reflect the change of the comprehensive safety factor of the mine wall. According to the orthogonal experimental calculation results in Table 2, a multiple regression equation is constructed by the Matlab data processing software and the expression of the regression equation can be obtained as:

$$y = 1.5513 - 0.0227x_1 + 0.3452x_2 + 0.0425x_3 - 0.0495x_4 + \varepsilon \tag{16}$$

In Equation (16): the dependent variable y represents the comprehensive safety factor of the mine wall; x_1 represents the width of the mine room (m); x_2 represents the width of the mine wall (m); x_3 represents the friction angle of the backfill (°); x_4 represents the bulk density of the backfill (KN·m^{-3}); ε represents the residual.

The complex correlation coefficient of the regression equation in Equation (16) $R = 0.8848$ shows that the regression equation fits well, and $F = 38.4076$, $p = 0.0000$ shows that the explanatory variable (x_i) is significant for the coefficient test. Figure 7 is the residual of the regression equation. It can be seen that there is only one abnormal point, which also shows that the regression effect is good.

According to the independent variable coefficient sign of the regression equation of the comprehensive safety factor, the comprehensive safety factor of the mine wall changes in the same direction as the width of the mine wall and the friction angle of the backfill. The increase in the width of the mine wall or the friction angle of the backfill will increase the comprehensive safety factor of the mine wall. The comprehensive safety factor of the mine wall changes inversely with the width of the mine room and the bulk density of the backfill, which is the same as the above quantitative analysis results. According to the size of the independent variable coefficient of the expression of the regression equation, the important factors affecting the stability of the mine wall are: the width of the mine wall, the bulk density of the backfill, the friction angle of the backfill, and the width of the mine room, which are the same as those shown in Table 3. At the same time, according to the independent variable coefficient value of the expression of the regression equation, the influence weight of each factor on the comprehensive safety factor of the mine wall can be quantitatively analyzed.

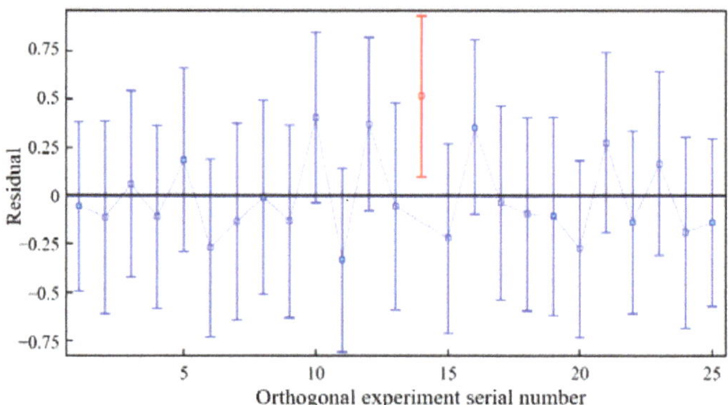

Figure 7. Residual diagram of the regression equation of the comprehensive safety factor of the mine wall.

5. Engineering Examples

According to the design scheme of isolated pillar mining in the Dongguashan Copper Mine, the width of the mine wall is 4 m, the width of the mine room is 78 m, and the filling scheme of the mine room uses full tailings cement filling. Experiments show that the bulk density of the backfill is 24 KN·m^{-3} and the friction angle is about 34.6°. According to Equation (15), the comprehensive safety factor of the mine wall is 1.39, and the comprehensive safety factor of the mine wall calculated by the regression equation of Equation (16) is 1.46; the error rate is less than 6%. Furthermore, the comprehensive safety factors of the mine wall obtained by the two equations both exceed the critical safety factor of 1.00. Theoretical calculation shows that under this mining design scheme, the mine wall is in a stable state.

According to the above design scheme, the isolated pillar is mined. After the mining is completed, the CMS goaf detection technology is used to scan the pillar back to the mined area to observe the retention of the mine wall [35], and use the BGK-A3 displacement meter to monitor the deformation of the roof rock during the mining process [36]. The triaxial stress meter is used to monitor the stress change of the mine wall. Figure 8 is a cross-sectional view of the scanning results of the open area after the mining of the 52-line isolated pillar G5 is completed.

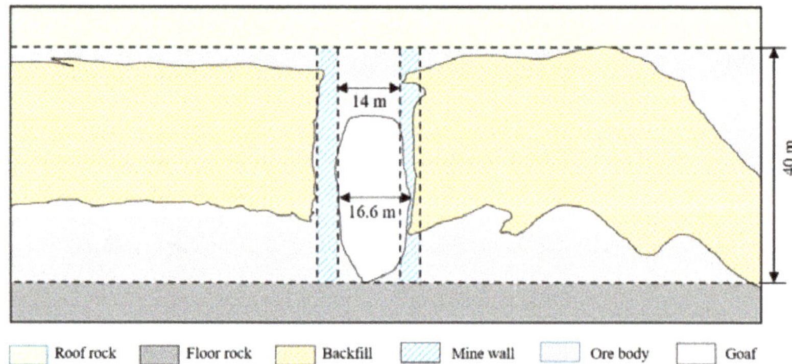

Figure 8. Vertical cross section view of empty area scan results.

It can be seen from Figure 8 that after the G5 isolation pillar is mined, the integrity of the top of the mine wall remains relatively good, no overturning failure occurs, no fracture occurs at the bottom of the mine wall and the pressure bearing performance is good. The integrity of the left mine wall is good, but there is over-mining in the middle of the right mine wall. The maximum span of the goaf is 16.6 m, and there is no large amount of backfill mixed in the gob scanning and ore mining. The maximum displacement of the roof surrounding rock measured by the BGK-A3 displacement meter is only 32 mm, and the deformation is in a controllable range. During the mining of the isolated pillar, the relative change of the maximum principal stress of the mine wall is 1.16 Mpa, the change of the intermediate principal stress is 1.17 Mpa, and the minimum change of the principal stress is 2.05 Mpa. The change in the stress value of the mine wall is small and it is in a stable state. The reliability of the mine wall is good, which is consistent with the theoretical calculation results and meets the mining requirements. In addition, according to the existing research results [15,37,38], this article analyzes the stability of the mine wall under this parameter by the energy variation method, the cusp catastrophe theory and the thin plate theory. The calculation results are consistent with the above situation, which further verifies the theoretical analysis results and improves the reliability of the research results.

6. Conclusions

In this study, according to the mining situation of an isolated pillar in the Dongguashan Copper Mine, the bearing mechanism of the backfill–mine wall system is analyzed, and the influence of backfill mechanical parameters on the stability of the mine wall is considered. Based on the failure forms of the mine wall under multi-directional loads, the comprehensive evaluation index of the mine wall stability is proposed, the calculation equation of the comprehensive safety factor is deduced, the orthogonal test is designed to analyze the sensitivity of the influencing factors of the mine wall stability, and the following research results are obtained:

1. The important factors affecting the stability of the mine wall are the width of the mine wall, the bulk density of the backfill, the friction angle of the filling body and the width of the mine room. Among them, the width of the mine wall mainly affects the bearing strength and the stress distribution state of the mine wall. The bulk density of the backfill and the friction angle mainly affect the lateral load of the mine wall, and the width of the mine room affects both the vertical load and the lateral load of the mine wall.
2. The main forms of mine wall failure are brittle compression failure and overturning failure. Increasing the width of the mine wall can significantly improve the mine wall compression safety factor. Reducing the bulk density of the backfill and increasing the friction angle of the backfill can improve the mine wall overturning safety factor. The increase of the mine width increases the compression safety factor and reduces the overturning safety factor of the mine wall.
3. According to the regression equation calculation, the comprehensive safety factor of the 52-line G5 mine wall is about 1.4, which is close to the theoretical equation and the actual situation of engineering exploration. It provides ideas for the optimization of mine wall design and filling scheme in the process of deep isolation pillar mining.

Author Contributions: Conceptualisation, J.G. and J.L.; methodology, X.C.; validation, X.X.; writing—original draft preparation, X.C. and Y.Z.; writing—review and editing, X.C. and Y.Z. All authors have read and agreed to the published version of the manuscript.

Funding: Project supported by the National Natural Science Foundation of China Received (No. 52174140); Project supported by the National Natural Science Foundation of China Received (No. 52074351); Postgraduate Scientific Research Innovation Project of Hunan Province (No. QL20210054); Supported by Postgraduate Innovative Project of Central South University "Research and practice of comprehensive reinforcement technology of wellbore in water-rich broken soil layer" (No. 2021XQLH066).

Acknowledgments: The team of authors express their gratitude to the editors and reviewers for valuable recommendations that have been taken into account to improve significantly the quality of this article.

Conflicts of Interest: The authors declare no conflict of interest.

References

1. Singh, R.; Mandal, P.K.; Singh, A.K.; Kumar, R.; Sinha, A. Coal pillar extraction at deep cover: With special reference to Indian coalfields. *Int. J. Coal Geol.* **2011**, *86*, 276–288. [CrossRef]
2. Wagner, H. Deep mining: A rock engineering challenge. *Rock Mech. Rock Eng.* **2019**, *52*, 1417–1446. [CrossRef]
3. Chen, X.J.; Li, L.Y.; Wang, L.; Qi, L.L. The current situation and prevention and control countermeasures for typical dynamic disasters in kilometer-deep mines in China. *Saf. Sci.* **2019**, *115*, 229–236. [CrossRef]
4. He, M.C.; Xie, H.P.; Peng, S.P.; Jiang, Y.D. Study on rock mechanics in deep mining engineering. *Chin. J. Rock Mech. Eng.* **2005**, *24*, 2803–2813.
5. Xie, H.P.; Gao, F.; Ju, Y. Research and development of rock mechanics in deep ground engineering. *Chin. J. Rock Mech. Eng.* **2015**, *34*, 2161–2178.
6. Dong, L.J.; Hu, Q.C.; Tong, X.J.; Liu, Y.F. Velocity-free MS/AE source location method for three-dimensional hole-containing structures. *Engineering* **2020**, *6*, 827–834. [CrossRef]
7. Qi, C.; Fourie, A. Cemented paste backfill for mineral tailings management: Review and future perspectives. *Miner. Eng.* **2019**, *144*, 106025. [CrossRef]
8. Fall, M.; Célestin, J.C.; Pokharel, M.; Touré, M. A contribution to understanding the effects of curing temperature on the mechanical properties of mine cemented tailings backfill. *Eng. Geol.* **2010**, *114*, 397–413. [CrossRef]
9. Lei, Z.; Rougier, E.; Knight, E.E.; Munjiza, A. A framework for grand scale parallelization of the combined finite discrete element method in 2d. *Comput. Part. Mech.* **2014**, *1*, 307–319. [CrossRef]
10. Das, A.J.; Mandal, P.K.; Bhattacharjee, R.; Tiwari, S.; Kushwaha, A.; Roy, L.B. Evaluation of stability of underground workings for exploitation of an inclined coal seam by the ubiquitous joint model. *Int. J. Rock Mech. Min. Sci.* **2017**, *93*, 101–114. [CrossRef]
11. Basarir, H.; Sun, Y.; Li, G. Gateway stability analysis by global-local modeling approach. *Int. J. Rock Mech. Min. Sci.* **2019**, *113*, 31–40. [CrossRef]
12. Ibishi, G. Stability assessment of post pillars in cut-and-fill stoping method at trepça underground mine. *Geomech. Eng.* **2019**, *28*, 463–475.
13. Zhang, G.C.; Chen, L.J.; Wen, Z.J.; Chen, M.; Tao, G.Z.; Li, Y.; Zuo, H. Squeezing failure behavior of roof-coal masses in a gob-side entry driven under unstable overlying strata. *Energy Sci. Eng.* **2020**, *8*, 2443–2456. [CrossRef]
14. Wei, G.; Ge, M.M. Stability of a coal pillar for strip mining based on an elastic-plastic analysis. *Int. J. Rock Mech. Min. Sci.* **2016**, *100*, 23–28.
15. Huang, Z.G.; Dai, X.G.; Dong, L.J. Buckling failures of reserved thin pillars under the combined action of in-plane and lateral hydrostatic compressive forces. *Comput. Geotech.* **2017**, *87*, 128–138. [CrossRef]
16. Xu, Y.Y.; Li, H.Z.; Guo, G.L.; Liu, X.P. Stability analysis of hyperbolic coal pillars with peeling and high temperature effects. *Energy Explor. Exploit.* **2020**, *38*, 1574–1588. [CrossRef]
17. Wang, X.R.; Guan, K.; Yang, T.H.; Liu, X.G. Instability mechanism of pillar burst in asymmetric mining based on cusp catastrophe model. *Rock Mech. Rock Eng.* **2021**, *54*, 1463–1479. [CrossRef]
18. Idris, M.A.; Saiang, D.; Nordlund, E. Stochastic assessment of pillar stability at Laisvall mine using artificial neural network. *Tunn. Undergr. Space Technol.* **2015**, *49*, 307–319. [CrossRef]
19. Liu, Z.X.; Luo, T.; Li, X.; Li, X.B.; Huai, Z.; Wang, S.F. Construction of reasonable pillar group for undersea mining in metal mine. *Trans. Nonferrous Met. Soc. China* **2018**, *28*, 757–765. [CrossRef]
20. Ding, H.X.; Li, G.H.; Dong, X.; Lin, Y. Prediction of pillar stability for underground mines using the stochastic gradient boosting technique. *IEEE Access* **2018**, *6*, 69253–69264. [CrossRef]
21. Zhang, L.Y.; Deng, K.Z.; Zhu, C.G.; Xing, Z.Q. Analysis of stability of coal pillars with multi-coal seam strip mining. *Trans. Nonferrous Met. Soc. China* **2011**, *21*, s549–s555. [CrossRef]
22. Zhang, P.; Heasley, K.A. Elimination of boundary effect for laminated overburden model in pillar stability analysis. *J. Cent. South Univ.* **2016**, *23*, 1468–1473. [CrossRef]
23. Yang, R.S.; Zhu, Y.; Li, Y.L.; Li, W.Y.; Lin, H. Coal pillar size design and surrounding rock control techniques in deep longwall entry. *Arab. J. Geosci.* **2020**, *13*, 453. [CrossRef]
24. Chen, S.M.; Wu, A.X.; Wang, Y.M.; Chen, X. Analysis of influencing factors of pillar stability and its application in deep mining. *J. Cent. South Univ.* **2018**, *49*, 2050–2057.
25. Li, C.; Zhou, J.; Armaghani, D.J.; Li, X. Stability analysis of underground mine hard rock pillars via combination of finite difference methods, neural networks, and Monte Carlo simulation techniques. *Undergr. Space* **2021**, *6*, 379–395. [CrossRef]
26. Zhu, C.M.; Zhang, W.P. *Structural Mechanics*; Higher Education Press: Beijing, China, 2009; Volume 1, pp. 168–194.
27. Zhao, W.; Cao, P.; Zhang, G. *Rock Mechanics*; Central South University Press: Changsha, China, 2010; pp. 162–166.
28. Jiang, L.C.; Wang, Y.D. Comprehensive safety factor of residual mining pillar under complex loads. *J. Cent. South Univ.* **2018**, *49*, 1511–1518.
29. Dong, L.J.; Chen, Y.C.; Sun, D.Y.; Zhang, Y.H. Implications for rock instability precursors and principal stress direction from rock acoustic experiments. *Int. J. Min. Sci. Technol.* **2021**, *31*, 789–798. [CrossRef]
30. Ji, W.D. *Mine Rock Mechanics*; Metallurgical Industry Press: Beijing, China, 1991; pp. 36–41.
31. Guo, J.; Zhao, Y.; Zhang, W.X.; Dai, X.G.; Xie, X.B. Stress analysis of mine wall in panel barrier pillar-stope under multi-directional loads. *J. Cent. South Univ.* **2018**, *49*, 3020–3028.

32. Jiang, L.C.; Su, Y. Theoretical model and application of critical blasting vibration velocity for instability of cemented backfill pillar. *Chin. J. Nonferrous Met.* **2019**, *29*, 2663–2670.
33. Cao, S.; Du, C.F.; Tan, Y.Y.; Fu, J.X. Mechanical model analysis of consolidated filling pillar using stage-delayed backfill in metal mines. *Rock Soil Mech.* **2015**, *36*, 2370–2376.
34. Bieniawski, Z.T.; Van Heerden, W.L. The significance of in situ tests on large rock specimens. *Int. J. Rock Mech. Min. Sci. Geomech. Abstr.* **1975**, *12*, 101–113. [CrossRef]
35. Dong, L.J.; Tong, X.J.; Hu, Q.C.; Tao, Q. Empty region identification method and experimental verification for the two-dimensional complex structure. *Int. J. Rock Mech. Min. Sci.* **2021**, *147*, 104885. [CrossRef]
36. Dong, L.J.; Tong, X.J.; Ma, J. Quantitative investigation of tomographic effects in abnormal regions of complex structures. *Engineering* **2021**, *7*, 1011–1022. [CrossRef]
37. Guo, J.; Feng, Y.F. Cusp catastrophe theory analysis and width optimization on instability of pillar in deep mining. *J. Saf. Sci. Technol.* **2017**, *13*, 111–116.
38. Xue, B.X.; Li, D.X.; Kong, L.Y. Stability analysis model of ore wall based on elastic thin plate theory and its application. *J. Min. Saf. Eng.* **2020**, *37*, 698–706.

Article

Risk Assessment of Mining Environmental Liabilities for Their Categorization and Prioritization in Gold-Mining Areas of Ecuador

Bryan Salgado-Almeida [1], Daniel A. Falquez-Torres [1], Paola L. Romero-Crespo [1], Priscila E. Valverde-Armas [1], Fredy Guzmán-Martínez [2] and Samantha Jiménez-Oyola [1,*]

[1] Escuela Superior Politécnica del Litoral, ESPOL, Facultad de Ingeniería en Ciencias de la Tierra, Campus Gustavo Galindo km 30.5 Vía Perimetral, Guayaquil P.O. Box 09-01-5863, Ecuador; bryjosal@espol.edu.ec (B.S.-A.); dfalquez@espol.edu.ec (D.A.F.-T.); plromero@espol.edu.ec (P.L.R.-C.); priesval@espol.edu.ec (P.E.V.-A.)

[2] Mexican Geological Survey (SGM), Boulevard Felipe Angeles Km. 93.50-4, Pachuca 42083, Mexico; fredyguzman@sgm.gob.mx

* Correspondence: sjimenez@espol.edu.ec

Abstract: Mining environmental liabilities (MEL) are of great concern because of potential risks to ecosystems and human health. In this research, the environmental risk (R_I) related to MEL existing in three artisanal and small-scale gold-mining areas of Ecuador was evaluated. For this purpose, data of 167 MEL including landfills, mining galleries, tailing deposits, and mineral processing plants from Macuchi, Tenguel–Ponce Enriquez, and Puyango mining areas, were analyzed. The risk assessment related to the presence of waste deposits was carried out based on the methodology proposed by the Spanish Geological Survey. Moreover, the procedure outlined in the Environmental Risk Assessment Guide of the Ministry of Environment of Peru for nonwaste deposits was applied. The highest R_I values were identified in Puyango and Tenguel–Ponce Enriquez. Thus, they were both categorized as priority control areas requiring intervention and rehabilitation plans. The MEL that require a high level of intervention include waste deposits and mine entrances associated with potentially toxic elements. Moreover, the point risk maps showed that rivers in the studied areas have a potential pollution risk. This study provides risk levels associated with MEL in mining areas from Ecuador. This information could be used for environmental management and pollution mitigation.

Keywords: mining pollution; potentially toxic elements; risk management; abandoned mining areas; mining waste deposits

Citation: Salgado-Almeida, B.; Falquez-Torres, D.A.; Romero-Crespo, P.L.; Valverde-Armas, P.E.; Guzmán-Martínez, F.; Jiménez-Oyola, S. Risk Assessment of Mining Environmental Liabilities for Their Categorization and Prioritization in Gold-Mining Areas of Ecuador. *Sustainability* **2022**, *14*, 6089. https://doi.org/10.3390/su14106089

Academic Editors: Longjun Dong, Yanlin Zhao and Wenxue Chen

Received: 11 April 2022
Accepted: 13 May 2022
Published: 17 May 2022

Publisher's Note: MDPI stays neutral with regard to jurisdictional claims in published maps and institutional affiliations.

1. Introduction

Mining environmental liabilities (MEL) are elements such as facilities, infrastructures, surfaces affected by spills, disturbed watercourses, machine shops, tool storages, ore storages, mining waste deposits or stockpiles currently located in abandoned mines that pose a permanent potential risk for human health and the environment [1–3].

The absence of clear regulations has led the accumulation of MEL in mining zones worldwide. There are polluted areas due to the mining activities that have been carried out for centuries in different countries such as Peru [1–4], Mexico [5], Chile [6], South Africa [7], Ghana [8], Slovakia [9], Korea [10], Spain [11], and Ecuador [12–16]. Moreover, numerous MEL have been reported in Latin American countries such as Bolivia, Chile, Colombia, and Peru [17].

In most cases, tailings dams are MEL that represent significant ecological risk because it is feasible that they release potentially toxic elements (PTE) such as As, Cd, Cu, Zn, and Mn into the nearby water sources. These contaminated water resources are often used for crop irrigation or for human consumption, creating a serious pollution issue.

Tailing dams pose a potential risk to the environment and the safety of people due to the characteristics of the wastes and the intensive storage [18]. The poor management of mining waste has triggered the failure of dams with catastrophic consequences for the environment, as reported in Minas Gerais—Brazil [19,20], Aznalcóllar—Spain [21], and Karamken—Russia [22].

Several papers have been written about tailings management and mining disposal in recent years, describing the soil contamination, groundwater, and sediment problems [23,24]. These studies provide information about the evaluation of the possible side effects of the contamination of a mining site and its potential influence on the quality of the environment and people's health [7,25,26]. However, despite the associated problems of MEL, there is scarce research in Latin America that aims to provide the risk levels derived from the vicinity of abandoned mining facilities.

A timely risk assessment may help to identify problematic MEL by estimating the probability and severity of the consequences in several scenarios [25], helping to prioritize the MEL intervention and the strategies for environmental management [27]. Accordingly, the environmental risk evaluation is a process that comprises the measurement of the possible negative impacts resulting from the exposition of one or several factors of environmental stress. These stress factors correspond to physical, chemical, and biological entities that may cause damage to the ecosystem [28].

According to Alberruche del Campo et al. [29], the methods of the environmental risk analysis that use evaluation matrices are the ones that best adapt to the objectives of establishing action priorities in territories with a large number of MEL. Nevertheless, for the vast majority of MEL information is scarce, preventing detailed assessments. As an alternative to this problem, the Spanish Geological Survey has developed a methodology called Simplified Risk Assessment. This method is based on an MEL inventory and is supported by available technical and scientific information, i.e., topographic maps, aerial images, geological maps, unit maps, hydrogeological, geochemical maps, land use, distance to population centers and the state of conservation of the ecosystem among others that rank MEL. Once the MEL that requires further attention have been identified, fieldwork should be conducted to collect detailed information so that the remediation or rehabilitation project is realistic and appropriate for the site of interest.

There is a considerable quantity of MEL in Ecuador in three well-known mining districts where the MEL were inventoried, such as the Macuchi, Tenguel–Ponce Enriquez and Puyango. These areas are considered as mining zones of special relevance in terms of artisanal and small-scale gold-mining activity in Ecuador. Nevertheless, there is no available data about the risk level these waste facilities cause to the environment and the exposed population's health. In this context, this research aims: (a) to estimate the associated risk of MEL in each mining district and (b) to categorize and prioritize MEL that require immediate attention based on the contamination they can pose and their effects on the environment and the population. This information could be utilized in future management strategies to mitigate the environmental contamination related to the mining sites' abandonment and contribute to the sustainable development of the population and the environment.

2. Materials and Methods

2.1. Study Area

The research focuses on three significant artisanal and small-scale gold-mining (ASGM) districts in Ecuador, which belong to Macuchi, Tenguel–Ponce Enriquez and the Puyango River basin (Figure 1). Macuchi is in the Cotopaxi province and comprises an area of 16 km^2 (Figure 1a). Tenguel–Ponce Enriquez is situated between the Guayas and Azuay provinces; it has an area of 192 km^2 (Figure 1b). The Puyango River basin is located in the El Oro and Loja provinces and has a surface of 1924 km^2 (Figure 1c). The Puyango-Tumbes basin is one of the most critical watersheds in South America [30].

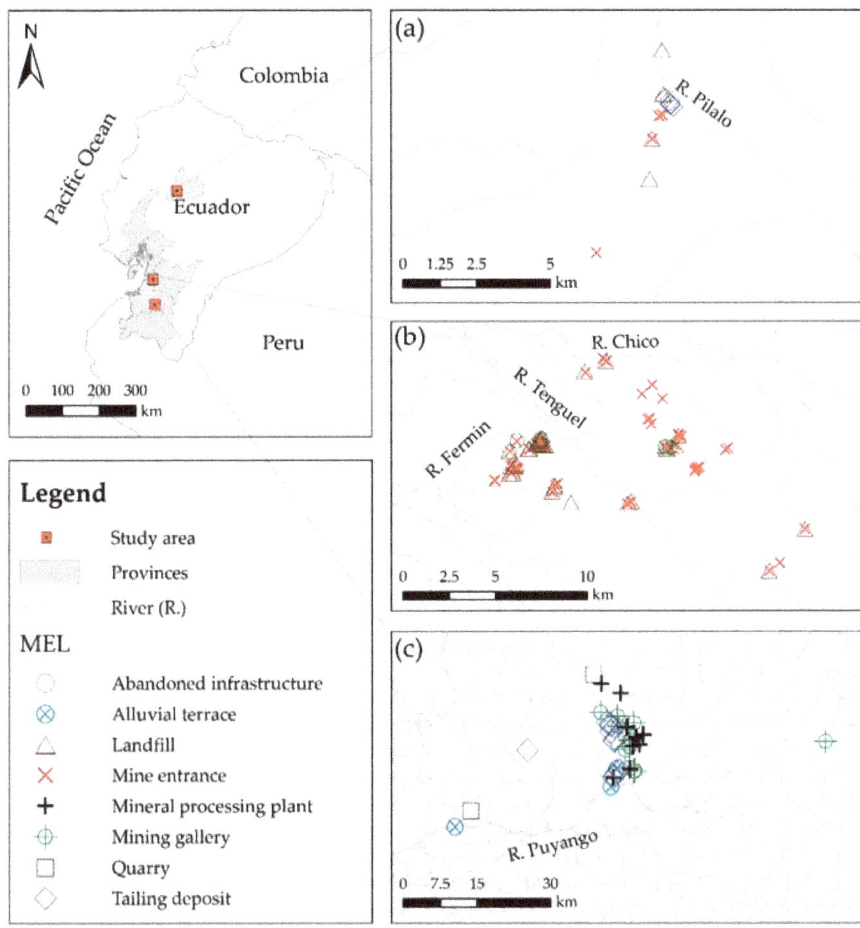

Figure 1. Location map of the study area and position of mining environmental liabilities (MEL). The MEL were in the mining districts of: (**a**) Macuchi; (**b**) Tenguel–Ponce Enriquez; (**c**) Puyango.

Intense mining and mineral processing activities (mainly located near rivers) have been reported in the three study areas [31], in addition to the agriculture and livestock activities [13,32]. The improper management of mining disposal, mainly from illegal mining, has derived from the accumulation of MEL and has significantly environmentally affected the ecosystem [13,33,34].

Previous research performed in the areas of interest reported the presence of PTE in several environmental facilities (Table S1, Supplementary Materials) as well as the potential risk of damage to the ecosystem and the population [35]. For example, in Macuchi, concentrations of PTE that exceed the maximum allowable limit (MAL) established in the Ecuadorian standard for water and soil have been reported [36]. At the same time, high levels of PTE, mainly As, in shallow water and river sediments have been reported in Tenguel–Ponce Enriquez and Puyango [12,15,34,37]. The PTE present in the contaminated areas, besides degrading the ecosystem, can enter the human body through different routes of exposure [38,39], affecting the area's residents' health [40,41].

2.2. Data Collection

A total of 167 MEL reported by the Ministry of Environment of Ecuador [30,36,42] were used in this study. The MEL included mine dumps, tailing dams, treatment plants, mining works, and abandoned facilities (Figure 1). Table 1 shows the number of MEL identified in each studied area. One of the most important MEL located in Macuchi and Tengel—Ponce Enriquez correspond to abandoned landfills and mine entrances. On the other hand, galleries and abandoned tailings deposits predominate within the Puyango area.

Table 1. Mining environmental liabilities reported in the studied areas.

MEL	Macuchi (n = 14)	Tenguel–Ponce Enriquez (n = 111)	Puyango (n = 42)
Landfills	4	34	-
Mining galleries (Mines)	-	-	14
Mine entrances	5	64	-
Tailings deposits	5	-	12
Abandoned infrastructure	-	13	-
Mineral processing plants	-	-	11
Alluvial terrace	-	-	3
Quarries	-	-	2
Landfills	4	34	-
Mining galleries (Mines)	-	-	14

n = number of MEL reported in the study area [30,36,42].

2.3. Risk Assessment

A simplified risk assessment was performed for the MEL following the procedure shown in Figure 2. For this purpose, the probability (I_P) and severity indices (I_S) associated with each MEL were calculated. In the case of mining waste deposits, the risk assessment was carried out using the methodology proposed by the Spanish Geological Survey [29]. For the rest of the MEL (nonwaste deposits), the procedure outlined in the Environmental Risk Assessment Guide of the Ministry of the Environment of Peru was followed [43]. The cartographic information was processed using Geographic Information Systems (GIS) through ArcMap 10.8.1 to identify the areas that represent a greater risk of affecting the population and ecosystem. This information allows proposing management actions for the MEL identified as significant concerns in each evaluated area.

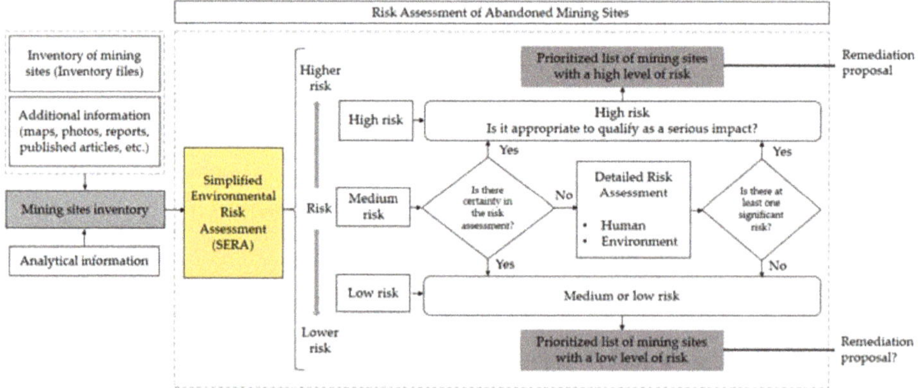

Figure 2. General scheme for simplified risk assessment of abandoned mining sites proposed by the Spanish Geological Survey [29].

2.3.1. Risk Scenarios

The risk characterization was conducted for four scenarios (S1–S4) identified as a priority in the present study (Figure 3). Two scenarios were considered for the risk assessment associated with the mining waste deposits: (i) S1—shallow waters are affected by the possible generation of acid mine drainage; and (ii) S2—population health is affected due to direct contact with MEL. In addition, two additional scenarios were considered for the MEL, such as galleries, mine entrances, infrastructure, processing plants, alluvial terraces, and abandoned quarries. Both scenarios potentially impact water bodies, urban areas, and agricultural areas: (iii) S3—contaminated water transportation or acid mine drainage; and (iv) S4—the drag of contaminated sediments.

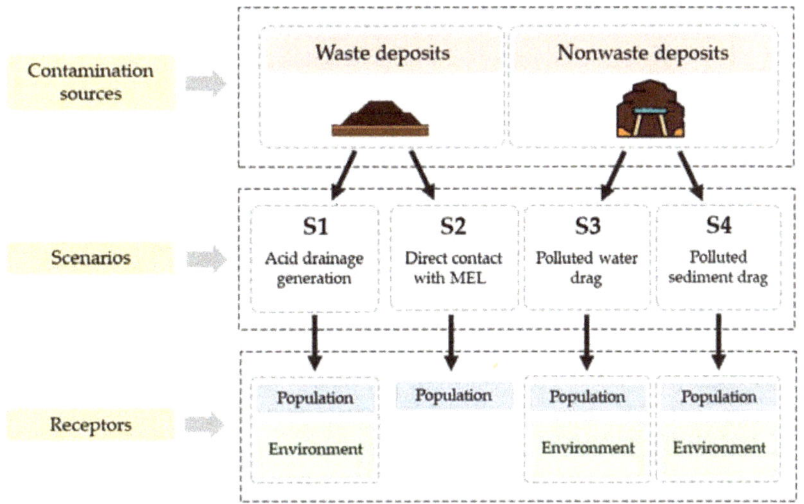

Figure 3. Scheme of risk assessment scenarios.

The probability (I_P) and severity (I_S) indices were rated on a scale from zero to five, being: very low ($\geq$0 and <1), low ($\geq$1 and < 2), medium ($\geq$2 and <3), high ($\geq$3 and <4), and very high ($\geq$4 and $\leq$5). The environmental risk estimation was obtained by multiplying the probability of occurrence (I_P) by the severity of the consequences (I_S). The last one included studying the types of receptors: the population effects and the natural environment [27]. Therefore, based on the $I_P \times I_S$ products, three grades of qualification for the risk (R_I) were distinguished: low risk ($0 \leq R_I \leq 5$), medium risk ($6 \leq R_I \leq 15$), and significant risk ($16 \leq R_I \leq 25$) [27,43].

2.3.2. The Probability Index (I_P)

The probability index for S1 was calculated considering the proximity factor to water bodies (P_R), the toxicity factor of the generated waste (F_{TOX}), and the factor of unprotected surface (F_{SD}), according to Equation (1) [29].

$$I_P\ (S1) = P_R \times F_{TOX} \times F_{SD} \quad (1)$$

The valuation criteria of P_R are shown in Table S2. F_{TOX} = 5 was established for the deposits with pH < 6.5 and the presence of acid mine drainage or heavy metal(loid)s. For the waste deposits with pH between 6.5 and 8.5 in its leachate, an F_{TOX} = 0.5 was rated. Moreover, F_{SD} was calculated by multiplying the exposed area of the waste structure (in ha) by a 0.5 factor, according to the methodology used. The exposed area considered for each deposit was 1 ha.

For the S2 scenario, I_P was quantified using Equation (2), which includes the pollutants direct contact concentration factor (F_{CCD}), waste deposits accessibility factor (F_{ACC}), and residential area proximity factor (P_{RR}).

$$I_P\ (S2) = F_{CCD} \times F_{ACC} \times P_{RR} \tag{2}$$

For the S2 evaluation, the total concentration values of heavy metal(loid)s (As, Cd, Cr, Hg, Ni, and Pb) reported in previous studies in soils and sediments (Table S1) were considered. For the MEL where the heavy metal(loid) concentration in soils was below the MAL of the Ecuadorian legislation [44], a concentration factor $F_{CCD} = 0$ was assigned, and for those which exceeded the MAL, $F_{CCD} = 5$ was allotted. Additionally, the valuation of P_{RR} and F_{ACC} is presented in Table S2. The criteria from Table 2 were used for the probability estimation (I_P) of S3 and S4. Both scenarios measure the effects of MEL presence in the population and natural environment.

Table 2. Assessment criteria for probability index (I_P) determination of scenario S3 and S4.

Criteria	I_P Value
Occurrence of the scenario continuously or daily	5
Scenario can happen within a week to a month	4
Scenario can happen within a month to a year	3
Scenario can happen within one to five years	2
Scenario can happen within a period greater than five years	1

2.3.3. The Severity Index (I_S)

The severity index (Is) for the S1 scenario evaluated the effects on the population I_S(S1PO) (Equation (3)) and in the natural environment I_S(S1NA) (Equation (4)).

$$I_S\ (S1PO) = 0.5P_{EX} + 0.5(F_{SUP-PO} \times V_P) \tag{3}$$

$$I_S(S1NA) = F_{SUP-NA} \times V_E \tag{4}$$

A population ($n > 50$) exposed to toxic elements by water consumption was considered ($P_{EX} = 5$). The exposition factor (F_{SUP}) was established by considering the receivers' distance to the mine waste deposit using the criteria in Table S3 (F_{SUP-PO} for the population and F_{SUP-NA} for the natural surroundings). Finally, the used criteria for evaluating population vulnerability (V_P) to the ingestion and/or direct contact with contaminated surface water and the natural environment to contamination with effluent from reservoirs (V_E) are presented in Table S3 with the F_{SUP} valuation criteria. On the other hand, in the S2 scenario, the impact on the population I_S(S2) was estimated according to the criteria shown in Table 3.

Table 3. Assessment criteria for severity index (I_S) determination of scenario S2.

Criteria	I_S Value
Uses with very high associated severity: marginal villages, children's parks	5
Uses with high associated severity: intensive recreational use (sports activities), isolated single-family homes	4
Uses with moderate associated severity: urbanized residential areas, nonintensive recreational use (trails, viewpoints)	3
Uses with low associated severity: agricultural and forestry activities	2
Uses with very low associated severity: other uses (commercial, industrial) with very low exposure	1

For the S3 and S4 scenarios, the severity index (I_S) for the population (I_S(S3PO) and I_S(S4PO)) and natural environment (I_S(S3NA) and I_S(S4NA)) was estimated according to the impact classification for the population and environment ($G_{PO/NA}$) (Equation (5)).

$$G_{PO/NA} = C + 2P + E + V_{PO/NA} \tag{5}$$

where C is the number of pollutants emitted to the environment, P is the dangerousness of the residue, E is the extension of the environmental impact, V_{PO} is the vulnerability to the affected population, and V_{NA} is the conservation state of the assessed environment. The values of the factors used in the I_S calculation for scenarios S3 and S4 are shown in Table S4.

3. Results

3.1. Risk Characterization

Calculated risk values results for S1–S2 and S3–S4 are presented in Tables S5 and S6, respectively.

3.1.1. Macuchi

Figure 4 shows the point risk map for the presence of MEL in the Macuchi area. In total, 93% of the MEL were identified in this region, representing a medium risk for the population and the natural environment. The risk values ranged from $6 \leq R_I \leq 15$. Figure 4a shows the abandoned landfills located at distances of >500 m from waterways, residential areas, and territories of environmental interest with a low probability of affectation. Nevertheless, in the surroundings of the Pilalo river, where the tailings are found, the affectation risk to the population from direct contact and contaminant effluents increases. This rise is because there exists a higher probability of affectation to the water bodies and the health of the people who receive that water. The highest risk valuation of affectation was obtained for the population in scenario S1 ($10 \leq R_I \leq 15$), however, these values only reach a medium-risk classification and not a high-risk one. This risk is caused by the mentioned tailings dams with acid drainage and heavy metal(loid)s, which exceed the MAL of Ecuadorian legislation. On the other hand, in scenario S2, 11% of the MEL presented a low risk, while 89% reported a medium risk for the population ($6 \leq R_I \leq 9$), see Figure 4b. Finally, the S3 and S4 scenarios resulted in a medium affectation risk for the people ($9 \leq R_I \leq 15$) and for the natural environment ($6 \leq R_I \leq 15$), as shown in Figure 4c,d.

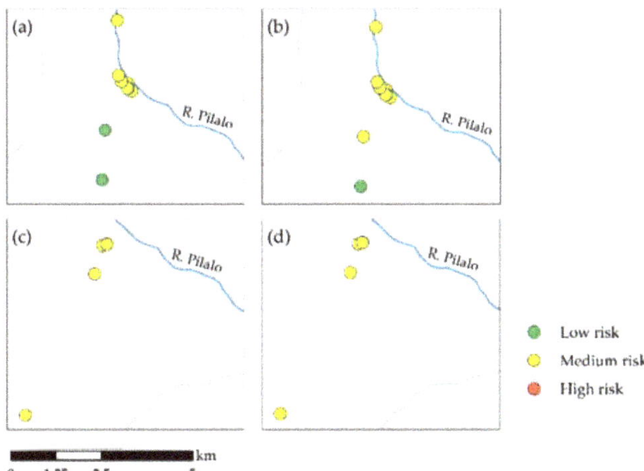

Figure 4. Point risk map for scenarios: (**a**) S1 for population and environment; (**b**) S2 for population; (**c**) S3 for population and environment; (**d**) S4 for population and environment in Macuchi area.

According to the MEL inventory [36], 60% of mine entrances studied in the sector do not present water and sediment accumulation. In this area, tailings deposits are the MEL of major concern for the population and ecosystem due to the potential risk of contamination with heavy metal(loid)s to the Pilalo river, which is used as a water supply by the residents of the nearby communities.

3.1.2. Tenguel–Ponce Enriquez

In the zone of Tenguel–Ponce Enriquez, no tailings deposits were reported. Approximately 31% of the MEL correspond to landfills, from which 79% represented a medium risk of affectation for the population and the environment in scenario S1 (Figure 5a). The risk ranges of this scenario were medium for the residents of the sector ($8 \leq R_I \leq 12$) and for the natural environment ($6 \leq R_I \leq 9$). The sites that presented a low risk corresponded to landfills located between 250–800 m from the waterways.

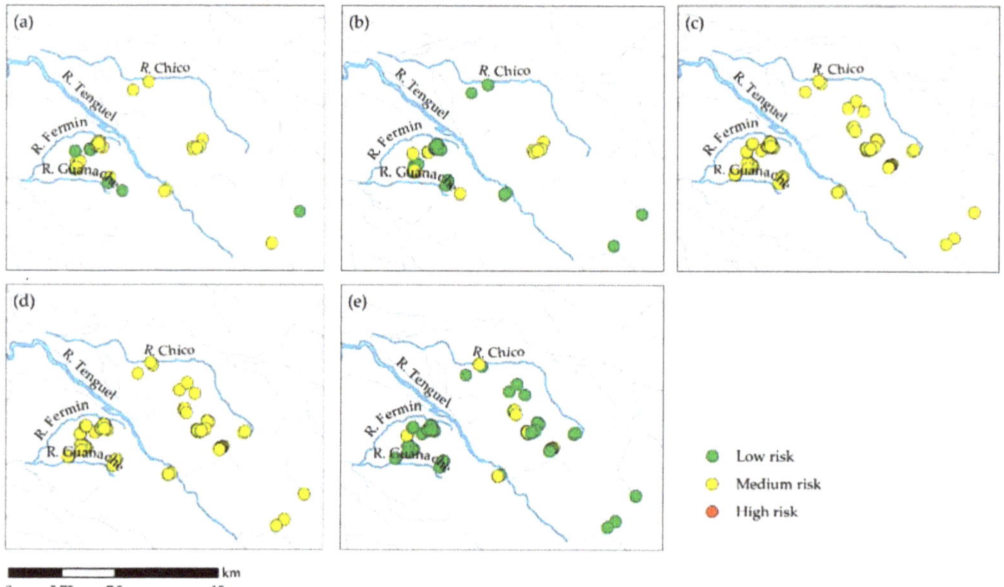

Figure 5. Point risk map for scenarios: (**a**) S1 for population and environment; (**b**) S2 for population; (**c**) S3 for population and environment; (**d**) S4 for population; (**e**) S4 for environment in Tenguel–Ponce Enriquez area.

In the S2 scenario (Figure 5b), approximately one-third (35%) of the landfills represented a medium affectation risk for the population with an $R_I = 6$ index. The MEL with a low-risk index were found at distances greater than 2 km from the communities of the sector. On the contrary, in the S3 scenario, the affectation risk was medium for both the population and the environment (Figure 5c). The values were $R_I \geq 9$ and $R_I \leq 15$ for the population and $R_I \geq 6$ and $R_I \leq 15$ for the natural environment. Similar results were obtained in the S4 scenario, in which all evaluated MEL presented a medium affectation risk for the population (Figure 5d). The risk for the natural environment (Figure 5e) was also medium ($R_I = 12$) in 18% of the analyzed sites, while 82% of the MEL showed a low affectation risk ($R_I = 3$). Moreover, the risk analysis allowed the identification of a significant quantity of MEL that are in the proximities of the Chico, Tenguel, Fermin, and Guanache rivers, a fact that represents a persistent threat of contamination by heavy metal(loid)s to the water sources, besides being a potential risk for the people and ecosystem.

3.1.3. Puyango

In Puyango, a significant quantity of abandoned tailings was reported (72% of the identified tailings in the three studied mining districts). As a result, 77% of the tailing deposits of the S1 scenario showed a medium affectation risk for the people and ecosystem (Figure 6a), with values of $8 \leq R_I \leq 12$ for the population and with a range of $6 \leq R_I \leq 9$ for the natural environment. A total of 33% of the tailings resulted in low-risk affectation values for the population and the environment with $R_I \leq 4$ due to the distances between these MEL and the river channels of the area being greater than 300 m. In the S2 scenario (Figure 6b), 100% of the tailings resulted in a medium affectation risk to the population's health, with a value of $R_I = 6$. The evaluation of S3 and S4 scenarios showed a medium risk for the people and the environment, as presented in Figure 6c,d. The risk index values of the S3 scenario were $R_I = 15$, while for the S4 scenario they were $R_I = 12$. For the assessment of each proposed scenario, all the MEL of the area that contained heavy metal(loid)s in concentrations that exceeded the Ecuadorian environmental standard were considered. The point risk maps indicated that approximately 98% of the MEL are located near water bodies such as the Amarillo, El Salado, and Puyango rivers and the Arcapamba creek.

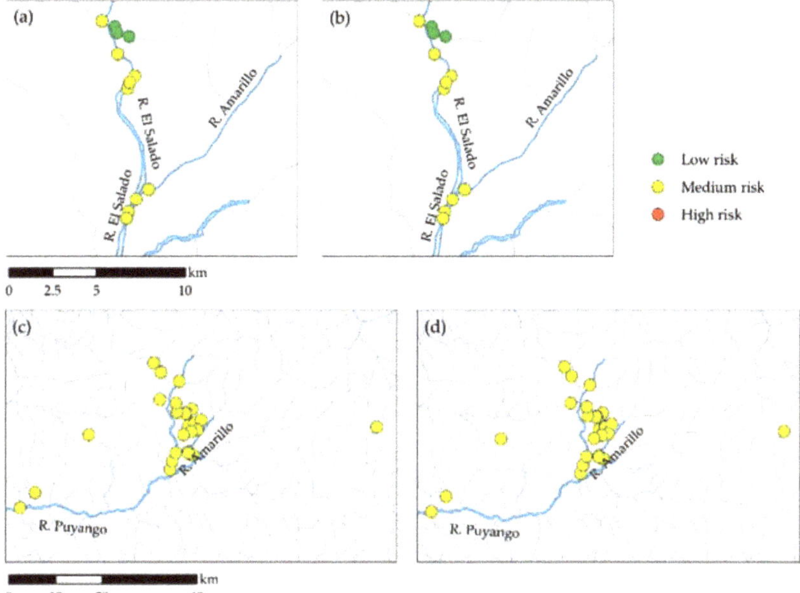

Figure 6. Point risk map for the scenario: (**a**) S1 for population and environment; (**b**) S2 for population; (**c**) S3 for population and environment; (**d**) S4 for population and environment in the Puyango area.

The results show that the pollution sources of major concern for the population and the environment are the mine entrances, landfills, and tailings deposits associated with the presence of acid drainage with a high content of heavy metal(loid)s. Moreover, the population near the tailings deposits and mine entrances may have health problems due to exposure to pollutants, released and mobilized from MEL, in the supply water.

4. Discussion

In the present study, the areas with the presence of deposits of abandoned waste (tailings and landfills) are Macuchi and Puyango, which reported a medium affectation risk in most of the analyzed sites in the S1 and S2 scenarios, which points out a potential risk that exposes the ecosystem and population. The presence of toxic elements such as heavy metal(loid)s in the MEL (that are released through different mechanisms to the environmental compartments) is causing the risk in the evaluated districts (Figure 7). Consequently, the ecosystem is highly affected by the presence of heavy metal(loid)s in shallow waters, soils, and sediments [15,45,46].

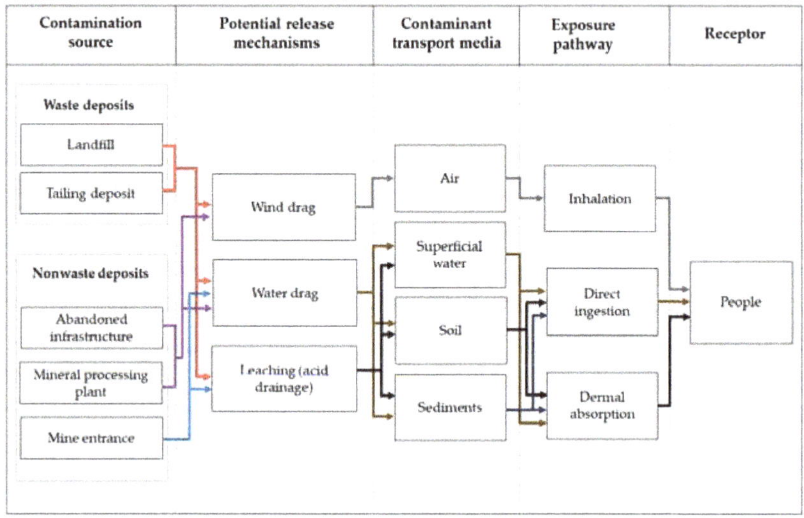

Figure 7. Conceptual site model for human health risk.

It is widely known that in the areas where tailing deposits exist, the concentration of heavy metal(loid)s in the soil is high due to the dispersion of the material particles and the drag of contaminants by the erosion processes [47]. The concentration of heavy metal(loid)s in the water in the three mining districts exceeds the MAL established by the Ecuadorian legislation. Additionally, the pH of the soils in the studied areas present acidity characteristics [30,42]. For example, effluents in tailing deposits with pH values of 2 and with concentrations of As (1.1 mg/L), Cd (4.9 mg/L), Cu (1103 mg/L), Fe (8422 mg/L), Pb (1.1 mg/L), and Zn (364 mg/L) are located in Macuchi [36]. Additionally, according to different studies, the detection of heavy metal(loid)s in shallow water and sediments is considered alarming in the zones of Tenguel–Ponce Enriquez and Puyango; a product of the direct or indirect discharges coming from the gold mining activities to the rivers [15,37,48,49]. This fact, therefore, increases the risk in the population and natural environment. The Puyango area may be particularly impacted since most of its MEL (98%) were found near water bodies, which may increase the presence of heavy metal(loid)s in surface waters, and therefore affect nearby populations and river users.

In Macuchi, the Pilalo river presents As, Pb, Cd, Cu, and Zn that exceed the MAL established in the Ecuadorian legislation of water and soils. In the area of Tenguel–Ponce Enriquez, the Tenguel river has Cu as its main pollutant and slight increases in As, Cd, Sb, Pb and Zn derived from gold-mining activities [50]. Moreover, the Fermin River has high concentrations of As, possibly due to the discharges made by the mineral processing plants of the area [37]. Jiménez-Oyola et al. [51] reported a severe ecological risk from As in the river sediments of Ponce Enriquez. On the other hand, Quishpe [46] identified that the Chico River presents average values of As higher than the MAL established in the environmental regulations for conserving aquatic wildlife in freshwater. In Puyango, the water bodies with greater affectation are the Calera and Amarillo rivers, where mineral processing has been carried out [45].

On an international level, a diverse study that evaluated the risk associated with the presence of MEL reported similar results to the ones obtained in the present research work. In San Luis Potosí, Mexico, the principal sources of toxic elements in the contaminated sites by MEL correspond to tailings deposits, dams, and slag deposits from which the pollutants can mobilize or be transported through the wind or runoff water and affect great soil extensions [5,52,53]. Moreover, in Zlata Idka–Slovenian, a study reported a high carcinogenic risk for the residents due to the high concentrations of As in soils, water, and vegetables [9]. Another case is Hualgayoc of Cajamarca, a mining district in Peru widely affected by the presence of MEL, which caused the contamination of agricultural soils with high contents of Pb (4683 mg/kg), Zn (724.2 mg/kg), Cu (511.6 mg/kg), As (3611 mg/kg), and Ag (33.4 mg/kg), exceeding the MAL established in the Peruvian legislation for agricultural soils [4]. In addition, around an abandoned mine of Hg in Asturias–Spain, exorbitant values of Hg were detected in the air, with concentrations up to 203.7 $\mu g/Nm^3$ [54]. Another example of MEL consequences is in South Korea, where most metallic mines have been abandoned, causing severe contamination cases [10]. Similarly, Ngole-Jeme and Fantke [7] reported high levels of heavy metal(loid)s such as As, Cd, Co, and Ni in soils in areas with the presence of abandoned tailing deposits in the mining city of Krugersdorp.

4.1. Potential Impacts of MEL on Human Health

The present study identified that the major affectation related to MEL presence is produced on the population by the direct or indirect exposition to toxic elements contained in waste deposits. In general, in the evaluated sites, the MEL contaminants are released through several mechanisms: leaching and air and water drag and transportation through air, shallow water, soil, and sediments, which finally come into contact with receptors through different routes of exposure such as inhalation, accidental ingestion or dermal absorption (Figure 7). The primary means of exposure is generally contaminated water because it is used for consumption or recreational use. The approximate number of people exposed to the contamination in the evaluated mining areas is between 2400 and 7000 inhabitants. In Macuchi, 59% of families (approximately 2470 people) consume water from the rivers and springs [55]. In Tenguel–Ponce Enriquez, around 25% of the homes (approximately 5500 people) take water from the rivers and springs from the Gala, Tenguel, and Siete basins [10]. In Puyango, nearly 46% of the population (approximately 7135 people) does not have water from a public network [56]. This situation forces the inhabitants to obtain water from wells, rivers, channels, and rainwater. Moreover, the identified MEL in the study zone are without surveillance or protection. People can easily have direct contact with the contaminated waste, especially children, during recreative activities. Figure 7 shows a site conceptual model pointing to the sources of contamination, the transport mechanism of pollutants, and the routes of exposure of significant concern.

Various kinds of human effects caused by exposure to MEL have been reported by several authors worldwide [1,5,8]. Heavy metal(loid)s such as As and Pb, contained in MEL, are potentially toxic for human health and can cause carcinogenic effects in the population at long exposures [57]. Diverse adverse effects for health, such as bladder, kidney, lung, liver, and colon cancer, cardiovascular and neurological diseases, and diabetes, are attributed to As exposure in humans [58]. On the other side, intoxication by Pb has centered on children because they are more susceptible to the present adverse effects during growth and neurobehavioral development [57–59]. Therefore, the risk for human health is strongly related to human behavior, which depends on the activities carried out in the contaminated site. Thus, the risk for human health depends on the pollutants of the environmental system, the heavy metal(loid) toxicity or metalloid, the routes of exposure, and the risk receptors [60]. In this sense, limiting exposure is a key factor in diminishing the risk.

4.2. Mining—Environmental Regulation

The environmentally responsible development of mineral resources is vital for the mining industry [61]. It is essential to consider the regulations and laws that guide the activities to be developed in an environmentally, economically, and socially sustainable way to promote sustainable development in favor of protecting the soil and land [62].

Adler Miserendino et al. [63] mention that the impacts, principally in the degradation of water quality and aquatic ecosystems caused by illegal mining and ASGM, result from deficiencies of the national regulatory framework. Moreover, factors such as the lack of capacity of local actors and internal political struggles contribute to the lack of integrated planning that ensures a postmining economy [64].

Furthermore, the lack of information and knowledge has had a significantly adverse effect on ASGM practices [65]. At the same time, the lack of technological investment in mining processes has potentiated significant impacts on hydric, soil, and air resource [66]. On an international level, the interventions to address the impacts of ASGM have commonly focused on the technological changes in mineral processing through eliminating the use of toxic elements [63]. For example, since 2010, the use of Hg in Ecuador has been forbidden in the gold recovery process. Nevertheless, informal miners still use amalgamation with Hg, and some of them even burn the amalgams inside their homes, exposing all of the members of the family to noxious fumes [67].

In so far as the environmental mining regulations in Ecuador, these are relatively new and still present deficiencies. For example, concerning MEL, unlike other countries such as Brazil, Peru, or Chile, Ecuador does not count with a legal instrument that allows the integral management of them. Additionally, regulation with a preventive and corrective approach is necessary. Moreover,, so is resource assignation permitting control on time and minimizing the generation of MEL in the mining areas [17].

The Mining Law of Ecuador [68] establishes that the mining concession holders and beneficiation plants should incorporate environmental management plans until closure and definitive abandonment of the mining project. In addition, it is found that the closing of operations and rehabilitation of affected areas should be a progressive process in the different stages of the project life cycle [69]. Therefore, the objective of rehabilitating mining zones is to make them safe, stable, nonpolluting, and self-sustainable for soil use after the activities [70,71]. Furthermore, closure and rehabilitation of the mines is an obligation for all the mining holders to prevent future contamination sites. Nevertheless, the subject of the environmental management plan is relatively new for the ASGM sector [72] and in practice still presents certain limitations. Finally, to limit the presence of illegal mining (closely related to environmental degradation), in 2014, illegal mining activities and environmental pollution were categorized as a felony according to the Integral Organic Criminal Code of Ecuador [73].

Despite the different legal instruments of the Ecuadorian environmental standard, there is a notorious deficiency in the associated regulations to recover degraded areas and manage environmental liabilities. Moreover, the ineffective application of the law is evident in many cases. This leads to thinking that an essential factor in the aspect of preservation of the environmental rights of nature and environment is apart from the legislation and regulations of the system, the ethical and moral quality of the system itself, and the people involved to correctly apply and obey the law so that the environmentally, socially, and economically sustainable development of the mining activity is not compromised. It is also important to consider that the main problem of artisanal miners is that there is no guide, such as a simplified rehabilitation plan, nor guidelines with simple techniques that avoid the costly requirements of rehabilitation [72].

4.3. Management and Public Policy: Suggested Actions

The identified impacts in the present study are a legacy from the past. They are associated mainly to ASGM and the illegal mining carried out in the study areas since the end of the XIXth century [31]. The evident lack of regulation in mining activities and the lack of recovery actions (on time) of degraded areas have caused MEL accumulation. The same accumulation has propitiated pollutants' transport in the different environmental compartments, resulting in severe anthropogenic contamination [66].

Table 4 shows the strategies and proposed actions for the management of MEL that represent a high risk for the population and the environment. Puyango and Tenguel–Ponce

Enriquez were identified as priority control areas, that is, they require urgent intervention as well as restoration and rehabilitation plans. MEL that require a high level of intervention include waste deposits, mine entrances, and abandoned processing plants. To achieve ecological restoration of the sites, some actions can be adopted, such as those presented in some research that propose phytoremediation [46,74–76] or the creation of geoparks in areas that do not represent a significant risk for the population [77].

Table 4. Categorization of priority areas and proposed strategies for pollution control.

MEL	Intervention Level			Proposed Actions
	High	Medium	Low	
Landfills	+		×	Covering, sealing, and revegetation of deposits Chemical and physical stability control and monitoring plan Water, soil, sediment, biotic component and stability control, and monitoring plan
Mining galleries		+		Chemical and physical stability control and monitoring plan Implementation of geotourism (museums and geoparks) in rehabilitated areas with low impact Plugging of higher risk mine entrances or galleries
Mine entrances	+	×		
Tailings Deposits	× *			Reuse/reuse/valorization of mining tailings Covering, sealing, and revegetation of tailings deposits Restoration plan and revegetation near the riverbanks Water, soil, sediment, biotic component and stability control, and monitoring plan
Abandoned infrastructure			+	Construction of a community meeting place Water, soil, sediment, biotic component and stability control, and monitoring plan
Mineral processing plants	*			Dismantling infrastructures Chemical stabilization of soils Control and monitoring of chemical stability of soils
Alluvial terrace		*		Treatment of the waters of affected rivers. Restitution of flora and fauna Physical and chemical stabilization of riverbanks
Quarries		*		Physical and chemical stabilization of soils Revegetation of the areas Restitution of fauna Water, soil, sediment, biotic component and stability control, and monitoring plan

(×) Macuchi MEL; (+) Tenguel–Ponce Enriquez MEL; (*) Puyango MEL.

Besides the restoration and rehabilitation plans, it is necessary to have continuous control and monitoring plans for the areas to ensure the correct recovery and restitution of the land. It is important to know that the restoration processes of the degraded mining areas are complex and require joint work between different institutions: mining companies, planners, investors, institutions, and local communities [78], with a common vision focused on sustainability. A lack of common vision can produce inefficacy in management plans [64]. Moreover, it is necessary to point out the importance of having clear public policies that encourage those involved to put social and environmental responsibility into practice during the execution of mining projects.

Regarding abandoned tailings, a sustainable alternative to manage these MEL is their reutilization and reuse in the construction industry [79,80]; tailings have been reused to fabricate bricks, ceramic materials, and cement [81–83]. Therefore, efforts should be focused on searching for the sustainable utilization of mining waste, promoting a circular economy. Finally, a risk communication plan must be in place to minimize the population's exposure to the potentially dangerous areas identified in this assessment.

5. Conclusions

This work provides a preliminary assessment of the risk associated with the presence of mining environmental liabilities in three artisanal and small-scale gold-mining areas of Ecuador. According to the results, Puyango and Tenguel–Ponce Enriquez appears to be the most affected mining areas by the presence of MEL, mainly waste deposits and mine entrances. Landfills, tailing, and mine entrances present a medium risk for the people and the environment due to the content of potentially toxic elements. Around 2% of MEL are far away from the population and water bodies, therefore, the result of the risk was low ($R_I < 3$) for the people and the ecosystem. The point risk maps showed that the rivers with major risk of contamination are: Pilalo in Macuchi; Chico, Tenguel, Fermin, and Guanache in

Tenguel–Ponce Enriquez and Amarillo, El Salado, and Puyango in the Puyango River basin. To our knowledge, the impacts caused in the studied areas are a product of the inherent contamination due to illegal mining activities since the end of the XIXth century. In this sense, it is necessary to remediate the polluted sites and practice continuous monitoring to restore the ecosystems properly. In addition, the population's exposure must be restricted from high-risk areas through a communication plan of the risks. Additionally, it is urgently necessary to investigate the bioavailability of heavy metal(loid)s in the environmental compartments, as well as their impact on the environment and people's health. This study highlights the need to implement regulations for the management of mining environmental liabilities in Ecuador to protect the environment and residents of mining communities, ensuring sustainability for the ecosystem, populations, and the mining industry.

Supplementary Materials: The following supporting information can be downloaded at: https://www.mdpi.com/article/10.3390/su14106089/s1, Table S1: Maximum and minimum concentrations of heavy metals in water, soil, and sediment samples from each study area, Table S2: Assessment criteria of parameters for probability index determination of scenario S1 and S2, Table S3: Assessment criteria of the severity index (Is) of scenario S1, Table S4: Assessment criteria of parameters for the severity index determination of scenarios S3 and S4, Table S5: Risk assessment results: scenarios S1 and S2, Table S6: Risk assessment results: scenarios S3 and S4.

Author Contributions: Conceptualization, B.S.-A. and S.J.-O.; data curation, B.S.-A.; formal analysis, B.S.-A. and S.J.-O.; investigation, P.L.R.-C., P.E.V.-A. and D.A.F.-T.; methodology, B.S.-A., S.J.-O. and F.G.-M.; supervision, P.L.R.-C., S.J.-O. and F.G.-M.; writing—original draft, B.S.-A., D.A.F.-T., P.E.V.-A. and S.J.-O.; writing—review and editing, P.L.R.-C. and F.G.-M.; project administration, P.L.R.-C. and S.J.-O. All authors have read and agreed to the published version of the manuscript.

Funding: This research received no external funding.

Institutional Review Board Statement: Not applicable.

Informed Consent Statement: Not applicable.

Data Availability Statement: Not applicable.

Conflicts of Interest: The authors declare no conflict of interest.

References

1. Corzo, A.; Gamboa, N. Environmental Impact of Mining Liabilities in Water Resources of Parac Micro-Watershed, San Mateo Huanchor District, Peru. *Environ. Dev. Sustain.* **2018**, *20*, 939–961. [CrossRef]
2. Guzmán-Martínez, F.; Arranz-González, J.C.; Fidel-Smoll, L.; Collahuazo, L.; Calderón, E.; Otero, O.; Arceo y Cabrilla, F. *Pasivos Ambientales Mineros: Manual Para El Inventario de Minas Abandonadas o Paralizadas*; ASGMI: Madrid, Spain, 2020.
3. González-Valoys, A.C.; Esbrí, J.M.; Campos, J.A.; Arrocha, J.; García-Noguero, E.M.; Monteza-Destro, T.; Martínez, E.; Jiménez-Ballesta, R.; Gutiérrez, E.; Vargas-Lombardo, M.; et al. Ecological and Health Risk Assessments of an Abandoned Gold Mine (Remance, Panama): Complex Scenarios Need a Combination of Indices. *Int. J. Environ. Res. Public Health* **2021**, *18*, 9369. [CrossRef] [PubMed]
4. Cruzado-Tafur, E.; Torró, L.; Bierla, K.; Szpunar, J.; Tauler, E. Heavy Metal Contents in Soils and Native Flora Inventory at Mining Environmental Liabilities in the Peruvian Andes. *J. S. Am. Earth Sci.* **2021**, *106*, 103107. [CrossRef]
5. Fernández-Macías, J.C.; González-Mille, D.J.; García-Arreola, M.E.; Cruz-Santiago, O.; Rivero-Pérez, N.E.; Pérez-Vázquez, F.; Ilizaliturri-Hernández, C.A. Integrated Probabilistic Risk Assessment in Sites Contaminated with Arsenic and Lead by Long-Term Mining Liabilities in San Luis Potosi, Mexico. *Ecotoxicol. Environ. Saf.* **2020**, *197*, 110568. [CrossRef]
6. Lam, E.J.; Gálvez, M.E.; Cánovas, M.; Montofré, I.L.; Rivero, D.; Faz, A. Evaluation of Metal Mobility from Copper Mine Tailings in Northern Chile. *Environ. Sci. Pollut. Res.* **2016**, *23*, 11901–11915. [CrossRef]
7. Ngole-Jeme, V.M.; Fantke, P. Ecological and Human Health Risks Associated with Abandoned Gold Mine Tailings Contaminated Soil. *PLoS ONE* **2017**, *12*, e0172517. [CrossRef]
8. Bempah, C.K.; Ewusi, A. Heavy Metals Contamination and Human Health Risk Assessment around Obuasi Gold Mine in Ghana. *Environ. Monit. Assess.* **2016**, *188*, 261. [CrossRef]
9. Rapant, S.; Dietzová, Z.; Cicmanová, S. Environmental and Health Risk Assessment in Abandoned Mining Area, Zlata Idka, Slovakia. *Environ. Geol.* **2006**, *51*, 387–397. [CrossRef]
10. Min, H.-G.; Kim, M.-S.; Kim, J.-G. Effect of Soil Characteristics on Arsenic Accumulation in Phytolith of Gramineae (*Phragmites Japonica*) and Fern (*Thelypteris Palustris*) Near the Gilgok Gold Mine. *Sustainability* **2021**, *13*, 3421. [CrossRef]

11. Fernández-Caliani, J.C.; Rosa, J.; Sánchez, A.M.; González-Castanedo, Y.; González, I.; Romero, A.; Galán, E. Datos Químicos y Mineralógicos Preliminares de las Partículas Atmosféricas Sedimentables en la Cuenca Minera de Riotinto (Huelva). *Macla: Revista de la Sociedad Española de Mineralogía*. 2010, pp. 79–80. Available online: http://hdl.handle.net/10272/7846 (accessed on 10 April 2022).
12. Appleton, J.D.; Williams, T.M.; Orbea, H.; Carrasco, M. Fluvial Contamination Associated with Artisanal Gold Mining in the Ponce Enríquez, Portovelo-Zaruma and Nambija Areas, Ecuador. *Water Air Soil Pollut.* **2001**, *131*, 19–39. [CrossRef]
13. Betancourt, Ó.; Barriga, R.; Guimarães, J.R.D.; Cueva, E.; Betancourt, S. Impacts on Environmental Health of Small-Scale Gold Mining in Ecuador. In *Ecohealth Research in Practice*; Springer: New York, NY, USA, 2012; pp. 119–130.
14. Tarras-Wahlberg, N.H.; Flachier, A.; Lane, S.N.; Sangfors, O. Environmental Impacts and Metal Exposure of Aquatic Ecosystems in Rivers Contaminated by Small Scale Gold Mining: The Puyango River Basin, Southern Ecuador. *Sci. Total Environ.* **2001**, *278*, 239–261. [CrossRef]
15. Carling, G.T.; Diaz, X.; Ponce, M.; Perez, L.; Nasimba, L.; Pazmino, E.; Rudd, A.; Merugu, S.; Fernandez, D.P.; Gale, B.K.; et al. Particulate and Dissolved Trace Element Concentrations in Three Southern Ecuador Rivers Impacted by Artisanal Gold Mining. *Water Air Soil Pollut.* **2013**, *224*, 1415. [CrossRef]
16. Mestanza-Ramón, C.; Cuenca-Cumbicus, J.; D'Orio, G.; Flores-Toala, J.; Segovia-Cáceres, S.; Bonilla-Bonilla, A.; Straface, S. Gold Mining in the Amazon Region of Ecuador: History and a Review of Its Socio-Environmental Impacts. *Land* **2022**, *11*, 221. [CrossRef]
17. Oblasser, A. *Estudio Sobre Lineamientos, Incentivos y Regulación Para El Manejo de Los Pasivos Ambientales Mineros (PAM), Incluyendo Cierre de Faenas Mineras: Bolivia (Estado Plurinacional), Chile, Colombia y El Perú*; Naciones Unidas: Santiago, Chile, 2016.
18. Dong, L.; Deng, S.; Wang, F. Some Developments and New Insights for Environmental Sustainability and Disaster Control of Tailings Dam. *J. Clean. Prod.* **2020**, *269*, 122270. [CrossRef]
19. Do Carmo, F.F.; Kamino, L.H.Y.; Junior, R.T.; de Campos, I.C.; do Carmo, F.F.; Silvino, G.; de Castro, K.J.d.S.X.; Mauro, M.L.; Rodrigues, N.U.A.; de Miranda, M.P.S.; et al. Fundão Tailings Dam Failures: The Environment Tragedy of the Largest Technological Disaster of Brazilian Mining in Global Context. *Perspect. Ecol. Conserv.* **2017**, *15*, 145–151. [CrossRef]
20. Garcia, L.C.; Ribeiro, D.B.; Oliveira Roque, F.; Ochoa-Quintero, J.M.; Laurance, W.F. Brazil's Worst Mining Disaster: Corporations Must Be Compelled to Pay the Actual Environmental Costs. *Ecol. Appl.* **2017**, *27*, 5–9. [CrossRef]
21. Paniagua-López, M.; Vela-Cano, M.; Correa-Galeote, D.; Martín-Peinado, F.; Garzón, F.J.M.; Pozo, C.; González-López, J.; Aragón, M.S. Soil Remediation Approach and Bacterial Community Structure in a Long-Term Contaminated Soil by a Mining Spill (Aznalcóllar, Spain). *Sci. Total Environ.* **2021**, *777*, 145128. [CrossRef]
22. Glotov, V.E.; Chlachula, J.; Glotova, L.P.; Little, E. Causes and Environmental Impact of the Gold-Tailings Dam Failure at Karamken, the Russian Far East. *Eng. Geol.* **2018**, *245*, 236–247. [CrossRef]
23. Cuentas Alvarado, M.; Velasquez Viza, O.; Arizaca Avalos, A.; Huisa Mamani, F. Evaluación de Riesgos de Pasivos Ambientales Mineros En La Comunidad de Condoraque—Puno. *Rev. Medio Ambient. Y Min.* **2019**, *4*, 43–57.
24. Peña-Carpio, E.; Menéndez-Aguado, J.M. Environmental Study of Gold Mining Tailings in the Ponce Enriquez Mining Area (Ecuador). *DYNA* **2016**, *83*, 237–245. [CrossRef]
25. Arranz-González, J.C.; Rodríguez-Gómez, V.; Fernández-Naranjo, F.J.; Vadillo-Fernández, L. Assessment of the Pollution Potential of a Special Case of Abandoned Sulfide Tailings Impoundment in Riotinto Mining District (SW Spain). *Environ. Sci. Pollut. Res.* **2021**, *28*, 14054–14067. [CrossRef] [PubMed]
26. Guzmán-Martínez, F.; Arranz-González, J.C.; García-Martínez, M.J.; Ortega, M.F.; Rodríguez-Gómez, V.; Jiménez-Oyola, S. Comparative Assessment of Leaching Tests According to Lixiviation and Geochemical Behavior of Potentially Toxic Elements from Abandoned Mining Wastes. *Mine Water Environ.* **2021**, *41*, 265–279. [CrossRef]
27. Neira, J.J.C.; Quito Quilla, S.J. Evaluación de Riesgo Ambiental Generado Por Pasivo Ambiental Minero En La Calidad de Agua Superficial. *Natura@Economía* **2020**, *5*, 1–14. [CrossRef]
28. USEPA. *Ecological Risk Assessment Guidance for Superfund: Process for Designing and Conducting Risk Assessments*; USEPA: Wahington, DC, USA, 1998.
29. Alberruche del Campo, M.E.; Arranz-González, J.C.; Rodríguez-Pacheco, R.; Vadillo-Fernández, L.; Rodríguez-Gómez, V.; Fernández-Naranjo, F.J. *Manual Para La Evaluación de Riesgos de Instalaciones de Residuos de Industrias Extractivas Cerradas o Abandonadas*, 1st ed.; Ministerio de Agricultura, Alimentación y Medio Ambiente España, Instituto Geológico y Minero de España: Madrid, Spain, 2014; ISBN 9788478409341.
30. MAE-PRAS. *Programa de Reparación Ambiental y Social—Plan de Reparación Integral de la Cuenca del Río Puyango*; MAE-PRAS: Quito, Ecuador, 2015; Available online: http://pras.ambiente.gob.ec/documents/228536/737569/LIBRO_PRI_PUYANGO.pdf/94bcfdb4-bf26-4d3a-afa3-d5e87cf7398b (accessed on 15 February 2022).
31. Rivera-Parra, J.L.; Beate, B.; Diaz, X.; Ochoa, M.B. Artisanal and Small Gold Mining and Petroleum Production as Potential Sources of Heavy Metal Contamination in Ecuador: A Call to Action. *Int. J. Environ. Res. Public Health* **2021**, *18*, 2794. [CrossRef] [PubMed]
32. Betancourt, O.; Narváez, A.; Roulet, M. Small-Scale Gold Mining in the Puyango River Basin, Southern Ecuador: A Study of Environmental Impacts AndHuman Exposures. *Ecohealth* **2005**, *2*, 323–332. [CrossRef]
33. Vilela-Pincay, W.; Espinosa-Encarnación, M.; Bravo-González, A. La Contaminación Ambiental Ocasionada Por La Minería En La Provincia de El Oro. *Estud. Gestión. Rev. Int. Adm.* **2020**, *8*, 215–233. [CrossRef]

34. Tarras-Wahlberg, N.H.; Flachier, A.; Fredriksson, G.; Lane, S.; Lundberg, B.; Sangfors, O. Environmental Impact of Small-Scale and Artisanal Gold Mining in Southern Ecuador. *Ambio* **2000**, *29*, 484–491. [CrossRef]
35. Escobar-Segovia, K.; Jiménez-Oyola, S.; Garcés-León, D.; Paz-Barzola, D.; Navarrete, E.C.; Romero-Crespo, P.; Salgado, B. Heavy Metals in Rivers Affected by Mining Activities in Ecuador: Pollution and Human Health Implications. *WIT Trans. Ecol. Environ.* **2021**, *250*, 61–72. [CrossRef]
36. MAE-PRAS. *Programa de Reparación Ambiental y Social—Plan de Reparación Integral de Macuchi*; MAE-PRAS: Quito, Ecuador, 2015; Volume 1. Available online: http://pras.ambiente.gob.ec/documents/228536/737569/PRI+Macuchi.pdf/ac1ccb5b-0f80-4114-81e5-8e004a434274 (accessed on 15 February 2022).
37. Sierra, C.; Ruíz-Barzola, O.; Menéndez, M.; Demey, J.R.; Vicente-Villardón, J.L. Geochemical Interactions Study in Surface River Sediments at an Artisanal Mining Area by Means of Canonical (MANOVA)-Biplot. *J. Geochem. Explor.* **2017**, *175*, 72–81. [CrossRef]
38. Jiménez-Oyola, S.; Escobar Segovia, K.; García-Martínez, M.-J.; Ortega, M.; Bolonio, D.; García-Garizabal, I.; Salgado, B. Human Health Risk Assessment for Exposure to Potentially Toxic Elements in Polluted Rivers in the Ecuadorian Amazon. *Water* **2021**, *13*, 613. [CrossRef]
39. Xu, Z.; Lu, Q.; Xu, X.; Feng, X.; Liang, L.; Liu, L.; Li, C.; Chen, Z.; Qiu, G. Multi-Pathway Mercury Health Risk Assessment, Categorization and Prioritization in an Abandoned Mercury Mining Area: A Pilot Study for Implementation of the Minamata Convention. *Chemosphere* **2020**, *260*, 127582. [CrossRef]
40. Castilhos, Z.; Rodrigues-Filho, S.; Cesar, R.; Rodrigues, A.P.; Villas-Bôas, R.; de Jesus, I.; Lima, M.; Faial, K.; Miranda, A.; Brabo, E.; et al. Human Exposure and Risk Assessment Associated with Mercury Contamination in Artisanal Gold Mining Areas in the Brazilian Amazon. *Environ. Sci. Pollut. Res.* **2015**, *22*, 11255–11264. [CrossRef] [PubMed]
41. Liang, C.P.; Chen, J.S.; Chien, Y.C.; Chen, C.F. Spatial Analysis of the Risk to Human Health from Exposure to Arsenic Contaminated Groundwater: A Kriging Approach. *Sci. Total Environ.* **2018**, *627*, 1048–1057. [CrossRef] [PubMed]
42. MAE-PRAS. *Programa de Reparación Ambiental y Social—Plan de Reparación Integral de la Zona de Estudio Tenguel—Camilo Ponce Enríquez*; MAE-PRAS: Quito, Ecuador, 2015; Volume 1, Available online: http://pras.ambiente.gob.ec/documents/228536/737569/PRI_Tenguel.pdf/58596e7c-d3aa-4380-b0c8-dfe9fde6ff2b (accessed on 15 February 2022).
43. MINAM. *Guía de Evaluación de Riesgos Ambientales*; MINAM: Magdalena del Mar, Peru, 2010.
44. TULSMA, Ministerio de Ambiente. Texto Unificado de Legislación Secundaria de Medio Ambiente: Norma de Calidad Ambiental del Recurso Suelo y Criterios de Remediación para Suelos Contaminados. 2015. Available online: https://www.gob.ec/sites/default/files/regulations/2018-09/Documento_Registro-Oficial-No-387-04-noviembre-2015_0.pdf (accessed on 20 February 2022).
45. Garcia, M.E.; Betancourt, O.; Cueva, E.; Guimaraes, J.R.D. Mining and Seasonal Variation of the Metals Concentration in the Puyango River Basin—Ecuador. *J. Environ. Prot.* **2012**, *3*, 1542–1550. [CrossRef]
46. Quishpe, Á. *Remoción de Arsénico de Efluentes Líquidos de Plantas de Beneficio de Oro y Cuerpos Hídricos, de La Zona Minéra de Ponce Enríquez, Por Rizofiltración Con Pasto Azul (Dactylis Glomerata)*; Escuela Politécnica Nacional: Quito, Ecuador, 2020.
47. Guzmán-Martínez, F.; Arranz-González, J.C.; Ortega, M.F.; García-Martínez, M.J.; Rodríguez-Gómez, V. A New Ranking Scale for Assessing Leaching Potential Pollution from Abandoned Mining Wastes Based on the Mexican Official Leaching Test. *J. Environ. Manag.* **2020**, *273*, 111139. [CrossRef] [PubMed]
48. Sandoval, F. Small-Scale Mining in Ecuador. *Min. Miner. Sustain. Dev.* **2001**, *75*, 28.
49. PRODEMINCA. *Monitoreo Ambiental de Las Áreas Mineras En El Sur de Ecuador 1996–1998; R-Ec-E-9.46/3.1-9810-069;* 1st ed.; UCP Prodeminca: Quito, Ecuador, 1998; ISBN 9978-40-872-x.
50. GAD. *Cantonal Camilo Ponce Enríquez Plan de Desarrollo y Ordenamiento Territorial Del Cantón Camilo Ponce Enríquez—Administración 2014–2019;* GAD: Azuay, Ecuador, 2013; Volume 53.
51. Jiménez-Oyola, S.; García-Martínez, M.-J.; Ortega, M.F.; Chavez, E.; Romero, P.; García-Garizabal, I.; Bolonio, D. Ecological and Probabilistic Human Health Risk Assessment of Heavy Metal(Loid)s in River Sediments Affected by Mining Activities in Ecuador. *Environ. Geochem. Health* **2021**, *43*, 4459–4474. [CrossRef]
52. Razo, I.; Carrizales, L.; Castro, J.; Díaz-Barriga, F.; Monroy, M. Arsenic and Heavy Metal Pollution of Soil, Water and Sediments in a Semi-Arid Climate Mining Area in Mexico. *Water Air Soil Pollut.* **2004**, *152*, 129–152. [CrossRef]
53. Martínez-Toledo, Á.; Montes-Rocha, A.; González-Mille, D.J.; Espinosa-Reyes, G.; Torres-Dosal, A.; Mejia-Saavedra, J.J.; Ilizaliturri-Hernández, C.A. Evaluation of Enzyme Activities in Long-Term Polluted Soils with Mine Tailing Deposits of San Luis Potosí, México. *J. Soils Sediments* **2017**, *17*, 364–375. [CrossRef]
54. Loredo, J.; Soto, J.; Álvarez, R.; Ordóñez, A. Atmospheric Monitoring at Abandoned Mercury Mine Sites in Asturias (NW Spain). *Environ. Monit. Assess.* **2007**, *130*, 201–214. [CrossRef]
55. GAD. *Parroquial Rural El Tingo Plan de Desarrollo y Ordemamintо Territorial de La Parroquia El Tingo;* GAD: Azuay, Ecuador, 2019; Volume 53, pp. 1689–1699.
56. GAD. *Municipal de Puyango Plan de Desarrollo y Ordenamiento Territorial Del Cantón Puyango;* GAD: Azuay, Ecuador, 2013.
57. RAIS Toxicity Profiles. Risk Assessment Information System. Available online: https://rais.ornl.gov/tools/tox_profiles.html (accessed on 4 March 2022).
58. Zhou, Y.; Niu, L.; Liu, K.; Yin, S.; Liu, W. Arsenic in Agricultural Soils across China: Distribution Pattern, Accumulation Trend, Influencing Factors, and Risk Assessment. *Sci. Total Environ.* **2018**, *616–617*, 156–163. [CrossRef]
59. IARC. *Evaluation of Carcinogenic Risk for Humans;* IARC: Lyon, France, 1987; Volume 1–42.

60. USEPA. *Exposure Factors Handbook: 2011 Edition*; USEPA: Washington, DA, USA, 2011.
61. Adrien Rimélé, M.; Dimitrakopoulos, R.; Gamache, M. A Stochastic Optimization Method with In-Pit Waste and Tailings Disposal for Open Pit Life-of-Mine Production Planning. *Resour. Policy* **2018**, *57*, 112–121. [CrossRef]
62. Monteiro, N.B.R.; Bezerra, A.K.L.; Moita Neto, J.M.; da Silva, E.A. Mining Law: In Search of Sustainable Mining. *Sustainability* **2021**, *13*, 867. [CrossRef]
63. Adler Miserendino, R.; Bergquist, B.A.; Adler, S.E.; Guimarães, J.R.D.; Lees, P.S.J.; Niquen, W.; Velasquez-López, P.C.; Veiga, M.M. Challenges to Measuring, Monitoring, and Addressing the Cumulative Impacts of Artisanal and Small-Scale Gold Mining in Ecuador. *Resour. Policy* **2013**, *38*, 713–722. [CrossRef]
64. Marais, L. Resources Policy and Mine Closure in South Africa: The Case of the Free State Goldfields. *Resour. Policy* **2013**, *38*, 363–372. [CrossRef]
65. Zvarivadza, T.; Nhleko, A.S. Resolving Artisanal and Small-Scale Mining Challenges: Moving from Conflict to Cooperation for Sustainability in Mine Planning. *Resour. Policy* **2018**, *56*, 78–86. [CrossRef]
66. SENAGUA. *Informe Técnico Muestreo y Análisis de La Calidad Del Agua En La Cuenca Del Río Puyango*; SENAGUA: Quito, Ecuador, 2011.
67. Gonçalves, A.O.; Marshall, B.G.; Kaplan, R.J.; Moreno-Chavez, J.; Veiga, M.M. Evidence of Reduced Mercury Loss and Increased Use of Cyanidation at Gold Processing Centers in Southern Ecuador. *J. Clean. Prod.* **2017**, *165*, 836–845. [CrossRef]
68. Asamblea Nacional Constituyente del Ecuador. *Mining Law of Ecuador*; Asamblea Nacional Constituyente del Ecuador: Quito, Ecuador, 2009; p. 65.
69. Ministerio de Ambiente. Reglamento Ambiental de Actividades Mineras. 2014, p. 54. Available online: https://www.ambiente.gob.ec/wp-content/uploads/downloads/2019/01/REGLAMENTO-AMBIENTAL-DE-ACTIVIDADES-MINERAS-MINISTERTIO-AMBIENTE.pdf (accessed on 10 March 2022).
70. Doley, D.; Audet, P. Identifying Natural and Novel Ecosystem Goals for Rehabilitation of Postmining Landscapes. In *Responsible Mining: Case Studies in Managing Social and Environmental Risks in the Developed World*; Society for Mining, Metallurgy and Exploration (SME): Englewood, CO, USA, 2013; pp. 609–638.
71. Lechner, A.M.; Kassulke, O.; Unger, C. Spatial Assessment of Open Cut Coal Mining Progressive Rehabilitation to Support the Monitoring of Rehabilitation Liabilities. *Resour. Policy* **2016**, *50*, 234–243. [CrossRef]
72. United Nations Environment Programme. *Analysis of Formalization Approaches in the Artisanal and Small-Scale Gold Mining Sector Based on Experiences in Ecuador, Mongolia, Peru, Tanzania and Uganda*; United Nations Environment Programme: Nairobi, Kenya, 2012.
73. Asamblea Nacional Constituyente del Ecuador. Código Orgánico Integral Penal. 2014, pp. 1–268. Available online: https://www.defensa.gob.ec/wp-content/uploads/downloads/2021/03/COIP_act_feb-2021.pdf (accessed on 10 March 2022).
74. Lam, E.J.; Cánovas, M.; Gálvez, M.E.; Montofré, Í.L.; Keith, B.F.; Faz, Á. Evaluation of the Phytoremediation Potential of Native Plants Growing on a Copper Mine Tailing in Northern Chile. *J. Geochem. Explor.* **2017**, *182*, 210–217. [CrossRef]
75. Vela-García, N.; Guamán-Burneo, M.C.; González-Romero, N.P. Efficient Bioremediation from Metallurgical Effluents through the Use of Microalgae Isolated from the Amazonic and Highlands of Ecuador. *Rev. Int. Contam. Ambient.* **2019**, *35*, 917–929. [CrossRef]
76. Yildirim, D.; Sasmaz, A. Phytoremediation of As, Ag, and Pb in Contaminated Soils Using Terrestrial Plants Grown on Gumuskoy Mining Area (Kutahya Turkey). *J. Geochem. Explor.* **2017**, *182*, 228–234. [CrossRef]
77. Franco, G.H.; Mero, P.C.; Carballo, F.M.; Narváez, G.H.; Bitar, J.B.; Torrens, R.B. Strategies for the Development of the Value of the Mining-Industrial Heritage of the Zaruma-Portovelo, Ecuador, in the Context of a Geopark Project. *Int. J. Energy Prod. Manag.* **2020**, *5*, 48–59. [CrossRef]
78. Popović, V.; Miljković, J.; Subić, J.; Jean-Vasile, A.; Adrian, N.; Nicolăescu, E. Sustainable Land Management in Mining Areas in Serbia and Romania. *Sustainability* **2015**, *7*, 11857–11877. [CrossRef]
79. Di Maria, A.; Van Acker, K. Turning Industrial Residues into Resources: An Environmental Impact Assessment of Goethite Valorization. *Engineering* **2018**, *4*, 421–429. [CrossRef]
80. Capasso, I.; Lirer, S.; Flora, A.; Ferone, C.; Cioffi, R.; Caputo, D.; Liguori, B. Reuse of Mining Waste as Aggregates in Fly Ash-Based Geopolymers. *J. Clean. Prod.* **2019**, *220*, 65–73. [CrossRef]
81. Gomes-Pimentel, M.; Rubens Cardoso da Silva, M.; de Cássia, S.; Viveiros, D.; Picanço, M.S. Manganese Mining Waste as a Novel Supplementary Material in Portland Cement. *Mater. Lett.* **2022**, *309*, 131459. [CrossRef]
82. Veiga Simão, F.; Chambart, H.; Vandemeulebroeke, L.; Cappuyns, V. Incorporation of Sulphidic Mining Waste Material in Ceramic Roof Tiles and Blocks. *J. Geochem. Explor.* **2021**, *225*, 106741. [CrossRef]
83. Da Silva, M.R.C.; Malacarne, C.S.; Longhi, M.A.; Kirchheim, A.P. Valorization of Kaolin Mining Waste from the Amazon Region (Brazil) for the Low-Carbon Cement Production. *Case Stud. Constr. Mater.* **2021**, *15*, e00756. [CrossRef]

Article

Optimization and Stability of the Bottom Structure Parameters of the Deep Sublevel Stope with Delayed Backfilling

Mochuan Guo [1,2,*], Yuye Tan [1,2,*], Da Chen [1,2], Weidong Song [1,2] and Shuai Cao [1,2]

[1] School of Civil and Resources Engineering, University of Science and Technology Beijing, Beijing 100083, China; chenda202205@163.com (D.C.); songwd@ustb.edu.cn (W.S.); luobiha435@163.com (S.C.)

[2] State Key Laboratory of High-Efficient Mining and Safety of Metal Mines of Ministry of Education, University of Science and Technology Beijing, Beijing 100083, China

* Correspondence: guomochuan@163.com (M.G.); tanyuye@ustb.edu.cn (Y.T.)

Abstract: This study analyzes the stability and optimizes the parameters of the bottom structure in sublevel stoping with the delayed backfilling method, improves production efficiency, and increases the ore recovery ratio under the premise of ensuring safe production. Theoretical formulas are used to calculate the stability of the pillar with the bottom structure. Numerical simulation is used to study the stability of muck slash during excavation. Finally, the optimization parameters of the bottom structure are obtained by combining a similar physical experimental model and numerical simulation. The results show that the excavation of the muck slash caused different degrees of deformation at the roof and floor of the roadway. The largest stress occurred at the roadway crossing, whereas the smallest stress was in the middle area. The excavation also caused the secondary stress concentration at the adjacent bottom structure but did not significantly impact its stability. During the mining process, the largest displacement deformation occurred at the roadway crossing, and the influence of mining disturbance on the stability of the bottom structure involves timeliness and periodicity. Considering the recovery ratio, dilution ratio, and stability, the spacing of the extracted ore drift is recommended to be 9 m. This study ensures the stability of the bottom structure in the mining process and obtained reasonable parameters of the extracted ore drift, which provides a scientific way for the mines that use sublevel stoping with the delayed backfilling method.

Keywords: bottom structure; numerical simulation; parameter optimization; stability

Citation: Guo, M.; Tan, Y.; Chen, D.; Song, W.; Cao, S. Optimization and Stability of the Bottom Structure Parameters of the Deep Sublevel Stope with Delayed Backfilling. *Minerals* **2022**, *12*, 709. https://doi.org/10.3390/min12060709

Academic Editors: Longjun Dong, Yanlin Zhao, Wenxue Chen and Abbas Taheri

Received: 8 April 2022
Accepted: 30 May 2022
Published: 1 June 2022

1. Introduction

The bottom structure of the sublevel stoping with the delayed backfill method is a chamber group composed of a series of muck slash, drilling drift, and extracted ore drift [1]. The layout and parameters of the bottom structure have a great influence on the efficiency of the mining method, labor productivity, ore blocked out, ore dilution and loss, and drawing work safety. Its stability is the key to ensuring the safety and efficiency of ore mining [2,3]. However, in the mining process, it will be subjected to the stress redistribution caused by excavation disturbance and the rock burst effect caused by stope blasting and ore falling [4]. Especially after entering deep mining, given complex factors such as high ground stress and dynamic disturbance, the deformation of the bottom structure roadway is significantly increased, and the stability problem is prominent, which poses a great threat and trouble to the safety of stope production and the efficient recovery of ore. Furthermore, the shallow mining method is challenging to apply to the subsequent deep mining of the mine [5].

In recent years, research on the bottom structure of stope has mainly optimized the main parameters and determined the optimal bottom structure form of stope using theoretical formulas and mathematical analysis. It predicted the failure of the roof, floor, and surrounding rock [6–16] and analyzed the stability of the bottom structure in the process of excavation disturbance combined with numerical simulation [17–23]. Previous studies have mainly focused on the parameter design of the muck slash in the caving method and the distribution and change of stress in excavation disturbance. Different muck slash was selected according to the technical and economic indicators, such as the ore recovery ratio. The stability of the stope bottom structure was analyzed from the surrounding rock of the bottom structure, the support of the roadway, and the interaction between ore and rock [24–31]. However, studies on the bottom structure of the sublevel stoping with the delayed backfill method are limited, and the parameter optimization of the extracted ore drift is also less involved. Therefore, studying the relevant parameters of the extracted ore drift is crucial to the loss and dilution of the ore in the mining process.

Empirical formula and FLAC3D numerical simulation software are used to study the stability, stress variation characteristics, displacement characteristics, and failure characteristics of the bottom structure in different periods of the sublevel stoping with the delayed backfill method and evaluate the stability of the bottom structure. A similar physical model test and PFC3D particle flow software are also used to examine the influence of the spacing between the extracted ore drift in the bottom structure of the stope on the ore loss and dilution level. A set of optimal parameters are here estimated to provide a basis for designing the bottom structure in the sublevel stoping with the delaying backfilling mining method.

2. Engineering Background

A gold mine adopts sublevel stoping with the delayed backfill method. The orebody level of a gold mine is −600–155 m. In order to study the stability of the bottom structure in the mining, the first and second sublevel of −616 m is selected for analysis. The maximum horizontal thickness of the orebody in this area is 105 m, and the minimum thickness is 57 m. The dip angle of the orebody close to the hanging wall is 25°, and the dip angle of the footwall is 48°. According to the actual geological work on site, the mine rock is mainly composed of granitic rock and has good stability. The physical and mechanical parameters of mine rock are shown in Table 1.

Table 1. Physical and mechanical parameters of ore and rock.

Mechanical Parameter	Hanging Wall	Footwall	Ore Body	Main Fracture Surface
natural density g/cm^3	2.7	2.72	2.82	2.73
Cohesion C/MPa	4.6	4.8	4.7	2.2
angle of internal friction $\varphi/(°)$	35	39	37	34
p-wave velocities m/s	4943	5401	5029	5014
tensile strength MPa	2.40	2.50	2.45	1.02
compressive strength MPa	72.0	85.0	70.0	42.0
elastic modulus E/GPa	16	21	20	13
Poisson ratio γ	0.22	0.21	0.24	0.22

In the mining, the ore body is mined from bottom to top, in the order of room to pillar. The width of the room and pillar in the stope is 10 m, the sublevel height is 15 m, the bank height is 60 m, and the length of the room is the horizontal thickness of the ore body. The section size of the stope track is 3.2 m × 3.0 m, and the section size of the air connection is 2.0 m × 2.0 m. The stope model is established through SURPAC, as shown in Figure 1.

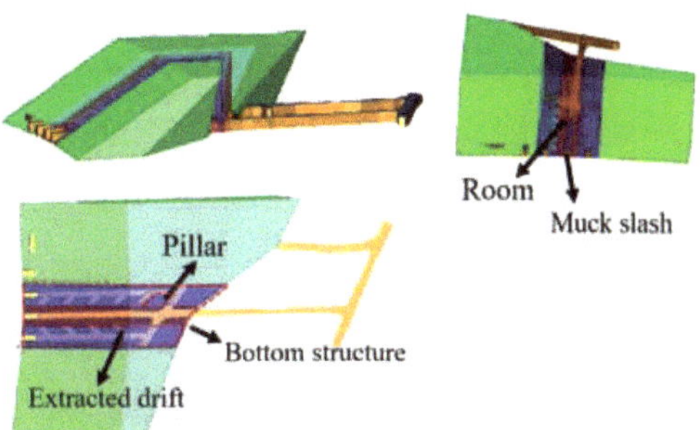

Figure 1. Stope solid model.

3. Stability Analysis of Bottom Structure

3.1. Study on Size of Bottom Structure

In the sublevel stoping with the delayed backfill method, the bottom structure of the muck slash can be simplified to a single pillar. In theory, the tangential stress on the sidewall surface of the pillar is the largest. The pillar's surface is affected by dynamic loads, such as mining blasting in practice. The fissure is abnormally developed, forming a ground stress fracture layer of 0.4–1 m, and the maximum stress value is transferred to the center of the pillar [32]. According to structural mechanics theory, when calculating the strength of pillars, the stress distribution in continuous or discontinuous pillars is uniform, and the stress mainly causes the failure of pillars in the vertical direction [33]. The pillar calculation generally does not consider the pillar weight unless it is very high. The required cross-sectional area to ensure pillar strength can be calculated according to the allowable bearing strength equation, as in Equation (1) [34]:

$$\frac{S\gamma Hk}{s} \leq \frac{\sigma_0 k_f}{n} \tag{1}$$

where S represents pillar support area (m^2), γ represents rock bulk density (kg/m^3), H represents stope distance to surface depth (m), k represents load coefficient, s represents the cross-sectional area of the pillar (m^2), σ_0 represents the UCS of the pillar (MPa), and n represents safety factor (3–5 when the surface does not allow collapse or long-term retention).

The equation of relative pillar area is obtained from Equation (1), as shown in Equation (2):

$$\frac{s}{S} \leq \frac{\gamma Hk}{\sigma_0 k_f} \tag{2}$$

When calculating the strength of the pillar, ensuring that the pillar is not affected by blasting shock and avoiding the longitudinal bending of the pillar are necessary. Therefore, the condition that the unknown parameter k_f needs to meet is $k_f \geq 2\omega$, where ω is the minimum resistance line of the room blasting. According to the mechanical parameters of mine rock in Table 1, at the level of −616 m, k_f = 1.2, n = 2, k = 0.6, and the length of the pillar edge is a. The pillar size obtained from Equation (2) satisfies the following conditions:

$$a \geq 4.29 \text{ m},$$

Field investigation reveals that, at −616 m level, the pillar with a small size and less than 4 m has different degrees of damage, but the size greater than 4 m, there is basically no failure form. Figure 2 is the typical failure form of the pillar in the field; most of the failures occurred at the root of the pillar. Figure 2a shows that three obvious parallel joints run through the pillar, and shear failure occurs along the joint surface under pressure. Over time, the failure will expand from the surface to the interior and eventually lose the support capacity. Figure 2b shows poor pillar stability. The root has been broken, which is prone to instability. In the later stage, the corresponding types of cross-section pillars should be strengthened.

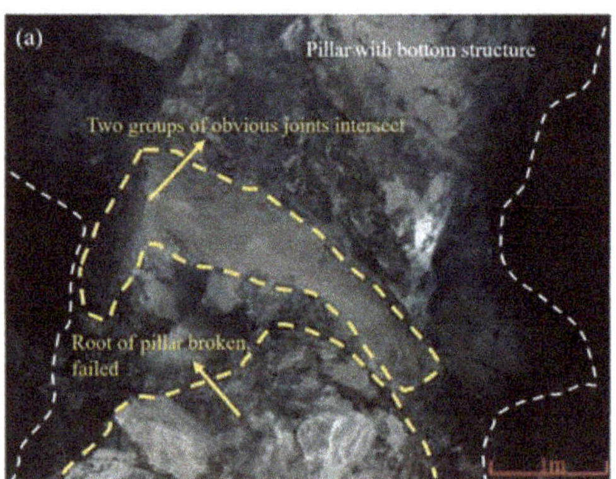

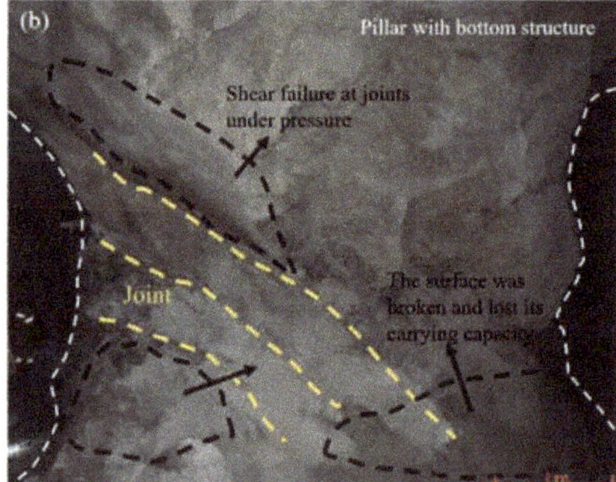

Figure 2. Typical pillar damage. (**a**) Pillar joint failure (**b**) Pillar root failure.

3.2. Numerical Simulation

3.2.1. Establish Model

For the stability analysis of the bottom structure, given the complex ground stress, the general theoretical calculation cannot be completed. Therefore, this study uses numerical analysis to simulate the ultimate span between the extracted ore drift of deep stope under the condition of no support. According to the actual situation of the mine, considering the Saint-Venant principle, the influence of boundary on the excavation of the central stope is reduced, and the calculation range is selected to be more than three times that of the excavation area. Considering the ore body and surrounding rock and the pitch angle, the length, width, and height are determined to be 300 m × 270 m × 120 m. After the model is established by ANSYS software (ANASYS2019, American ANSYS company), it is calculated by FLAC3D according to different lithological groups. The model is meshed by four-node tetrahedral elements. The calculation model is composed of 308,130 zones and 417,809 grid points. The mine rock mechanics parameters used in the calculation are shown in Table 2.

Table 2. Parameter of bulk modulus and shear modulus.

Mechanical Parameter	Hanging Wall	Footwall	Ore Body	Main Fracture Surface
volume modulus K/Kpa	9.52	12.07	12.82	7.74
shear modulus G/Mpa	13.11	17.36	16.13	10.66

The stress field of underground engineering existed before excavation. In this simulation, the two-stage elastic–plastic solution was used to obtain the initial stress field of the stope in sedimentary consolidation through the gravity effect of the overlying formation. The stress state of rock mass at a depth of H from the surface is expressed in Equations (3) and (4). The Mohr–Coulomb criterion is used for the mechanical model of mine rock, as shown in Equation (5) [33].

$$\sigma_x = \sigma_y = \frac{\mu}{1-\mu}\sigma \tag{3}$$

$$\sigma_2 = \gamma H, \tag{4}$$

$$f_s = \sigma_1 - \sigma_3 \frac{1+sin\varphi}{1-sin\varphi} - 2c\sqrt{\frac{1+sin\varphi}{1-sin\varphi}} \tag{5}$$

where γ represents Poisson ratio, M represents average rock weight, H represents depth from surface (m), c represents cohesion, φ represents the internal friction angle, and f_s represents the destroy judgment coefficient.

Considering the constitutive characteristics of mine rock, the continuity of medium, and boundary conditions, the stress loaded on the top of the model is found to be 19.46 MPa, and the horizontal stress loaded on the model's boundary is 24.3 MPa. Figure 3a shows the stress nephogram in the vertical and horizontal directions of the initial stress field, and Figure 3b shows the research area division of the model. This study selects the scheme with the minimum spacing between extracted ore drift of 7 m for numerical simulation. By analyzing the stability of the small spacing, the parameter basis is provided for the design of a larger spacing of the extracted ore drift, and the recovery ratio of the ore is improved under the premise of ensuring safety.

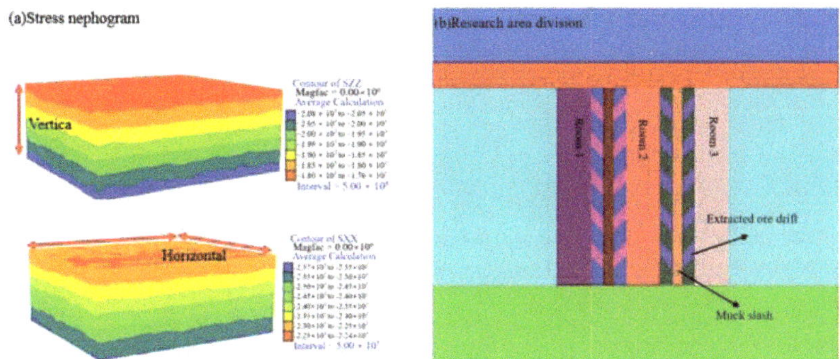

Figure 3. Initial diagram of the calculation model.

3.2.2. Simulation Result Analysis

(1) Stress analysis during excavation

Figure 4 illustrates the stress distribution nephogram of the bottom structure during excavation. The maximum principal stress of the rock mass after excavation has been redistributed. The two sides of the roadway and the shallow part of the roof and floor are in the pressure relief area. The deep part of the roof and floor of the roadway and the shallow part of the two bottom angles of the roadway are in the stress concentration area. Figure 4a shows that the disturbance stress caused by excavation is concentrated on the bottom structure, and the stress value is the largest at the intersection tip of the roadway. After excavation, the stress on the two sides of the extracted ore drift and muck slash is significantly greater than that outside the bottom structure. The stress decreases gradually from the bottom structure area to the outside. According to the process of room excavation in Figure 4b–d, in the process of room excavation, the influence area is larger than that of the roadway excavation. The excavation of the room causes the secondary concentration of the structural stress near the bottom, and the stress concentration area of the bottom structure changes obviously. The stress concentration value is the highest at the tip and gradually decreases from the outside to the inside, showing an elliptical distribution. The excavation of the three rooms does not significantly impact each other, indicating that the bottom structure is stable under this parameter.

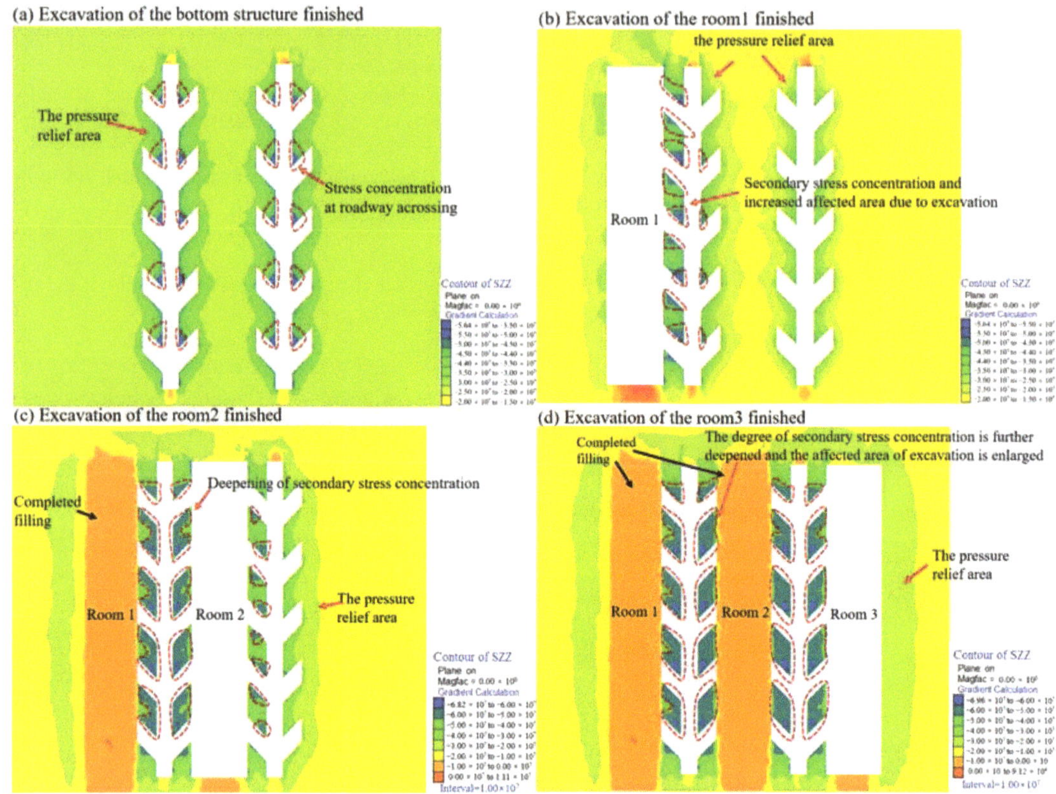

Figure 4. Stress distribution of bottom structure.

(2) Displacement analysis during excavation

Figure 5 presents the displacement nephogram of the roof and floor in the mining process shown in the section diagram of the bottom structure of the pillar. After excavating the bottom structure, the roof and floor of the roadway have different degrees of deformation, and the bottom structure also has a small amount of settlement. With the mining of the ore body, the hanging area of the rock mass increases, and the initial stress balance system is destroyed. Under the action of the internal unbalanced stress, the surrounding rock seeks a new balance through deformation. The deformation process of the rock mass makes the accumulated stress of the surrounding rock release and transfer continuously. Simultaneously, the deformation of the bottom structural displacement also means that the energy is absorbed inside. The above figures' comparison shows that the bottom structure's displacement caused by room 1 mainly concentrates on the roof and floor. When rooms 2 and 3 are stoping, the displacement of the bottom structure changed greatly, and the displacement increased by three times. With the mining of the room, the area with large displacement continues to expand to the surrounding, and the bottom structure bears more stress transfer caused by mining.

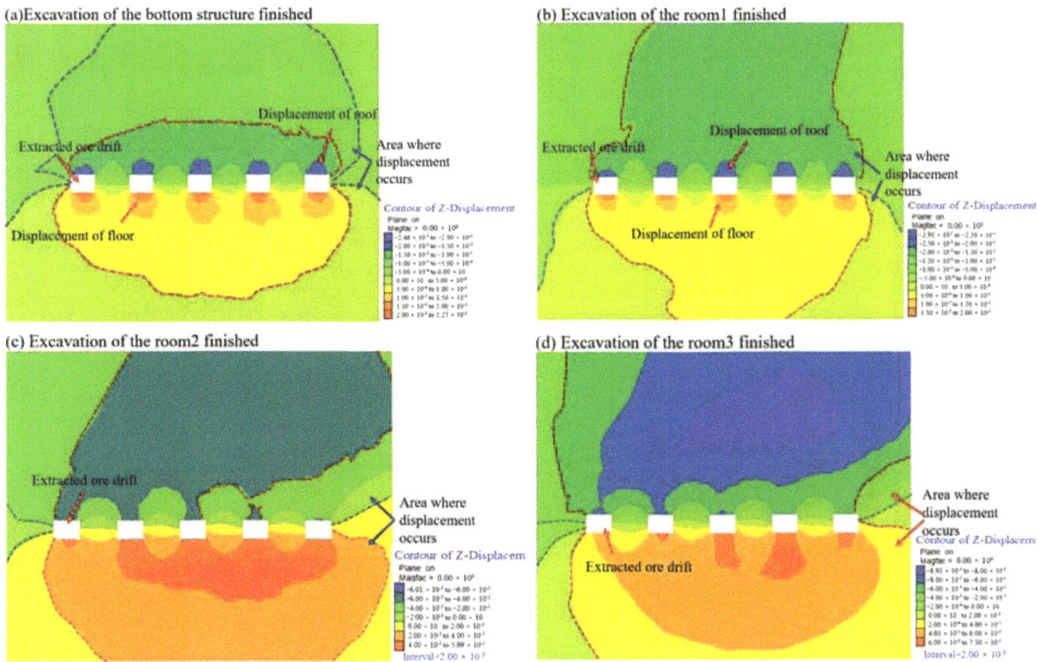

Figure 5. Displacement of the excavation.

(3) Plastic-zone analysis during excavation

Figure 6 shows the plastic area where the deformation of the soil cannot be recovered during the excavation of the room; Figure 6a shows that, after the excavation of the bottom structure, only a small area of failure zone is generated near the hanging wall of the roadway. The main fracture surface is close to the hanging wall, and the rock mass is mainly selvage. However, the poor stability leads to the failure of the rock mass. Figure 6b–d shows that the stress of the adjacent bottom structure of room 1 increases significantly after excavation, and the stress concentration at the intersection of the muck slash and the extracted ore drift is more pronounced. The stress concentration of both sides of the bottom structure is caused again after the excavation of room 2, which has a certain influence on the bottom structure of room 1. The failure zone of the ore body of the bottom structure on both sides is further increased, and a circle of plastic zone is formed in the periphery. Only the middle elliptical area has not been destroyed. After the excavation of room 3, the stress of the bottom structure is further increased, and the plastic zone area of the bottom structure is close to half of the total area. Only the middle part does not occur plastic deformation, and the bottom structure is in a critical state of failure. Throughout the development of the plastic zone, the influence of mining disturbance on the stability of the bottom structure has timeliness and stages.

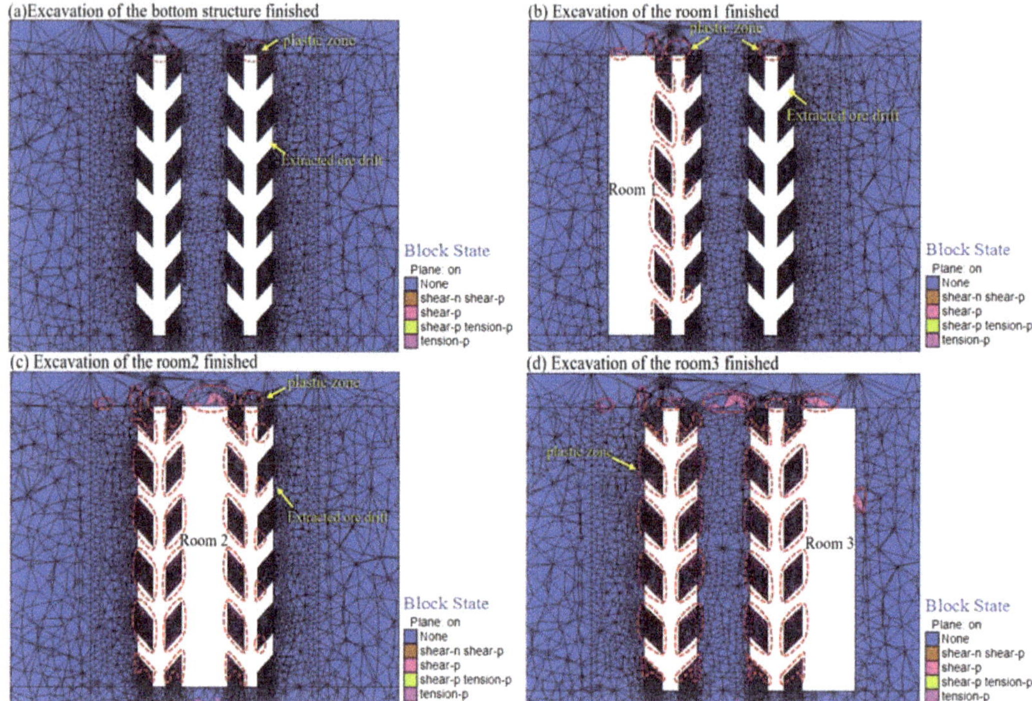

Figure 6. Distribution of plastic zone.

Figure 7 illustrates that the failure range of the bottom structure is in the shape of a funnel, and the middle failure area is the largest. The main reason is that, after the formation of the bottom structure, the pillar bears the coating pressure on the excavation area, forming the compression-shear failure, which causes the local failure of the bottom structure. However, the bottom structure is not fully yielding and is still stable.

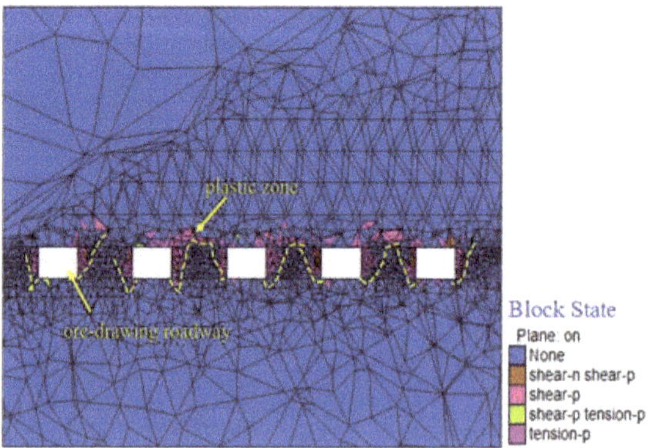

Figure 7. Left view after excavation.

(4) Displacement monitoring analysis

Figure 8 shows the distribution of displacement monitoring points in the bottom structure. MP represents the monitoring point. According to the displacement change of the roof and floor, 24 monitoring points are arranged in the bottom structure. The first group of MPs is located at the intersection of the roadway roof, numbered 11–14. The second group of MPs is located in the center of the roof of the extracted ore drift, numbered 21–24. The third group of MPs is located in the center of the pillar of the bottom structure, numbered 31–34. The layout points of the fourth, fifth, and sixth groups are located on the bottom floor at the same position as the first three groups.

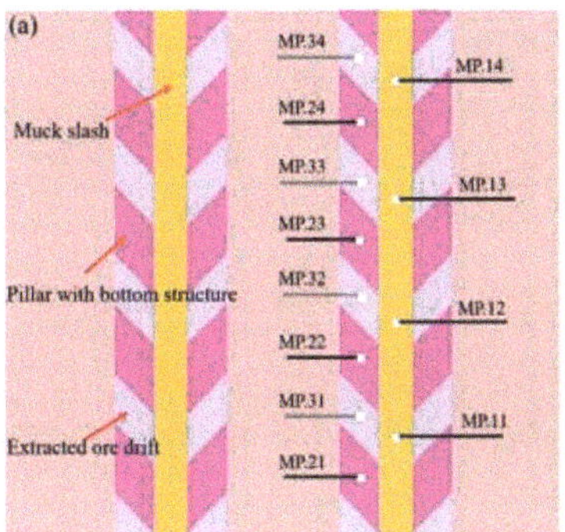

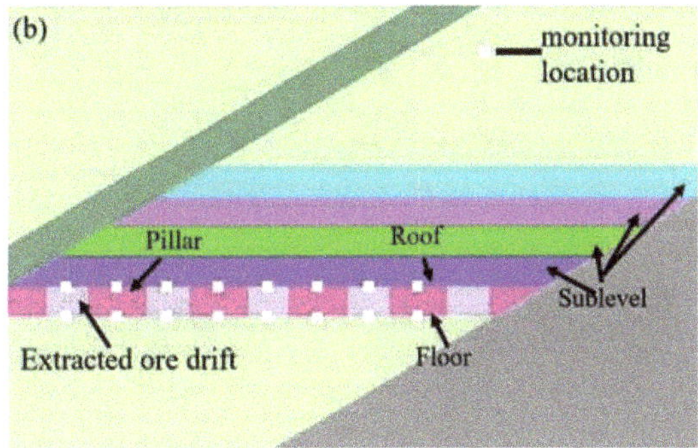

Figure 8. Distribution of monitoring points. (**a**) Top view of MP layout (**b**) Side view of MP layout.

Figure 9 shows each monitoring point's displacement process in the bottom structure. The displacement variation of the roof and floor in the bottom structure is similar. The displacement and deformation of the roof and floor at the roadway intersection are the largest, and most of the roof subsidence and floor heave occur in the mining process. A small amount of displacement is caused after excavation, but the deformation is limited. During room excavation, the displacement of the roof and floor changes significantly and gradually stabilizes with the mining.

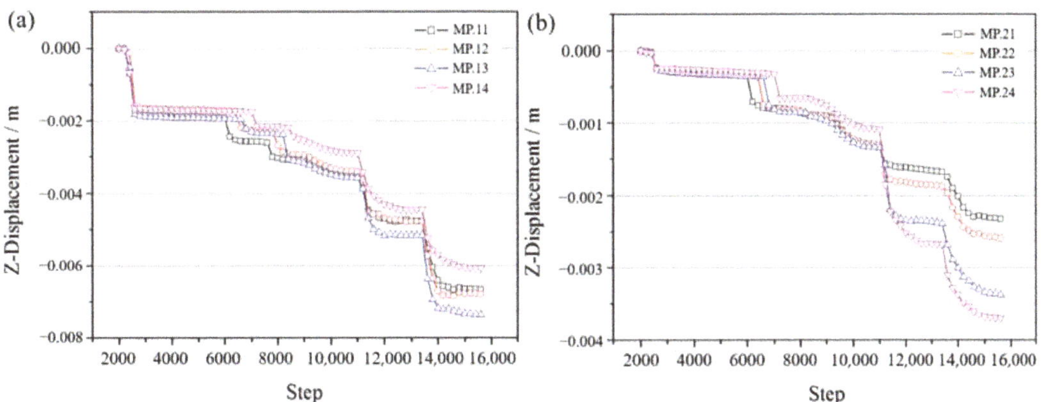

Figure 9. *Cont.*

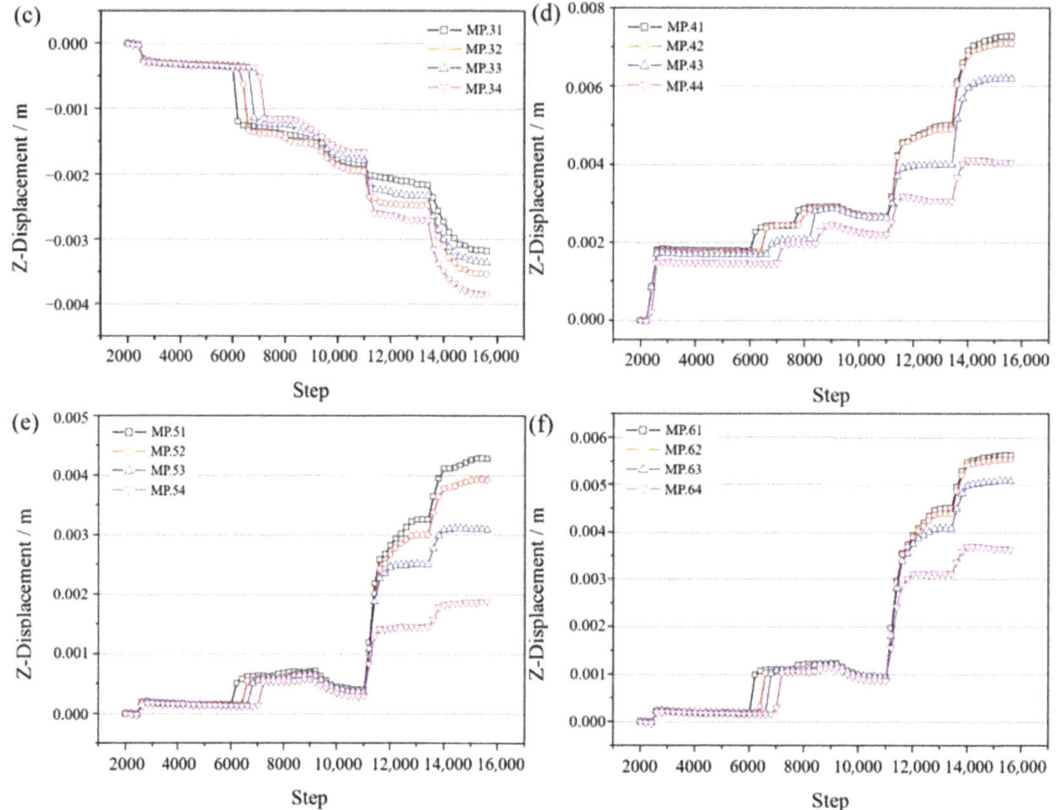

Figure 9. Displacement curve of each monitoring point. (**a**) The first group of MPs (**b**) The second group of MPs (**c**) The third group of MPs (**d**) The fourth group of MPs (**e**) The fifth group of MPs (**f**) The sixth group of MPs.

4. Spacing Optimization of Extracted Ore Drift with Bottom Structure

The extracted ore drift is the key channel for drawing. When the spacing is too wide, the stress superposition is reduced, and the stability of the roadway is improved. However, the residual ore in the ridge and the footwall between the roadway increase, and the ore recovery ratio cannot be guaranteed [35]. Therefore, the spacing of extracted ore drift involves ore recovery and roadway stability. Its essence is to achieve a reasonable balance between ore recovery and roadway stability and maximize ore recovery to ensure roadway stability.

4.1. Physical Modeling Experiment

4.1.1. Model Design and Material Selection

This experiment adopts a 3D ore model. According to the stope size and roadway layout, the model ratio was 1:66.6, the model size was 750 × 150 × 225 mm, extracted ore drift size was 45 × 45 mm. The roadway had a symmetrical layout. The experimental model is shown in Figure 10. The bulk density of ore and waste rock selected in this experiment was 1.98 g/cm^3 and 1.68 g/cm^3, respectively. The ore block size range of the stope in the gold mine was 400~800 mm, and the waste rock block size range was 600~1000 mm. According to the geometric similarity in the similarity theory, the particle size of ore and rock in the experiment is ground according to the geometric size ratio of 1:66.6. The ore and waste rock used in the experiment were obtained from mine production.

The particle size of ore and waste rock were ground to 6~12 mm and 9~15 mm, respectively; after screening by lattice sieve, they were mixed according to the experiment's proportion of particle size distribution.

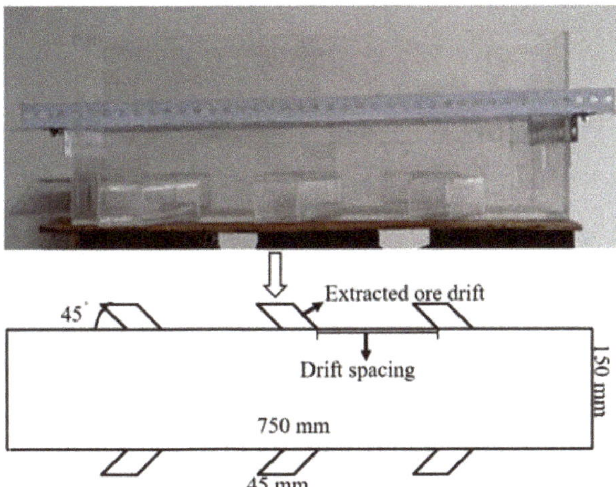

Figure 10. Experimental model diagram.

4.1.2. Experimental Design and Steps

According to the requirements of the mining method, the spacing was set to 7, 8, 9, 10, 11, 12, and 13 m, taking the roadway spacing as variable parameters. During the experiment, each group was filled with one waste rock and four ore layers, and three to five groups of extracted ore drifts were arranged. After mucking was completed, the drawing was started. In the experiment, a small spoon was used for ore drawing. According to the one-time ore drawing amount in actual production and the proportion of the physical model, the ore drawing amount per shovel was about 25 g. When the ore drawing was similar to that of waste rock, the drawing was stopped, and the drawing of the next ore drift was carried out after weighing. In order to ensure the uniform ore drawing of each ore drift, the next drift for ore drawing was changed after every 20 shovels. This drawing was terminated when all of the extracted ore drift was unable to be drawn. Then the ore was sorted as in the drawing model, the spacing of the extracted ore drift was adjusted, the ore and rock were screened, the total ore waste rock was counted, and the ore drawing experiment for the next scheme was prepared. The experimental process is shown in Figure 11.

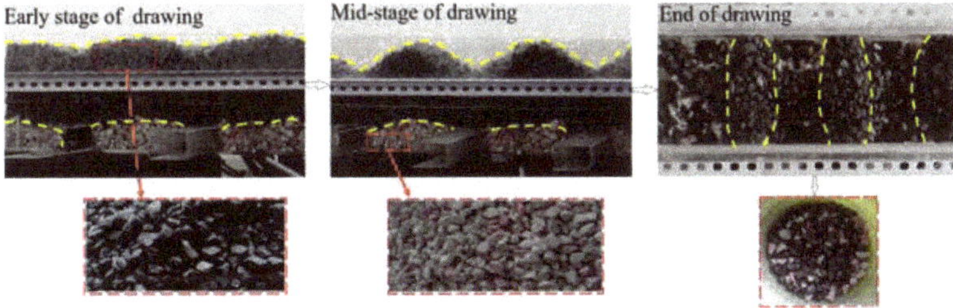

Figure 11. The drawing process.

4.1.3. Experimental Result Analysis

The drawing experiment was carried out according to the above experimental scheme. Under different experimental schemes, ore quantity, waste rock quantity, total mine rock quantity, dilution ratio, and ore recovery ratio were obtained. The relationship between ore loss dilution ratio and ore drawing roadway spacing was obtained by fitting the results of the different ore drawing roadway spacing. The experimental results are shown in Figure 12.

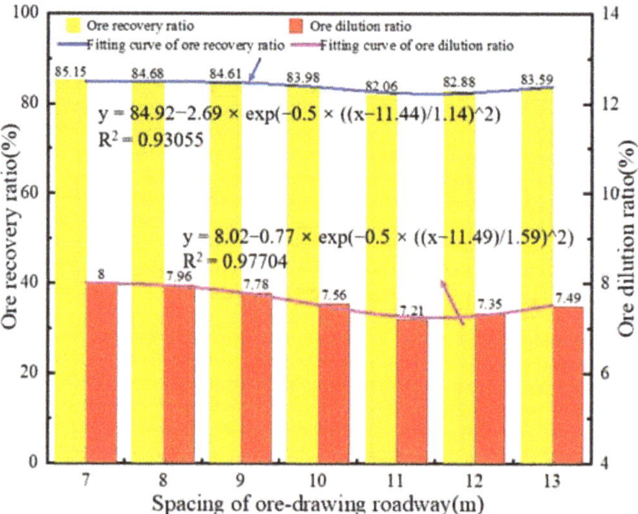

Figure 12. Effects of different mining roadway spacings on recovery ratio and dilution ratio.

Figure 12 shows that the change trends of ore recovery ratio and dilution ratio are the same, and the two establish a slight downward trend with the increase in the spacing of the mining roadway. When the spacing is 7–10 m, the change is gentle and suddenly drops at 10 m. When the spacing is 11 m–13 m, it is gradually stable, and the overall trend is flat–steep–flat. Combined with experimental observation and analysis of ore–rock movement law, when the spacing becomes larger, the number of outlets decreases, and the recovery ratio decreases. When the spacing is 10 m, the recovery ratio decreases sharply because of the decline of one outlet.

4.2. Numerical Simulation Analysis

4.2.1. Establishment of Mining Model

According to the indoor rock mechanics test and field data, the simplified stope model is established using PFC3D numerical simulation software. The size of the model is $X \times Y \times Z = 55\ m \times 10\ m \times 15\ m$, the size of the extracted ore drift is $X \times Y = 3\ m \times 3\ m$, the angle is 45°, the radius of the ore is 18 mm, and the waste rock is 19 mm. The mechanical parameters of the ore and rock are shown in Table 3. The spacing of 7 m is selected as the research object for numerical simulation analysis. The initial mining model is shown in Figure 13.

Table 3. Parameter table of numerical simulation.

Material	Tangential Stiffness ($N{\cdot}m^{-1}$)	Normal Stiffness ($N{\cdot}m^{-1}$)	Density (kg/m^3)	Friction Coefficient
surrounding rock	1.2 × 108	1.2 × 108	/	0.35
ore	1.0 × 108	1.0 × 108	2700	0.3
rock	1.0 × 108	1.0 × 108	3000	0.3

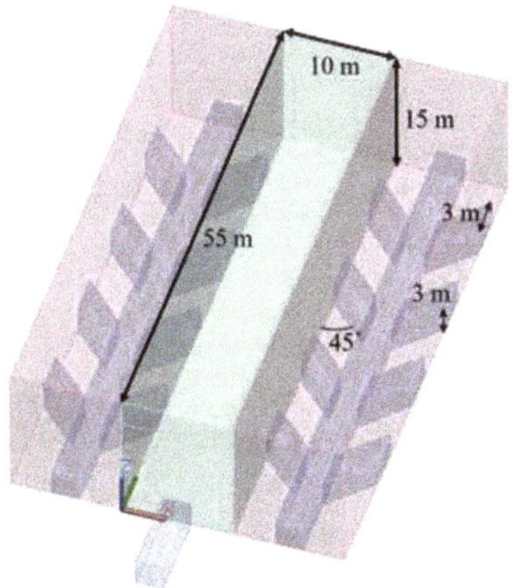

Figure 13. Drawing of ore drawing model.

4.2.2. Drawing Simulation Results Analysis

The following is a comprehensive analysis of the flow of ore and rock in the ore body. The study examines the outflow of ore and waste rock and the distribution of ore and rock in the ore body in each period of the drawing process. It intuitively shows the whole drawing process, obtains the output of ore and waste rock after completing the drawing process and the residual parts of ore and waste rock in the ore body, and calculates the ore recovery ratio and dilution ratio.

Figure 14 illustrates the schematic diagram of each stage in the drawing. In Figure 14a, the radius expansion method fills the particles. After the particle filling is completed, the initial balance is carried out. Figure 14b shows that the movement of the upper orebody is not evident in the early stage of drawing. Most of the mine rock mixtures released for the first time are waste rocks, and only a small amount of ore is released. Figure 14c is the drawing process. A small movement occurs in the middle ore body, and the ore flow law is the same as the laboratory physical model experiment. Figure 14d is the final drawing. The ore flow diagram of each stage shows that the mine rock mixture released for the first time is mostly waste rock, and only a small amount of ore is released. After drawing, the ore in the upper part of the outlet is released first and gradually forms a funnel shape. Only a small amount of intermediate ore moves. After the drawing is cut off, a large amount of waste rock is found at both spont mouth ends. The ore in the upper is less, and most of the ore is in the lower part. Finally, more ore is seen in the middle of the spont mouth and at the top of the hanging wall.

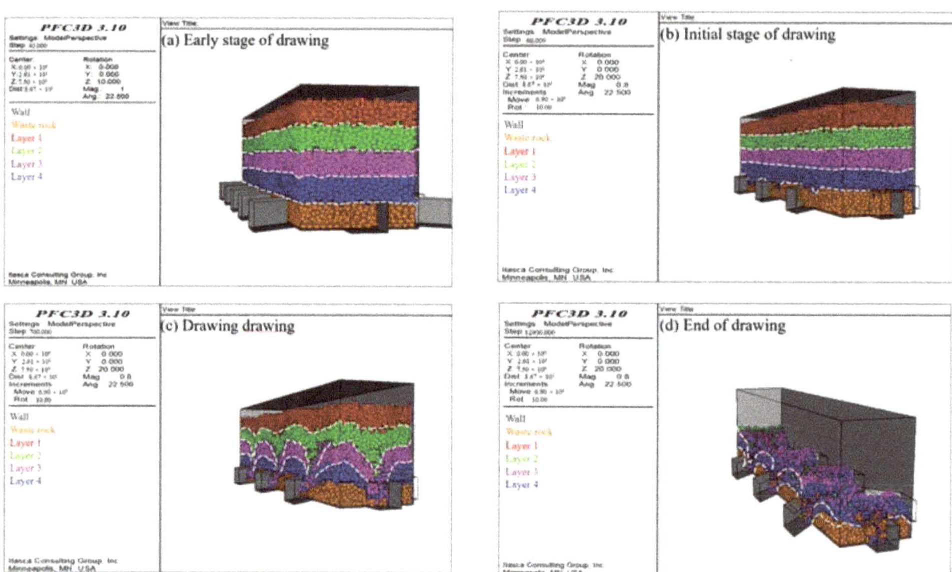

Figure 14. Schematic diagram of the drawing process.

Figure 15 shows the relationship between the extracted ore drift spacing and the ore recovery and dilution ratio in the numerical simulation. Its change trend is consistent with the experimental trend of a similar physical model when the dilution ratio is constant in the number of spont mouths. With the increase in spacing, the residual ore and rock increase, the residual waste rock at the bottom increases, and the dilution ratio decreases. When the number of the spont mouth increases, ore and rock drawing increase, and the dilution ratio increases. Considering the recovery ratio and dilution ratio, using the parameter of 9–10 m is recommended for the spacing between the extracted ore drift.

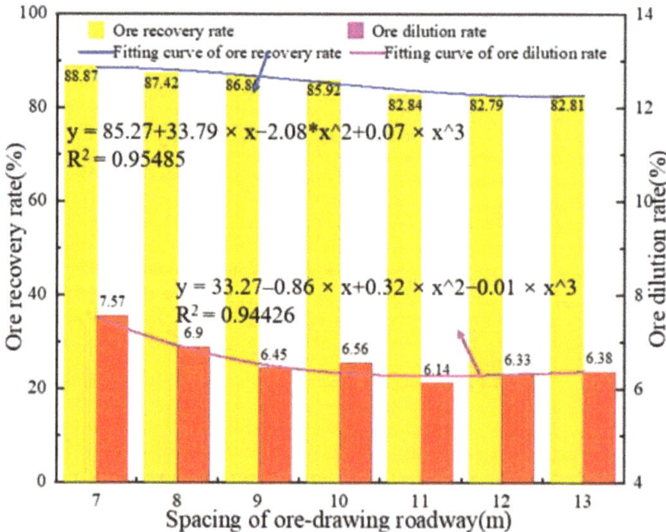

Figure 15. Influence of different spacing of extracted ore drift on recovery ratio and dilution ratio.

5. Conclusions

(1) The numerical analysis shows that the excavation of the muck slash caused different degrees of deformation at the roof and floor of the roadway. The roadway has two sides. The shallow part of the roof and floor are in the pressure relief area. The deep and shallow strata of the two bottom corners of the roadway are in the stress concentration area. The excavation of the adjacent two roadways does not affect each other. The largest stress occurred at the roadway crossing, whereas stress is the smallest in the middle area. Stress is symmetrically distributed, and only a small area of the failure zone is generated near the hanging wall at the roadway.

(2) During the excavation of the room, the secondary stress concentration at the adjacent bottom structure and the stress concentration area of the bottom structure are significantly changed but did not significantly impact its stability. It only affects the near bottom structure to bear more stress transfer.

(3) In mining, the displacement deformation of the roof and floor at the intersection is the largest and gradually tends to be stable with the mining. The influence of mining disturbance on the stability of the bottom structure has timeliness and periodicity.

(4) In the stope of this study, the bottom structure of the mining roadway is simplified as a single pillar when the sublevel stoping with delayed backfill method is used. When the actual size of the pillar is less than the theoretical, the pillar will be destroyed along the joint or at the root. Considering the recovery ratio, dilution ratio, and stability, the spacing of the extracted ore drift is recommended to be 9 m, which can be increased to 10 m if stability is poor.

Author Contributions: Conceptualization, M.G. and Y.T.; methodology, Y.T.; software, M.G.; validation, Y.T. and D.C.; formal analysis, M.G.; investigation, D.C.; resources, Y.T. and S.C.; data curation, M.G.; writing—original draft preparation, M.G.; writing—review and editing, Y.T. and W.S.; funding acquisition, Y.T. and W.S. All authors have read and agreed to the published version of the manuscript.

Funding: This work was financially supported by the National Natural Science Foundation of China, grant number 52004019 and Fundamental Research Fund for the Central Universities(FRF-BD-20-03A).

Data Availability Statement: Not applicable.

Acknowledgments: The author gratefully acknowledge all the professors for their guidance and help during testing and writing.

Conflicts of Interest: The authors declare no conflict of interest.

References

1. Changbin, Y.; Guoyuan, X. Numerical Simulation Analysis of Dynamic Stability of Vertically Arranged Underground Chambers under dynamic load. *J. Cent. South Univ.* **2006**, *37*, 593–599.
2. Delentas, A.; Benardos, A.; Nomikos, P. Analyzing Stability Conditions and Ore Dilution in Open Stope Mining. *Minerals* **2021**, *11*, 1404. [CrossRef]
3. Li, S.; Zhao, Z.; Yu, H.; Wang, X. The Recent Progress China Has Made in the Backfill Mining Method, Part II: The Composition and Typical Examples of Backfill Systems. *Minerals* **2021**, *11*, 1362. [CrossRef]
4. Lu, J.; Chu, H.; Lin, S.; Gong, Y. Research on the stability of bottom structure based on FLAC3D numerical simulation. *Min. Res. Dev.* **2020**, *40*, 19–23.
5. Li, X.; Yao, J.; Gong, F. Dynamic problems in deep mining of hard rock metal mines. *Chin. J. Nonferrous Met.* **2011**, *21*, 2551–2563. [CrossRef]
6. Cheng, J.; Zhang, Q.; Xue, X.; Wang, S.; Cao, R. Optimization of stope structure parameters based on AHP and TOPSIS method. *Min. Metall. Eng.* **2014**, *34*, 1–5.
7. Ma, M.; Guo, Q.; Pan, J.; Ma, C.; Cai, M. Optimal Support Solution for a Soft Rock Roadway Based on the Drucker–Prager Yield Criteria. *Minerals* **2022**, *12*, 1. [CrossRef]
8. Bai, X.; Marcotte, D.; Simon, R. Underground stope optimization with network flow method. *Comput. Geosci.* **2013**, *52*, 361–371. [CrossRef]
9. Sameera, S.D.S.; Erkan, T.; Ali, A.M.W. Designing an optimal stope layout for underground mining based on a heuristic algorithm. *Int. J. Min. Sci. Technol.* **2015**, *25*, 767–772.

10. Yin, S.; Wu, A.; Li, X. Orthogonal range analysis on sensitivity of influencing factors of pillar stability. *J. Coal* **2012**, *37*, 48–52. [CrossRef]
11. Shi, Y.; Ye, Y.; Hu, N.; Jiao, Y.; Wang, X. Physical Simulation Test on Surrounding Rock Deformation of Roof Rockburst in Continuous Tunneling Roadway. *Minerals* **2021**, *11*, 1335. [CrossRef]
12. Lan, M.; Liu, Z.; Li, X. Multi-objective multi-attribute optimization of structural parameters of subsequent filling stope. *J. Cent. South Univ.* **2019**, *50*, 375–383.
13. Qi, C.; Chen, Q. Evolutionary Random Forest Algorithms for Predicting the Maximum Failure Depth of Open Stope Hangingwalls. *IEEE Access* **2018**, *6*, 72808–72813. [CrossRef]
14. Yu, Y.; Lu, J.; Chen, D.; Pan, Y.; Zhao, X.; Zhang, L. Study on the Stability Principle of Mechanical Structure of Roadway with Composite Roof. *Minerals* **2021**, *11*, 1003. [CrossRef]
15. Nan, S.; Qian, G.; Liu, Z. Numerical Simulation of Fluid-Solid Coupling in Surrounding Rock and Parameter Optimization for Filling Mining. *Int. Symp. Mine Saf. Sci. Eng.* **2011**, *26*, 1639–1647.
16. Feng, X.; Ding, Z.; Hu, Q.; Zhao, X.; Ali, M.; Banquando, J.T. Orthogonal Numerical Analysis of Deformation and Failure Characteristics of Deep Roadway in Coal Mines: A Case Study. *Minerals* **2022**, *12*, 185. [CrossRef]
17. Zhang, D.; Bai, J.; Yan, S.; Wang, R.; Meng, N.; Wang, G. Investigation on the Failure Mechanism of Weak Floors in Deep and High-Stress Roadway and the Corresponding Control Technology. *Minerals* **2021**, *11*, 1408. [CrossRef]
18. Zhang, Z.; Shi, X.; Qiu, X. Using Mathews synthetic chart and dynamic and static load numerical simulation to evaluate stope stability of inclined ore body. *J. Nonferrous Met. China* **2022**, 1–11. Available online: http://kns.cnki.net/kcms/detail/43.1238.TG.20210831.1316.005.html (accessed on 12 February 2022).
19. Wang, X.; Wang, C.; Zhang, Q.; Yuan, J.; Chen, B. Analysis of stope stability based on ANSYS program. *Met. Mine* **2008**, *08*, 17–20.
20. Liu, D.; Shao, A.; Jin, C.; Ding, S.; Fan, F. Construction of numerical calculation model of underground broken orebody and optimization of stope structure parameters. *J. Cent. South Univ.* **2019**, *50*, 437–444.
21. Xu, W.; Song, W.; Du, J.; Mei, L.; Liu, B. Stability Analysis of Excavation Disturbance of Bottom Structure Roadway Group in Metal Mine. *Undergr. Space Eng. J.* **2014**, *10*, 689–696.
22. Huang, Y.; Yang, W.; Li, Y.; Guo, W. Spatial Distribution Characteristics of Plastic Failure and Grouting Diffusion within Deep Roadway Surrounding Rock under Three-Dimensional Unequal Ground Stress and Its Application. *Minerals* **2022**, *12*, 296. [CrossRef]
23. Zhu, D.; Wang, J.; Gong, W.; Sun, Z. Model Test and Numerical Study on Surrounding Rock Deformation and Overburden Strata Movement Law of Gob-Side Entry Retaining via Roof Cutting. *Minerals* **2020**, *10*, 458. [CrossRef]
24. Wenjie, G. Optimization of stope structure parameters based on MIDAS-GTS. *Min. Res. Dev.* **2017**, *37*, 1–5. [CrossRef]
25. Milic, V.; Radovanovic, M. Determination of the main parameters of semi-level induced caving method with lateral loading. *J. Min. Sci.* **2021**, *57*, 76–85. [CrossRef]
26. Wang, S.; Li, X. Grade Distribution Modeling within the Bauxite Seams of the Wachangping Mine, China, Using a Multi-Step Interpolation Algorithm. *Minerals* **2017**, *7*, 71. [CrossRef]
27. Huang, Y.; Dai, X.; Bai, Y.; Wang, L. Optimization of stope bottom structure and improvement of ore extraction process in high-stage stope. *Min. Metall. Eng.* **2013**, *33*, 21–24, 29.
28. Wu, P.; Chen, L.; Li, M.; Wang, L.; Wang, X.; Zhang, W. Surrounding Rock Stability Control Technology of Roadway in Large Inclination Seam with Weak Structural Plane in Roof. *Minerals* **2021**, *11*, 881. [CrossRef]
29. Zhao, H.; Liu, Y.; Li, J.; Xu, J. Analysis of damage process and zoning failure characteristics of floor rock mass under isolated island coal pillar. *J. China Univ. Min. Technol.* **2021**, *50*, 963–974.
30. Zhang, C.; Song, W.; Fu, J.; Li, Y. Rock mass stability analysis under seabed mining disturbance. *J. China Univ. Min. Technol.* **2020**, *49*, 1035–1045.
31. Xie, X.; Xie, H.; Tian, H.; Xiong, H.; Li, J. Study on meso-mechanism of pillar damage and fracture instability under excavation disturbance. *Min. Metall. Eng.* **2019**, *39*, 30–36.
32. Wang, G.; Yuan, Y.; Gao, Y.; Sun, M.; Ge, Y. Optimization of bottom structure of large diameter deep hole stage open stope and subsequent filling mining method. *World Nonferrous Met.* **2021**, *22*, 31–32.
33. Zhao, K.; Yan, H.; Feng, X.; Wang, X.; Zhang, J.; Zhao, K. Pillar stability analysis based on energy method. *J. Mech.* **2016**, *48*, 976–983.
34. Song, W.; Cao, S.; Fu, J.; Jiang, G.; Wu, F. Sensitivity analysis of influencing factors of pillar stability and its application. *Geotech. Mech.* **2014**, *35*, 271–277.
35. Guo, J.; Li, M.; Wang, X. Study on the spacing of ore drawing roadway of broken ore body in Xujiagou copper mine, Shaanxi. *Met. Mine* **2018**, *1*, 58–62.

Article

Long- and Short-Term Strategies for Estimation of Hydraulic Fracturing Cost Using Fuzzy Logic

Hyunjun Im [1,*], Hyongdoo Jang [1], Erkan Topal [1] and Micah Nehring [2]

[1] Mining Engineering, Western Australian School of Mines: Minerals, Energy and Chemical Engineering, Curtin University, Perth, WA 6845, Australia; hyongdoo.jang@curtin.edu.au (H.J.); e.topal@curtin.edu.au (E.T.)

[2] School of Mechanical and Mining Engineering, University of Queensland, Brisbane, QLD 4072, Australia; m.nehring@uq.edu.au

* Correspondence: hyunjun.im@postgrad.curtin.edu.au

Abstract: Over two decades, block caving mining has developed the application of hydraulic fracturing as a preconditioning method. This study aims to estimate hydraulic fracturing costs in block caving operations and suggests the base case of specified costs based on the U.S. Energy Information Administration (EIA) report. Furthermore, it applies cavability factors to develop the long- and short-term strategies through the fuzzy inference system. In the long-term strategy, we suggest three possible scenarios for reducing the long-term strategy's uncertainty by considering the association for the advancement of cost engineering (AACE)'s contingency rate. Moreover, each fuzzy membership function of the three possible redeveloped scenarios was analysed through arithmetic operations over independent/dependent fuzzy numbers for comparing each scenario. The outcome of flexible cost estimation suggested deciding on the scale of infrastructure and ore production by facilitating undercut propagation and controlling block height of block caving operation including additional fragmentation processes. The result of this study also illustrated that systematic fuzzy cost engineering could help estimate the initial stage of budgeting. In addition, through solving the uncertainty of fuzzy calculation values, the project schedule identification is presented by recognising the dependence on each scenario's common characteristic of the cavability parameter and cost contingency rate.

Keywords: fuzzy logic; hydraulic fracturing; cost optimisation; fuzzy cost engineering; mining planning

Citation: Im, H.; Jang, H.; Topal, E.; Nehring, M. Long- and Short-Term Strategies for Estimation of Hydraulic Fracturing Cost Using Fuzzy Logic. *Minerals* **2022**, *12*, 715. https://doi.org/10.3390/min12060715

Academic Editors: Longjun Dong, Yanlin Zhao and Wenxue Chen

Received: 17 March 2022
Accepted: 1 June 2022
Published: 4 June 2022

1. Introduction

Block caving is a mass underground mining method that suitably applies for massive cavable orebodies. Due to the capability of large extraction in massive orebodies, the block caving method has the superiority of operating cost by economies of scale over any other underground mining method. Furthermore, including the facilitation of block caving mechanism helps obtain small fragmentation and reduce the entire cost of the comminution process. Even though block caving could induce a low-cost mining method through an automatic flow system, it is limited in application to uncavable hard rock masses that highly rely on the influence of gravitational force. Thus, the application of an effective hard rock-preconditioning method is vitally required to achieve the desired depth, production rate, and caving performance, such as cave initiation, cave propagation, and draw rates.

Preconditioning is a prerequisite and comprehensive process applied to perturb a rock mass property before cave initiation by altering/modifying its in situ geomechanical properties through destressing and fracturing [1]. These processes comprise treating the rock mass either by drilling a single hole and pressurising it with fluid, called Hydraulic Fracturing (HF) or by drilling up holes into the rock mass and packing them with explosives to break the in situ rock masses. The extensive use of preconditioning methods nowadays suggests that it may become a fundamental part of block caving.

Over the past two decades, applying the preconditioning technique used in the caving method has evolved to extract massive, deeper, and low-grade ore bodies. Various engineering and academic studies have been conducted considering the extraction of ore from larger and deeper mines, including Ridgeway Deeps [2,3], Northparkes mine [4–6], Cadia East mine [7], Salvador mine [8], and the El Teniente mine [9] which has a product range from 6 to 35 Mtpa with plans to increase production in future.

HF, before its application in the mining industry, has been widely used in the petroleum industry to recover trapped oil from low-porosity and low-permeability rock. It was used for the first time in mining operations at Northparkes mine (Australia) in 1997 [4]. The potential of its practical use in the full-scale mining industry is to mitigate the risks associated with underground mass mining, such as rockburst, by inducing micro-seismic events by injecting a fluid to cause a fault slippage [10]. For controlling the outburst in the coal mine, the massive HF method is recently used through surface HF in underground coal rock [11]. In addition, the contribution of the natural fracture is identified for flow channels when the HF technique is applied [12].

In the HF application, new fractures are created by inducing tensile stresses along the borehole boundary by applying pressure in an isolated section of the hole. Water or a water-based cross-linked polymer gel is injected in an isolated section between the two straddle packers. The fracture initiates along the borehole boundary, mostly below the borehole collar, after the pressure exceeds the rock mass tensile strength and propagates the fracture length by increasing the injection pressure. This results in a reduction of connection time between the two fractures during the caving process. The HF method could enhance the cavability, draw rate, and fragmentation size in block caving operations by connecting existing natural fractures and new fractures.

Even though there is a growth in demand for applying such preconditioning methods, there are difficulties estimating the economic cost/benefit. Since preconditioning techniques have just been applied to block caving operations, there are uncertainties in analysing costs in its practical application. In addition, a lack of knowledge for estimating the cost of preconditioning techniques causes hardship to the decision-makers. Thus, decision-makers should often rely on experts' knowledge. Since the conventional approach of the mining industry flows with decision trees, it often has caused trouble with the imprecision of prediction because of the deterministic way of decision. To avoid these uncertainties, which in an operating situation always incur major problems in the mining industry, this study applied the fuzzy inference system (FIS) to develop a pragmatic decision-making support tool that enables a more smooth and practical approximation in a probabilistic way.

This study addresses the short-and long-term strategies of the HF cost estimation in block caving, where the proposed approach is not only reducing the uncertainty in the economic assessment of HF by using fuzzy modelling, but also addressing how to identify the dependence on cost measurement due to engineering variables and order the scenario between several cases. For fuzzy modelling in the short-term and long-term based on the characteristics of block caving, the input parameters would be selected from the cavability's natural factors and induced factors, respectively. In a short-term strategy, the cost of HF was measured for one borehole by using the natural factor of rock mass, such as uniaxial compressive stress, in situ stress, water, and discontinuity properties. In a long-term strategy, it was modelled for a year-based project by induced factors that indicate the character of block caving, such as block height, draw rate, fragmentation, and propagation of caving. Through induced factors, we make different scenarios and differentiate both correlated cases and uncorrelated cases by considering the variable's dependency on how to decide the contingency rate.

The rest of the paper is organised as follows. Section 2 describes the related case studies for fuzzy cost estimation. Section 3 provides a brief explanation based on the fuzzy set theory and discusses HF cost and cavability index. In Section 4, the results of fuzzy modelling in short-term strategy and long-term strategy suggest the application of contingency rate on the results of the defuzzified long-term strategy's result. Finally, some conclusions are provided in Section 5.

2. Fuzzy Logic Applications into the Mining and Cost Estimation

The application of fuzzy logic in the mining industry is well explored. For handling the complex mine design and implementing the techniques, the trends of using fuzzy logic suggest a new approach to solving the issues related to traditional economic activity in mineral sectors [13]. Bandopadhyay [14] analysed the roof and floor conditions for mining method selection with fuzzy aggregation procedures. Den Hartog MH [15] developed a performance prediction of a rock-cutting trencher with a knowledge-based fuzzy model. In this paper, the developed model uses six input variables (rock strength, joint spacing of three joint sets, joint orientation, and trench dimensions) for estimating the performance of a trencher (bit consumption and production rates). Jiang, Y.M. [16] applied the fuzzy set theory in the evaluation of roof categories in longwall mining. Cabseoy, T. [17] and Bascetin, A. [18] used fuzzy set theory for the selection of optimum equipment combinations in surface mines. Yao, Y., Li, X., Yuan, Z. [19] developed fuzzy logic and a neural network approach by utilising data obtained from cutting tests. Wei, X., Wang, C.Y., Zhoub, Z.H. [20] suggested the fuzzy ranking system for evaluating the sawability of granite and for selecting a suitable tool and determining the optimum sawing parameters. Li, W. et al. [21] applied the fuzzy probability measures to the analysis on actual cases of excavation, mining, ground surface movement, and subsidence. M. Iphar [22] applied fuzzy sets to the diggability index (DI) rating method for surface mine equipment selection. They selected the inputs as weathering degree, rock strength (UCS), joint spacing, bedding spacing and put them into a fuzzy inference system for DI rating. Edrahim Ghaesei and Mohammad Ataei [23] proposed the roof fall rate, which was applied to fuzzy logic by using the input parameters as CMRR (coal mine roof rating), PRSUP (primary roof support), intersection span, and depth of cover. Dong, L.J. et al. [24] developed the evaluation index of a man–machine–environment system and established it for the clean and safe production grade of a phosphorous mine by using the grid-based fuzzy Borda method (GFB).

The fuzzy logic for the case of cost estimation analysis has had a few applications: Muñoz, M., & Miranda, E. [25] introduced a FIS for estimating the premium cost of an options exchange. It is applied in the derivative market in Mexico, and its results are compared with the Black–Sholes theoretical model. Kasie, F.M. et al. [26] proposed a decision support system that stabilises the flow of fixtures in manufacturing systems. This paper justifies the cost-effectiveness of the decision support system and strengthens the decision proposed using a similarity measure. Fayek, A.R., & Rodriguez Flores, J.R. [27] developed a fuzzy model to assess the quality of infrastructure projects at the conceptual cost estimating stage. In this paper, the fuzzy expert system provides a systematic approach for assessing the quality and cost of components to perform a fitness. Alshibani, A., & Hassanain, M.A. [28] applied the fuzzy set theory for estimating the maintenance cost of constructed facilities. Tony Chen [29] proposed an entropy-consensus agent-based fuzzy collaborative intelligence (FCI) for estimating the minimum unit cost of a semiconductor product, which has critical weakness due to insufficient consensus. Anthony, K. et al. [30] illustrated the construction of cost estimating relationships (CERs) using fuzzy sets and fuzzy inferencing procedures. Petley, G.J., & Edwards, D.W. [31] developed fuzzy matching to use in capital cost estimation about the plant materials of constructions. Zima & Krzysztof [32] suggested fuzzy case-based reasoning (CBR) for estimating costs in the early phase of the construction project. Shaheen, A. A. et al. [33] addressed the fuzzy range estimating model that utilises the major cost packages identified by experts. Through this method, the paper would like to overcome the shortcomings of Monte Carlo Simulations. Plebankiewicz, E. et al. [34] performed the analysis for the validity of the model structure assumptions by comparing the selected initially fuzzy approach and probabilistic approach to calculate life cycle cost (LCC). Cocodia, E. [35] illustrated fuzzy analysis for the development of cost estimation reasoning (CER)s risk for use in conceptual cost estimates for futures Floating Production, Storage and Off (PRSO) developments.

Moreover, fuzzy engineering economic analysis is researched by many authors. Kahraman, C. et al. [36] developed engineering economy techniques under fuzziness which are the new extensions of the fuzzy set such as intuitionistic, hesitant and type-2 fuzzy sets to be applied to environmental problems. Kaino, T et al. [37] described fuzzy sensitivity analysis and its application. Çoban, V., & Onar, S.Ç. [38] applied the Pythagorean fuzzy sets for dealing with uncertainty and imprecision inherent in production levels and energy price for life-cycle cost (LCC) and levelled cost of energy (LCOE) methods. Dimitrovski, A.D. & Matos, M.A. [39] applied the fuzzy sets approach in utility economic analysis. In this paper, they considered the dependence that may exist between the fuzzy variables and the influence of this dependence on the results. They also considered mathematical operations over independent fuzzy numbers.

In this paper, a new systematic strategy with the FIS of HF method is developed with fundamental fuzzy rules based on the engineering index of cavability. Furthermore, we propose a method that considers contingency costs to reduce risk and uncertainty for long term HF cost estimation.

3. Fuzzy Inference System Development

3.1. Fuzzy Set

Fuzzy set theory is designed for representing the formalised tools to deal with, hardly representing the probabilistic model. Fuzzy set theories could provide better rational decision-making tools with specified membership functions than probabilistic methods. However, no answer is always correct when rules or facts are based on experience and past data which are often ambiguous. The fuzzy set theory also can be used in uncertain economic decisions to deal with vagueness. The methodology of the paper, for dealing with the mathematical expression of cavability index (CI), which is a classified rating system, utilises fuzzy set theory and gives the rating system the quantitative number through a membership function of fuzzy numbers. The fuzzy modelling method does not exclude statistical methods but becomes a tool when other approaches for HF cost optimisation are not pertinent.

3.1.1. Fuzzy Arithmetic Operation

In general, a financial calculation is difficult to convert into a probability model due to uncertainty. However, the fuzzy financial calculation can address this uncertainty because it can show the decision-maker what can happen. For this calculation, arithmetic calculations of fuzzy numbers can be considered, synthesised in the equation for generic operation, and described by the extension principles [40]. It is noteworthy that if the computation of fuzzy numbers is the same number of arithmetic calculations, the calculation can be solved when both variables are independent, such as the normal number of arithmetic calculations. Still, when both variables are dependent, the arithmetic calculations cannot be applied, and the extension principle should be applied. In other words, an extension principle is a method of extending an ordinary number of operations to fuzzy concepts.

When applying the extension principle, it is necessary to consider dependency and independency in probabilistic calculations. In most cases, however, these dependent features tend to be neglected without proper consideration. For example, although the tendency of rock characteristics in mines is heterogeneous, it does not mean all variables of cavability are independent of each other characteristic of the rock mass. This could reveal the correlation between the two variables expressed as fuzzy numbers, as it implies that the variables will behave in the same way. In other words, if one of the variables takes a certain value from its support set, the other variables could share precedent values in proportion to the support set. Thus, the more information with common characteristics, the more uncertainty can be remained by considering dependence. In this situation, the dependence between variables of cavability also works in cost estimation because some uncertain values directly or indirectly affect the elements of HF cost estimation through the motivated use of common uncertain parameters. Therefore, this paper excludes the tendency to exacerbate the uncertainty of long-term operation. That can occur from the induced cavability factor, expressed by the fuzzy set in block caving from a long-term perspective. It has modelled three scenarios for consideration of the dependence.

3.1.2. Fuzzy Inferences System

Fuzzy Inferences System (FIS), based on fuzzy propositions and fuzzy sets, is commonly referred to fuzzy reasoning as the computing framework. FIS is constituted of a fuzzifier, rule-based structure, and defuzzifier. For using FIS between cavability index and HF cost estimation, it is fuzzified through the Mamdani fuzzy algorithm in this paper. Through a fuzzified algorithm, a rule-based structure drives reasonable conclusions from soft computing between obscure information and other information when the defuzzification processes ended.

3.2. Cost Optimisation in Mining

Cost optimisation is a continuous process aiming to minimise operating costs whilst improving business values without significantly impacting other associated factors. Cost optimisation has become a vital part of the success of mining projects. However, there are several parameters associated with the uncertainty which can affect the overall cost optimisation process. Therefore, the selection of parameters is key to the success of the cost optimisation process.

Before considering the Hydraulic Fracturing (HF) costs, it is required to review and analyse the newest creative techniques of HF in block cave mining. Firstly, the problem of using either a cased borehole or an open borehole is questionable. Usually, in the oil and gas industry, the cased borehole is applied to the HF operations to retrieve the fluids and initiate the HF process by making perforations link up a single axial HF. After the perforation process, the HF may initiate from its preferred propagation plane outside the cased borehole. However, in cave mining, for the HF initiation process, directional HF could be introduced for inducing lower breakdown pressure and promoting HF transverse initiation [4,7]. Unlike the cased borehole case, the initiation of the HF process in the open borehole is replaced through the cutting machine, which creates an initial notch around an open borehole as an artificial weakness for the hydraulic fractures to initiate [41].

Next, using proppant material for HF in cave mining is questionable. According to recent research, HF for preconditioning in cave mining is suggested for using proppants to enhance the induced stress shadow effect. Usually, in the oil and gas industry, proppant is injected into HF to induce greater length and width of the fracture. The formed fractures are prevented from being closed due to in situ stress and are kept semi-permanent by injecting the proppant. In the geotechnical view, fractures caused by HF can be re-orientation in the preferred direction due to the effects of stress shadow. Because HF causes stress in the rock due to the liquid injected into the rock, the stress shadow effect induces deformation of the rock and forms the stress reversal region. In other words, since the injection rate exceeds the flow capacity of the rock formation, the stress created here is called the stress shadow effect, which deforms the existing state of stress due to the superimposition of in situ stress. The stress shadow effect depends on whether a stress reversal region occurs around an existing fracture or between existing fractures and the extent of this stress reversal region [41]. Because the preconditioning of HF used in block caving is to make the rock mass more blocky, it is preferred that the formed fractures become networks or connect with existing cracks. Therefore, it becomes more important to form a joint network by maintaining the crack and how cracks occur easily. However, HF, usually used in conventional block-caving, does not use proppant (mainly using sand), but it is suggested to induce stress shadow effects to each crack by keeping the gap between the cracks by injecting proppant into the cracks, which are caused by HF.

Lastly, the cost estimation of HF to be applied in block caving mines is relatively unprecedented compared to that of the oil and gas industry. Thus, this study refers to a report by the U.S Energy Information Administration (EIA) that performed an analysis of upstream drilling and production costs conducted by IHS Global Inc. [42]. However, the cost of HF used in the oil industry cannot be directly applied to block cave mining as the above-mentioned paper discussed. This is because of the differences in HF techniques between the oil and gas industry and the block cave mining, which were summarised in table 1 in Q. He et al. [41].

According to Bunger et al. [43] and Adams and Rowe [44], the most significant difference between the gas & oil industry and the mining industry is the fracture size. Table 1 shows that the amount of injection rates used in the oil industry differs by a factor of 25 from those used in the block cave mining industry and, in terms of injection volumes, they differ by a factor 12.5 to 50 from those used in the block cave mining industry. Above all, the reason that caused the quantitative difference between the two industries in HF is the size of the fracture, and similarly, distances between fractures create the difference in the quantitative factors for HF.

Table 1. Differences between hydraulic fracturing in the cave mining industry and hydraulic fracturing in the shale gas industry, revised by Bunger et al. [43] and Adams and Rowe [44].

Application	HF Size	Injection Volume (m^3)	Injection Rate (L/s)	Addictive	Proppant	Distances between Fractures (m)	σ_3 Orientation
Cave mining industry	About 30 m in radius	8–20	5–10	No	Some	1.25	Mostly vertical in Australia
Shale gas industry	Hundreds of meters in half-length	135–1000	75–250	Yes	Yes	About 100	Mostly horizontal

3.3. Cost Variables of Hydraulic Fracturing Costs

Many factors may also contribute to the fluctuation of HF costs. For example, HF cost is influenced by drilling cost (rig rates and drilling), fluid cost, proppant cost, pump cost, and equipment cost heavily impacted by more significant market conditions and the time required for drilling the total well depth. In addition, HF costs depend on rock mass conditions and the cost of minerals. Furthermore, HF costs are influenced by geological properties for geo-steering, which is required to optimise the preconditioning efficiency by orienting the wellbore perpendicular to the minimum in situ stress orientation and inducing the interaction between hydraulic fractures and natural fractures for the growth of hydraulic fractures. Therefore, HF costs might need a comprehensive function of budgeting estimation.

Based on the listed reasons above, the elements of the HF cost estimation used for block caving should include the following: the HF uses an open borehole rather than a cased borehole, and by using proppant to form stress reversal regions and to utilise the effect of stress shadow effect, the use of proppant to expand the width of the fracture and to proceed in the preferred direction. In addition, the study rectifies the EIA's HF cost estimation for the difference in HF size and HF distances between fractures. Furthermore, HF cost factors are discussed that may contribute to formulating an estimation. As a result, the values are rectified and supplemented for all the illustrated cost aspects described in Table 2. In Table 2, the P10, P50, and P90 stand for percentiles such as P denotes the percentiles in the distribution, P10 denotes the lower 10% of the observations and P50 denotes a value close to the mean value.

Table 2. P10, P50, and P90 of one borehole cost estimation with 500 m length for HF in block cave mining.

HF Cost Component	Items	P10	P50	P90
Frack cost	Fluid cost	$31,250	$68,750	$125,000
Proppant cost	Water and sand cost	$31,250	$68,750	$131,250
Drilling cost		$97,500	$125,000	$200,000
Frack pumping cost	Stage cost	$15,000	$16,250	$16,875
	Break pressures cost	$137,500	$187,500	$225,000
	Injection rate cost	$4750	$7500	$11,750
Others		$118,750	$137,500	$175,000

3.3.1. Drilling Related Costs

Drilling related costs make up roughly 20% to 25% of the total cost of HF in block caving. This cost is associated with borehole depth, drilling rates, drilling penetration, and drilling efficiency. Drilling costs had many parameters which are likely linked to quantifiable costs, such as drilling crews' wages and drilling equipment. It also includes intangible costs such as the drill bits consumption, equipment rental fees, drilling costs, and other service costs.

3.3.2. Frack Pumping Costs

The frack pumping cost is a highly volatile variable. It consists of 20% to 25% of total HF costs. Between the frack pumping variables, only the injection rate is selected for considering the difference of application between block caving and the oil-gas industry. Furthermore, the break pressure, which must be higher than the intact rock strength, would also consist in fracking pumping costs. Moreover, in a block caving HF operation, there is no need to take an option for several stages. Thus, this study just assumed the operation would be set as a single stage.

3.3.3. Fluid and Proppant Costs

The fluid and proppant costs make up 15% to 30% of the total HF borehole cost. As discussed above, the purpose of the fluid and proppant in HF is to initiate and propagate fractures. Several substances could make up the proppant, such as artificial proppant and coated proppant, etc. However, this study only considers the fluid and proppant as water and sand, respectively. Even though water has a low viscosity and can be injected at a high or low rate, it could initiate micro-fractures and propagate in the target area. It is also available in abundance and cheaper than any other fluid used for HF.

3.4. *Cavability*

Cavability is the ability to unravel the in situ rock mass when developing the undercut process and consider all three stages of caving: initiation, propagation, and continuous caving, which is the vital feature of the caving process [45]. In addition, the cavability of rock mass will have an essential role in controlling the mine design at the initial stage of economic issues in a given geological environment. On the other hand, the preconditioning task has the ultimate goal of weakening the rock mass or developing discontinuities. Preconditioning has been used for enhancing the cave initiation and propagation to achieve the desired draw rates. Because of the effects that can be achieved by preconditioning are the same as the purpose of recognising the rock mass's cavability in cost-effective utility, the study assumed preconditioning and cavability share direct intermediary characteristics. In other words, if cavability ultimately represents the degree of capability for caving in situ rock mass in the orebody, then preconditioning is intended to proceed with this process for improving the ability into the formed rock to be caved. Since the natural factors of cavability can directly or indirectly affect boreholes required for preconditioning, this value is presented through fuzzy logic at a short-term cost estimation. The induced factors of cavability propose a strategy to estimate costs for long-term planning of preconditioning. For instance, if obtained from the rock with weak geology of the orebody, preconditioning can maximise its effectiveness by reducing the strength of the pressure that can be seen as an effect, and later transferred to a part made of strong rock that would be estimated through cost estimation fuzzy logic in the long-term plan.

3.4.1. Cavability Index

The Cavability index (CI) is referred to the degree of cavability in the caving mining method. CI is an essential geomechanics factor for conducting caving by forming an ellipsoid shape in the draw column using the gravitational separation process. In addition to this importance, CI helps address uncertainty in the design and planning process phases of the caving operation because the engineer's decision is incomplete with information and data on the rock mass. Therefore, CI is often obtained through structural analysis, numerical modelling, and an empirical chart estimated using rock mass rating (RMR) and IRMR [46,47]. In addition, CI obtained from these empirical formulas ultimately helps infer the dimension of the undercut which is the most important component of block caving design, and to set up the hydraulic radius (HR). HR is necessary to ensure the direction of caving propagation for the unsupported area of the back of the cave. That is, the determination of HR is used by CI to decide the required dimension for caving and sustain mining operations.

CI of the orebody is influenced by the rock mass of its natural properties. It is also influenced by induced factors resulting from the mining process [48,49]. These properties have been demonstrated by a review of RMR classification [45].

Natural Factor

The natural factors in cavability generally use the physical properties of rock mass classification. Such a natural factor includes uniaxial compressive strength (UCS), in situ stress, water, and rock mass discontinuity properties. UCS shows the characteristics of rock material strength. In addition, in situ stress serves as an important factor in the natural factor of cavability, as it informs the magnitude and orientation of regional stress in the caving mine [48]. Water reduces the effective stress of discontinuous joints in the rock mass, weakening their shear strength. Water can increase the capability of cavability by reducing the friction between the joints. Lastly, the natural factors also include discontinuity properties. This factor also distinguishes six ways of spacing, orientation, aperture, persistence, roughness, and filling of joints as used in representative rock mass classifications (RMR, IRMR, Q systems).

Induced Factor

The induced factor represents the structural characteristics required by the caving process. These induced factors are made up by engineering decision-makers and are regarded as a way for mining the orebody. Induced factors are included in an undercut, HR, block height, and fragmentation. For identifying the uncertainty of mine, cavability is specified by each of the criteria.

First, undercut is favoured to choose the advanced undercut method, which is the development of undercut ahead of the partially developing extraction level. This method helps ramp up the production material and destress the extraction level for reducing the abutment stress, which could damage the extraction level. HR is the area divided by the perimeter. The effects of HR are included as the induced factor of cavability, as they play a significant role in the design of caved material channels and in determining the overall development. Block height is one of the critical parts of induced factors in cavability. After cave initiation, cave material falls off from the caprock as an ellipsoid shape. The vertical heights of the ellipsoid as the height of draw (HOD) influence secondary fragmentation. Since as long as HOD applies, secondary fragmentation is a much finer fragmentation during the separate gravitational work and facilitates the transition of primary fragmentation to secondary fragmentation. In this sense, block height is also included in induced factors of cavability. Lastly, fragmentation of ore in block caving influences the overall operation process due to its impact reaching the comminution process, equipment selection, production rates, determination of drawbell plan and size, and continuous production schedule. Since the fragmentation of a deposit could change the entire mine life, diagnosing the result of the preconditioning method makes the right decision more important.

4. Cost Estimation with a Strategic View

4.1. Long- and Short-Term Strategy for Estimation of HF Cost Using Fuzzy Logic

The study not only plans to model the cost estimation of HF using fuzzy functions but also proposes correlated and uncorrelated cost estimation to avoid ignoring the effect of the interdependence of certain variables on the cost estimation. To consider these features, this paper referred to the work of Dimitrovski, A.D., & Matos, M.A [50]. Even though that study had an analysis of recognising and designing dependent differences as cash flows parameter which usually consists of interest rates and time values, this study turns our attention to the engineering parameters' degree of the dependent and independent relationship since inadequate values of engineering variables can also lead to additional uncertainties in cost estimation. It can occur mainly in the industries on which upstream operations (oil and mining sector) are based because the physical properties of the underground are invisible and the exact estimate of the material being produced is almost impossible. To overcome those uncertainties, the study proposes a model that considers the extent to which variables are caused by their dependence, as well as probabilistic estimation methods.

Before investigating in the light of the strategic approach using fuzzy logic, the study we discuss the basic definition of the strategy. A term of periodic strategy in mining operations ultimately exists for the optimisation of the cost. In other words, this is to make the utilisation of the budget to induce the best efficiency through maximised cash flow with low capital cost and low operating costs. Thus, the project's duration can be divided into short-term, medium-term, and long-term. The short-term can be defined as day-to-day, but in this paper, the concept of the shift-to-shift is addressed, and the long-term defines each other period within a year's range if the operation of HF lasts just about three years. However, the medium-term was not discussed because the HF fracturing project had a relatively smaller magnitude of size than other projects.

Furthermore, the short-term strategy is modelled to estimate the HF cost with the natural factors of cavability as input parameters. The output of this model is modelled based on "the fracture cost", "proppant cost", "drilling cost", "fracking pumping cost", and "others" (such as labour costs and other related operating costs). The short-term strategy will propose that those assigned orientations determine the timing of expansion (adjusting the rate of undercut process and initial production), information on the scale of infrastructure, and direction of preconditioning operation. Although the short-term strategy proposed in this paper may not address the complexity of all phenomena occurring in mining operations, it is assumed that the criterion of the cavability parameter would provide an optimal approach with sufficient logical information due to the cost per borehole of HF preconditioning. In addition, it will induce not only to be practical but also provide the opportunity to create periodic updates and action plans for real mine sites through achievable scenario analysis. It will also help decide the equipment selection and borehole spacing of the HF and drilling diameter of the borehole.

The long-term strategy of mining planning would generally propose to solve the problem of large-scale optimisation, which is aimed at finding block extraction sequences. Since the sequence problem is directly the order to derive the maximum NPV, it performs the optimal strategy, incorporating physical and economic constraints and future changes in mineral prices. However, this paper will set the induced factors of cavability in optimising HF cost estimation as constraints of the long-term strategy. In this case, it suggests three possible scenarios for HF cost. Thus, according to those assumptions, the analysis was made to changes in three scenarios, as the measurement of HF costs from a long-term perspective can vary. Those modelled scenarios are integral to the optimisation of HF operation; these inform the development of long-term planning parameters and contain the consideration of contingency planning for practical operations. To sum up, all these processes through HF preconditioning aim to maximise rock mass's cavability assessment, the efficiency of capital costs, and the scheduling problem. The outcome of maximised cost estimation would help decide on the scale of infrastructure and ore production by facilitating undercut propagation and controlling block heights of block caving operation with inducing additional fragmentation.

4.2. Short-Term Strategy for HF Operation

In this study, the Mamdani FIS was applied to construct short-term HF costs. The essential component of this model is made up of the input-output sets and if-then rules. The fuzzy set of input-output sets is developed by triangular and trapezoidal membership functions, which can minimise their uncertainty. After setting the input-output sets, all the consequent fuzzy sets will be integrated into a new fuzzy set by using the maximum operator as the if-then rules construction process. Once the construction of if-then rules is finalised, it requires a procedure of defuzzification that converts the form of output-fuzzy sets into a crisp numerical value. The defuzzification method is processed by the method of COA defuzzification, and it would perform to get the crispy value of each HF cost variable.

For the short-term strategy, it is necessary to restrict the range of the scale to one borehole operation. This model is assigned to the 9 input variables (UCS, in situ stress, joint spacing, joint orientation, water, joint aperture, joint persistence, joint roughness, joint filling), and five output variables (stage and others costs, fluid and proppant costs, Drilling costs, break pressures, injection rates). Those fuzzified sets are illustrated in Figure 1. Before specifically describing this model, there is a questionable relationship between input variables (cavability) and "stage and other costs", which is the subset of HF cost estimation. Due to the property of cavability, it seems not to relate to the stage cost and other costs associated to the operation's maintenance and labour cost. However, the lower the CI, the longer the time to shift the operation of HF, so the other and stage costs were set to output variables together because the time, cost, and labour cost would also increase due to the chain of actions.

Specifically, input and output variables are illustrated with trapezoidal and triangular membership functions in Figures 1 and 2. The membership function of input variables is classified into five classes. The fuzzy input sets were designed to intersect each class based on the membership degree of 0.5, respectively. The output variable is also classified into five classes. However, it was determined that the three classes (P10, P50, P90) that were originally calculated alone would not be more detailed to calculate the value but would bring up the result of the broad and wide value. Thus, this research goes a little further and classifies it into five classes by adding more P32 and P68 cases. This helps to flexibly cope with the estimation of output variables by applying one standard deviation. The fuzzy output sets, categorised into five classes, are designed to intersect each class based on the membership degree of 0.4 (Figure 2).

The next stage of fuzzy modelling is the construction of the if-then rules. As discussed before, the CI is classified into the five parameters. Thus, the number of if-then rules will have the 5 powers of 9 rules. However, among them, the if-then rule is applied as itself, the number of rules would be too large. Thus, the generality of the rule is given in consideration of the weighted value of the CI of Rafiee [44]. Moreover, the natural characteristic of rock will ignore some impossible rules. When constructing the if-then rules, the logical rules set that the higher the CI, the lower the HF cost, contrary to the lower the CI, the higher the HF cost. As the last process of this methodology, the COA method is employed for the defuzzification process due to the calculation simplicity. The fuzzy set output was translated into a crisp numerical value through the COA defuzzification method, which would lead to the final HF cost estimation. Following the determination of the cost estimation, its membership degree is obtained by using the fuzzy sets, which represent the output variables.

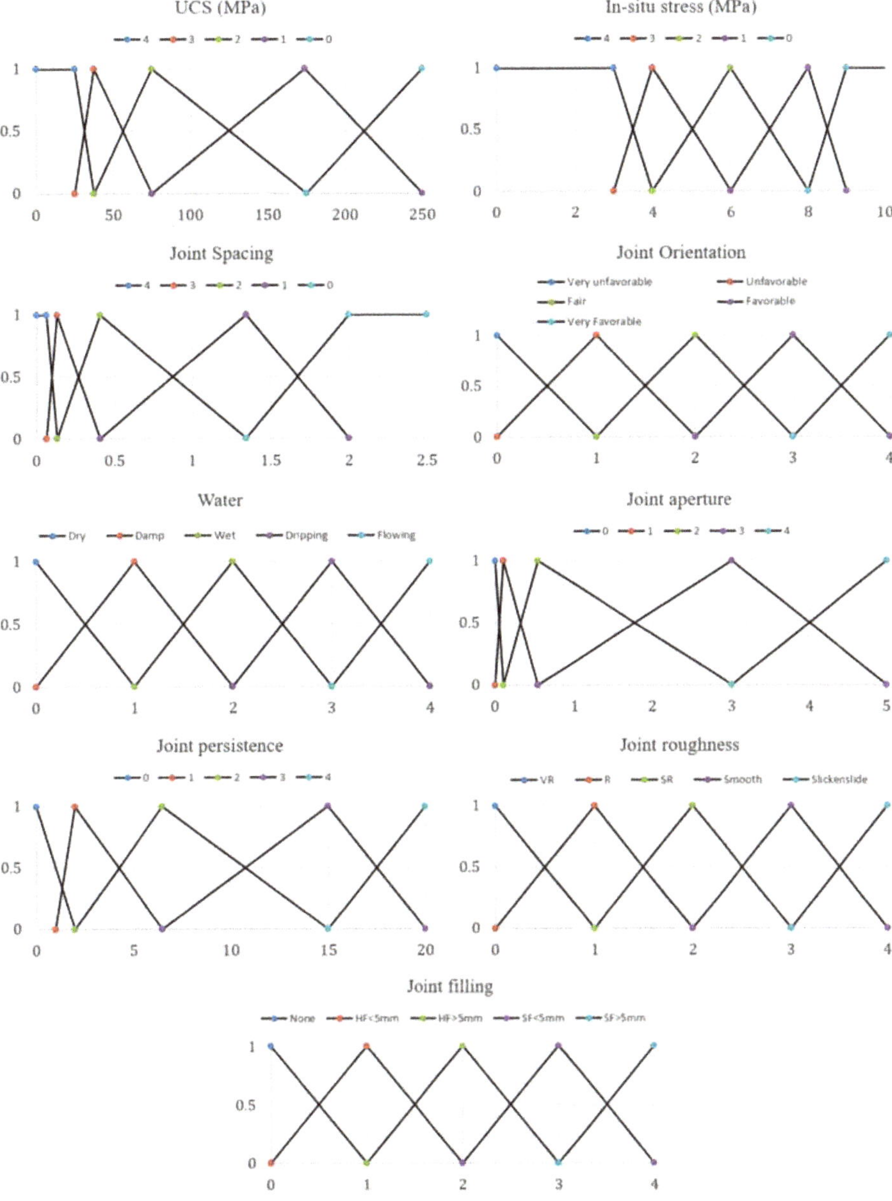

Figure 1. Illustrations of cavability natural factors as input variables (UCS, in situ stress, joint spacing, joint orientation, water, joint aperture, joint persistence, joint roughness, joint filling) membership function for each variable (0: Very Low, 1: Low, 2: Medium, 3: High, 4: Very High).

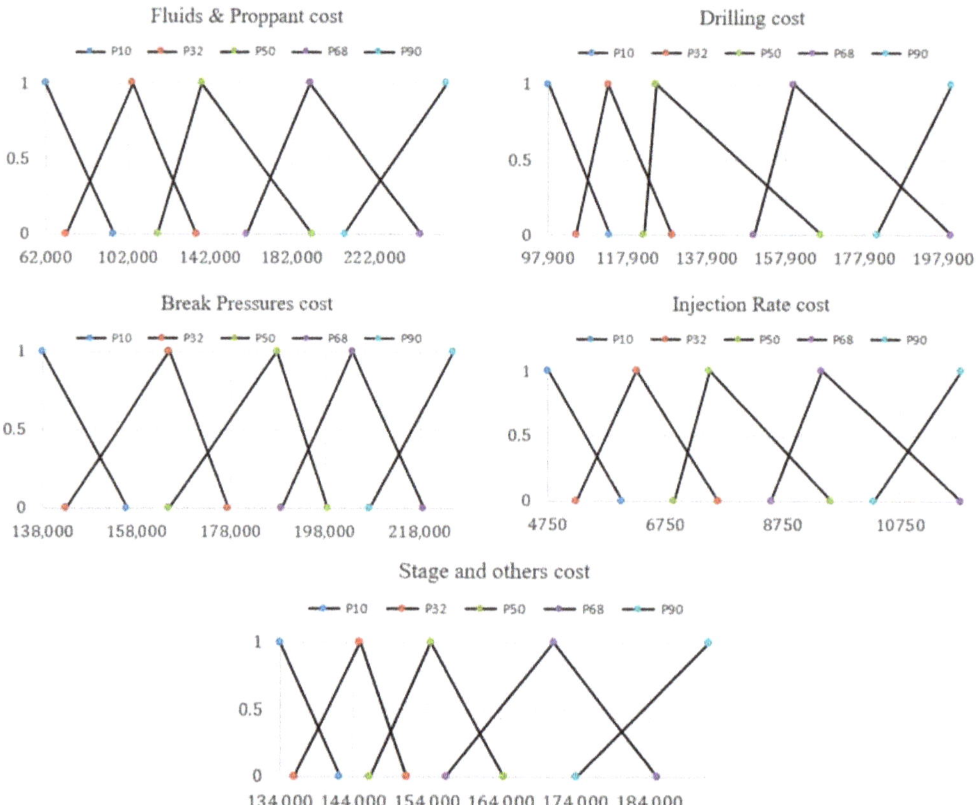

Figure 2. Illustrations of HF cost estimation as output variables (fluid and proppant cost, drilling cost, break pressures cost, injection rate cost, stage and others cost), membership function for each variable (P10: Very Low, P32: Low, P50: Medium, P68: High, P90: Very High).

4.3. Long-Term Strategy for HF Estimation

In this section, the study tries to prioritise the uncertainties encountered when executing the HF long-term planning in the block caving mine. First of all, mines in block caving are exposed to environments where the characteristics of rocks can be distributed variously. Therefore, long-term strategies for HF should be considered heterogeneous characteristics of rock for mining planning and constructing essential infrastructure rather than a homogenous characteristic of rock. Thus, the study makes a choice to measure the cavability of the induced factors for dealing with the heterogeneous features of the block caving sector as constraints of the long-term strategy. Then, according to the nature of rock characteristics, three scenarios in the block caving sector have been suggested. Those cases are illustrated in Table 3.

Secondly, those scenarios are modelled with FIS in the same way as the short-term strategy. Fuzzy modelling for long-term strategy is constructed based on four induced factors of cavability to extract defuzified long-term HF cost estimation with the range of 1 year. For further explanation, induced factors are made of HR, fragmentation size, block height, and undercut direction as the input variables in fuzzy sets. Finally, those defuzzified results are compared and analysed by presenting the correlated and uncorrelated cases of the differentiated long-term budget, according to the composition of scenarios A, B, and C. For comparison of analysis, the rectified extension principle has been used rather than the applied arithmetical operation mentioned above.

Table 3. Three scenarios are illustrated for long-term strategies for hydraulic fracturing costs.

Rating	0	1	2	3	4
		Scenario A			
Hydraulic radius		18.5 m			
Fragmentation		512			
Block height		125 m			
Undercut direction			Fair		
		Scenario B			
Hydraulic radius				50 m	
Fragmentation					0.2
Block height			180 m		
Undercut direction			Fair		
		Scenario C			
Hydraulic radius		25			
Fragmentation			10		
Block height				225 m	
Undercut direction			Fair		

Those modelled scenarios are integral to the optimisation of HF operation, and these are not only informing the development of long-term planning parameters but also containing the consideration of contingency planning for practical operations. Measuring the amount of uncertainty would help make a decision maker construct the scenario's consecutive sequence in the mining operation. To sum up, all these processes through HF preconditioning aim to maximise rock mass's cavability assessment, the efficiency of capital costs, and the scheduling problem. The outcome of flexible cost estimation would help decide on the scale of infrastructure, ore production by facilitating undercut propagation, and controlling block heights of block caving operation by inducing an additional fragmentation process.

After assuming that the induced cavability factor is an input variable, it is compounded to each output result for recalculation of fuzzy correlation. Furthermore, this strategy utilises the value of contingency cost, which is referred to as association for the advancement of cost engineering (AACE)'s contingency table. Through this way, the long-term strategy for HF operations could conclude to validate "cost estimation" through the fuzzy economic analysis.

The scenarios assumed in Table 3 imply that each of them has a different rock property in one massive ore body. However, this scenario also implies uncertainty in scheduling decisions or budgeting about which part of the rock mass to approach first. Thus, to design the long-term strategy, we followed the way we used in the short-term strategy method with fuzzy input and output variables. After the following results are obtained from the FIS, each of which is defuzzified by the elements of the COA shown in Table 4. The basic logical configuration is the same the lower the score of cavability, the higher the cost of HF.

The defuzified values were available according to the assigned fuzzy rules shown in Table 4. However, although the values have tentatively been obtained through long-term cost estimation for each scenario through an intuitive and uncertainty-tackling FIS by considering engineering parameters, it is still too vague to apply the estimated cost. Here, the study is going to focus a little more on the nature of estimation. Universally, cost estimation should not represent individual numbers but the range of risks and possible costs. Although the study used the data based on the cost estimation for HF in the oil industry through the EIA HF report, there are limitations to applying the environmental constraints of the underground mining industry and unstable operation, so cost estimation must be extended into those uncertainties.

Table 4. After defuzzified results of long-term cost estimation of hydraulic fracturing.

Scenarios/Cost	Others and Stage Cost	Fluid and Proppants Costs	Drilling Costs Costs	Break Pressures Costs	Injection Rate Costs
scenario A	$4,950,000	$5,780,000	$4,770,000	$5,800,000	$280,000
scenario B	$4,380,000	$3,900,000	$3,810,000	$4,840,000	$208,000
scenario C	$4,460,000	$4,290,000	$3,870,000	$5,160,000	$225,000

If the values are concluded only with the individual values shown above in Table 4, one will not be able to respond appropriately within the range sanctioned in the project when risk events occur because it is easy to become a time range of cost estimation. Therefore, if possible, it appropriate cost estimation is considered to derive the optimal condition cost through risk analysis before the funding decision.

Before taking the concept of contingency cost, too much or too little range of cost should be considered as it may cause problems. Since the initial starting point of a project may tend to schedule HF projects quickly and safely, it derives from overestimating the budget calculated for the project. In this case, the overrun cost estimation will lock up the capital, which may adversely affect the budget for the next schedule. On the other hand, the more detailed and strategic the project is, the more pressure it may create for lower cost estimation, which increases the likelihood that the outcome of the project will be disruptive. In general, to quantify the risk, detailed distribution of the contingency could be decisive, but it is beyond the scope of this paper to discuss how to quantify the risk.

There are two representative methods for quantifying the risk, the parametric estimating method and the expected value, which are introduced in RP 42R-08 and RP 44R-08 on AACE's website, respectively. However, parametric estimating is chosen for this study because a method is used to quantify systemic risk because it helps analyse the impact of risk through empirical data. Before using this quantification method, it is necessary to classify the project definition of HF application in block cave operation by referring to AACE's class estimate. For this reason, it is summarised with estimate criteria in Table 5 for deciding the scope of the project range, which describes the AACE's estimate class of deliverable status and target status by referring to RP 18R-97 [51].

Through Table 5, the application of HF in the block cave mine was assumed to be the initial control estimate against which all costs were assigned. Therefore, as this criterion, the study considers that the HF operation could belong to Class 3.

However, before taking the parametric estimation in this paper, the systemic contingency allowance table, proposed by Hollmann, J.K. [52], is applied to the fuzzy model for CPI research (which is publicly available on AACE's website) since the in-depth analysis of parametric estimating in this study is beyond the scope of this paper. This table is introduced in Table 6. As followed by research, the contingency percentage was determined by the technology rating with the complexity of the project for each class in the systemic contingency allowance table.

Accordingly, the study considered the cavability to represent the complex operation and assumed the distribution of contingency cost. For example, the higher the cavability, the more likely it is that accidental risk events in the underground mine could be created, so a high percentage of contingency cost was allocated. Conversely, if the cavability is lowered, the HF cost will be higher, but the accidental events in the underground mine will be less. As those assumptions followed, it is assumed that the complexity of the contingency rate is proportional to the degree of cavability, and the HF technology criterion is tentatively set at the medium level.

Table 5. AACE estimate Classes, the Maturity level of project definition deliverables and the Expected accuracy range [51].

Estimate Class	Primary Characteristic	Secondary Characteristic
	Maturity level of project definition deliverables	Expected Accuracy range
Class5	Key deliverables and target status: block flow diagram by key stakeholders	P10: −20% to −50% P90: +30% to +100%
Class 4	Key deliverables and target status: process flow diagrams (PFDS) issued for design.	P10: −15% to −30% P90: +20% to +50%
Class 3	Key deliverables and target status: piping and instrumentation diagrams (P& IDs) issued for design.	P10: −10% to −20% P90: 10% to +30%
Class 2	Key deliverables and target status: All specifications and datasheet complete including for instrumentation.	P10: −5% to −15% P90: +5% to +20%
Class 1	Key deliverables and target status: All deliverables in the maturity matrix complete.	P10: −3% to −10% P90: +3% to +15%

Table 6. Typical systemic contingency allowances based on project attributes by Hollmann, J.K. [52].

Systemic Contingency as a Percentage of the Unexpended Base Estimate										
Scope Definition		Class 3			Class 4			Class 5		
Complexity		Low	Medium	High	Low	Medium	High	Low	Medium	High
Tech.	Low	3%	8%	12%	10%	15%	20%	19%	24%	29%
	Medium	6%	11%	15%	13%	18%	23%	22%	27%	32%
	High	15%	20%	25%	22%	27%	32%	32%	47%	42%

Systemic contingency cost would be set up with the crispy costs, which would take a task of fuzzified parameters as HF cost contingency. According to those criteria, the contingency cost of scenarios a, b and c will be assigned to the systemic contingency percentage of 6%, 12.5%, and 11.5%, respectively. After this contingency cost is allocated to each scenario, the crispy values are assumed as the median values of the corresponding fuzzy number for the fuzzification modelling. Then, it is redistributed to the trapezoidal fuzzy model through each value setting with 0-cut intervals and 1-cut intervals of the fuzzy numbers.

As follows from Table 7, the contingency costs are associated with each scenario, with the contingency cost rate, which contains the uncertainty in this model. Finally, the corresponding trapezoidal fuzzy intervals are multiplied with Table 4 (defuzzified model) and the result of this calculation is shown in Table 8. Thus, Figure 3 could illustrate the considered form of both trapezoidal shapes of graphs. It can be regarded as a final representation of the long-term cost estimation of HF for each scenario, including the contingency rate.

Table 7. Contingency rate considered with fuzzy sets.

Contingency Rate/α Cut Intervals	α = 0	α = 1	α = 1	α = 0
Scenario A	3.5%	5%	7%	8.5%
Scenario B	9.5%	11%	14%	15.5%
Scenario C	8.5%	10%	13%	14.5%

Table 8. Fuzzified cost estimation included in the range of contingency costs.

Scenarios/Cost Estimation as per Contingency Rate	α = 0	α = 1	α = 1	α = 0
A	3.5%	5%	7%	8.5%
	$22,335,300	$22,659,000	$23,090,600	$23,414,300
B	9.5%	11%	14%	15.5%
	$18,766,110	$19,023,180	$19,537,320	$19,794,390
C	8.5%	10%	13%	14.5%
	$19,535,425	$19,805,500	$20,345,650	$20,615,725

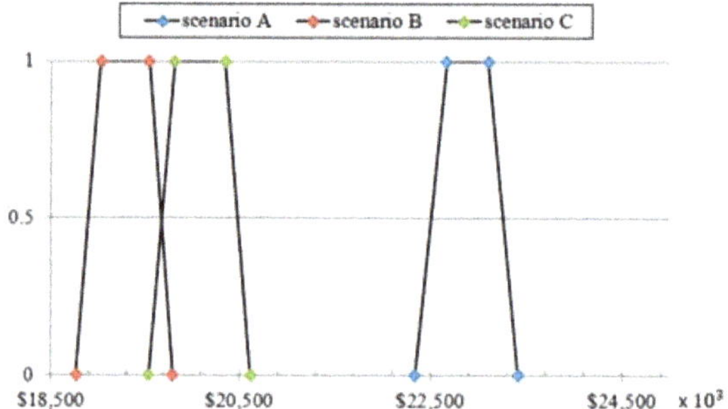

Figure 3. Fuzzy model of each scenario for hydraulic fracturing cost from base estimate.

For comparing the cost estimation of HF from the strategic perspective, however, the uncertainty of the fuzzy equation should consider the relationship between the engineering parameter and the cost contingency rate. Firstly, as illustrated in Figure 3, it is evident that scenario A is the last option to consider if ordered in terms of cost. However, scenarios B and C cannot be confident about which one is clearly the better option. Because scenarios B and c overlap each other in the fuzzy trapezoidal function, the right extreme of scenario B and the left extreme of scenario C are overlapping. Thus, it is impossible to find a definite answer as to which scenario should be worked on first. To address this, the calculation is done through the arithmetic operation of the fuzzy number introduced earlier in Section 4. The difference between scenarios B and C in values through the calculation method using Zadeh's extension principle is presented in Figure 4. However, in the case of calculation, the value of the assumption that the variables of cost estimation included in scenarios b and c are independent of each other is implied.

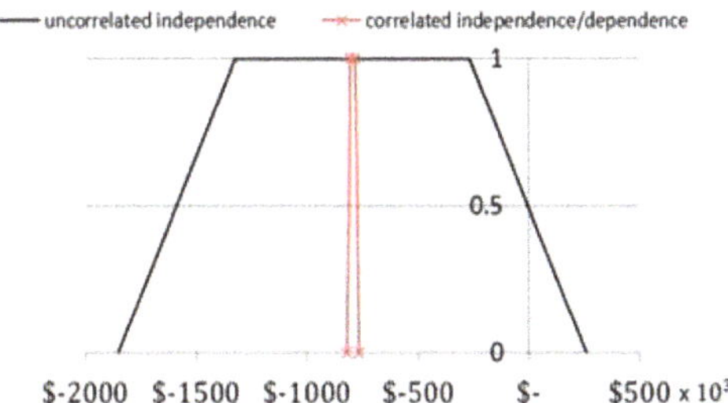

Figure 4. Independence case of difference of scenarios B and C.

More specifically, the subtraction between independent fuzzy numbers located at both extremes naturally implicates the best estimate, which tends to simplify the various variables used to derive the calculation. However, the simplification of the calculation may induce results that ignore both recognition of cavability variables commonly used for cost estimation and contingency rate newly introduced for long-term strategy. In other words, if the dependence between variables is not considered, the result of fuzzy arithmetic operation causes a result of exaggerated uncertainty. As a result, the fuzzy arithmetic operation by the extension principle is applied. In the light of the characteristic of fuzzy arithmetic operation, the use of fuzzy logic models allows the computation of variable dependencies. For optimisation of decision making through these considerations, calculations that consider dependencies for each scenario are presented in Figures 5 and 6.

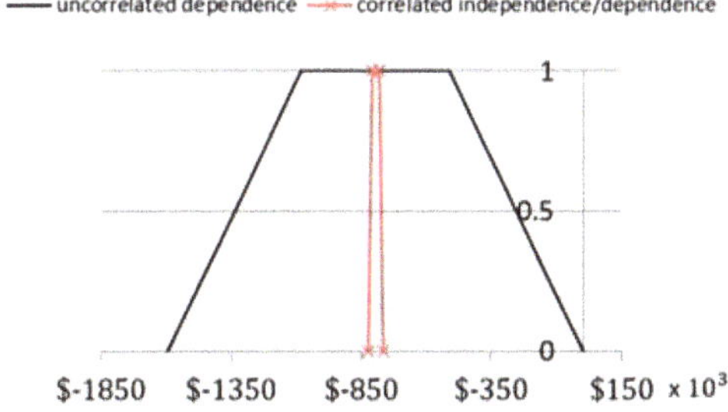

Figure 5. Uncorrelated dependence costs recognised with the difference of B and C.

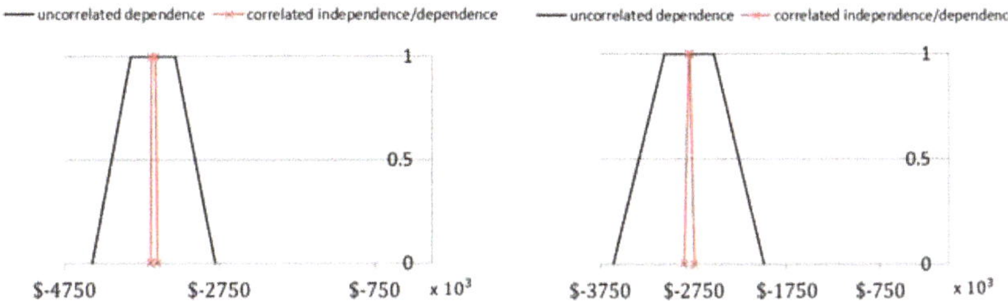

Figure 6. Dependence recognised with the difference of B and A (**Left**) and C and A (**Right**).

As presented in Figure 4, according to this perspective, it noted that the difference between scenarios b and c has a positive value at the right extreme when independent of each other. In this case, it implies that there is a possibility that scenario C will be better. However, if the two variables are dependent on each other, the right extreme value of the difference between the two scenarios appears as approximate 0. It implies that the value of scenario B has a lower cost than choosing scenario C in any case.

25. Muñoz, M.; Miranda, E. A fuzzy system for estimating premium cost of option exchange using mamdani inference: Derivates market of Mexico. In Proceedings of the 2016 IEEE International Conference on Fuzzy Systems (FUZZ-IEEE), Vancouver, BC, Canada, 24–29 July 2016. [CrossRef]
26. Kasie, F.M.; Bright, G.; Walker, A. An intelligent decision support system for on-demand fixture retrieval, adaptation and manufacture. *J. Manuf. Technol. Manag.* **2017**, *28*, 189–211. [CrossRef]
27. Fayek, A.R.; Flores, J.R.R. Application of fuzzy logic to quality assessment of infrastructure projects at conceptual cost estimating stage. *Can. J. Civ. Eng.* **2010**, *37*, 1137–1147. [CrossRef]
28. Alshibani, A.; Hassanain, M.A. Estimating facilities maintenance cost using post-occupancy evaluation and fuzzy set theory. *J. Qual. Maint. Eng.* **2018**, *24*, 449–467. [CrossRef]
29. Chen, T. Estimating unit cost using agent-based fuzzy collaborative intelligence approach with entropy-consensus. *Appl. Soft Comput.* **2018**, *73*, 884–897. [CrossRef]
30. Mason, A.K.; Kahn, D.J., Sr. Estimating costs with fuzzy logic. *AACE Int. Trans.* 1997, p. 122. Available online: https://www.proquest.com/scholarly-journals/estimating-costs-with-fuzzy-logic/docview/208181169/se-2?accountid=10382 (accessed on 16 March 2022).
31. Petley, G.; Edwards, D. Further developments in chemical plant cost estimating using fuzzy matching. *Comput. Chem. Eng.* **1995**, *19*, 675–680. [CrossRef]
32. Zima, K. The use of fuzzy case-based reasoning in estimating costs in the early phase of the construction project. *AIP Conf. Proc.* **2015**, *1648*, 600010. [CrossRef]
33. Shaheen, A.; Fayek, A.R.; Abourizk, S.M. Fuzzy Numbers in Cost Range Estimating. *J. Constr. Eng. Manag.* **2007**, *133*, 325–334. [CrossRef]
34. Plebankiewicz, E.; Meszek, W.; Zima, K.; Wieczorek, D. Probabilistic and Fuzzy Approaches for Estimating the Life Cycle Costs of Buildings under Conditions of Exposure to Risk. *Sustainability* **2020**, *12*, 226. [CrossRef]
35. Cocodia, E. Risk Based Fuzzy Modeling of Cost Estimating Relationships for Floating Structures. In Proceedings of the International Conference on Offshore Mechanics and Arctic Engineering, Estoril, Portugal, 15–20 June 2008; Volume 48180, pp. 129–142. [CrossRef]
36. Kahraman, C.; Sarı, İ.U.; Onar, S.C.; Oztaysi, B. Fuzzy economic analysis methods for environmental economics. In *Intelligence Systems in Environmental Management: Theory and Applications*; Springer: Cham, Switzerland, 2017; pp. 315–346. [CrossRef]
37. Kaino, T.; Hirota, K.; Pedrycz, W. Fuzzy Sensitivity Analysis and Its Application. In *Fuzzy Engineering Economics with Applications*; Springer: Berlin, Germany, 2008; pp. 183–216. [CrossRef]
38. Çoban, V.; Onar, S.Ç. Pythagorean fuzzy engineering economic analysis of solar power plants. *Soft Comput.* **2018**, *22*, 5007–5020. [CrossRef]
39. Dimitrovski, A.; Matos, M. Fuzzy engineering economic analysis [of electric utilities]. *IEEE Trans. Power Syst.* **2000**, *15*, 283–289. [CrossRef]
40. Zadeh, L.A. Fuzzy sets. *Inf. Control.* **1965**, *8*, 338–353. [CrossRef]
41. He, Q.; Suorineni, F.T.; Oh, J. Strategies for Creating Prescribed Hydraulic Fractures in Cave Mining. *Rock Mech. Rock Eng.* **2016**, *50*, 967–993. [CrossRef]
42. Trends in U.S. Oil and Natural Gas Upstream Costs. 2016. Available online: https://www.eia.gov/analysis/studies/drilling/pdf/upstream.pdf (accessed on 16 March 2022).
43. Bunger, A.P.; Jeffrey, R.G.; Kear, J.; Zhang, X.; Morgan, M. Experimental investigation of the interaction among closely spaced hydraulic fractures. In Proceedings of the 45th US Rock Mechanics/Geomechanics Symposium, San Francisco, CA, USA, 26–29 June 2011.
44. Adams, J.; Rowe, C. Differentiating applications of hydraulic fracturing. In Proceedings of the ISRM International Conference for Effective and Sustainable Hydraulic Fracturing, Brisbane, Australia, 20–22 May 2013.
45. Rafiee, R.; Ataei, M.; KhalooKakaie, R.; Rafiee, R.; Ataei, M.; KhalooKakaie, R. A new cavability index in block caving mines using fuzzy rock engineering system. *Int. J. Rock Mech. Min. Sci.* **2015**, *77*, 68–76. [CrossRef]
46. Bieniawski, Z.T. *Engineering Rock Mass Classifications: A Complete Manual for Engineers and Geologists in Mining, Civil, and Petroleum Engineering*; John Wiley & Sons: Hoboken, NJ, USA, 1989.
47. Laubscher, D.H. *Cave Mining Handbook*; De Beers: Johannesburg, South Africa, 2003; 138p.
48. Lorig, L.; Board, M.P.; Potyondy, D.; Coetzee, M.J. Numerical modeling of caving using continuum and micro-mechanical models. In Proceedings of the CAMI'95 Canadian Conference on Computer Applications in the Mining Industry, Montreal, QC, Kanada, 22 October 1995; pp. 416–424.
49. Mawdesley, C.A. Predicting Rock Mass Cavability in Block Caving Mines. Ph.D. Thesis, The University of Queensland, Brisbane, Australia, 2002.
50. Dimitrovski, A.; Matos, M. Fuzzy present worth analysis with correlated and uncorrelated cash flows. In *Fuzzy Engineering Economics with Applications*; Springer: Berlin, Germany, 2008; pp. 11–41. [CrossRef]
51. RP18R-97. Available online: https://online.aacei.org/aacessa/ecssashop.show_category?p_category_id=RP&p_cust_id=&p_order_serno=&p_promo_cd=&p_price_cd=&p_session_serno=&p_trans_ty= (accessed on 16 March 2022).
52. Hollmann, J.K. Improve your contingency estimates for more realistic project budgets: Reliable risk-analysis and contingency-estimation practices help to better manage costs in CPI projects of all sizes. *Chem. Eng.* **2014**, *121*, 36–44.

Article

Study of Key Technology of Gob-Side Entry Retention in a High Gas Outburst Coal Seam in the Karst Mountain Area

Zhenqian Ma [1,*], Dongyue Zhang [1], Yunqin Cao [1], Wei Yang [2] and Biao Xu [3]

1 School of Mining, Guizhou University, Guiyang 550025, China; 17885906234@163.com (D.Z.); cyq@gzu.edu.cn (Y.C.)
2 Guizhou Panjiang Coal and Electricity Group Technology Research Institute Co., Ltd., Guiyang 550081, China; pjjmtzbxhf@163.com
3 Guizhou Qianxi Honglin Mining Co., Ltd., Bijie 551500, China; z1727389800@126.com
* Correspondence: zqma@gzu.edu.cn

Abstract: In the gob-side entry retaining by roof cutting (GERRC) technique, pressure is offloaded via directional roof cutting, and a roadway is automatically formed due to the ground pressure and rock-breaking expansion. To improve the application of the theory and technical system of GERRC in the Karst area in Southwest China, this research studies the key technology of GERRC in a high gas outburst coal seam, based on the engineering background of the 39114 working face of the Honglin coal mine. According to the geological conditions of the 39114 working face, by means of formula calculation, UDEC numerical modeling, and on-site drilling peeping, the optimal roof-cutting parameters suitable for the 39114 working face were determined: the roof cutting height was 7 m, the roof cutting angle was 15°, and the spacing of pre-splitting blasting holes was 600 mm. Additionally, the above roof-cutting parameters have achieved good results in the engineering practices of the 39114 transportation roadway, which shows that the technology of GERRC is feasible in high gas outburst mines and achieves the goal of safe and efficient mining.

Keywords: high gas outburst coal seam; pre-splitting blasting; gob-side entry retaining; roof cutting; UDEC simulation

Citation: Ma, Z.; Zhang, D.; Cao, Y.; Yang, W.; Xu, B. Study of Key Technology of Gob-Side Entry Retention in a High Gas Outburst Coal Seam in the Karst Mountain Area. *Energies* **2022**, *15*, 4161. https://doi.org/10.3390/en15114161

Academic Editor: Adam Smoliński

Received: 11 April 2022
Accepted: 13 May 2022
Published: 6 June 2022

1. Introduction

As the world's largest energy production and consumption country, coal occupies the main position in China's energy consumption structure. With the depletion of easy-to-mine coal resources, gob-side entry retaining technology has become the primary research and application direction of coal-efficient mining because of its advantages of maximizing resource recovery, avoiding coal loss, and reducing roadway excavation. Since its application in China in the 1970s, gob-side entry retaining (GER) technology has made outstanding contributions to the field of non-pillar mining [1–3].

On the basis of gob-side entry retaining technology (GER), He Manchao's team proposed the technology of gob-side entry retaining by roof cutting (GERRC). In gob-side entry retaining, the crushing and expansion characteristics of caving waste rock are used to allow waste rock to fill the mined-out areas and form roadway sides, and further cancel the filling body [4–7]. Because of its advantages of low cost, easy operation, and fast lane keeping, this technology has been tested and popularized under different geological and mining conditions, such as different coal thickness [8–10], different buried depth [11,12], and different roof conditions [13,14], and produced great economic and social benefits. On this basis, domestic and foreign scholars have conducted in-depth research on GERRC.

Scholars in various countries have carried out in-depth research on GERRC. For example, Wang Q [14] performed a field engineering test in the Ningtiaota mining field, and proved the feasibility both of no coal-pillar mining schemes, and the technology of automatically forming gob-side entry retention by roof cutting and pressure relief. He M [15] used mechanical analysis and numerical simulation to analyze the structural characteristics of composite roofs and the effect of GERRC under composite roofs. Shao L [16] studied the pressure-relief law of secondary gob-side entry retention after double-side roof cutting in a thin coal seam, which provides an important reference value for secondary gob-side entry retention in the future. Zhu Z [17] used numerical simulation to analyze key problems, such as abutment pressure and stress concentration shells, and study the distribution and evolution characteristics of macroscopic stress fields of surrounding rocks in GERRC, and the differences from traditional pillar-retaining mining methods. Ma X [18] studied the design of GERRC by means of mechanical analysis, geometric analysis, and numerical simulation. Chen T [19] established a mechanical model of hard roofs, and analyzed the process of pressure relief by roof cutting; the key stage of roadway-stability control in the process of pressure relief by roof cutting was determined. Based on key block theory, Sun B [20] established the mechanical model of basic roof fracture structures, and the roof fracture mechanism of gob-side entry retention under the conditions of roof cutting and pressure relief was studied.

In addition, high gas mines are widely distributed in China and around the world [21–23]. To explore the applicability of the GERRC technology to high gas mines, the haulage roadway of the 39114 working face in the Honglin Coal Mine of Guizhou Province was taken as the research background. The key technology and advanced influence range of GERRC and pressure relief in gas outburst mines were studied by means of theoretical analysis, numerical simulation, and field tests. The research results have certain significance for improving the theory and technology systems of GERRC.

2. Engineering Survey

Honglin Coal Mine is located in Qianxi City, Guizhou province, China. The 9# and 15# coal seams are mainly mined in the Honglin Coal Mine. The 9# coal seam is unstable, with a thickness of 1.08–3.67 m, and the average thickness is 2.21 m.

The 39114 working face mainly mines the 9# coal seam, and the 9# coal seam is located in the weak aquifer in the middle of the Longtan formation. The upper limestone aquifer of the Changxing formation is about 60 m, and the lower is about 65 m from the strong aquifer of the Maokou formation. The water insulation of the coal seam is good, where only a small amount of bedrock fissure water is contained, and there is no water-conducting fault in the area.

The immediate roof of the coal seam is argillaceous siltstone, with a thickness of 3.0–4.5 m, and the average thickness is 3.38 m. The main roof is fine-grained sandstone with a thickness of 5.3–7.6, and the average thickness is 6.97 m. The main floor is argillaceous siltstone with a thickness of 6.72 m. The column diagram of the roof and floor lithology is shown in Figure 1.

Histogram	Coal seams number	Rock name	Lithological description
		Muddy siltstone	Gray, thin - medium thick layered, microwave - shaped
	7#	Coal	Black, lumpy, bright coal with dark coal
		Fine-sandstone	Light gray, thin-thick layered, fine-grained structure
		Muddy siltstone	Gray, medium thick layered, parallel joints
		Fine-sandstone	Black, lumpy, bright coal with dark coal
	9#	Coal	Black, lumpy, bright coal with dark coal
		Muddy siltstone	Gray, thin - medium thick layered, parallel joints

Figure 1. Column diagram of the roof and floor lithology.

The 39114 haulage roadway is a rectangular section with a width of 4.5 m and a height of 2.8 m. The original support section is as shown in Figure 2.

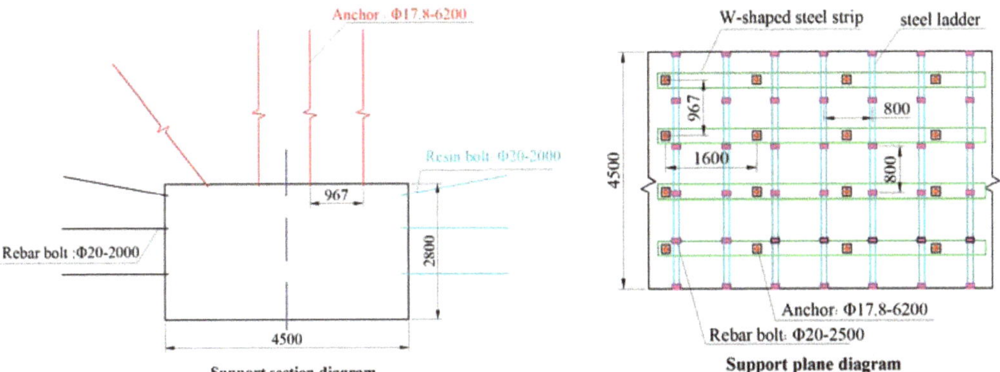

Figure 2. Original roadway support parameters.

3. Gas Prevention and Control Measures

The 9# coal seam belongs to a high gas outburst coal seam; the gas content at the elevation of +1435 m to +1540 m is 13.9374 m^3/t, the gas pressure is 0.8 MPa, and the firmness is 1.3. To ensure the safe and efficient production of the mine, it is necessary to take corresponding gas outburst prevention measures during the excavation and mining of the 39114 working face. Therefore, the method of mining the liberated seam was adopted as the regional outburst prevention measure during the roadway excavation of the 39114 working face. During the mining of the working face, pre-draining coal seam gas was used as the regional outburst prevention measure.

3.1. Mining the Liberated Seam

When the liberated seam is mined, the roof or floor of the protective coal seams can be damaged and cracks can develop. At the same time, the gas in the protective coal seams changes from an adsorption state to a free state under the influence of mining. This free gas moves to the liberated seam through the cracks, so as to achieve the effect of treating the protective coal seam gas [24].

The 39114 working face is situated in the 9# coal seam, which is adjacent to the 7# and 15# coal seams. The average interval between the 7# coal seam and the 9# coal seam is 11 m, the average interval between the 15# coal seam and the 9# coal seam is 55 m, and the positional relationship between the three coal seams is shown in Figure 3. The original gas content and pressure of the 7#, 9#, 15# coal seams are shown in Table 1. According to the above data, the threat of gas disaster in the 15# coal seam is relatively small, so the 15# coal seam was selected as the liberated seam, and the 9# coal seam was selected as the protective seam.

Table 1. Gas parameters of the 7#, 9#, 15# coal seams.

Coal Seam	Original Gas Content (m^3/t)	Original Gas Pressure (MPa)
7#	12.46	0.08–0.58
9#	13.9374	0.12–0.69
15#	3.6	0.05–0.4

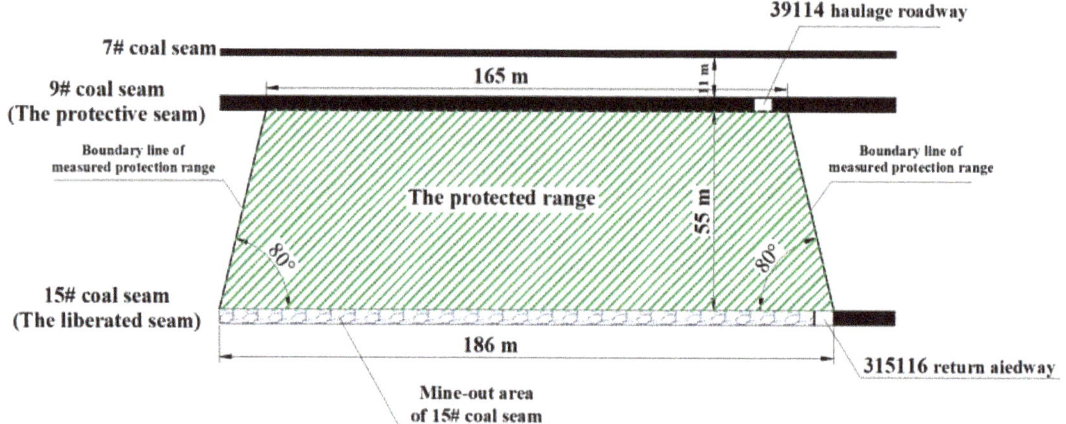

Figure 3. Positional relationship and protected range of the liberated and protective seams.

Before mining the 15# coal seam, the pre-drainage of gas per 80 m roadway could reach the standard level for about 25 days during the excavation of roadways in the 39114 working face. The single-hole gas pre-drainage concentration was about 8–36%, and the average excavation length per month was 30–50 m.

After mining the 15# coal seam, within the protection range of the liberated seam, the pre-drainage of gas per 80 m roadway was about 15 days, which was about 40% shorter than it was before mining the liberated seam. The single hole extraction concentration was 15–48%, and the average excavation length per month could be increased to 90 m.

To further verify the gas prevention and control effect of the liberated seam, 6 inspection holes and 18 inspection points were arranged in the 39114 haulage roadway. The maximum gas content was 6.831 m^3/t, with an average of 6.13 m^3/t, which was 55.9% lower than the original average gas content of the 9# coal seam, and the protected range as shown in Figure 3.

The monitoring results show that mining the 15# coal seam as the liberated seam has a better gas-control effect on the 9# coal seam.

3.2. Gas Pre-Drainage

After the excavation of the 39114 mining roadway was completed, in order to ensure the safe and efficient advancement of the working face, pre-drainage boreholes needed to be arranged in the 9# coal seam. In addition, the overlying coal seam of the 9# coal seam is the 7# coal seam, and the distance between the two coal seams is only 11 m. Therefore, when arranging gas pre-drainage boreholes in the 9# coal seam, the gas pre-drainage boreholes of the 7# coal seam also needed to be arranged.

After the gas pre-drainage method was determined, the gas in the 9# coal seam was pre-extracted by drilling along the seam in the 39114 return airway, the 39114 haulage roadway, and the setup entry, as shown in Figure 4. The drilling parameters of gas pre-drainage through the 7# coal seam are shown in Table 2.

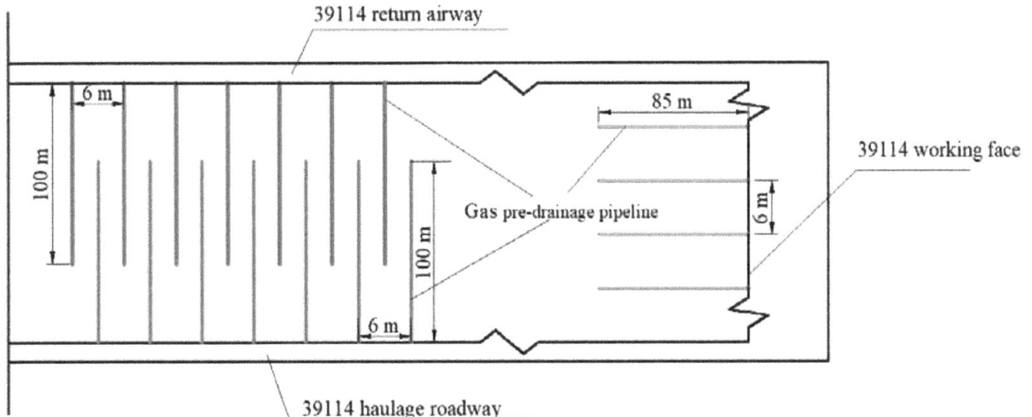

Figure 4. Diagram of the arrangement of pre-drainage boreholes.

Table 2. Design parameters of gas pre-drainage boreholes.

		39114 Return Airway		
Number	**Azimuth (°)**	**Dip Angle (°)**	**Design Length (m)**	**Spacing (m)**
1	89	+28	47	0.38
2	110	+25	48	0.38
3	126	+20	55	0.38
4	138	+15	64	0.38
5	146	+11	76	0.38
6	152	+8	88	0.38
		39114 Haulage Roadway		
Number	**Azimuth (°)**	**Dip Angle (°)**	**Design Length (m)**	**Spacing (m)**
1	77	+29	48.5	0.3
2	60	+28	54.2	0.3
3	47	+26	63	0.3
4	38	+23	74	0.3
5	31	+21	85.5	0.3
6	27	+19	98.2	0.3

3.3. *Evaluation of Gas Pre-Drainage Effect*

In order to verify the gas pre-drainage effect of the 9# coal seam, the gas pre-drainage rate and residual gas content were analyzed in the range of 300 m outward from the setup entry of the 39114 working face. In this area, there were a total of 146 horizontal boreholes and 168 through-layer boreholes.

3.3.1. Gas Pre-Drainage Rate

According to the estimation formula of the total amount of gas in the original coal seam:

$$Q_o = S \times q \times h \times \gamma \tag{1}$$

In this formula, Q_o is the original coal seam gas volume in the gas pre-drainage borehole control area, unit: m^3; S is the area of borehole control range, unit: m^2; q is the original gas content, unit: m^3/t; h is the thickness of the coal seam, unit: m; γ is the bulk density of the coal seam, unit: t/m^3.

After calculation, the original gas volume of the 9# coal seam Q_{o9} was 2.3733 million m^3, and the original gas volume of the 7# coal seamQ_{o7} was 0.2533 million m^3.

Because the residual gas content after pre-drainage should be lower than 8 m^3/t, the calculation formula of minimum gas pre-drainage amount can be obtained:

$$Q_{min} = T \times (q - 8) \quad (2)$$

In this formula, Q_{min} is the minimum gas pre-drainage volume, unit: m^3; T is the total coal weight within the control range of the borehole, unit: t. It is calculated that the minimum gas pre-drainage volume from the 7# and 9# coal seams was 1.1017 million m^3. Thus, the minimum gas pre-drainage rate is 1.1017/(0.2533 + 2.3733) = 41.94%.

In order to detect the gas pre-drainage effect, the GD_3 pipeline automatic metering system and the manual CZJ-70 gas comprehensive parameter tester were used to monitor the gas pre-drainage data in the 39114 working face. The total gas pre-drainage volume of the GD_3 pipeline automatic metering system is 1.1653 million m^3; therefore, the gas pre-drainage rate is 1.1653/(0.2533 + 2.3733) = 44.37%. According to the statistics of the measured data of each of the pre-drainage boreholes by the CZJ-70 gas comprehensive parameter tester, the total gas pre-drainage volume is 1.1842 million m^3; therefore, the gas pre-drainage rate is 1.1842/(0.2533 + 2.3733) = 45.10%. The above two gas pre-drainage rates are greater than the minimum gas pre-drainage rate, which is 41.94%, indicating that the pre-drainage effect is remarkable, achieving the purpose of gas outburst elimination.

3.3.2. Residual Gas Content and Residual Gas Pressure

To further verify the accuracy of the gas outburst elimination effect in the 39114 working face, 6 boreholes in the 9# coal seam were respectively arranged in the 39114 return airway and haulage roadway, and the residual gas pressure and residual gas content were measured. In addition, 4 boreholes through the 7# coal seam were respectively arranged in the 39114 return airway and haulage roadway to determine the residual gas pressure and residual gas content of the 7# coal seam. The parameters of the boreholes after on-site measurement and calculation are shown in Table 3.

Table 3. Parameters of boreholes.

Location	Number	Depth (m)	Azimuth(°)	Residual Gas Pressure (MPa)	Residual Gas Content (m^3/t)
39114 return airway (along the 9# coal seam)	1	44	180	0.092	6.1033
		87	180	0.094	6.1695
	2	44	180	0.101	6.3613
	3	87	180	0.081	5.7999
	4	44	180	0.098	6.2805
	5	87	180	0.093	6.1316
	6	46	180	0.091	6.0843
39114 haulage roadway (along 9# coal seam)	1	54.2	0	0.08	5.7639
	2	80.56	0	0.088	5.9996
	3	48	0	0.101	6.3538
	4	82	0	0.089	6.0343
	5	53	0	0.095	6.1963
	6	87	0	0.095	6.204
	Average value			0.092	6.114
39114 return airway (through the 7# coal seam)	1	48.6	89	0.123	6.1639
	2	76	146	0.13	6.3053
39114 haulage roadway (through the 7# coal seam)	1	45	35	0.121	6.1187
	2	101.84	35	0.128	6.2756
	Average value			0.126	6.2159

The residual gas content in all of the boreholes measured above was less than 8 m^3/t. The residual gas pressure was less than 0.74 MPa, which meets the requirements of coal-seam outburst elimination. To further evaluate the outburst elimination effect, 12 boreholes which measured 42 mm in diameter and 10 m in depth were constructed in the 39114 working face, and the WTC-2 gas outburst parameter instrument was used to verify the gas desorption index of the drilling ($K1$) and drilling cuttings weight (S). After measurement, the maximum value of $K1$ was 0.23, and the maximum value of S was 1.8, which are all less than the critical index ($K1 < 0.50$, $S < 6.0$). Therefore, the above two measurement results show that the outburst elimination effect of the 39114 working face is noteworthy, which created good conditions for the implementation of gob-side entry retention.

4. Determination of Key Technology of Roof Cutting

4.1. Numerical Simulation

As shown in Figure 5, α refers to the cutting angle between the roof cutting line and the plumb line. H is the roof cutting height, which refers to the maximum vertical height of the roof cutting seam.

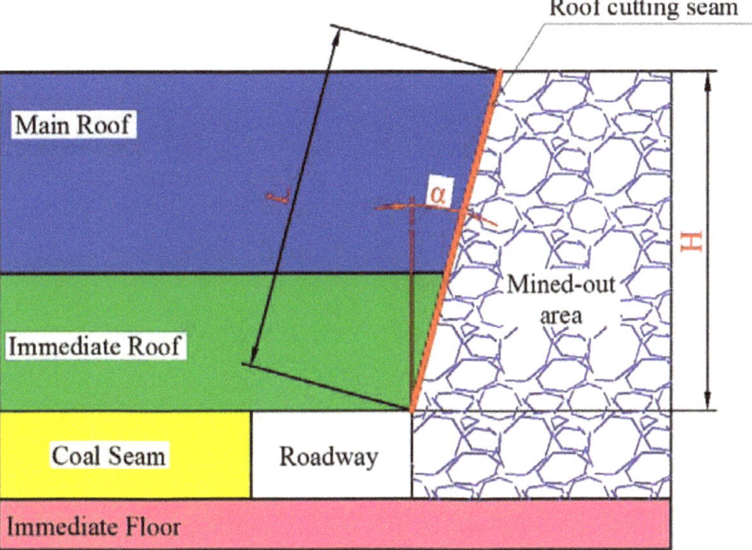

Figure 5. Roof cutting diagram.

To determine the reasonable roof cutting height and angle, the UDEC numerical simulation model was established according to the engineering geological conditions of the 39114 working face. The model was 100 m long and 70 m high, and fixed constraints were imposed on the left, right, and bottom boundaries. A vertical load of 4.0 MPa was applied to the top of the model to simulate the 160 m thick overlying strata. Subsequently, the numerical models with different roof cutting heights and angles were established and calculated. Based on the above, the collapse morphology and vertical displacement variation of roadway-surrounding rock for different roof cutting parameters were studied.

4.1.1. Roof Cutting Height

During the simulation of roof cutting height, the roof cutting heights were 5.5 m, 7 m and 8.5 m. The other parameters remain unchanged, and the roof cutting angle was always 15°, which was biased towards the mined-out areas. The excavation method of the working face is gradual excavation, and half of the model was intercepted for analysis after calculation and balance. Collapse morphology and vertical displacement of the model for different roof cutting heights are shown in Figure 6. At the same time, the monitoring points were arranged on the roadway roof; these were used to monitor the maximum deformation of the roadway roof, and the results are shown in Figure 7.

Analysis of the collapse morphology of roadway-surrounding rock for different roof cutting height shows that roof cutting can effectively cut off the structural connection between the roof of mined-out areas and the roadway roof. The roof of the mined-out area broke and fell down along the roof cutting seam, and the roof of the roadway formed a short-arm beam structure. After the basic roof sunk, it was stable after contacting the collapsed gangue in the mined-out areas, providing bearing support for the roof of the roadway.

It can be seen from Figure 6a that when the roof cutting height is 5.5 m, the roof of mined-out areas is cut down along the roof cutting seam at the height of 5.5 m. In the process of roof cutting, there is friction between the roof of the mined-out areas and the roadway roof, which exerts a downward force on the short-arm beam structure of the roadway roof, resulting in the subsidence of the roadway roof; the maximum deformation of the roadway roof was 167 mm. In addition, there is a large unfilled space between the crushed gangue and the basic roof, as shown in Figure 6a.

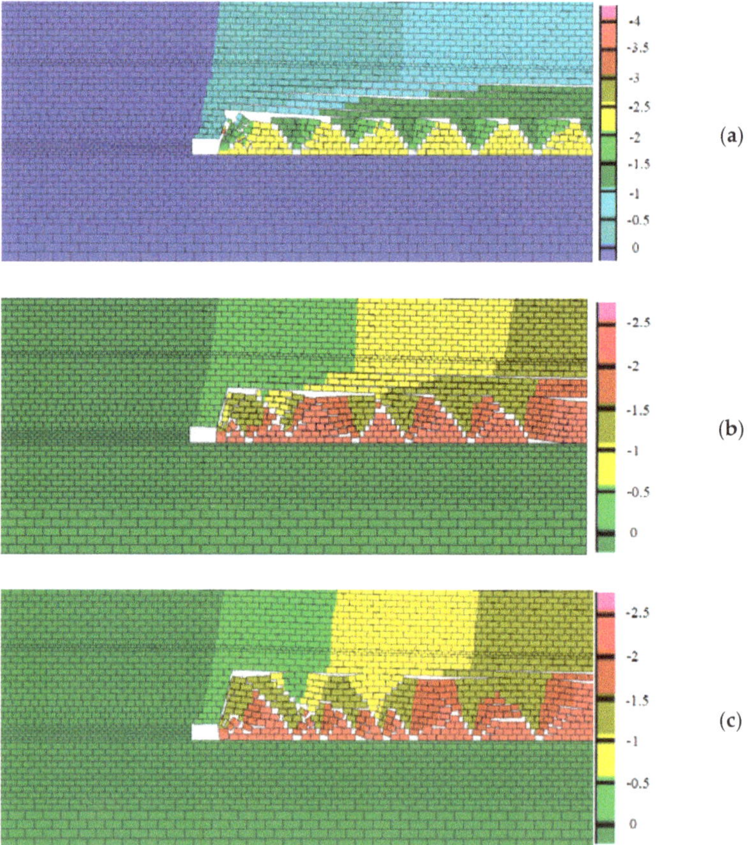

Figure 6. Collapse morphology and vertical displacement model (unit: m). (**a**) Roof cutting height 5.5 m, (**b**) Roof cutting height 7 m, (**c**) Roof cutting height 8.5 m.

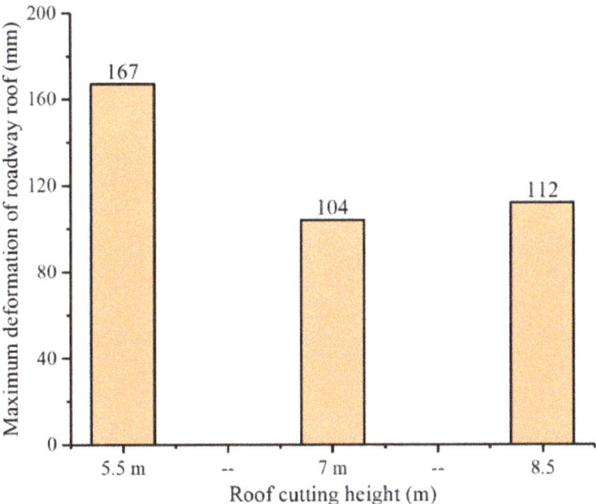

Figure 7. Maximum deformation of the roadway roof for different cutting heights.

When the roof cutting height is increased to 7 m, the roof of the mined-out areas is cut down along the roof cutting seam at a height of 7 m. It can be seen from Figure 6b that with the increase in roof cutting height, the roof cutting range of the mined-out area increases, and the degree to which the mined-out area is filled increases, after the collapse of rock fragmentation and expansion. Meanwhile, the deformation of the roadway roof is effectively controlled, and the maximum roadway roof deformation is 104 mm, which is smaller than that when the cutting height is 5.5 m.

As can be seen from Figure 6c, when the roof cutting height is further increased to 8.5 m, the unfilled space between the main roof and mined-out areas is further reduced. However, due to the increased roof cutting height, the force exerted on the short-arm structure of the roadway roof also increases, resulting in the greater deformation of the roadway roof, with a maximum deformation of 112 mm.

In summary, the height of roof cutting affects both the degree to which the mined-out areas are filled, and the force of the roof cutting on the short-arm structure of the roadway. A reasonable roof cutting height is needed to ensure that the expanded gangue fills the mined-out areas as much as possible and the cutting force on the roof of the roadway is as small as possible. From the numerical simulation results, it can be seen that increasing the roof cutting height within a certain range can improve the expansion volume of gangue and reduce the unfilled space of mined-out areas. However, further increases in roof cutting height may not be conducive to the stability of roadway roof, and also increase construction costs and difficulty. The numerical simulation results show that when the cutting height is 7 m, the roof cutting effect is preferable.

4.1.2. Roof Cutting Angle

A reasonable cutting angle can reduce the friction between the gob roof and the reserved roadway roof, enhance the cutting effect and controlling the roadway roof deformation. In the roof cutting angle simulation, the roof cutting angles of 0°, 10°, 15° and 20° were simulated to observe the surrounding rock collapse morphology and vertical displacement distribution characteristics of the roadway. Except for the cutting angle, the other parameters were kept unchanged, and the cutting height was 7 m. The simulation results are shown in Figure 8, and the maximum deformation of the roadway roof for different cutting angles is shown in Figure 9.

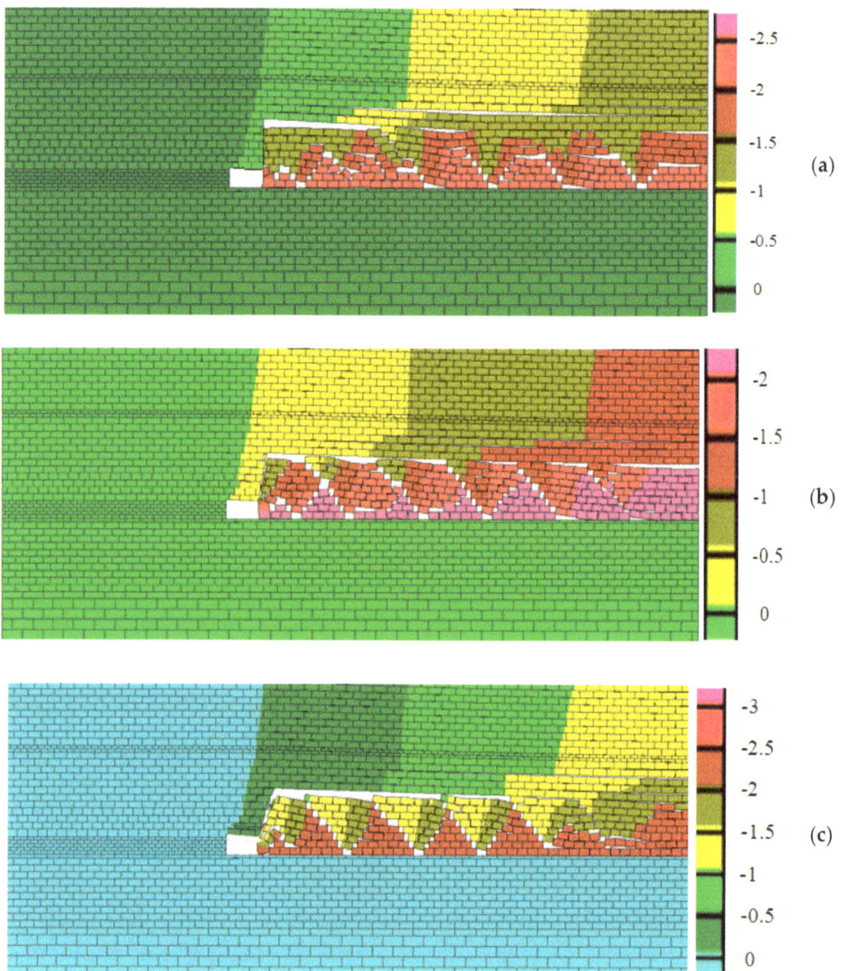

Figure 8. Collapse morphology and vertical displacement model (unit: m). (**a**) Roof cutting angle 0°, (**b**) Roof cutting angle 10°, (**c**) Roof cutting angle 20°.

As shown in Figure 8a, when the roof cutting line is perpendicular to the roadway roof, that is, the cutting angle is 0°, the force of the roof of the mined-out areas on the short-arm beam structure of the roadway roof is significantly increased, resulting in a significant deformation of the roadway roof. The cutting seam line is perpendicular to the roof, and the crushed and expanded gangue only plays a vertical support role for the direct roof and does not play a support role for the short-arm beam structure of the roadway roof, which is also one of the direct reasons for the large deformation of the roadway. When the roof cutting angle is 0°, the maximum roadway roof deformation is 164 mm, as shown in Figure 9. As shown in Figure 8b, the friction between the roof of the mined-out areas and the roadway roof decreases after the roof cutting angle is increased to 10°. The gangue collapse is close to the short-arm beam structure of the roadway, which provides a certain support for it. Increasing the cutting angle can effectively control the roof deformation. When the cutting angle is 10°, the maximum deformation of the roadway roof is 116 mm, which is 29.3% lower than the deformation when the cutting angle is 0°, as shown in Figure 9.

When the roof cutting angle is 15°, the force between the roof of the mined-out areas and the roadway roof is further decreased, and the collapsed rock is also close to the short-arm beam structure of the roadway, providing greater support. Compared with the roof cutting angle of 10°, the maximum deformation of the roadway roof is slightly reduced to 104 mm, which is a reduction of 10.3%. As shown in Figure 8c, when the roof cutting angle continues to increase to 20°, due to the large roof cutting angle, the collapsed rock is more inclined to the mined-out areas, resulting in increased space between the collapsed rock and the short-arm structure of the roadway roof, which reduces the supporting effect of collapsed rock on the roadway roof. Therefore, it is not conducive to roadway stability, and the maximum roadway roof deformation is 137 mm.

In summary, there is an appropriate value for the roof cutting angle. When the roof cutting angle exceeds this value, continuing to increase the roof cutting angle is detrimental to the stability of the roadway. It can be seen from the numerical simulation results that when the cutting angle is ~15°, the roof cutting effect is preferable.

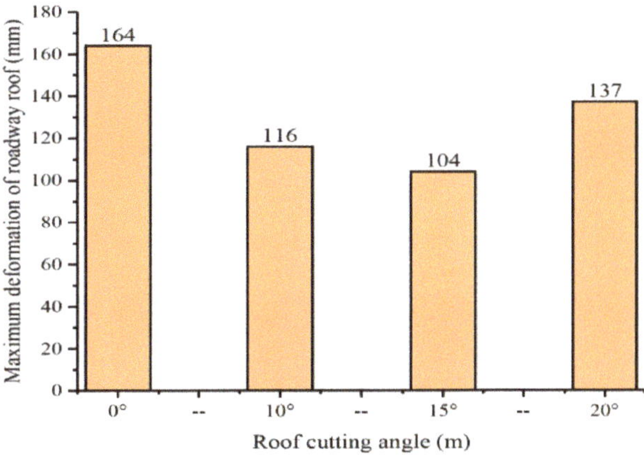

Figure 9. Maximum deformation of the roadway roof for different cutting angles.

4.2. Calculation of Roof Cutting Seam Parameters

To further determine the roof cutting parameters of the Honglin Coal Mine, the previous formula was used, combined with the numerical simulation results, to calculate the parameters that are conducive to the design of the optimal roof cutting parameters.

4.2.1. Calculation of Roof Cutting Height

To ensure that the roof collapse within the roof cutting range of the mined-out areas can effectively support the overburden, the cutting height H can be calculated according to the following formula [25]:

$$H = \frac{m - \Delta H_1 - \Delta H_2}{K - 1} \tag{3}$$

In this formula, m is the coal seam mining height, which is 2.2 m; ΔH_1 is roof subsidence, unit: m; ΔH_2 is floor heave, unit: m; K is the roof expansion coefficient, taken as 1.35. Regardless of roof subsidence and floor heave, the roof cutting height H is calculated to be 6.3 m.

In a comprehensive comparison of the 39114 haulage roadway roof lithology histogram, the roadway immediate roof lithology was determined to be argillaceous siltstone, with an average thickness of 3.38 m. The main roof lithology was fine-grained sandstone, with an average thickness of 6.97 m. To facilitate on-site construction and make the roof of the mined-out area more prone to collapse along the roof cutting seam, and combined with the numerical simulation results, the height of roof cutting was determined to be 7 m, which was adjusted according to the drilling peep results in the later stage.

4.2.2. Calculation of the Roof Cutting Angle

He Manchao, an academic at the Chinese Academy of Science, obtained this formula for calculating the roof cutting angle, based on the theory of masonry beams and the S-R stability principle of the surrounding rock structure [26]:

$$\alpha \geq \phi - \arctan\frac{2(h - \Delta S)}{L} \tag{4}$$

In the formula: α is the roof cutting angle, unit: °; φ is the internal friction angle between rock blocks, which is 21°; h is the thickness of the main roof, is 6.3 m; L is the lateral span of the main roof block, unit: m; ΔS is the subsidence of the rock block, which is 4 m. Additionally, it was calculated that the roof cutting angle $\alpha \geq 14.4°$.

To reduce the amount and difficulty of roof cutting construction, combined with the numerical simulation results, the cutting angle was determined to be 15°. In summary, the roof cutting height of the 39114 haulage roadway in the Honglin Coal Mine was determined to be 7 m, and the roof cutting angle was determined to be 15°.

4.3. Pre-Splitting Blasting Parameters

4.3.1. Spacing of Blast Holes

The blast hole spacing in hard rock roofs is usually set in the range of 450–550 mm. Therefore, the preliminary blast hole spacing was set at 500 mm. To further determine the blast hole spacing, pre-split blasting experiments were carried out in the field on blast holes with a spacing of 500 mm, 600 mm, and 700 mm. The borehole peeping instrument was used to observe cracks in the holes, and the peeping image is shown in Figure 10.

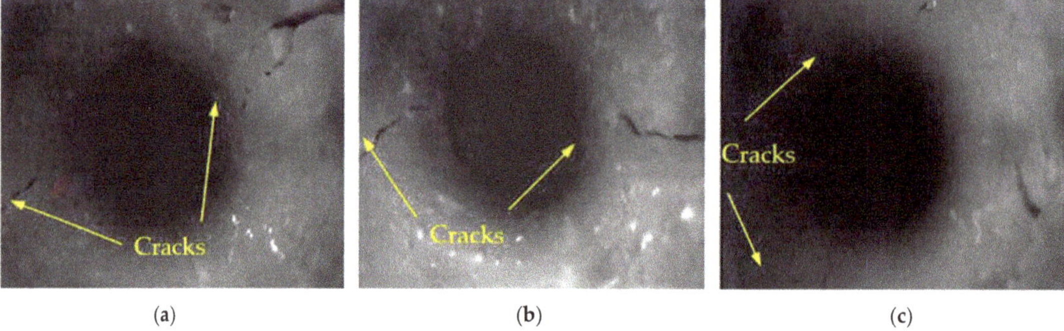

Figure 10. Blast hole drilling peep with different spacing. (**a**) Spacing 500 mm (**b**) Spacing 600 mm (**c**) Spacing 700 mm.

As shown in Figure 10a,b, when the blast hole spacing was 500 mm and 600 mm, two obvious cracks appeared in the blast holes, indicating that spacings of 500 mm and 600 mm were appropriate, and achieved the effect of pre-splitting blasting. As shown in Figure 10c, there were also two cracks in the blast hole, but the crack width was smaller, indicating that the spacing of 700 mm between the blast holes was too large and the cutting effect was poor.

To sum up, it was determined that the blast hole spacing of the roof cutting construction should be 600 mm.

4.3.2. Blasting Parameters

After engineering practice, the blasting parameters were determined as follows. The blast hole depth was 7 m, and the spacing was 600 mm. Three binding energy tubes were used for each hole, and the charge structure was 3–2–1. The sealing mud length is 2.5 m, based on previous field experience, as shown in Figure 11.

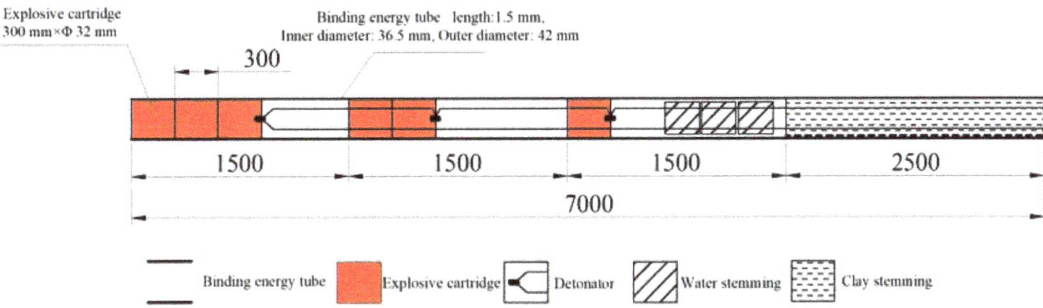

Figure 11. Explosive charge structure.

5. Engineering Practice

The whole process of gob-side entry retaining in the high gas outburst coal seam is shown in Figure 12. The mining of the liberation seam eliminates the danger of a coal seam outburst and provides satisfactory conditions for roadway excavation. After that, the working face began to advance, and the coal seam gas pre-drainage ensured the safety and efficiency in the process of coal seam mining.

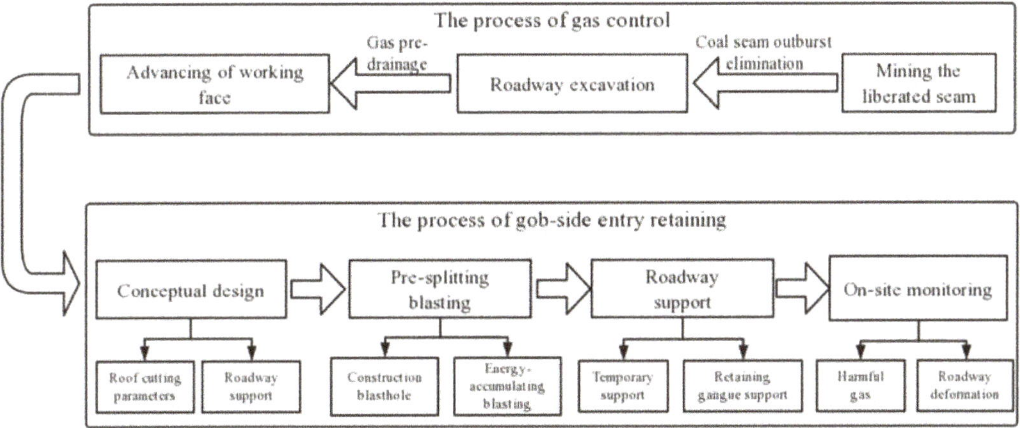

Figure 12. Whole process of gob-side entry retention in a high gas outburst coal seam.

The outburst danger of the coal seam gas was eliminated by mining liberated seam and pre-drainage coal seam gas, which created satisfactory conditions for gob-side entry retention. To keep the roadway along the mined-out area, the roof cutting parameters and the roadway support parameters were first determined. Then, blastholes were constructed and pre-splitting blasting was conducted, and support and temporary support in the process of roof cutting were deployed in a timely manner for gangue retention and to ensure the stability of the roadway.

5.1. Roadway Support

5.1.1. Retaining Gangue Support

As the 9# coal seam of the Honglin coal mine is a high gas outburst coal seam, in the design of the 39114 transport roadway gangue support mode, a layer of air duct cloth was installed in the mined-out area and fixed with iron wire mesh. This can prevent gas from flowing into the roadway from mined-out areas, effectively prevent air leakage, and preclude the spontaneous combustion of the coal seam caused by fresh air entering the mined-out areas.

The retaining gangue support in the 39114 working face is shown in Figure 13, and the on-site retaining support effect is shown in Figure 14.

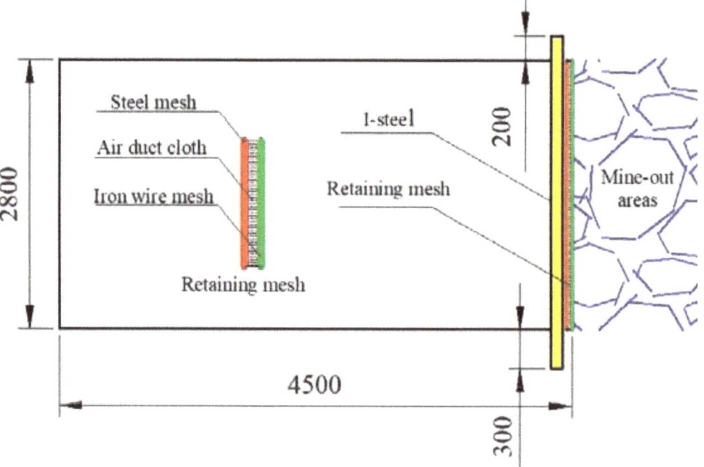

Figure 13. Retaining gangue support schematic diagram.

Figure 14. On-site retaining support effect.

5.1.2. Advanced Temporary Support

To prevent the influence of advanced stress concentration on the haulage roadway, single hydraulic props and hinged roof beams were used as advanced temporary support. The row spacing of the single hydraulic props was 1000 mm, which was 500 mm away from the roadway rib, as shown in Figure 15. The range of the advanced temporary supports was 50 m. The working face was advanced in the range of 0–20 m, and two rows of single hydraulic props were assembled along the roadway; then, when the working face was advanced in the range of 20–50 m, one row of single hydraulic props was assembled on the mined-out area side along the roadway, as shown in Figure 15. The effect of the on-site advanced temporary support is shown in Figure 16.

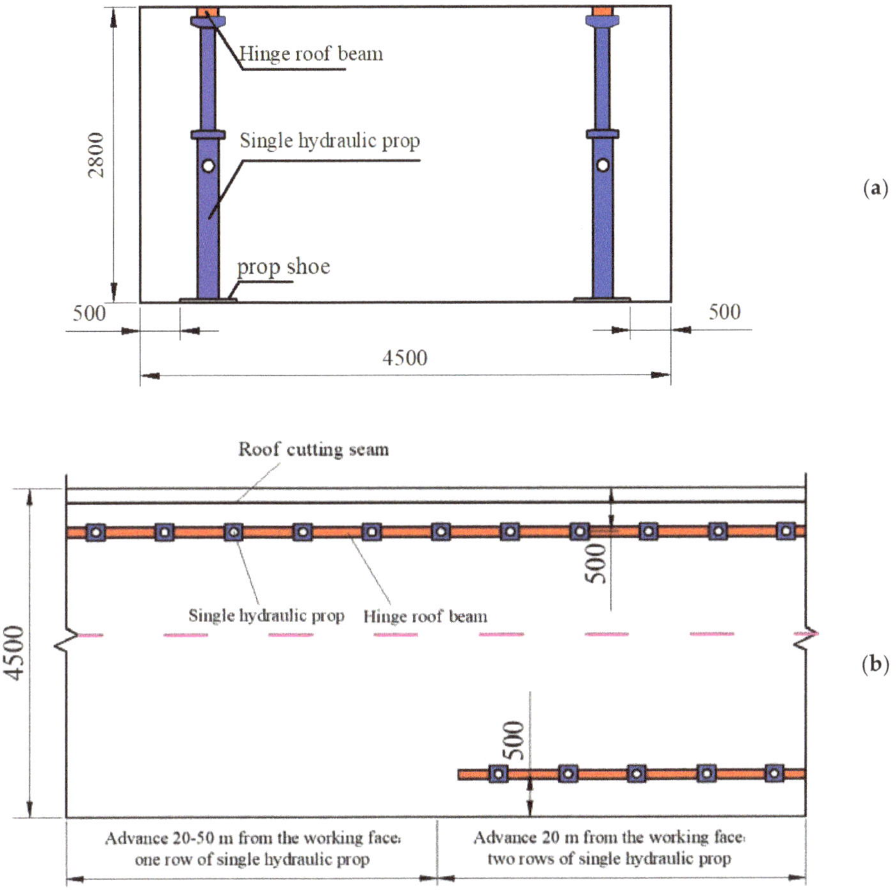

Figure 15. Schematic diagram of the advanced temporary support (unit: mm). (**a**) Support section, (**b**) Support planar.

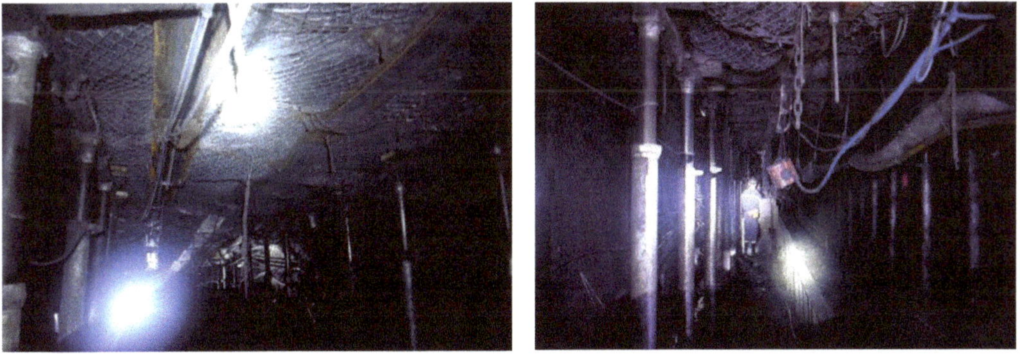

Figure 16. Effect of on-site advanced temporary support.

5.1.3. Delayed Temporary Support

Three rows of single hydraulic props were arranged along the roadway as delayed temporary support. On the roof cutting side, a single hydraulic prop with a hinged roof beam was used, and the spacing of the single hydraulic props was 1000 mm. In addition, a row of single hydraulic props was arranged on both sides of the roadway center-line, 500 mm away from the roadway center-line, and the row spacing of these single hydraulic props was 2000 mm. Figure 17 is the schematic diagram of the delayed temporary support, and the effect of the on-site delayed temporary support is shown in Figure 18.

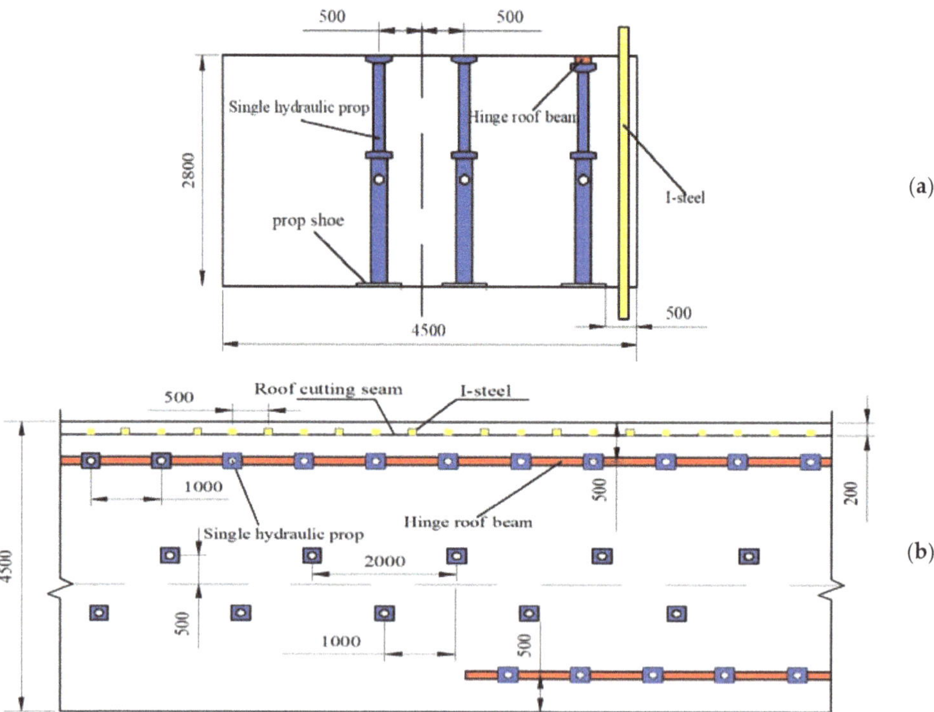

Figure 17. Schematic diagram of the delayed temporary support. (**a**) Support section, (**b**) Support planar.

Figure 18. Effect of on-site delayed temporary support.

5.2. On-Site Monitoring

As the 39114 working face did not have Y-line ventilation conditions, to prevent fresh air from flowing into the mined-out areas, sandbag walls were arranged behind the working face along the mined-out areas. This resulted in the inability to establish measuring stations in the retaining roadway. Therefore, ground pressure monitoring was only carried out in the roof cutting section.

5.2.1. Advanced Influence Range of Roof Cutting

To determine the advanced influence range of roof cutting, three stations were established along the 39114 haulage roadway, and one bolt in the middle of the roadway roof was selected to have a bolt dynamometer installed in it to monitor the working force. The monitoring stations were 60 m, 160 m, and 260 m away from the setup entry, and the monitoring results are shown in Figure 19.

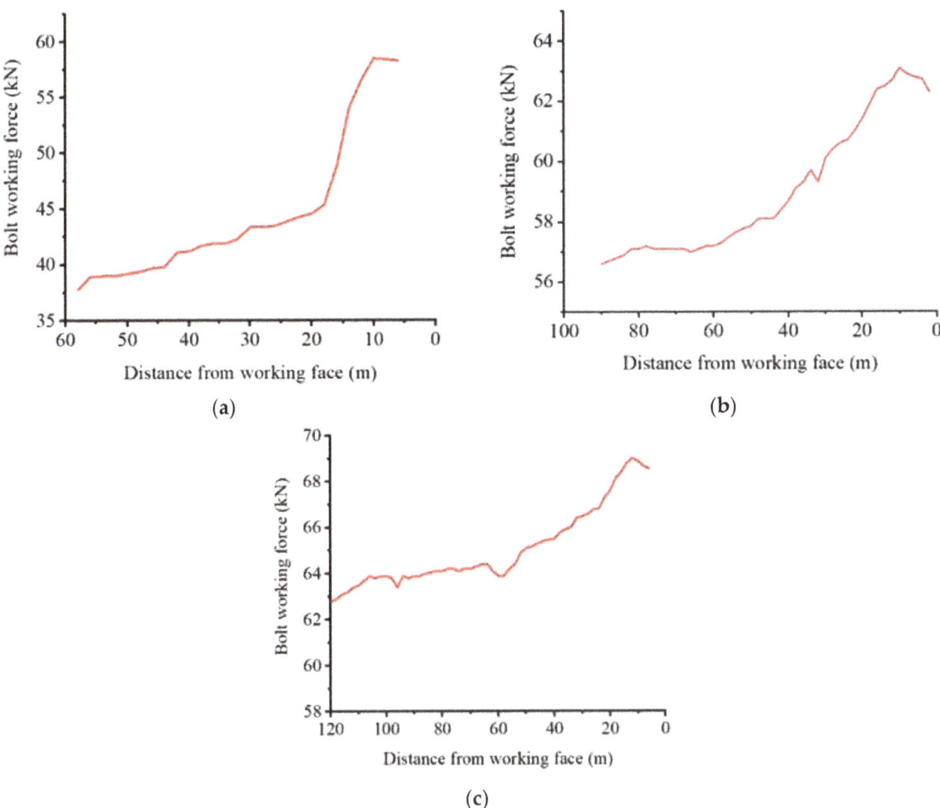

Figure 19. Force of bolts. (**a**) Station 1# (**b**) Station 2# (**c**) Station 3#.

As shown in Figure 19a, when the distance from the working face was 40–60 m, the working force of the bolts increased slowly, and when the distance from the working face was less than 20 m, the working force of bolts increased sharply. As shown in Figure 19b, the variation trend of the bolt working force at station 2# was basically consistent with that at station 1#. When the distance from the working face was greater than 50 m, the bolt working force increased slowly. When the distance was less than 40 m, the bolt working force increased rapidly. As shown in Figure 19c, when the distance from the working face was greater than 60 m, the working force of the bolt increased gently, and increased rapidly when the distance from the working face was less than 60 m.

It can be seen from Figure 19 that, with an increase in the advancing length of the working face, the working force of the bolts at the three stations showed an increasing trend. This also showed that, with the advance of the working face, the advance pressure of the roadway gradually increased and the deformation of the roadway continued to increase.

While monitoring the working force of the bolt, the deformation of the roof-floor and two ribs of the three stations were also monitored. The monitoring results are shown in Figure 20.

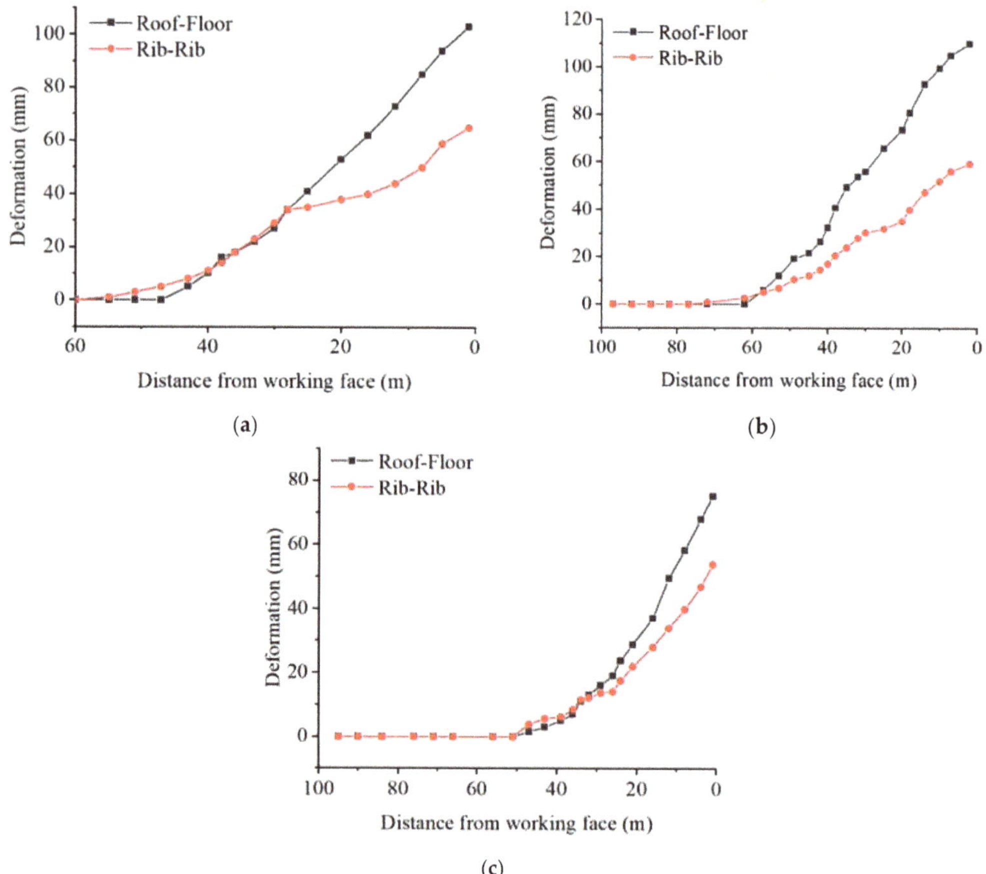

Figure 20. Deformation of the roof-floor and two ribs. (**a**) Station 1# (**b**) Station 2# (**c**) Station 3#.

5.2.2. Harmful Gas Monitoring in Mined-Out Areas

The 9# coal seam in the Honglin Coal Mine is a high gas outburst coal seam. Although outburst elimination treatment was carried out before mining, residual gas accumulation may still occur in mined-out areas. In addition, due to the isolation of fresh air flow in the mined-out areas, the residual coal in the mined-out areas is prone to anaerobic oxidation to produce CO. Therefore, it is necessary to use a JSG-8 monitoring system to monitor the CH_4 and CO concentration in the gas extraction pipeline in the mined-out areas. The JSG-8 is a mine fire beam tube monitoring system, which is an electronic product for monitoring and predicting underground natural fires. The system extracts the gas in the underground monitoring beam pipe to the ground through microcomputer control, and uses the gas chromatograph for analysis, which allows on-line monitoring of the content of CO, CO_2, CH_4, and other gases.

It can be seen from Figure 21 that the CH_4 concentration and CO concentration in the extraction pipeline fluctuated slightly, but the CH_4 concentration fluctuated around 4% and the CO concentration fluctuated around 0.06%. This shows that the concentration of harmful gases in the mined-out areas was relatively small, and the prevention and control of gas and fire in the mined-out areas was within a controllable range when gob-side entry retention was used.

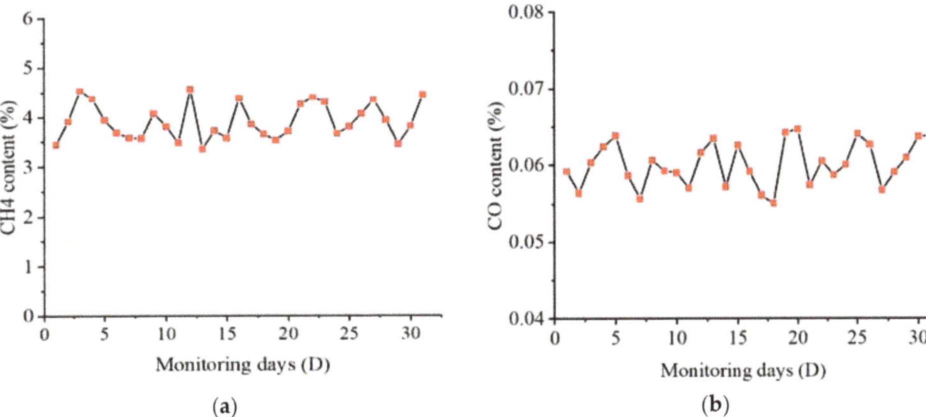

Figure 21. Harmful gas monitoring over a period of one month. (**a**) CH_4, (**b**) CO.

6. Conclusions

(1) The 39114 working face of the Honglin coal mine adopts the methods of mining the protective layer and gas pre-drainage to eliminate outbursts in the coal seam. After coal seam outburst elimination, the CH_4 content of adjacent coal seams in the 39114 working face decreased by 50.2% and 56.1%, respectively, indicating that the mining of the liberated seam and gas pre-drainage can effectively control coal seam gas.

(2) UDEC numerical simulation software was used to simulate different roof cutting heights and cutting angles. It was found that increasing the roof cutting height or cutting angle within a certain range can improve the volume of gangue fragmentation and reduce the unfilled space in the mined-out area, indicating that there are optimal values for cutting height and cutting angle.

(3) After engineering practice was carried out on the 39114 working face by the use of determined optimal roof cutting parameters, the average maximum deformation of the roof was 98.3 mm and that of the two ribs was 62.3 mm. This shows that the technology of roof cutting, and keeping pressure-relief retaining roadways along mined-out areas, can effectively control roadway deformation.

(4) The concentrations of CH_4 and CO in the mined-out areas were monitored. The gas concentration in the mined-out areas fluctuated up and down around 0.3% and the CO concentration fluctuated up and down around 0.06%, indicating that retaining gangue support with an air duct cloth can effectively control the generation and escape of harmful gases.

(5) The technology of gob-side entry retaining by roof cutting was feasible in the high gas outburst mine, and achieves the goal of safe and efficient mining.

Author Contributions: Formal analysis, D.Z.; investigation, Y.C.; resources, Z.M. and B.X.; writing—original draft preparation, D.Z.; writing—review and editing, Z.M.; project administration, W.Y. All authors have read and agreed to the published version of the manuscript.

Funding: The National Natural Science Foundation of China (Grant No. 51904080), Grant to Z.M.; and the Science and Technology Program of Guizhou Province of China (Grant No. [2021] general 352 No. [2021]3001) during the research.

Institutional Review Board Statement: The study did not require ethical approval.

Informed Consent Statement: Informed consent was obtained from all subjects involved in the study.

Data Availability Statement: This study did not report any data.

Acknowledgments: The authors of this paper would like to express their gratitude for the National Natural Science Foundation of China (No. 51904080) and the Science and Technology Project of Guizhou Province (No. 2021 general 352 and No. [2021]3001) during the research.

Conflicts of Interest: The authors declared that they have no conflict of interest.

References

1. Xu, Y.; Bai, J.B.; Chen, Y.; Yue, J. Development of Scientific Mining: A Case Study of Gob-Side Entry Retaining Technology in China. *Adv. Mater. Res.* **2011**, *255–260*, 3786–3792.
2. Zhang, N.; Yuan, L.; Han, C.; Xue, J.; Kan, J. Stability and deformation of surrounding rock in pillarless gob-side entry retaining. *Saf. Sci.* **2012**, *50*, 593–599. [CrossRef]
3. Yang, H.; Cao, S.; Li, Y.; Sun, C.; Guo, P. Soft Roof Failure Mechanism and Supporting Method for Gob-Side Entry Retaining. *Minerals* **2015**, *5*, 707–722. [CrossRef]
4. He, M.; Gao, Y.; Yang, J.; Gong, W. An Innovative Approach for Gob-Side Entry Retaining in Thick Coal Seam Longwall Mining. *Energies* **2017**, *10*, 1785. [CrossRef]
5. Hu, J.; He, M.; Wang, J.; Ma, Z.; Wang, Y.; Zhang, X. Key Parameters of Roof Cutting of Gob-Side Entry Retaining in a Deep Inclined Thick Coal Seam with Hard Roof. *Energies* **2019**, *12*, 934. [CrossRef]
6. Zhen, E.; Gao, Y.; Wang, Y.; Wang, S. Comparative Study on Two Types of Nonpillar Mining Techniques by Roof Cutting and by Filling Artificial Materials. *Adv. Civ. Eng.* **2019**, *2019*, 5267240. [CrossRef]
7. Sun, X.; Liu, Y.; Wang, J.; Li, J.; Sun, S.; Cui, X. Study on Three-Dimensional Stress Field of Gob-Side Entry Retaining by Roof Cutting without Pillar under Near-Group Coal Seam Mining. *Processes* **2019**, *7*, 552. [CrossRef]
8. Yang, X.; Sun, Y.; Wang, E.; Mao, W. Study and Application of Gob-Side Entry Retaining Formed by Roof Cutting and Pressure Relief in Medium-Thickness Coal Seam with Hard Roof. *Geotech. Geol. Eng.* **2020**, *38*, 3709–3723. [CrossRef]
9. Wang, Q.; Jiang, B.; Wang, L.; Liu, B.; Li, S.; Gao, H.; Wang, Y. Control mechanism of roof fracture in no-pillar roadways automatically formed by roof cutting and pressure releasing. *Arab. J. Geosci.* **2020**, *13*, 274. [CrossRef]
10. Xiao-Ming, S.; Gan, L.; Peng, S.; Chengyu, M.; Chengwei, Z.; Qi, L.; Xing, X. Application Research on Gob-Side Entry Retaining Methods in No. 1200 Working Face in Zhongxing Mine. *Geotech. Geol. Eng.* **2018**, *37*, 185–200. [CrossRef]
11. Wang, Y.; Gao, Y.; Wang, E.; He, M.; Yang, J. Roof Deformation Characteristics and Preventive Techniques Using a Novel Non-Pillar Mining Method of Gob-Side Entry Retaining by Roof Cutting. *Energies* **2018**, *11*, 627. [CrossRef]
12. Ma, Z.; Wang, J.; He, M.; Gao, Y.; Hu, J. Key Technologies and Application Test of an Innovative Noncoal Pillar Mining Approach: A Case Study. *Energies* **2018**, *11*, 2853. [CrossRef]
13. Ma, X.; He, M.; Wang, J.; Gao, Y.; Zhu, D.; Liu, Y. Mine Strata Pressure Characteristics and Mechanisms in Gob-Side Entry Retention by Roof Cutting under Medium-Thick Coal Seam and Compound Roof Conditions. *Energies* **2018**, *11*, 2539. [CrossRef]
14. Wang, Q.; He, M.; Yang, J.; Gao, H.; Jiang, B.; Yu, H. Study of a no-pillar mining technique with automatically formed gob-side entry retaining for longwall mining in coal mines. *Int. J. Rock Mech. Min. Sci.* **2018**, *110*, 1–8. [CrossRef]
15. He, M.; Ma, X.; Yu, B. Analysis of Strata Behavior Process Characteristics of Gob-Side Entry Retaining with Roof Cutting and Pressure Releasing Based on Composite Roof Structure. *Shock Vib.* **2019**, *2019*, 2380342. [CrossRef]
16. Shao, L.; Huang, B.; Zhao, X. Secondary Gob-side Entry Retaining Technology with Double Side Roof Cutting and Pressure Relief in Thin Coal Seam. *Int. J. Oil Gas Coal Eng.* **2020**, *8*, 91. [CrossRef]
17. Zhu, Z.; Zhu, C.; Yuan, H. Distribution and Evolution Characteristics of Macroscopic Stress Field in Gob-Side Entry Retaining by Roof Cutting. *Geotech. Geol. Eng.* **2019**, *37*, 2963–2976. [CrossRef]
18. Xingen, M.; Manchao, H.; Jiandong, S.; Jie, H.; Xingyu, Z.; Jiabin, Z. Research on the Design of Roof Cutting Parameters of Non Coal Pillar Gob-side Entry Retaining Mining with Roof Cutting and Pressure Releasing. *Geotech. Geol. Eng.* **2019**, *37*, 1169–1184. [CrossRef]
19. Tian, C.; Wang, A.; Liu, Y.; Jia, T. Study on the Migration Law of Overlying Strata of Gob-Side Entry Retaining Formed by Roof Cutting and Pressure Releasing in the Shallow Seam. *Shock Vib.* **2020**, *2020*, 8821160. [CrossRef]

20. Sun, B.-J.; Hua, X.-Z.; Zhang, Y.; Yin, J.; He, K.; Zhao, C.; Li, Y. Analysis of Roof Deformation Mechanism and Control Measures with Roof Cutting and Pressure Releasing in Gob-Side Entry Retaining. *Shock Vib.* **2021**, *2021*, 6677407. [CrossRef]
21. Zhang, J.; Yang, F.; Wang, R.; Zheng, C.; Sun, N.; Lei, W.; Xu, T.; Zhang, J.; Miao, Z. Study on Optimum Design of Outburst Prevention Measures in Weijiadi Coal Mine. *IOP Conf. Series Earth Environ. Sci.* **2019**, *252*, 052098. [CrossRef]
22. Hudeček, V.; Vaněk, M.; Černý, I. Economic Assessment of Methods of Coal and Gas Outburst Prevention in the Extraction of Coalfaces. *New Trends Prod. Eng.* **2020**, *3*, 379–393. [CrossRef]
23. Hudecek, V.; Urban, P.; Stavtnoha, J. Problems of higher gas emission and coal and gas outbursts in the Czech Republic and other countries of the world. *J. Mines Met. Fuels* **2010**, *58*, 212–216.
24. Xiao, F.; Duan, L.; Ge, Z. Floor fracture law of coal mining face and gas pre-drainage application. *J. Coal* **2010**, *35*, 417–419. [CrossRef]
25. Chi, B.; Zhou, K.; He, M.; Yang, J.; Wang, Q.; Ma, X. Optimization of support parameters of roof cutting roadway in large mining height working face. *Coal Sci. Technol.* **2017**, *45*, 128–133. [CrossRef]
26. He, M.; Ma, X.; Niu, F.; Wang, J.; Liu, Y. Adaptability research and application of fast coal pillar-free self-forming roadway with composite roof in medium-thick coal seam. *J. Rock Mech. Eng.* **2018**, *37*, 2641–2654. [CrossRef]

Article

Study on the Improved Method of Wedge Cutting Blasting with Center Holes Detonated Subsequently

Bing Cheng [1,2], Haibo Wang [1], Qi Zong [1,*], Mengxiang Wang [1], Pengfei Gao [1] and Nao Lv [1]

[1] School of Civil Engineering and Architecture, Anhui University of Science and Technology, Huainan 232001, China; dzcx1995@163.com (B.C.); wanghb_aust@163.com (H.W.); mxwang@aust.edu.cn (M.W.); pfgao@aust.edu.cn (P.G.); lvnao1990@163.com (N.L.)
[2] School of Chemical Engineering, Anhui University of Science and Technology, Huainan 232001, China
* Correspondence: qzong@aust.edu.cn

Abstract: To acquire a satisfying cutting effect during medium-length hole blasting driving of rock tunnels, an improved wedge cutting blasting method with supplementary blasting of the center holes was proposed. Initially, the cavity forming mechanism of the improved cutting method was analyzed theoretically. The results suggested that cutting hole blasting could realize the ejection of rock within the range from free face to critical cutting depth, and hence reduce the restraining force of the center hole blasting, and the supplementary blasting of the center holes could further accomplish the expulsion of the residuary rock. Subsequently, simulation of the improved cutting method was implemented to exhibit the stress wave evolution and reveal the stress field distribution. The simulation results indicated that cutting hole blasting could cause the preliminary failure of the residuary rock, and center hole blasting could strengthen the stress field intensity in 1.8–2.5 m in order to aggravate the destruction of the residuary rock. Hence, the residuary rock could be broken into small fragments that were easy to expel out. Finally, a field application experiment was conducted in a coal mine rock tunnel. Using the improved wedge cutting method instead of the conventional wedge cutting method, the full-face blasting driving efficiency was obviously enhanced and the overall blasting driving expense was significantly reduced, which forcefully confirmed the engineering usefulness of the improved wedge cutting method in the medium-length hole blasting driving of rock tunnels.

Keywords: wedge cutting blasting method; center holes; supplementary blasting; cavity forming mechanism; numerical simulation; field application experiment

Citation: Cheng, B.; Wang, H.; Zong, Q.; Wang, M.; Gao, P.; Lv, N. Study on the Improved Method of Wedge Cutting Blasting with Center Holes Detonated Subsequently. *Energies* **2022**, *15*, 4282. https://doi.org/10.3390/en15124282

Academic Editor: Vamegh Rasouli

Received: 3 May 2022
Accepted: 9 June 2022
Published: 10 June 2022

1. Introduction

To date, the drilling and blasting method still occupies a dominant position in coal mine underground tunnel driving because of its salient features of convenient operation, strong adaptability, and economical investment, despite the continuous development of mechanical excavators [1–4]. During the blasting driving of rock tunnels, the main function of cutting blasting is creating an extra free face and rock bulking space for subsequent blasting procedures. Therefore, the selection and design of the cutting blasting method is the key to determining the overall blasting driving efficiency of coal mine rock tunnels [5–8]. According to the hole layout, the common cutting blasting methods are broadly categorized into parallel-hole and inclined-hole cutting blasting methods, and wedge cutting blasting is the most frequently employed among various inclined-hole cutting methods. Engineering practices prove that the prominent defects of parallel-hole cutting blasting are many cutting holes, high explosive consumption, and a small cavity size [9,10]. Conversely, the conspicuous advantages of the wedge cutting method are few cutting holes, low explosive consumption, and large cavity size [11,12]. Therefore, the wedge cutting method has been the most extensively used cutting blasting method in the full-face blasting driving of coal mine rock tunnels.

Recently, numerous experts and scholars have executed a great deal of research on wedge cutting blasting and have obtained fruitful academic achievements. For instance, based on full consideration of the failure modes of different cavity surfaces, Dai and Du [13] proposed a computing method of wedge cutting parameters, and further discussed the influences of rock hardness and explosive performance on the design of cutting parameters. According to the difference between blasting stress wave and detonation gas, Wang et al. [14] divided the cavity forming process into the rock damage stage and rock throwing stage, and investigated the cavity forming mechanism by using a simplified mechanical model. To investigate the influence of cutting hole angle symmetry on the wedge cutting results, Liang et al. [15] performed similar simulation experiments to analyze the cutting effect from the aspects of cavity size and blasting hole utilization, and concluded that the symmetrical arrangement of cutting hole angles would be conducive to the full utilization of blasting energy. Through laboratory model tests, Pu et al. [16] obtained the factors affecting the wedge cutting effect, and further computed and sequenced the gray relational grade of these factors. Then, they found that the cutting hole angle was the primary factor affecting the cutting blasting results, with a high gray relational grade of 2.39. Yang et al. [17] concretely investigated the effect of the cutting hole angle on the cutting cavity size and rock fragment size based on the model test data, and obtained the optimal cutting hole angle that could be used to guide the design of the cutting parameters. Because of the rapidity and complexity of the blasting process, the experimental research has some deficiencies such as a huge expense, long period, and low accuracy, and the theoretical calculation can only solve some simplified problems [18,19]. With the fast progress and development of computer numerical technology, the numerical technique has become a practicable means to study the cutting blasting mechanisms of rock tunnels. For example, using the nonlinear analysis platform LS-DYNA, Xie et al. [20] simulated the damage development process of cutting blasting under different crustal stress, and further optimized the design of cutting blasting in deep rock. Hu et al. [21] obtained the cutting cavity extension by introducing the rock damage criterion into the LS-DYNA, and thus demonstrated the effectiveness of wedge cutting blasting in the blasting driving of mine rock tunnels. Cheng et al. [22] also simulated the stress wave propagation of wedge cutting blasting under different explosive diameters, and further analyzed the influence of the explosive diameter on stress distribution features and cutting cavity formation. According to the above review, the current research mainly focuses on the blasting mechanism, charge parameters, and hole parameters of the conventional wedge cutting method.

Currently, the increasing demand for coal energy resources has put forward higher requirements for the driving speed of coal mine rock tunnels. However, the blasting diving efficiency of coal mine rock tunnels in China is always at a low speed of 75 m/month, which cannot maintain the balance between mining and driving and seriously restricts the productivity of underground coal mines. Accordingly, traditional shallow hole blasting with a hole depth of less than 1.8 m has become inopportune, and the medium-length hole blasting with a hole depth of more than 2.0 m has turned into an inevitable and judicious choice [23,24]. However, the restraining force from the surrounding rock will considerably increase with the hole depth. Because of the low power of the permitted explosives for coal mine, the powerful restraining force will inevitably deteriorate the cutting blasting effect; that is, only a small extra free face and insufficient rock bulking space can be created after cutting blasting [25], which will lead to small cyclical footage and low blasting hole utilization in the subsequent blasting driving procedures.

Consequently, in the present paper, an improved wedge cutting blasting method with supplementary blasting of the center holes was proposed to achieve a good cutting blasting effect in the medium-length hole blasting diving of coal mine rock tunnels. Initially, the cavity forming mechanism for the improved cutting method was analyzed theoretically. Subsequently, a numerical analysis was performed to display the evolution process of the stress wave and to reveal the distribution characteristics of the stress field. Ultimately, field application experiment was conducted in a coal mine rock tunnel to examine the engineering usefulness of the improved wedge cutting method.

2. Theoretical Analysis of Cavity Forming Mechanism

2.1. Analysis of Restraining Force

Existing research results have indicated that the restraining force from the surrounding rock would increase with the hole depth (or cutting depth) during single-face blasting such as cutting blasting. However, because of the limitations of research methods and test means, it is extremely difficult to evaluate the restraining force quantitatively and accurately. Previous studies [26,27] have pointed out that the increasing rate of the restraining force could increase with the hole depth during single-hole charge blasting. Therefore, with reference to their research achievements, the conventional wedge cutting blasting was equivalent to single-hole charge blasting, and the variation curve of the restraining force with the cutting depth was plotted and is shown in Figure 1. Here, the curve $A_0A_1A_2A_3$ corresponds to the change law of the restraining force for conventional wedge cutting blasting, and H_2 and F_2 represent the critical cutting depth and the critical restraining force, respectively. The interpretation of H_2 and F_2 is that if the actual cutting depth is less than or equal to the critical cutting depth H_2, the cavity forming force can always overcome the actual restraining force lower than or equal to the critical restraining force F_2, and then the rock in the cutting cavity can be completely ejected.

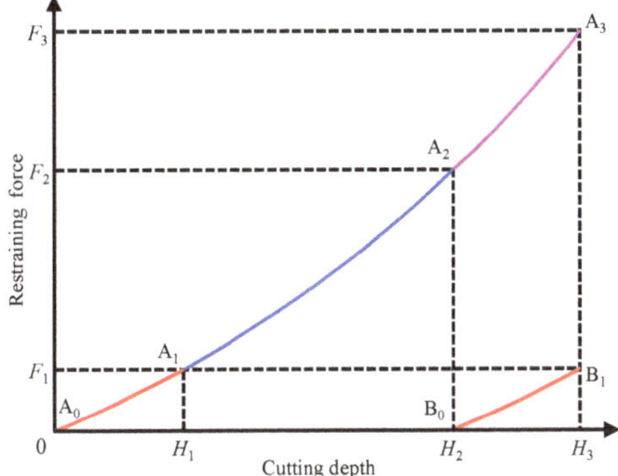

Figure 1. Variation curve of the restraining force with cutting depth.

As indicated in Figure 1, during conventional wedge cutting blasting, the restraining force would gradually increase with the cutting depth along the curve $A_0A_1A_2A_3$. Because the design cutting depth H_3 was greater than the critical cutting depth H_2, the restraining force was always higher than the critical restraining force F_2 in the range of H_2–H_3. Therefore, the rock in the range of 0–H_2 could be thrown out, while the rock within the range of H_2–H_3 could be stuck in the cutting cavity. Eventually, only a small extra free face and insufficient rock bulking space could be provided for the subsequent blasting procedures, which would exert a detrimental effect on the overall driving efficiency.

Subsequently, it was assumed that supplementary blasting was performed at the bottom of the cutting cavity. As the rock in the range of 0–H_2 had been fully ejected, a new free face could be formed at the cutting depth of H_2, which would lead to a redistribution of the restraining force in the range of H_2–H_3. The restraining force would return to zero at a cutting depth of H_2 and would continue to increase along the curve B_0B_1 in the range of H_2–H_3, and curve B_0B_1 was consistent with the curve A_0A_1, as illustrated in Figure 1. The residuary cutting depth in the range of H_2–H_3 was equal to H_1, and H_1 was less than the critical cutting depth H_2, which means that the maximum restraining force during supplementary blasting was equal to F_1, and F_1 was lower than the critical restraining force

F_2. Therefore, the residuary rock within the range of H_2–H_3 could be completely ejected after the supplementary blasting. Eventually, a large extra free face and sufficient rock bulking space could be provided for the subsequent blasting procedures, which would have a beneficial effect on the overall blasting tunneling efficiency.

2.2. Improved Cutting Method and Cavity Forming Process

Based on the analysis of the restraining force, an improved wedge cutting blasting method with supplementary blasting of the center holes could be developed. As illustrated in Figure 2, similar to the conventional wedge cutting method, cutting holes were symmetrically fixed on the left and right sides of the heading face, and the angle between the cutting holes and the heading face was within the range of 75–85°. Most of the explosives required for cutting blasting were filled in the cutting holes and were detonated by the first-stage detonators. Different from the conventional wedge cutting method, the center holes perpendicular to the heading face were arranged in the middle area of the heading face, and its depth was the same as the vertical depth of the cutting holes. A small amount of explosives were loaded at the bottom of the center holes and detonated by the second-stage detonators. The forming procedure of the cutting cavity is shown in Figure 3.

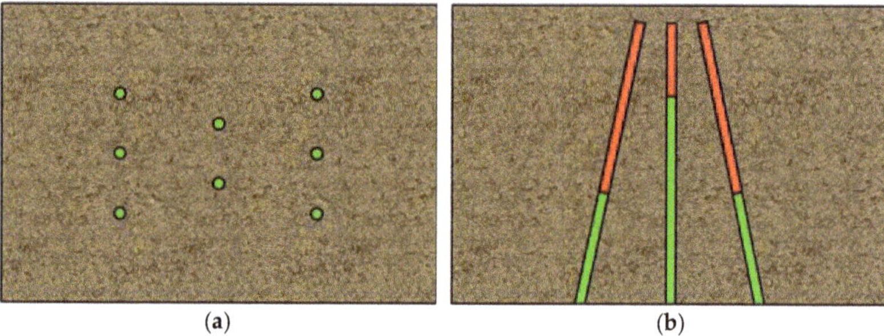

Figure 2. Hole layout of the improved wedge cutting method: (**a**) front view and (**b**) top view.

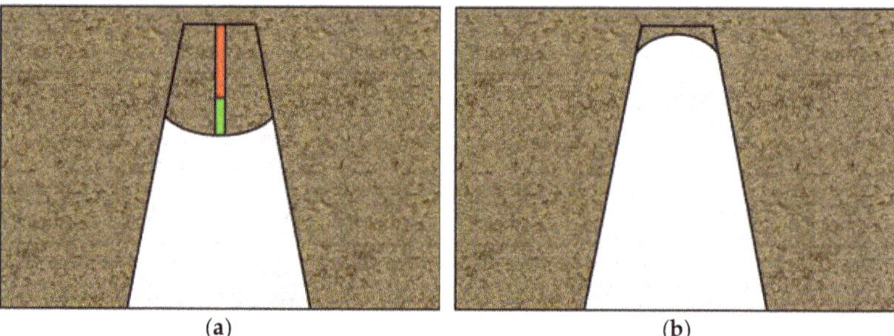

Figure 3. Cavity forming procedure: (**a**) cutting hole blasting and (**b**) center hole blasting.

After early blasting of the cutting holes, the rock in the range from the free face to the critical cutting depth formed rock fragments under the intensive blasting load, and then these rock fragments were ejected out from the cutting cavity under the push action of the detonation gas. During the medium-length hole blasting, the rock at the bottom of the cutting cavity could not be thrown out under the powerful restraining force from the surrounding rock, but it would be preliminarily destroyed by the blasting stress wave from the cutting holes. Furthermore, a new free face could be created to reduce the restraining force of the center hole blasting. The cavity created by the cutting hole blasting is shown in Figure 3a. After supplementary blasting of the center holes, because of the preliminary destruction and new free face caused by the early blasting of the cutting holes, the residuary rock easily formed rock fragments under the blasting stress wave of the center holes, and then these rock fragments could be ejected almost completely under the push action of the detonation gas. The cavity generated by the center hole blasting is presented in Figure 3b.

Compared with the conventional wedge cutting method, the process of cutting blasting was changed from a one-step blasting pattern to a two-step blasting pattern. The early blasting of the cutting holes and the supplementary blasting of the center holes were viewed as the first step blasting and second step blasting, respectively. For the first step, early blasting of the cutting holes could accomplish most of the cavity forming task and could reduce the restraining force of the center hole blasting. For the second step, supplementary blasting of the center holes could complete the rest of the cavity forming task to achieve a satisfying cutting effect.

3. Numerical Simulation and Analysis

3.1. Numerical Model Descriptions

A three-dimensional numerical model with dimensions of 4000 mm × 3000 mm × 3000 mm (length × width × height) was erected using an eight-node solid element SOLID164 in ANSYS/LS-DYNA, as shown in Figure 4. Two center holes perpendicular to the free face were arranged on the middle area of the heading face, and the space, depth, and charge length of the center holes were 400, 2500, and 660 mm, respectively. Six wedge cutting holes were symmetrically arranged on the left and right sides of the heading face; the angle between cutting holes and heading face was 77.5°; and the space, top space, bottom space, vertical depth, and charge length of the cutting holes were 400, 1500, 400, 2500, and 1320 mm, respectively. In addition, 40 mm diameter holes and 35 mm diameter explosive sticks were adopted in all of the blasting holes. The geometrical parameters of the numerical model were taken from the blasting parameters of the later field application experiments.

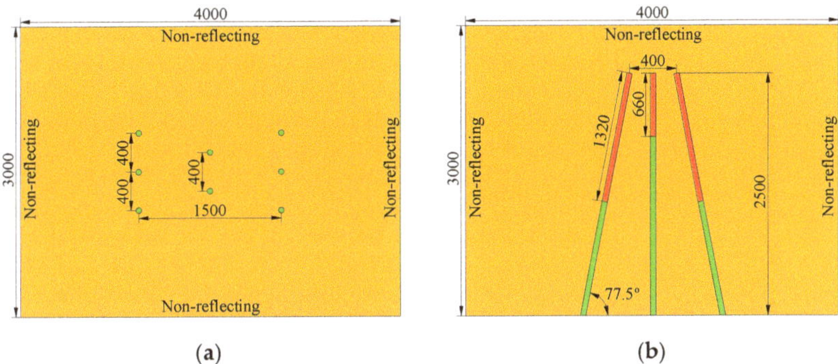

Figure 4. Numerical calculation model (unit: mm): (**a**) front view and (**b**) vertical view.

In this simulation, a reverse initiation pattern was performed at all of the blasting holes, and the cutting holes and center holes were initiated at 0 and 1000 μs, respectively. Through several tentative simulations, the time interval of 1000 μs ensured that the peak stresses induced by the cutting holes and the center holes were independent of each other. Setting the initiation points and initiation times was completed using the keyword *INITIAL_DETONATION. To effectively avoid the negative influence of stress wave reflection from the artificial boundary on the precision of the calculation results, non-reflecting boundaries were exerted into the rock model surfaces, except for the heading face [28–30]. In addition, the convergence tests of the mesh size were implemented before the formal simulation. During the tests, the number of elements continued to increase until the peak stress difference at the same stress monitoring point between two consecutive tests was less than 5% [31–33]. The meshing of the rock model is shown in Figure 5.

Figure 5. Meshing of the rock model.

3.2. Algorithms and Material Models

In ANSYS/LSDYNA, the algorithms used to describe the continuous media mainly include Euler, Lagrange, and ALE (Arbitrary−Lagrange−Euler) [34,35], and the latter two algorithms were used in this study. The Lagrange algorithm is more effective for the description of solid media [36,37]; thus, it was adopted to simulate the rock and stemming. The ALE algorithm is well suited to describing the fluid media [38]; thus, it was used to simulate the explosive and air. Furthermore, the transfer problem of mechanical information between the fluid and solid media was solved by defining the fluid−structure interaction algorithm.

It is well known that the determination of the blasting load is very important for simulating the blasting process. To achieve high accuracy, the *MAT_HIGH_EXPLOSIVE_BURN material model was adopted for the explosive, and the detonation pressure resulting from the release of explosive energy was described by the JWL state equation [39–41]. As for the other materials included in this study, the *MAT_NULL material model and linear polynomial state equation were selected for the air [42]. The dynamic response of brittle materials subjected to high energy explosive loads is a complex rate-dependent process [43]; thus, the *MAT_PLASTIC_KINEMATIC material model considering the rate dependence was adopted to simulate the rock [44–46]. Moreover, the nonlinear deformation process of stemming was described by the *MAT_SOIL_AND_FORM material model. In this study, except for the mechanical parameters of the rock that were tested and obtained through laboratory tests, the main parameters of the other three materials were selected from the existing literature [47–49]. The main constitutive parameters of the above four materials are given in Tables 1–4.

Table 1. Main constitutive parameters of the explosive.

ρ_e (kg·m^{-3})	D_e (m·s^{-1})	A_e (GPa)	B_e (GPa)	R_1	R_2	ω_e	E_e (GPa)
1100	3200	214	0.182	4.15	0.95	0.15	4.20

ρ_e is the density. D_e is the detonation velocity. A_e, B_e, R_1, R_2 and ω_e are the material constants. E_e is the specific energy.

Table 2. Main constitutive parameters of air.

ρ_a (kg·m^{-3})	C_0	C_1	C_2	C_3	C_4	C_5	C_6
1.25	0.00	0.00	0.00	0.00	0.40	0.40	0.00

ρ_a is the density. C_0–C_6 are the coefficient of state equation.

Table 3. Main constitutive parameters of rock.

ρ_r (kg·m^{-3})	E (GPa)	μ	σ_{sc} (MPa)	σ_{st} (MPa)	C	P	λ
2680	28.1	0.24	78.1	6.4	2.63	3.96	1.0

ρ_r is the density. E is the elasticity modulus. μ is Poisson ratio; σ_{sc} is the compressive strength; σ_{st} is the tensile strength. C and P are the stain rate parameters. λ is the hardening parameter.

Table 4. Main constitutive parameters of stemming.

ρ_s (kg·m^{-3})	G (GPa)	K (GPa)	Y_0	Y_1	Y_2	P_c (GPa)
1800	0.016	1.328	0.0033	1.31 × 10^{-7}	0.1232	0.0

ρ_s is the density. G is the shear modulus. K is the bulk modulus. Y_0–Y_2 are the yield parameters. P_c is the cut-off pressure.

3.3. Simulation Results and Discussions

3.3.1. Dynamic Evolution of Stress Wave

The keyword file containing the above model and material information was submitted to the solver LS-DYNA. Then, the rock model was partitioned along the horizontal symmetric plane by the post-processing software LS-PREPOST to clearly present the dynamic evolution process of the stress wave. The dynamic developments of the stress wave after the charge detonation in cutting holes and center holes are presented in Figures 6 and 7, respectively.

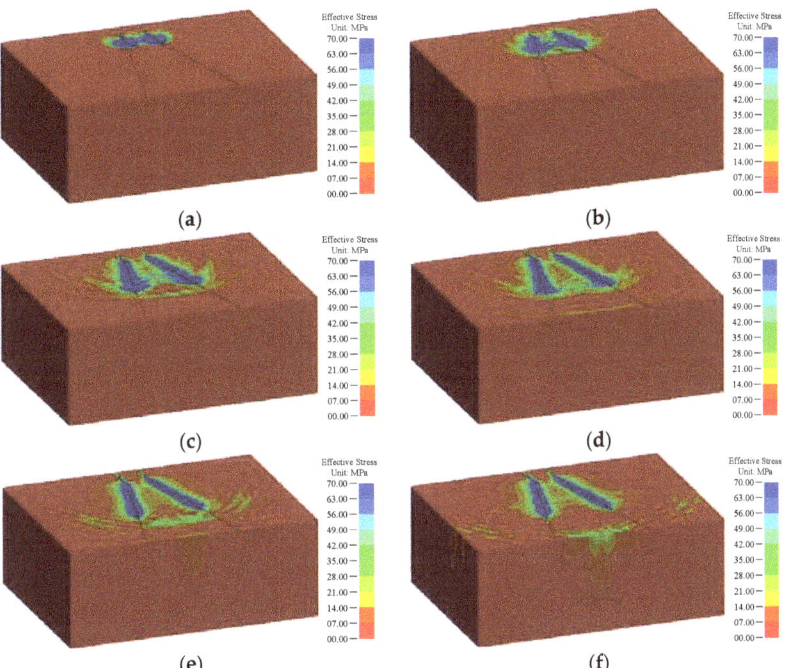

Figure 6. Dynamic evolution process of stress wave after charge detonation in cutting holes: (**a**) 080 μs, (**b**) 160 μs, (**c**) 320 μs, (**d**) 470 μs, (**e**) 600 μs, and (**f**) 800 μs.

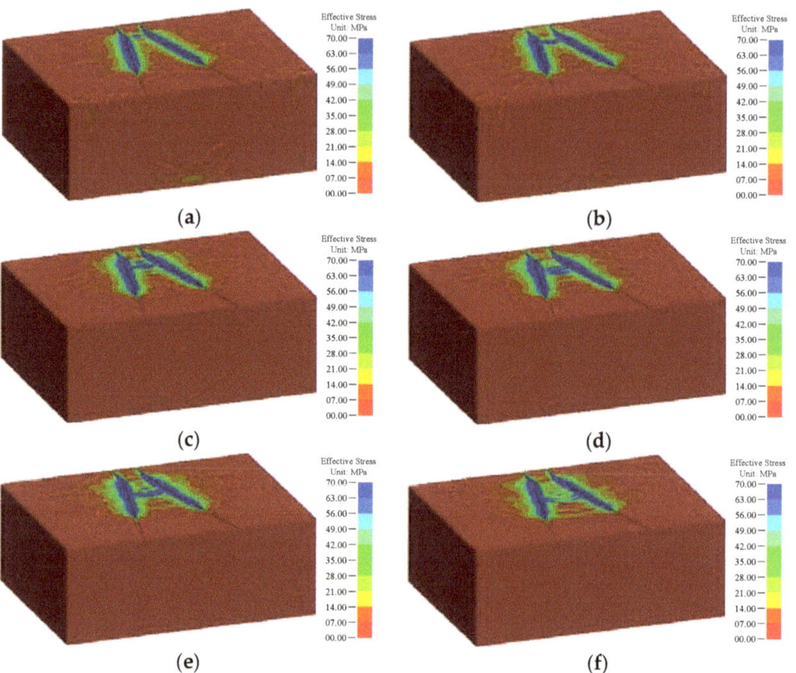

Figure 7. Dynamic evolution process of stress wave after charge detonation in center holes: (**a**) 1050 μs, (**b**) 1100 μs, (**c**) 1150 μs, (**d**) 1200 μs, (**e**) 1250 μs, and (**f**) 1300 μs.

As illustrated in Figure 6, when the explosive was detonated in the cutting holes, a huge amount of chemical energy was released in order to induce the stress wave. With the rapid detonation reaction of the explosive in the cutting holes, the induced blasting stress wave propagated towards the heading face. Subsequently, the stress waves derived from the cutting holes were superimposed on each other in the middle zone of the blasting model, which would promote the destruction of rock in the cutting cavity. At 320 μs, the detonation transmission of the explosive in each of the cutting holes was completed. However, the blasting stress wave continued to propagate to the heading face and was reflected as the tensile stress wave after reaching the heading face. Because of the low tensile strength of various brittle materials such as rock, the tensile stress wave would cause great tensile damage to the rock close to the heading face [50]. After that, the forward-propagation compression wave met the backward-propagation tensile wave to form a complex stress environment, which would increase the damage level of the rock in the cutting cavity. The above stress wave evolution after the charge detonation of cutting holes was basically in accordance with the relevant numerical simulation results of the conventional wedge cutting method [22].

As shown in Figure 7, when the explosive located in the center holes was detonated at 1000 μs, a large amount of chemical energy was also released to generate the stress wave. With the rapid detonation reaction of the explosive in the center holes, the generated blasting stress wave propagated towards the heading face along the center hole in a circular wave front, and was superimposed with the stress wave remaining from the cutting holes to aggravate the destruction of the rock at the bottom of the cutting cavity. At 1150 μs, the detonation transmission of the explosive in each center hole was completed, and the blasting stress wave still continued to propagate towards the heading face.

From Figures 6 and 7, the rock in the cutting cavity could be subjected to the blasting stress waves from the cutting holes and the center holes successively. Owing to the deficiency of the numerical simulation, the throwing process of the rock could not be obtained. However, in fact, after the cutting hole blasting, the rock in the range from the free face to critical cutting depth could be thrown out and would not be subjected to the stress wave of the center holes. In contrast, because of the powerful restraining force, the residuary rock in the range from the critical cutting depth to the design cutting depth could not be expelled in practice. Therefore, only the residuary rock would be subjected to the stress waves generated by the cutting holes and the center holes successively.

3.3.2. Distribution Features of Stress Field

In order to intuitively study the specific influence of center hole blasting on the distribution features of the stress field in the cutting cavity, a monitoring line perpendicular to the heading face was set in the middle of the rock numerical model. A monitoring point was fixed every 0.1 m on the monitoring line, and there were 26 stress monitoring points in the design cutting depth range of 0–2.5 m. Then, the stress−time curves of these stress monitoring points were output, and that of the typical stress monitoring points at cutting depths of 0.7, 1.3, 1.9, and 2.4 m are presented in Figure 8.

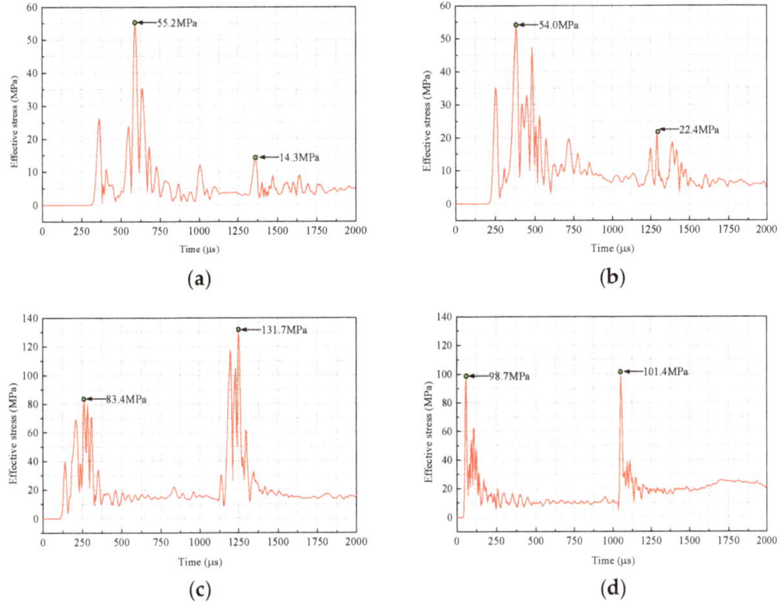

Figure 8. Stress−time curves of typical stress monitoring points: (**a**) 0.7 m, (**b**) 1.3 m, (**c**) 1.9 m, and (**d**) 2.5 m.

As shown in Figure 8, under the blasting stress waves caused by the cutting holes and the center holes, two peak stresses appeared successively on the stress−time curve of each stress monitoring point. Nevertheless, after early blasting of the cutting holes, the rock in the range from the free face to critical cutting depth could be thrown out and would not be subjected to the stress wave of the center holes, and the rock in the range from the critical cutting depth to the design cutting depth could not be expelled and would be subjected to two stress waves from the cutting holes and the center holes. Therefore, the second peak stress did not exist at the stress monitoring points in the range from the free face to critical cutting depth, and the first and second peak stresses actually existed at the stress monitoring points in the range from the critical cutting depth to the design cutting depth.

According to the subsequent field application experiment, an average cyclical footage of 1.79 m could be acquired using the conventional wedge cutting method, and the critical cutting depth was conservatively estimated to be about 1.8 m. Accordingly, after early blasting of the cutting holes, the rock in the range of 0–1.8 m was ejected out and would not be affected by the stress wave of the center holes, while the rock in the range of 1.8–2.5 m was confined in the cutting cavity and would be subjected to two stress waves from the cutting holes and the center holes. This also means that the second peak stress did not exist in 0–1.8 m, while both the first and second peak stresses existed within 1.8–2.5 m.

Both the first and second peak stresses were extracted to draw the variation curve of the peak stress with the cutting depth, as shown in Figure 9. It could be observed that the first peak stress substantially increased with the cutting depth and reached a maximum value of 102.9 MPa at 2.2 m. The second peak stress increased initially and decreased afterwards, and reached a maximum value of 149.0 MPa at 2.0 m. Nevertheless, because the rock within 0–1.8 m was not subjected to the stress wave of the center holes, only the range of 1.8–2.5 m was taken for further contrastive analysis.

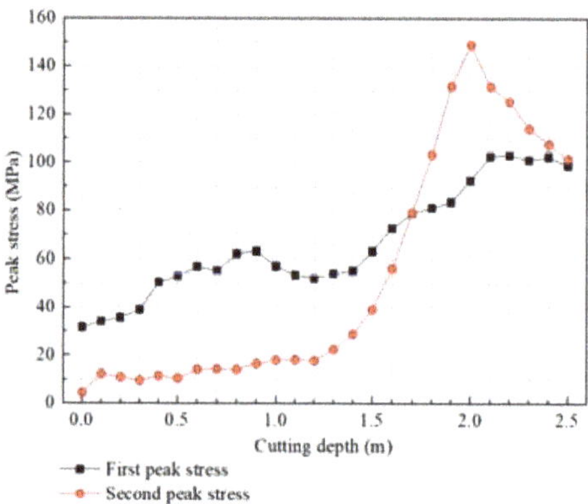

Figure 9. Variation curve of the peak stress with the cutting depth.

From Figure 9, in the range of 1.8–2.5 m, the first peak stresses were greater than the compressive strength of the rock. Although the residuary rock was stuck at the bottom of the cutting cavity as a result of the powerful restraining force after the cutting hole blasting, it would be preliminarily destroyed by the stress wave produced by the cutting holes, thus reducing the difficulty of the center hole blasting. Also in this range, the second peak stress was always greater than the first peak stress. The maximum values of the first peak stresses and second peak stresses were 102.9 and 149.0 MPa, respectively, and the latter was 1.45 times that of the former. The average values of the first peak stresses and the second peak stresses were 95.5 and 119.7 MPa, respectively, and the latter was 1.25 times of the former. It was believed that the stress field intensity in the range of 1.8–2.5 m could be obviously increased after center hole blasting, which was conductive to aggravating the destruction of the residuary rock. Consequently, under the successive action of two stress waves, the residuary rock could be broken into small fragments that were easy to throw out, thus forming a cutting cavity in line with the design size.

4. Field Application Experiment

4.1. Project Overview

To investigate the engineering effectiveness of the improved wedge cutting method with supplementary blasting of the center holes, the conventional wedge cutting method and improved wedge cutting method were applied into blasting driving in the same coal mine rock tunnel. The same rock tunnel could minimize the harmful influence of geological conditions on the reliability of test results as much as possible. The rock tunnel was located in the Pansan Coal mine, Huainan, China, which adopted a straight wall arch section, with a height of 3.65 m, a width of 4.7 m, and an area of 14.79 m^2. The rock strata that the tunnel passed through were sandstone, with a compressive strength of 78.1 MPa and a tensile strength of 6.4 MPa.

4.2. Design of Blasting Scheme

In the field application experiment, two types of drill bits with diameters of 40 mm and 32 mm were selected for the drilling operations. The permitted water-gel explosive for coal mine, including two specifications, 35 mm × 330 mm × 350 g (diameter × length × weight) and 29 mm × 430 mm × 310 g, were used as the blasting material. The center holes and cutting holes employed Φ 40 mm (Φ represents the diameter) blasting holes and Φ 35 mm explosive sticks, and Φ 32 mm blasting holes and Φ 29 mm explosive sticks were used for the other holes. In addition, the coal mine permitted electric detonator with five delay stages was adopted as the detonation device.

Prior to the research and development of the improved wedge cutting method, the conventional wedge cutting method had been adopted in blasting site for a long time. Therefore, the blasting design using the conventional wedge cutting method was regarded as the original scheme, and that using the improved wedge cutting method was considered as the optimized scheme. Then, the blasting parameters of the original and optimized schemes are provided in Tables 5 and 6, and the blasting hole layouts of the original and optimized schemes are indicated in Figures 10 and 11.

Table 5. Blasting parameters of the original scheme.

Hole Name	Hole No.	Hole Amount	Charge Weight (kg)		Detonating Sequence
			Per Hole	Sub-Total	
Cutting hole	1–6	6	1.40	8.40	1
Stoping hole	7–19	13	0.93	12.09	2
Stoping hole	20–33	14	0.93	13.02	3
Contour hole	34–56	23	0.62	14.26	4
Floor hole	57–65	9	0.93	8.37	4
Total	N/A	65	N/A	56.14	N/A

Table 6. Blasting parameters of the optimized scheme.

Hole Name	Hole No.	Hole Amount	Charge Weight (kg)		Detonating Sequence
			Per Hole	Sub-Total	
Center hole	1–2	2	0.70	1.40	2
Cutting hole	3–8	6	1.40	8.40	1
Stoping hole	9–21	13	0.93	12.09	3
Stoping hole	22–35	14	0.93	13.02	4
Contour hole	36–58	23	0.62	14.26	5
Floor hole	59–67	9	0.93	8.37	5
Total	N/A	67	N/A	57.54	N/A

4.3. Experimental Results

During the field experiment, because of the limitation of measuring means, it was very difficult to detect and evaluate the cutting effect separately without affecting the whole engineering schedule. Accordingly, the cutting blasting effect could usually be indirectly reflected by the final blasting results, such as the cyclical footage, blasting hole utilization, specific explosive consumption, and specific detonator consumption. The statistical data of the final blasting results under the two blasting schemes are shown in Tables 7 and 8.

As shown in Table 7, when the conventional wedge cutting method was employed for full-face control blasting, the average values of the cyclical footage, blasting hole utilization, specific explosive consumption, and specific detonator consumption were 1.79 m, 77.8%, 2.12 kg·m^{-3}, and 2.46 PCS·m^{-3}, respectively. As indicated in Table 8, when the improved wedge cutting blasting method was adopted for full-face blasting driving, the average values of cyclical footage, blasting hole utilization, specific explosive consumption, and specific detonator consumption were 2.19 m, 95.2%, 1.78 kg·m^{-3}, and 2.07 PCS·m^{-3}, respectively. Compared with the conventional wedge cutting method, the improved wedge cutting method increased the average cyclical footage and average blasting hole utilization by 0.40 m and 17.4%, and reduced the average specific explosive consumption and average specific detonator consumption by 0.34 kg·m^{-3} and 0.39 PCS·m^{-3}, respectively. The improvement in overall blasting driving efficiency and the reduction in overall blasting driving expense indirectly indicated that the improved wedge cutting method could achieve a satisfying cutting effect and had well engineering applicability in the medium-length hole blasting driving of rock tunnels.

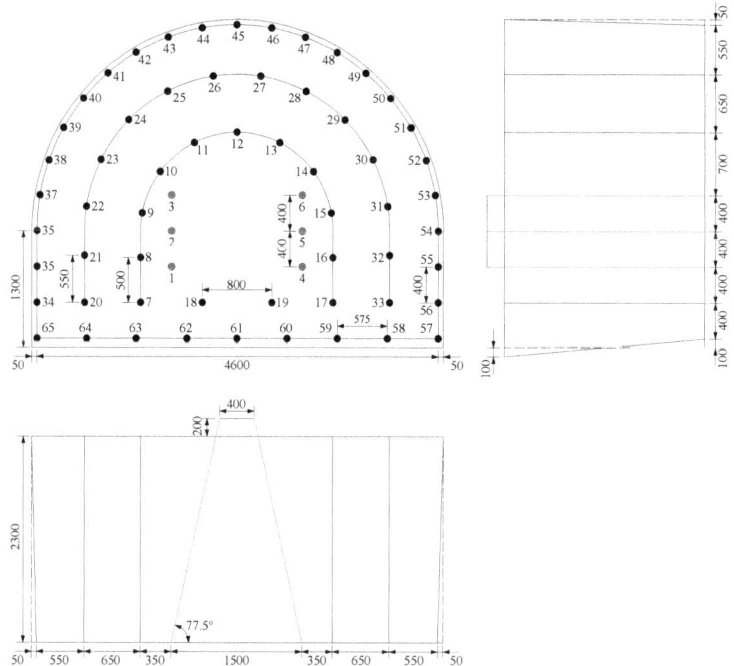

Figure 10. Blasting hole layout of the original scheme (unit: mm).

Table 7. Statistical data of the final blasting results for the original scheme.

No.	L (m)	U (%)	Q (kg·m^{-3})	R (PCS·m^{-3})
1	1.75	76.0	2.17	2.51
2	1.80	78.3	2.11	2.44
3	1.80	78.3	2.11	2.44
4	1.80	78.3	2.11	2.44
5	1.75	76.0	2.17	2.51
6	1.80	78.3	2.11	2.44
7	1.75	76.0	2.17	2.51
8	1.85	80.4	2.05	2.38
9	1.80	78.3	2.11	2.44
10	1.80	78.3	2.11	2.44
Average	1.79	77.8	2.12	2.46

L is the cyclical footage. U is the blasting hole utilization. Q is the specific explosive consumption. R is the specific detonator consumption.

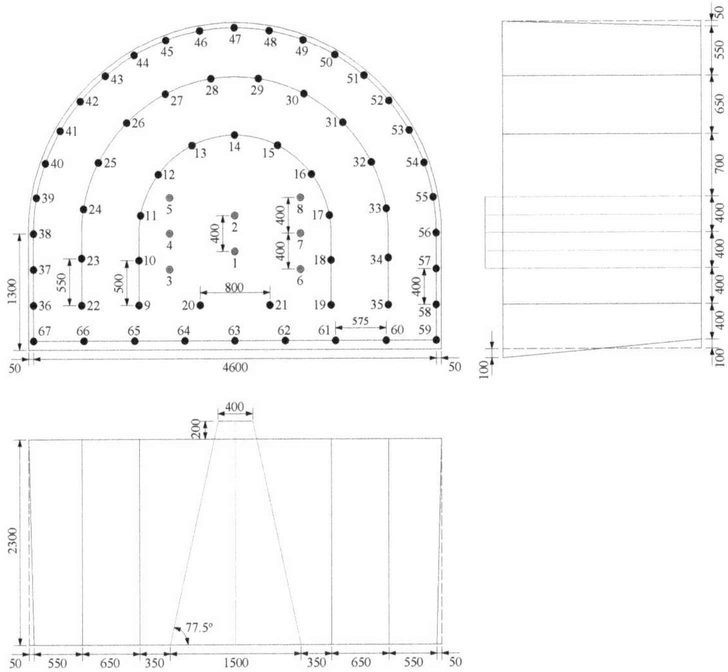

Figure 11. Blasting hole layout of the optimized scheme (unit: mm).

Table 8. Statistical data of the final blasting results for the optimized scheme.

No.	*L* (m)	*U* (%)	*Q* (kg·m^{-3})	*R* (PCS·m^{-3})
1	2.15	93.5	1.81	2.11
2	2.20	95.6	1.76	2.06
3	2.20	95.6	1.76	2.06
4	2.15	93.5	1.81	2.11
5	2.25	97.8	1.73	2.01
6	2.20	95.6	1.77	2.06
7	2.20	95.6	1.77	2.06
8	2.15	93.5	1.81	2.11
9	2.20	95.6	1.77	2.06
10	2.20	95.6	1.77	2.06
Average	2.19	95.2	1.78	2.07

L is the cyclical footage. *U* is the blasting hole utilization. *Q* is the specific explosive consumption. *R* is the specific detonator consumption.

5. Conclusions

(1) The theoretical analysis indicated that the rock in the range from the free face to critical cutting depth could be ejected out after the cutting hole blasting, which reduced the restraining force of the center hole blasting, and then the center hole blasting could further complete the expulsion of residuary rock to achieve a satisfying cutting effect.

(2) The numerical simulation visually presented the dynamic evolution of the stress wave for the improved wedge cutting method. In 1.8–2.5 m, the residuary rock suffered from the stress waves generated by the cutting holes and the center holes successively. The stress field with a low intensity induced by the cutting hole blasting could cause the preliminary failure of the residuary rock, and the stress field with a high intensity generated by the center hole blasting could aggravate the destruction of the residuary

rock, which was conductive to breaking the residuary rock into small fragments that were easily thrown out, hence forming a cutting cavity meeting the design size.

(3) In comparison with the conventional wedge cutting method, the improved wedge cutting method attained an increase in the average cyclical footage of 0.40 m and the blasting hole utilization of 17.4%, and a decrease in the average specific explosive consumption of 0.34 kg·m^{-3} and the average specific detonator consumption of 0.39 PCS·m^{-3}, which convincingly validated the engineering applicability of the improved wedge cutting method in the rock tunnel blasting driving.

Author Contributions: Conceptualization, B.C. and Q.Z.; methodology, B.C.; validation, B.C., H.W. and Q.Z.; formal analysis, B.C. and Q.Z.; investigation, B.C., H.W., Q.Z., M.W., P.G. and N.L.; data curation, Q.Z.; writing—original draft preparation, B.C.; writing—review and editing, B.C. All authors have read and agreed to the published version of the manuscript.

Funding: This study was funded by the National Natural Science Foundation of China (nos. 51374012 and 51404010).

Institutional Review Board Statement: Not applicable.

Informed Consent Statement: Not applicable.

Data Availability Statement: The data used to support the findings of this study are included within the article.

Conflicts of Interest: The authors declare no conflict of interest.

References

1. Zare, S.; Bruland, A.; Rostami, J. Evaluating D&B and TBM tunnelling using NTNU prediction models. *Tunn. Undergr. Space Technol.* **2016**, *59*, 55–64.
2. Costamagna, E.; Oggeri, C.; Segarra, P.; Castedo, R.; Navarro, J. Assessment of contour profile quality in D&B tunnelling. *Tunn. Undergr. Space Technol.* **2018**, *75*, 67–80.
3. Ji, L.; Zhou, C.B.; Lu, S.W.; Jiang, N.; Li, H.B. Modeling study of cumulative damage effects and safety criterion of surrounding rock under multiple full-face blasting of a large cross-section tunnel. *Int. J. Rock Mech. Min. Sci.* **2021**, *147*, 104882. [CrossRef]
4. Chandrakar, S.; Paul, P.S.; Sawmliana, C. Influence of void ratio on "Blast Pull" for different confinement factors of development headings in underground metalliferous mines. *Tunn. Undergr. Space Technol.* **2021**, *108*, 103716. [CrossRef]
5. Zare, S.; Bruland, A. Comparison of tunnel blast design models. *Tunn. Undergr. Space Technol.* **2006**, *21*, 533–541. [CrossRef]
6. Zhao, Z.Y.; Zhang, Y.; Bao, H.R. Tunnel blasting simulations by the discontinuous deformation analysis. *Int. J. Comput. Methods* **2011**, *8*, 277–292. [CrossRef]
7. Zuo, J.J.; Yang, R.S.; Xiao, C.L. Model test of empty hole cut blasting in coal mine rock drivage. *J. Min. Sci. Technol.* **2018**, *3*, 335–341.
8. Zhang, H.; Li, T.C.; Wu, S.; Zhang, X.T.; Gao, W.L.; Shi, Q.P. A study of innovative cut blasting for rock roadway excavation based on numerical simulation and field tests. *Tunn. Undergr. Space Technol.* **2022**, *119*, 104233. [CrossRef]
9. Cardu, M.; Seccatore, J. Quantifying the difficulty of tunneling by drilling and blasting. *Tunn. Undergr. Space Technol.* **2016**, *60*, 178–182. [CrossRef]
10. Ding, Z.; Jia, J.; Li, X.; Li, J.; Li, Y.; Liao, J. Experimental study and application of medium-length hole blasting technique in coal-rock roadway. *Energy Sci. Eng.* **2019**, *8*, 1554–1566. [CrossRef]
11. Shapiro, V.Y. Efficiency of cut configuration in driving tunnels with a set of deep blast holes. *Sov. Min. Sci.* **1989**, *25*, 379–386. [CrossRef]
12. Zhang, Z.R.; Yang, R.S. Multi-step cutting technology and its application in rock roadways. *Chin. J. Rock Mech. Eng.* **2019**, *38*, 551–559.
13. Dai, J.; Du, X.L. Research on blasting parameters of wedge-shaped cutting for rock tunnel. *Min. Res. Dev.* **2011**, *32*, 90–93.
14. Wang, Z.K.; Gu, X.W.; Zhang, W.L.; Xie, Q.K.; Xu, X.C.; Wang, Q. Analysis of the cavity formation mechanism of wedge cut blasting in hard rock. *Shock Vib.* **2019**, *2019*, 1828313. [CrossRef]
15. Liang, W.M.; Wang, Y.X.; Chu, H.B.; Yang, X.L. Study on effect of symmetry of wedge shaped cutting hole angle on cut blasting. *Met. Mine* **2009**, *38*, 21–24.
16. Pu, C.J.; Liao, T.; Xiao, D.J.; Wang, J.Q.; Jiang, R. Grey relation analysis of influence factors on rock tunnel wedge-shaped cut blasting. *Ind. Miner. Processing* **2016**, *45*, 34–38.
17. Yang, D.Q.; Wang, X.G.; Wang, Y.J.; An, H.M.; Lei, Z. Experiment and analysis of wedge cutting angle on cutting effect. *Adv. Civ. Eng.* **2020**, *2020*, 5126790. [CrossRef]

18. Sazid, M.; Singh, T.N. Two-dimensional dynamic finite element simulation of rock blasting. *Arab. J. Geosci.* **2013**, *6*, 3703–3708. [CrossRef]
19. Zehtab, B.; Salehi, H. Finite-element-based monte carlo simulation for sandwich panel-retrofitted unreinforced masonry walls subject to air blast. *Arab. J. Sci. Eng.* **2020**, *45*, 3479–3498. [CrossRef]
20. Xie, L.X.; Lu, W.B.; Zhang, Q.B.; Jiang, Q.H.; Chen, M.; Zhao, J. Analysis of damage mechanisms and optimization of cut blasting design under high in-situ stresses. *Tunn. Undergr. Space Technol.* **2017**, *66*, 19–33. [CrossRef]
21. Hu, J.H.; Yang, C.; Zhou, K.P.; Zhou, B.R.; Zhang, S.G. Temporal-spatial evolution and application of blasting cavity of single wedge cutting. *J. Cent. South Univ.* **2017**, *48*, 3309–3315.
22. Cheng, B.; Wang, H.B.; Zong, Q.; Xu, Y.; Wang, M.X.; Zheng, Q.Q.; Li, C.J. A study on cut blasting with large diameter charges in hard rock roadways. *Adv. Civ. Eng.* **2020**, *2020*, 8873412. [CrossRef]
23. Zong, Q.; Liu, Q.H. Application research on cutting technology of mid-deep hole blasting in coal mine rock tunnel. *Blasting* **2010**, *27*, 35–39.
24. Yang, R.S.; Zhang, Z.R.; An, C.; Zheng, C.D.; Ding, C.X.; Xiao, C.L. Discussion on ultra-deep depth problem of slot hole in blasting excavation of rock roadway in coal mine. *Coal Sci. Technol.* **2020**, *48*, 10–23.
25. Ding, C.X.; Yang, R.S.; Zheng, C.D.; Yang, L.Y.; He, S.L.; Feng, C. Numerical analysis of deep hole multi-stage cut blasting of vertical shaft using a continuum-based discrete element method. *Arab. J. Geosci.* **2021**, *14*, 1086. [CrossRef]
26. Zong, Q. Preliminary investigations into deep-hole millisecond delay-action blasting. *J. Huainan Min. Inst.* **1992**, *12*, 18–22.
27. Yang, R.S.; Zheng, C.D.; Yang, L.Y.; Zuo, J.J.; Cheng, T.L.; Ding, C.X.; Li, Q. Study of two-step parallel cutting technology for deep-hole blasting in shaft excavation. *Shock Vib.* **2021**, *2021*, 8815564. [CrossRef]
28. Liu, K.; Li, Q.Y.; Wu, C.Q.; Li, X.B.; Li, J. A study of cut blasting for one-step raise excavation based on numerical simulation and field blast tests. *Int. J. Rock Mech. Min. Sci.* **2018**, *109*, 91–104. [CrossRef]
29. Chen, M.; Ye, Z.W.; Lu, W.B.; Wei, D.; Yan, P. An improved method for calculating the peak explosion pressure on the borehole wall in decoupling charge blasting. *Int. J. Impact Eng.* **2020**, *146*, 103695. [CrossRef]
30. Jiang, N.; Zhu, B.; He, X.; Zhou, C.B.; Luo, X.D.; Wu, T.Y. Safety assessment of buried pressurized gas pipelines subject to blasting vibrations induced by metro foundation pit excavation. *Tunn. Undergr. Space Technol.* **2020**, *102*, 103448. [CrossRef]
31. Blair, D.P. The free surface influence on blast vibration. *Int. J. Rock Mech. Min. Sci.* **2015**, *77*, 182–191. [CrossRef]
32. Yang, J.H.; Lu, W.B.; Li, P.; Yan, P. Evaluation of rock vibration generated in blasting excavation of deep-buried tunnels. *KSCE J. Civ. Eng.* **2018**, *22*, 2593–2608. [CrossRef]
33. Hajibagherpour, A.R.; Mansouri, H.; Bahaaddini, M. Numerical modeling of the fractured zones around a blasthole. *Comput. Geotech.* **2020**, *123*, 103535. [CrossRef]
34. Park, D.; Jeon, S. Reduction of blast-induced vibration in the direction of tunneling using an air-deck at the bottom of a blasthole. *Int. J. Rock Mech. Min. Sci.* **2010**, *47*, 752–761. [CrossRef]
35. Kim, J.G.; Song, J.J. Abrasive water jet cutting methods for reducing blast-induced ground vibration in tunnel excavation. *Int. J. Rock Mech. Min. Sci.* **2015**, *75*, 147–158. [CrossRef]
36. Jang, H.; Handel, D.; Ko, Y.; Yang, H.S.; Miedecke, J. Effects of water deck on rock blasting performance. *Int. J. Rock Mech. Min. Sci.* **2018**, *112*, 77–83. [CrossRef]
37. Hu, Y.G.; Lu, W.B.; Chen, M.; Yan, P.; Yang, J.H. Comparison of blast-induced damage between presplit and smooth blasting of high rock slope. *Rock Mech. Rock Eng.* **2014**, *47*, 1307–1320. [CrossRef]
38. Cheng, B.; Wang, H.B.; Zong, Q.; Xu, Y.; Wang, M.X.; Zhu, N.N. Study on the novel technique of straight hole cutting blasting with a bottom charged central hole exploded supplementally. *Arab. J. Geosci.* **2021**, *14*, 2867. [CrossRef]
39. Li, H.C.; Zhang, X.T.; Li, D.; Wu, L.M.; Zhou, H.M. Numerical simulation of the effect of empty hole between adjacent blast holes in the perforation process of blasting. *J. Intell. Fuzzy Syst.* **2019**, *37*, 3137–3148. [CrossRef]
40. Guo, Y.; Li, Q.; Yang, R.S.; Xu, P.; Zhao, Y.; Wang, Y. Evolution of stress field in cylindrical blasting with bottom initiation. *Opt. Lasers Eng.* **2020**, *133*, 106153. [CrossRef]
41. Gao, Q.D.; Lu, W.B.; Yan, P.; Hu, H.R.; Yang, Z.W.; Chen, M. Effect of initiation location on distribution and utilization of explosion energy during rock blasting. *Bull. Eng. Geol. Environ.* **2019**, *78*, 3433–3447. [CrossRef]
42. Xu, X.D.; He, M.C.; Zhu, C.; Lin, Y.; Chen, C. A new calculation model of blasting damage degree—Based on fractal and tie rod damage theory. *Eng. Fract. Mech.* **2019**, *220*, 106619. [CrossRef]
43. Xia, W.J.; Lu, W.B.; Wang, G.H.; Yan, P.; Liu, D.; Leng, Z.D. Safety threshold of blasting vibration velocity in foundation excavation of Baihetan super-high arch dam. *Bull. Eng. Geol. Environ.* **2019**, *79*, 4999–5012. [CrossRef]
44. Yang, R.S.; Ding, C.X.; Yang, L.Y.; Lei, Z.; Zheng, C.D. Study of decoupled charge blasting based on high-speed digital image correlation method. *Tunn. Undergr. Space Technol.* **2019**, *83*, 51–59. [CrossRef]
45. Cheng, B.; Wang, H.; Zong, Q.; Xu, Y.; Wang, M.; Zheng, Q. Study of the double wedge cut technique in medium-depth hole blasting of rock roadways. *Arab. J. Sci. Eng.* **2021**, *46*, 4895–4909. [CrossRef]
46. Chen, Y.; Ma, S.; Yang, Y.; Meng, N.; Bai, J.B. Application of shallow-hole blasting in improving the stability of gob-side retaining entry in deep mines: A case study. *Energies* **2020**, *12*, 3623. [CrossRef]
47. Parviz, M.; Aminnejad, B.; Fiouz, A. Numerical simulation of dynamic response of water in buried pipeline under explosion. *KSCE J. Civ. Eng.* **2017**, *21*, 2798–2806. [CrossRef]

48. Wang, Z.L.; Wang, H.C.; Wang, J.G.; Tian, N.C. Finite element analyses of constitutive models performance in the simulation of blast-induced rock cracks. *Comput. Geotech.* **2021**, *135*, 104172. [CrossRef]
49. Liu, K.; Li, Q.Y.; Wu, C.Q.; Li, X.B.; Li, J. Optimization of spherical cartridge blasting mode in one-step raise excavation using pre-split blasting. *Int. J. Rock Mech. Min. Sci.* **2020**, *126*, 104182. [CrossRef]
50. Fourney, W.L.; Dick, R.D.; Wang, X.J.; Wei, Y. Fragmentation mechanism in crater blasting. *Int. J. Rock Mech. Min. Sci. Geomech. Abstr.* **1993**, *30*, 413–429. [CrossRef]

Essay

Parameter Optimization and Fragmentation Prediction of Fan-Shaped Deep Hole Blasting in Sanxin Gold and Copper Mine

Bo Ke [1,2], Ruohan Pan [1], Jian Zhang [2,*], Wei Wang [1,3], Yong Hu [3], Gao Lei [3], Xiuwen Chi [1], Gaofeng Ren [1] and Yuhao You [1]

1 School of Resources and Environmental Engineering, Wuhan University of Technology, Wuhan 430070, China; boke@whut.edu.cn (B.K.); panrh@whut.edu.com (R.P.); 15871178782@139.com (W.W.); xwchi@whut.edu.cn (X.C.); rgfwhut@163.com (G.R.); a1622078050@gmail.com (Y.Y.)
2 School of Urban Construction, Wuchang University of Technology, Wuhan 430223, China
3 Hubei Sanxin Gold Copper Limited Company, Huangshi 435199, China; scb@hbsanxin.com (Y.H.); 13971757475@139.com (G.L.)
* Correspondence: csuzhangjian@126.com

Abstract: For San-Xin gold and copper mine, deep blasting large block rate is high resulting in difficulty in transporting the ore out; secondary blasting not only increases blasting costs but is more likely to cause the top and bottom plate of the underground to become loose causing safety hazards. Based on the research background of Sanxin gold and copper mine, deep hole blasting parameters were determined by single-hole, variable-hole pitch, and oblique hole blasting tests, further using the inversion method to determine the optimal deep hole blasting parameters. Meanwhile, the PSO-BP neural network method was used to predict the block rate in deep hole blasting. The results of the study showed that the optimal minimum resistance line was 1.24–1.44 m, which was lower than 1.6–1.8 m in the original blasting design, which was one of the reasons for the higher blasting block rate. In addition, the PSO-BP deep hole blasting fragmentation prediction model predicts the block rate of the optimized blasting parameters and predicted a block rate of 6.83% after the optimization of hole network parameters. Its prediction accuracy is high, and the blasting parameter optimization can effectively reduce the block rate. It can reasonably reduce the rate of large pieces produced by blasting, improve blasting efficiency, and save blasting costs for enterprises. The result has wide applicability and can provide solutions for underground mines that also have problems with blasting large block rate.

Keywords: deep hole blasting parameters; blasting funnel; blasting fragmentation; neural network

Citation: Ke, B.; Pan, R.; Zhang, J.; Wang, W.; Hu, Y.; Lei, G.; Chi, X.; Ren, G.; You, Y. Parameter Optimization and Fragmentation Prediction of Fan-Shaped Deep Hole Blasting in Sanxin Gold and Copper Mine. *Minerals* **2022**, *12*, 788. https://doi.org/10.3390/min12070788

Academic Editor: Amin Beiranvand Pour

Received: 24 April 2022
Accepted: 15 June 2022
Published: 21 June 2022

Publisher's Note: MDPI stays neutral with regard to jurisdictional claims in published maps and institutional affiliations.

1. Introduction

As a large industrial country, mineral resources play an important role in social infrastructure construction. However, due to over-exploitation, shallow mineral resources have been facing depletion, affected by mining equipment and mining technology. There are a certain number of hidden resources in the process of recovery. In order to meet the sustainable development of mines, mining enterprises often need secondary mining of these resources, and these hidden resources often present extremely complex conditions, especially in the filling body for recovery, in the surrounding filling body strength, poor stability of the quarry, difficulty in recovery, etc. How to ensure the stability of the filling quarry under the conditions of dynamic explosive load is a key technical problem facing the mining industry.

Blasting is the most important part of mine production. Its quality and efficiency are not only related to the production efficiency of the entire mining process, but also determine the economic benefits of the enterprise. In the blasting process, the control of blasting fragmentation is very important. The fragmentation of blasting is an important indicator for evaluating the effect of blasting. Scholars have carried out a lot of research on the theory of fragmentation of rock blasting and have summarized a variety of control methods [1]. Shi Xiaopeng systematically studied the effect of blasting vibration on the filling body of large-diameter deep holes in Anqing copper mine using regression calculations and theoretical analysis, and then proposed technical measures to reduce the vibration intensity of large-diameter deep holes in Anqing copper mine based on the actual situation of blasting production [2]. In the case of Jinchuan's second mine, where the roof and the left and right helpers of the working face are filled bodies, Wang Xianlai made a series of elaborations on rock drilling, blasting material selection, hollowing method, number of holes, single shot dosage, and micro-difference time for blasting operations in the mining process, and concluded and optimized blasting parameters, which improved the safety of underground workers working under filled bodies and provided experience for reducing mine casualties [3]. However, the blasting process is very complicated, and the mining technology of different mines is different, resulting in a big difference between the prediction results of the fragmentation model and the actual ore fragmentation after blasting.

The degree of damage to the rock by explosives and the process control are mainly realized by blasting parameters. These parameters mainly include the nature of the explosives, the unit consumption of explosives, deep hole density factor, the minimum resistance line, the bottom distance of the hole, etc. These factors are in series with each other and determine the volume of the blasting funnel, which in turn affects the block size produced by blasting. Regarding the determination of blasting parameters, scholars have made a lot of explorations on the research of blasting parameters, optimizing blasting parameters from theoretical analysis, physical test (funnel test), engineering analogy, numerical simulation, and machine learning calculations. For example, Akande and Lawal examine optimization of blasting parameters for economic production of granite aggregates in Ratcon and NSCE quarries located at Ibadan using regression models [4]. Rezaeineshat et al. used robust techniques to design the blasting parameters in open-pit mine with the aim to reducing ground vibration [5]. Inanloo Arabi Shad and Ahangari proposed an empirical relation to calculate the proper burden in blast design of open-pit mines and promoted this method to other mines [6]. Wang et al. proposed a method combining UDEC and LS-DYNA to study the effects of parameters on the blasting effects [7]. Messaoud et al. classified the quality of rock mass through cluster analysis, and then studied the influence of different rock parameters on blasting through principal component analysis [8]. Monjezi and Dehghani used the neural network to analyze the relationship between the physical and mechanical properties of the rock mass, the performance of the explosive, the hole network parameters and the blasting backlash, optimize the blasting parameters, reduce the damage of the blasting backlash, and the optimization effect is obvious [9]. Sadollah et al. uses reverse neural network to optimize rock drilling and blasting parameters [10]. Similar research can be found in many other studies [11–13].

With the development of computer technology, in recent years, traditional theory combined with machine learning to predict the distribution of rock blasting fragmentation has gradually been recognized by everyone. Monjezi et al. established a 4-layer feed-forward artificial neural network (ANN) model to predict the distribution of fragmentation [14]. Ghiasi et al. predicted the large block rate generated in the open-pit blasting operation of Golegohar Iron Mine by using multiple regression methods and artificial neural network [15]. Ghaeini et al. and Ghaeini Hesarouieh et al. used the Mutual Information (MI) method to predict the blasting fragmentation of the Meydook Mine and compared it with the Kuz-Ram empirical model and statistical model. The results show that the MI model has higher accuracy than the Kuz-Ram and statistical models [16,17]. Based on meta-heuristics and machine learning algorithms, Xie et al. predicted the rock size distribution in mine blasting using various novel soft computing models [18]. Bahrami et al. implemented artificial neural network method to develop a model to predict rock fragmentation due to blasting in an iron ore mine [19]. Murlidhar et al. proposed a new hybrid imperialism competitive algorithm (ICA)-artificial neural network (ANN) and found the proposed ICA-ANN model can be implemented better in improving the performance capacity of ANN model in estimating rock fragmentation during blasting [20]. Ouchterlony and Sanchidrián reviewed the prediction equations for blast fragmentation [21]. These models include some early fragmentation models, first Kuz-Ram models [22], crush zone model (CZM) and the two-component mode l(TCM) [23–25], extended Kuz-Ram model [26] and Swebrec function [27]. Some other methods were also adopted to predict rock fragmentation by many studies, for example, statistical modelling approach [28], muck-pile model [29], KCO model [30], and deep learning approach [31].

In fan-shaped deep hole blasting, the boulder yield is relatively high. How to take effective measures to reduce the output rate of boulders in deep hole blasting and improve the economic benefits of the mine is an urgent problem. Take Sanxin gold-copper mine as an example, the boulder yield is higher than 12%, in some stopes boulder yield even reached more than 20%. High boulder yield is not good for safe production and increases the workload of secondary blasting. This study takes Sanxin gold-copper mine as the research background and aims at reducing the boulder yield of this mine. Funnel tests [32] were conducted to determine the optimal blasting parameters including hole distance, explosive unit consumption, and row space. A combined particle swarm optimization (PSO) and BP neural network model was established to predict the boulder yield [33]. A targeted study was conducted to explore the effect of the maximum single-section charge on the filling body, and the safe distance of the filling body at different charges from the perspective of safe vibration speed, which provides a basis for the optimization of blasting parameters in the field.

2. Blasting Funnel Test

2.1. Test Equipment

Measuring tools include: weighing instrument (platform scale with weighing capacity $\geq$ 100 kg, electronic scale with sensing capacity $\leq$ 20 g), 50 m soft leather ruler, 5 m steel tape, shovel, gun stick and lattice screen (50 mm) $\times$ 50 mm, 100 mm $\times$ 100 mm, 200 mm $\times$ 200 mm, 300 mm $\times$ 300 mm, 400 mm $\times$ 400 mm grid).

2.2. Selection of Test Site

The selection of the blasting funnel test site should take into account that the ore lithology of the test orebody should be same or as close as possible to the mined area. According to the field investigation, the test site is selected at 8–3 rock drilling roadway at -609 m level in Sanxin Gold and Copper Mine. The funnel test site is shown in Figure 1.

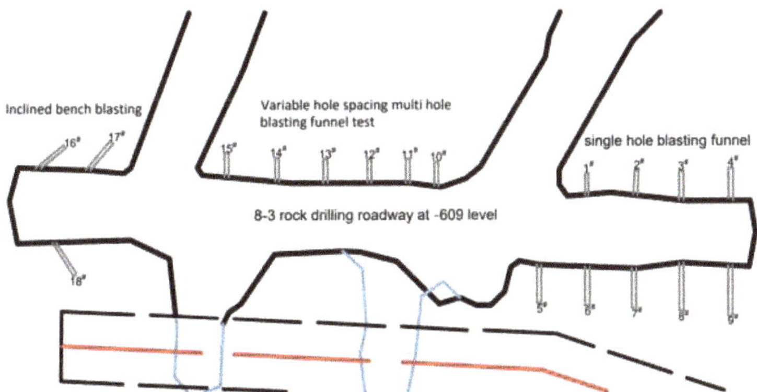

Figure 1. Plane layout of blast holes for series blasting funnel test. (In the figure, '#' is for the convenience of numbering the holes, $1^{\#}$ represents hole number one, and so on).

The physical and mechanical properties of the rock at the test site are shown in Table 1, and the performance parameters of the explosives used are shown in Table 2.

Table 1. Results of physical and mechanical properties of 609 horizontal ore and rock.

Lithology	Bulk Density (N/m^3)	Tensile Strength (MPa)	Elastic Modulus (GPa)	Poisson's Ratio	Cohesion (MPa)	Internal Friction Angle (°)
Ore bearing marble	32,300	6.01	41.007	0.24	17.88	50.5

Table 2. Main performance parameters of ANFO explosive.

Explosive Type	Density (g/cm^3)	Explosion Velocity (m/s)	Stiffness (mm)	Explosive Force (ML)	Explosive Weight (g)
ANFO explosive	1	3000	12	260	400

2.3. Single Hole Blasting Funnel Test

The designed blast hole depth is: there are 9 boreholes numbered 1–9 in 0.4 m, 0.5 m, 0.6 m, 0.7 m, 0.8 m, 0.9 m, 1.0 m, 1.1 m, and 1.2 m. Due to the errors in site construction and the sinking of the overlying rock layer caused by ground pressure, the actual depth of the blast hole is 0.4 m, 0.5 m, 0.6 m, 0.65 m, 0.7 m, 0.82 m, 0.9 m, 1.0 m, 1.1 m. The distance between holes is 2.4 m, the diameter of blast hole is 60 mm, and the charging length is 200 mm. Before charging, the individual deep hole is adjusted to the designed charge depth with gun mud, then the charge package is filled with gun stick, and finally the hole is plugged with 0.2 m gun mud. The ANFO explosive is initiated in reverse with the bottom of the detonator hole to reduce the blasting influence between the adjacent holes, and there are two non-adjacent holes in each blasting.

The layout of the blasthole is shown in Figure 2 below:

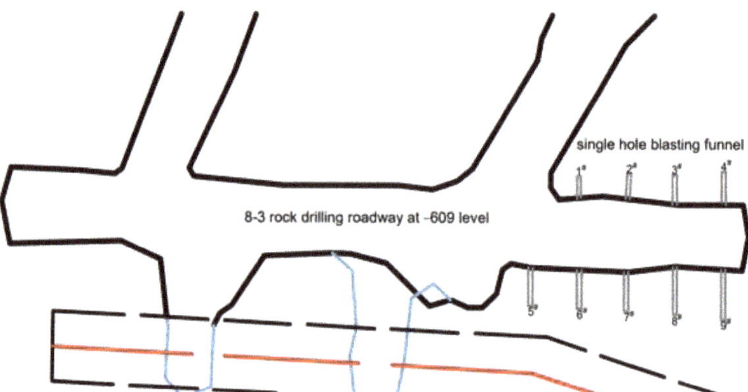

Figure 2. Schematic diagram of plane layout of blast holes in funnel test of single-hole blasting. (In the figure, '#' is for the convenience of numbering the holes, 1# represents hole number one, and so on).

2.4. Variable Hole Spacing Multi Hole Blasting Funnel Test

Six blastholes numbered 10–15 are arranged. The hole distance is 1.75, 2, 2.25, 2.5, and 2.75 times of the optimum funnel radius (the specific blasthole distance is 1.03 m, 1.18 m, 1.33 m, 1.48 m, 1.62 m), respectively. There are six holes in a group. The blasthole is arranged on the vertical free surface, and the hole diameter is 60 mm. The length of the charge is 200 mm. After blasting, the measuring process of single-hole blasting funnel test is repeated, and the funnel profile is drawn based on the paint on the grid.

The layout of the blasthole is shown in Figure 3 below:

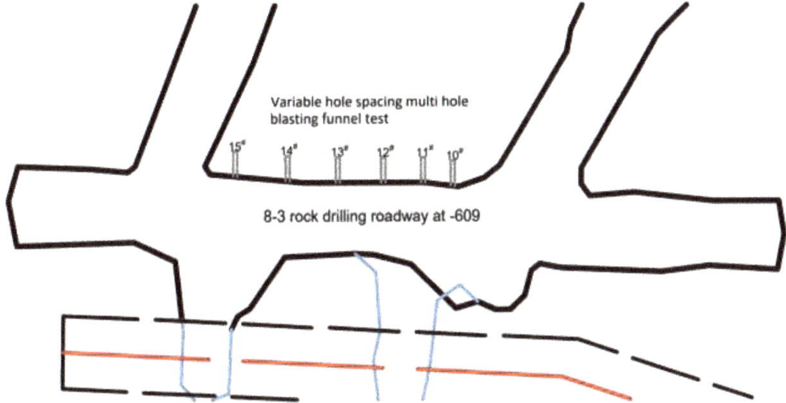

Figure 3. Schematic diagram of blast hole plane layout in funnel test of multi-hole blasting with variable hole distance. (In the figure, '#' is for the convenience of numbering the holes, $10^{\#}$ represents hole number one, and so on).

2.5. Inclined Bench Blasting Test

The design scheme is shown in Figure 4, and the blasthole layout is shown in Figure 5 below:

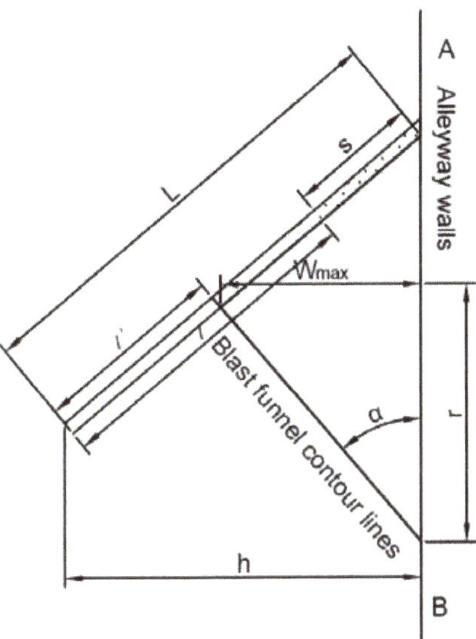

Figure 4. Schematic diagram of geometric parameters of inclined bench blasting.

In the drawing, AB is the roadway wall, L is the hole depth, m;

l—charge depth, m;

l' is the residual blasthole depth, m;

h—the distance between the bottom of the blasthole and the roadway wall (free surface), m;

r—blasting funnel radius, m;

s—blasthole clogging depth, m;
W_{max}—maximum resistance line of blasting, m;
α—is the angle between the blasthole and the free surface AB.

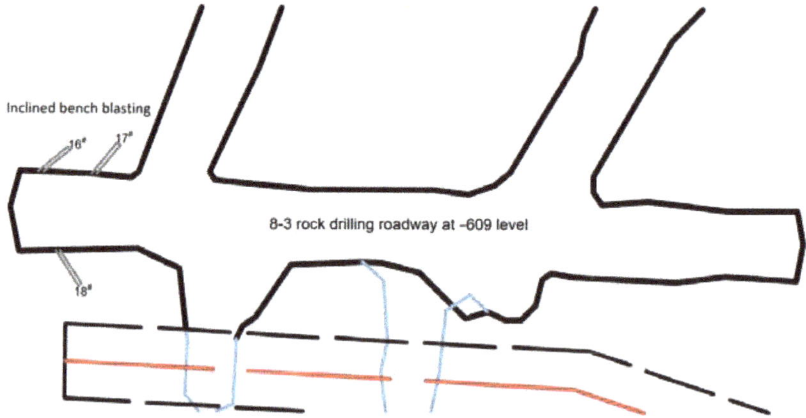

Figure 5. Schematic diagram of plane layout of hole in funnel test of inclined bench blasting. (In the figure, '#' is for the convenience of numbering the holes, 16# represents hole number one, and so on).

There are three blastholes in the inclined bench blasting experiment, numbered 16–18, drilling depth is the same, and the angles between them and the wall of the roadway are 30°, 40°, and 50°, respectively.

2.6. Test Data Collection

The measurements before and after blasting are as follows:

1. Before blasting, the blasthole diameter, blasthole depth, explosive filling depth and packing length are measured respectively with steel tape measure, the explosive burying depth is recorded, and the colored stripe cloth is laid at the bottom of the roadway to ensure that the ore falling after blasting falls on the colored stripe cloth.
2. After blasting, 200 mm × 200 mm grid is used to cover the blasting funnel surface, the grid is kept perpendicular to the roadway floor, the original free surface position was replaced with the grid plane, and the residual hole depth of the blasting funnel and the straight line distance between the bottom of the hole and the grid plane was measured.
3. Deducting the falling part of the rock around the funnel mouth, taking the center line of the blasthole as the axis, the distance between the funnel boundary and the axis of eight different directions is directly measured every 45°, and the average value is taken as the blasting funnel radius.
4. The grid is used to screen the ore after blasting, and the ores of different sizes after screening are weighed and recorded respectively.
5. The rock fragmentation of blasting funnel test is screened. According to the actual production situation of the mine, the bulk above 400 mm is defined as unqualified. By weighing the ores of different sizes, the percentage of the total rock mass in blasting is calculated, and the blasting effect of blasting funnel test is compared.
6. After inclined bench blasting, in order to ensure the accuracy of the measured data, a gun stick is inserted into the residual hole, and a tape measure is used to measure the resistance line at the opening formed by blasting, and the average value of the resistance line is the maximum resistance line W_{max}.

3. Data Processing and Blasting Parameter Inversion

3.1. Data Processing of Single-Hole Blasting Funnel

The hole 9 was blown up in the field blasting so that the data could not be measured; the single-hole blasting funnel test results were only obtained for single-hole blasting 1 to 8. The test results of single-hole blasting funnel are shown in Table 3.

Table 3. Funnel test results of single-hole blasting.

Blasthole Number	Blasthole Depth (m)	Explosive Package Length (m)	Explosive Package Depth (m)	Funnel Volume (m^3)	Funnel Depth (m)	Funnel Radius (m)	Buried Depth Ratio	Unit Explosive Blasting Volume (m^3)
1#	0.4	0.2	0.3	0.1305	0.27	0.371	0.29	0.3263
2#	0.5	0.2	0.39	0.1729	0.27	0.428	0.37	0.4322
3#	0.6	0.2	0.5	0.2166	0.33	0.606	0.48	0.5415
4#	0.65	0.2	0.55	0.2463	0.38	0.573	0.52	0.6157
5#	0.7	0.2	0.45	0.1963	0.3	0.523	0.43	0.4908
6#	0.82	0.2	0.72	0.1761	0.22	0.396	0.69	0.4402
7#	0.9	0.2	0.76	0.111	0.29	0.333	0.72	0.2775
8#	1	0.2	0.85	0.0916	0.16	0.298	0.81	0.229

The characteristic curve of the blasting funnel (V-L curve) and the curve of the relationship between the radius of the funnel and the burial depth of the explosive package (R-L curve) are generally characterized by polynomials, while the curve polynomials can be fitted based on the principle of least squares, and the accuracy of the test curve fit is mainly expressed by the sum of squares of the errors between the fitted and the test values of the polynomials. After the observation of the experimental data, the quartic polynomial is used to fit. MATLAB software was used to make the blasting funnel characteristic curve and the relationship curve between funnel radius and explosive package burial depth, as shown in Figures 6 and 7.

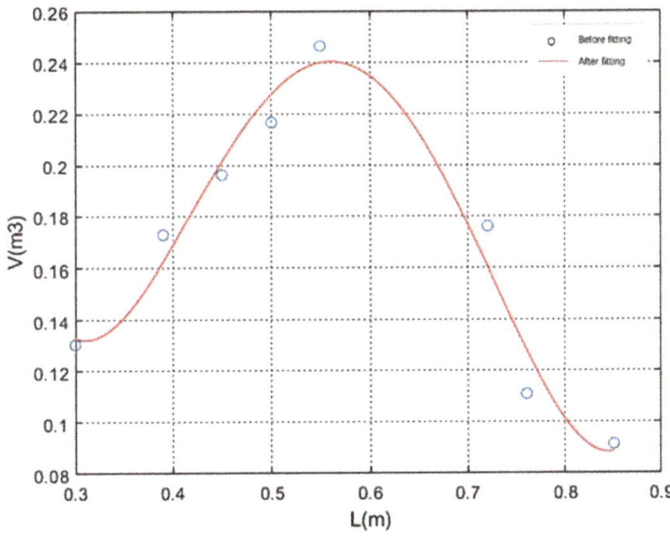

Figure 6. V-L blast funnel characteristic curve.

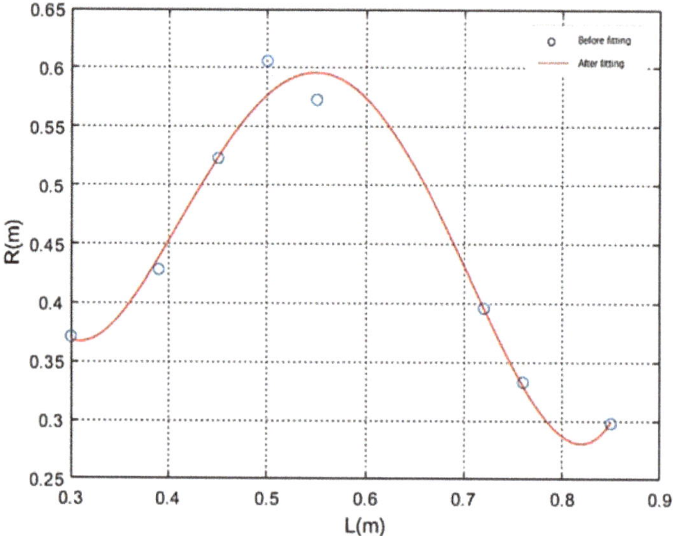

Figure 7. R-L funnel radius and explosive package burial depth relationship curve.

According to the principle of the least square method, the test data of the blasting funnel are regressed for four times, and the volume of the blasting funnel (V_j) and the buried depth of the explosive charge (L_e) are obtained. The multiple expressions between the radius of blasting funnel (R_j) and the buried depth of explosive charge (L_j) are as follows:

$$V = 25.44L^4 - 58.19L^3 + 46.29L^2 - 14.96L + 1.82 \quad (1)$$

$$R = 63.66L^4 - 142.36L^3 + 111.13L^2 - 35.41L + 4.32 \quad (2)$$

From Figures 7 and 8, we get V_j = 0.24 m^3 and R_j = 0.59 m. Substituting V_j and R_j into Equations (7) and (8) gives:

Optimum embedment depth: L_j = 0.56 m,

Critical buried depth of charge: L_e = 0.57 m

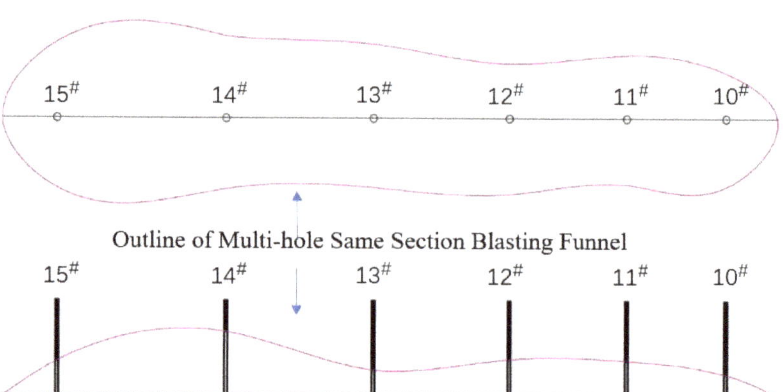

Figure 8. Outline of multi-hole same section blasting funnel.

3.2. Data Processing of Multi Hole Blasting Funnel with Variable Hole Spacing

See Figure 8 for the outline of blasting interface, and Table 4 for the test results:

Table 4. Test results of multi hole blasting funnel with variable hole spacing.

Blast Hole Number	Medicine Bag Buried Depth/m	Blast Hole Spacing/m	Unit Explosive Blasting Volume (m^3)	Volume of Blasting Funnel (m^3)	Description of Blasting Fragmentation
10~11#	0.56	1.05	0.9483	0.3793	Small and uniform lumpiness
11~12#	0.56	1.24	1.0478	0.4191	Small and uniform lumpiness
12~13#	0.56	1.43	1.0601	0.4243	Small and uniform lumpiness
13~14#	0.56	1.55	0.7503	0.3001	The lumpiness is uniform, and there is basically no large block
14~15#	0.56	1.80	1.1315	0.4526	Large fragmentation

From the blasting effect observed in the field, it is known that a group of blastholes with a distance of 1.80 m are not connected into grooves after blasting, and basically form independent blasting funnels. After blasting, the four groups of blastholes with hole spacing of 1.05 m, 1.24 m, 1.43 m, and 1.55 m are connected to form grooves along the center line of the blasthole.

The above results show that when the hole spacing is small, the explosive blasting energy can be superimposed on each other, and the blasting effect is good, and there is basically no inter-hole spine. With the increase of hole spacing, the explosive blasting energy becomes weaker and the blasting effect becomes worse gradually. The spine between the holes gradually appears. When the hole spacing increases to a certain value, the line in the center of the hole no longer communicates to form a strip blasting groove, but develops into an independent blasting funnel in the triangular spine between the two holes.

According to the test results of porous blasting funnel in the same section, if the blasting parameters such as reasonable blasting hole spacing and explosive unit consumption are selected, the triangular spine between adjacent blasting funnels can be broken better. In order to ensure the blasting crushing effect, it is recommended that the hole spacing is 2.5 times of the optimum blasting funnel radius according to the blasting funnel test results, and hole spacing is 1.55 m. According to the above situation, under the test conditions, the optimum hole bottom distance is 1.43 to 1.55 m, and the blasting volume per unit explosive is 0.7503 m^3.

3.3. Data Processing of Inclined Bench Blasting Test

In the experiment of inclined bench blasting, there are three blastholes, numbered 16#, 17#, and 18#. The destruction of the 16# blasthole during blasting, the data could not be measured. So only the data for blastholes 17# and 18# were obtained, and the angles between them and the wall of the roadway are 40° and 50°, respectively. Generally speaking, before blasting, the optimum burying depth obtained from the single-hole blasting funnel experiment is taken as the predicted median of the blasting funnel length. The hole length of the inclined bench blasting funnel experiment is 1.6 m and the explosive filling depth is 1.25 m. The layout and results of the test holes are shown in Table 5. The test results show that the average value of the best resistance line under the test conditions is 0.712 m.

Table 5. Test results of inclined bench blasting funnel.

Blasthole Number	Hole Depth/m	Inclination Angle/°	Orifice Resistance Line/m	Hole Bottom Resistance Line/m	Charge Length/m	Best Resistance Line/m
17#	1.6	40	0.8	1.02	0.36	0.804
18#	1.6	50	0.96	1.22	0.34	0.62
average	1.6	45	0.88	1.12	0.35	0.712

3.4. Inversion of Stope Deep Hole Blasting Parameters

On the basis of several blasting funnel tests in the field, the following blasting parameters are obtained when the blasthole diameter is 60 mm:

Optimum embedding depth: L_j = 0.56 m;

Optimum funnel radius: R_j = 0.59 m;

Critical buried depth of charge: L_e = 0.57 m

Volume of blasting funnel at optimum burial depth: V_j = 0.24 m^3;

Optimum hole bottom distance: a = 1.43~1.55 m;

Weight of spherical charge: Q_0 = 0.4 kg;

According to Livingston's blasting funnel theory and based on the blasting energy balance criterion and similarity principle, many parameters suitable for deep-hole sector blasting in Sanxin Gold and Copper Mine need to be further deduced so as to provide scientific basis for the design of deep-hole sector blasting in the future.

The strain energy coefficient E and the optimum burial depth ratio Δ_j can be determined by single-hole blast funnel tests:

$$E = \frac{L_e}{Q_0^{\frac{1}{3}}} \tag{3}$$

$$\Delta_j = \frac{L_j}{L_e} \tag{4}$$

where,

L_e—critical buried depth, m;

E—strain energy coefficient, which is constant for specific rocks and explosives;

Q_0—weight of spherical charge, kg;

L_j—optimum buried depth, m;

Δ_j—optimum burial depth ratio. Δ_j is constant for specific rocks and explosives.

By substituting the Parameters (3) and the Equation (4), the strain energy coefficient E = 0.77 and the optimum buried depth ratio $\Delta_j = 0.98$ are obtained.

3.4.1. Row Spacing

The row spacing (minimum resistance line) is an important basic parameter for the design of deep hole blasting, based on the principle that the amount of charge that can be loaded into a deep hole ($Q = \pi d^2 L\Delta/4$) and the amount of charge required for a deep hole (unit volume of explosive consumption multiplied by the amount of blasting square that the hole is burdened with ($Q = WaLq$) are equal, obtaining the equation:

$$W = d\sqrt{\frac{0.785\Delta\tau}{mq}} \tag{5}$$

In the equation:

d—aperture, d = 0.060 m:

Δ—charge density, Δ = 1.0 × 10^3 kg/m^3;

τ—deep hole charge coefficient, $\tau = \frac{\text{Charge length}}{\text{Gun hole length}} = 0.85$

m—deep hole density factor, $m = 1.0 \sim 1.2$

q—single consumption of explosives.

The effective explosive unit consumption approved by conventional blasting theory is:

$$q = \frac{Q}{f(n)W^3} \tag{6}$$

where: $f(n)$—exponential function of blasting action, $f(n)$ = 0.4 + 0.6n^3,

n—blasting action index, $n = \frac{R}{L_J}$.

Q—weight of spherical charge, 0.4 kg;

Combining the two Equations (5) and (6) yields the following equation:

$$W = \frac{mQ}{0.785d^2\Delta\tau f(n)} \tag{7}$$

Therefore, according to the above equation, the minimum resistance line W = 1.24~1.44 m can be determined.

It can be seen that the minimum resistance line of 1.6 to 1.8 m used in the original blasting parameters is too large, which may be one of the reasons for the high blasting block rate.

Vertical sector holes have a hole diameter of 75 mm, according to the slant hole blasting funnel test results and blasting cube root similarity law, when the blasting funnel test package in the best burial depth W_1 = 0.712 m, package radius r_1 = 30 mm. When the diameter of the blasthole is 75 mm (prototype), the radius of explosive charge is r_2 = 37.5 mm (prototype), and the blasting resistance line is W_2 (prototype). The Equation (13) should be satisfied:

$$\frac{W_1}{W_2} = \frac{r_1}{r_2} \tag{8}$$

Substitute the relevant data into the equation to obtain: b = $W_2 \approx 0.88$ m.

Therefore, the best minimum resistance line for the prototype spherical explosive package can be calculated by the spherical explosive package model test. According to other scholars' studies, the blasting effect of columnar charge in blastholes is different from that of spherical explosive packages. When the blasting effect is the same, the resistance line of the pillar charge package is greater than that of the spherical charge package if the amount of the charge is also the same.

$$W_z = K_w W_a \tag{9}$$

where: W_z—resistance line of cylindrical charge;

W_a—resistance line of spherical charge;

K_w—coefficient, K_w = 1.2~1.6, take K_w = 1.25

Substitute the relevant data into Equation (9) to obtain the minimum resistance line W (row distance b) W = b = 1.4 m.

3.4.2. Hole Bottom Distance

The bottom distance of the hole a and the minimum resistance line W (or row distance b) of the fan-shaped gun hole satisfy the Equation (10).

$$a = m \times W \tag{10}$$

where: m is the density coefficient, which is the ratio of the distance between the bottom of the hole and the minimum line of resistance.

As shown in Figure 9, W < R, a > R, the rock breaking area of a single hole is S (a, W), and the two-hole blasting fragmentation zone just covers all the rock mass in the single-hole blasting area, and the superposition zone of the two-hole blasting fragmentation zone eliminates the area where the crushing force cannot be reached. It can be considered that all the rock masses in the single hole blasting area have been fully broken at this time, which is the ideal condition of rock mass fragmentation, and it is also the situation that the design wants to achieve. For this reason, the function S (a, W) can be defined: 0 < W< R, 0 < a < 2R, make S obtain the maximum value as far as possible and the previously determined w to determine the optimal value of a. According to the geometric relationship in Figure 9, the explicit function expression of hole network area S (a, W) can be obtained:

$$S = 2\left(\frac{\pi R^2}{2} - R^2\arctan\frac{a}{2w} + \frac{aw}{4}\right) \tag{11}$$

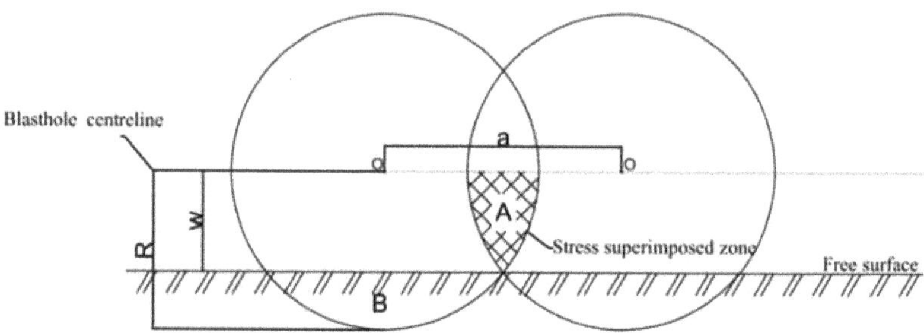

Figure 9. Stress superposition diagram at $0 < W < R$, $0 < a < 2R$.

Because: $\frac{1}{4}a^2 + W^2 = R^2$, in order to maximize the value of pore network area s, let the first derivative of s be zero, and according to the empirical calculation:

$$a = 1.23R \tag{12}$$

$$W = 0.789R \tag{13}$$

Then the deep hole density coefficient is: $m = a/W = 1.559$.

Based on the above equation calculation as well as the wide hole spacing multi-hole with section blasting funnel test, $m = 1.6$ is taken here, and small resistance line with large hole bottom distance blasting is implemented.

$$a = 1.6 \times 1.4 = 2.24 \approx 2.2\ \text{m} \tag{14}$$

3.4.3. Explosive Unit Consumption

From the single-hole and multi-hole with the same section blasting funnel test, it is known that when the explosive package is the best burial depth ratio and the blasting funnel volume is the largest, the explosive unit consumption is optimal. The best unit explosive consumption approved for single-hole and variable-pitch multi-hole tests by the optimal burial depth of the explosive package funnel blasting volume are $q_1 = 0.65\ \text{kg/m}^3$ ($q_1 = 0.4/0.6157 = 0.65\ \text{kg/m}^3$), $q_2 = 0.53\ \text{kg/m}^3$ ($q_2 = 0.4/0.7503 = 0.53\ \text{kg/m}^3$). However, the analysis shows that the test condition of explosive charge funnel blasting is the condition of full opening and free surface, and the cutting action of joints and fissures will increase the blasting ballast quantity to a certain extent, and the calculated explosive consumption index is low. At the same time, industrial production back blasting blocks up and down the disk for silica or quartz orthoclase diorite, the rock is loose and broken, its detonability is worse than the test lithology, and back blasting for extrusion blasting, and full free surface conditions are very different. Therefore, the test results must be corrected to ensure that the quality of recovery blasting and reduce the rate of large blocks.

Substitute relevant data into the Equation (6) to obtain: $q_3 = 0.706\ \text{kg/m}^3$.

The average explosive unit consumption obtained from the above three calculation methods is chosen as the average explosive unit consumption for the blasting design of the test section, which is taken as

$$q = \frac{q_1 + q_2 + q_3}{3} = 0.629\ \text{kg/m}^3 \tag{15}$$

4. Prediction of Blasting Fragmentation Based on PSO-BP Neural Network Algorithm

4.1. PSO Optimized BP Neural Network Algorithm

Due to the random selection of initial weights and thresholds during the application of BP networks, local convergence minima occur, which reduces the fitting effect, and to solve this problem, the initial weights and thresholds of the PSO optimized BP neural network

(PSO-BP) algorithm are used to solve the local minima problem and improve the prediction accuracy of the BP neural network algorithm. In PSO, D is the dimension of the whole search space. See Equation (16) for the position of the i particle.

$$X_i = [x_{i1}, x_{i2}, \cdots, x_{iD}]^T = \begin{bmatrix} \omega_{11} & \cdots & \omega_{nl} \\ \omega_{1m} & \cdots & \omega_{nm} \\ \theta_1{}^1 & \cdots & \theta_n{}^1 \\ \theta_1{}^2 & \cdots & \theta_n{}^2 \end{bmatrix}^T \tag{16}$$

$$D = (n+m)k + k + m \tag{17}$$

where: the elements of $\omega_{11} \ \cdots \ \omega_{nl} - \omega^1$
the elements of $\omega_{1m} \ \cdots \ \omega_{nm} - \omega^2$
the elements of $\theta_1{}^1 \ \cdots \ \theta_n{}^1 - \theta^1$
the elements of $\theta_1{}^2 \ \cdots \ \theta_n{}^2 - \theta^2$

D is the dimension of this search space as the sum of all weights and thresholds in the BP neural network.

A population is randomly initialized with each particle setting a maximum velocity V_{max} corresponding to an initial position maximum X_{max} and each particle setting a minimum velocity V_{min} corresponding to an initial position minimum X_{min}, and the positions and velocities of the particles are randomly selected in the interval $[-X_{max}, X_{max}]$, $[-V_{max}, V_{max}]$ for initialization, and each particle in the D-dimensional space in this population is represented by a set of weights and thresholds of the BP neural network. Set the termination condition of PSO optimization, the number of iterations N, the population size M, the initial value of inertia weight ω, the initial value of acceleration factor c_1, c_2, and so on. According to the initial position of each particle, the fitness value $F(i)$ of each particle is obtained by the fitness function, see Equation (18).

$$F(i) = k(\sum_{i=1}^{n} abs(y_i - o_i)) \tag{18}$$

where: n—number of output nodes;
y_i—expected output of the i node of the neural network;
o_i—predicted output of the i node;
k—coefficient.

Set the extreme value P_{gBest} of each particle population to the optimal value of the objective function; calculate the fitness value of each particle, if $F(i) > P_{iBest}$, $F(i)$ replaces P_{iBest} and optimize the group extreme value P_{gBest}; update the speed and position of particles according to Equations (17) and (18) in each iteration; after the termination of PSO optimization, get the optimal weight and threshold of BP neural network; replace the BP neural network to initialize the initial weight and threshold of BP neural network, and input sample data. The BP neural network trains the model in the way of error reverse transmission. When the error between the output value of the BP neural network and the real value is large, the reverse error is calculated to reverse forward to update the connection weight and threshold between the layers until the termination condition is reached. The algorithm flow of PSO-BP is shown in Figure 10.

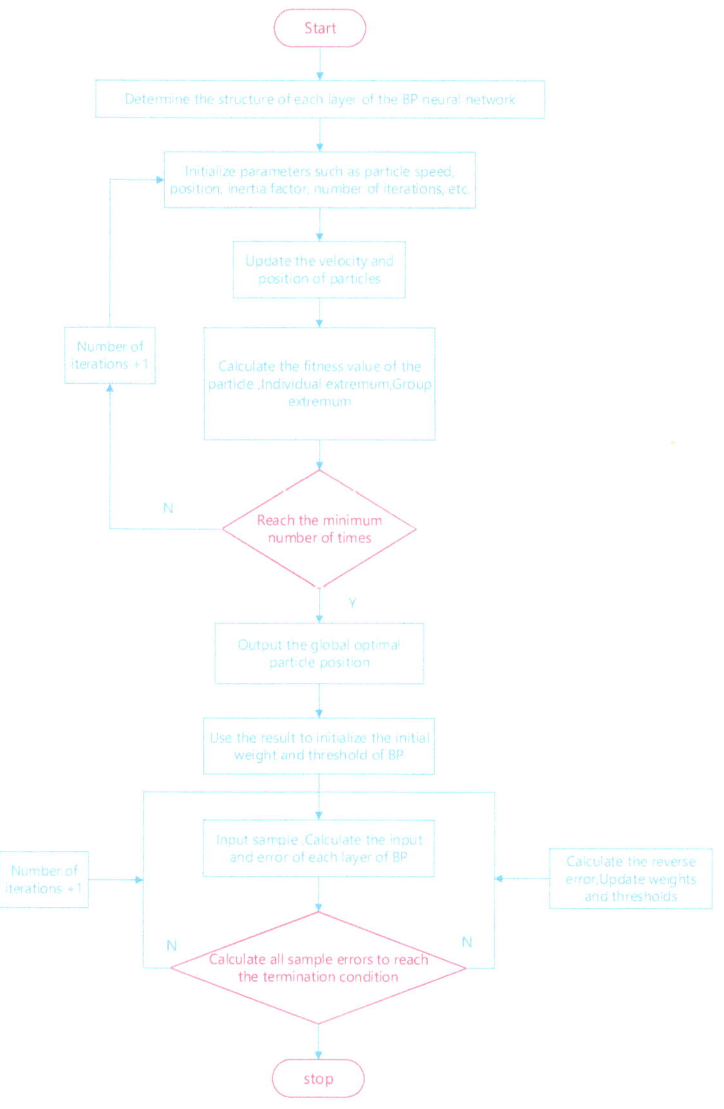

Figure 10. Algorithm flow of PSO-BP.

4.2. Neural Network Model Structure

There are many factors that affect the fragmentation distribution of blasting, and it is difficult to establish a mathematical expression to establish a relationship. How to highlight the principal contradiction from many influencing factors and select the main influencing factors to simplify the model as far as possible on the basis of ensuring the reliability of the data without reducing the prediction accuracy of the model is a problem that we should carefully consider before establishing the model.

1. There are many factors in the rock that can affect the blasting effect, and this model does not consider all the factors affecting the rock, but only selects some factors that have a significant impact on the blasting effect, such as the tensile strength of the rock, the compressive strength of the rock, and the spacing between the joints of the rock.

2. In terms of blasting technology and blasting parameters, combined with the existing equipment in the mine, this model only considers the simple blasting technology of regular hole layout, uniform charge and millisecond initiation with detonating cord, which is also the conventional production technology of the mine; in fact, the design of blasting technology is an art of flexible change.

3. There is no full consideration as to whether the charge is explosive coupled with blasting rock mass and whether the blasting energy can be fully utilized. At the same time, in the process of construction, many man-made uncertainties and equipment uncertainties cannot be fully considered.

4. The original data were normalized with $x' = 2 \times \frac{x - x_{\min}}{x_{\max} - x_{\min}} - 1$ because of different input factor magnitudes and large differences in order of magnitude, and the output data calculated by the network model are the normalized results, and the final data are obtained by reverse normalization for easy reading.

Determine the BP neural network structure of the deep hole blasting block degree prediction model; determine the initial position and velocity of the particle in space according to the BP neural network; set the termination condition of the particle population optimization, the number of iterations N, the population size, the initial value of inertia weight ω, maximum particle speed V_{max}, the initial value of acceleration factor c_1, c_2, etc. According to Equation (18) calculate the particle's $F(i)$; according to $F(i)$ update P_{iBest}, P_{gBest}; according to Equations (1) and (2) update the particle position and velocity; calculate the new position of the particle $F(i)$; if $F(i) > P_{iBest}$, then $F(i)$ replaces P_{iBest}; optimize the population extremum P_{gBest}; in each iteration according to Equations (1) and (2) update the velocity and position of the particle; determine whether to reach the minimum error. If the error requirement is satisfied then PSO optimization is terminated, and after PSO optimization is terminated, the global optimal particle position is output, the optimal initial weights and thresholds of the deformation prediction BP network are obtained, and the training of the highway plane vertical deformation PSO-BP network is started.

Selection of input layer neuron parameters: rock characteristic parameters (density, compressive strength, joint spacing); seven blasting design parameters (explosive unit consumption, hole diameter, hole bottom distance, minimum resistance line) are used as the input layer of the neural network model. The data of the orthogonal experiments are shown in the following Table 6.

Table 6. Field orthogonal experimental data.

No.	Density/(t·m^{-3})	Compressive Strength/Mpa	Joint Crack/ (Strip m^{-1})	Aperture/mm	Hole Bottom Distance/m	Minimum Line of Resistance/m	Unit Consumption of Explosives/(kg·m^{-3})	Test Block Rate/%
1#	2.94	92.7	1.07	65	1.6	1.4	0.579	14.8
2#	2.94	92.7	1.3	70	1.9	1.6	0.629	12.62
3#	2.94	92.7	1.07	65	2.2	1.8	0.579	18.24
4#	2.94	92.7	1.3	70	1.9	1.4	0.629	18.02
5#	3.14	98.4	1.07	65	1.6	1.8	0.579	13.34
6#	3.14	98.4	1.3	70	1.9	1.6	0.629	12.68
7#	3.14	98.4	1.76	75	2.2	1.4	0.679	17.36
8#	3.14	98.4	1.3	70	1.9	1.6	0.629	13.48
9#	3.23	104.8	1.07	65	2.2	1.8	0.579	14.95
10#	3.23	104.8	1.07	65	1.6	1.4	0.579	14.32
11#	3.23	104.8	1.3	70	1.9	1.6	0.629	11.35
12#	3.23	104.8	1.76	75	2.2	1.8	0.679	13.56
13#	2.94	92.7	1.76	75	2.2	1.8	0.679	15.28
14#	3.14	98.4	1.76	75	1.6	1.4	0.679	15.65
15#	3.23	104.8	1.76	75	2.2	1.4	0.679	15.46
16#	2.94	92.7	1.76	75	1.6	1.6	0.679	13.35
17#	2.94	92.7	1.07	65	1.6	1.8	0.579	16.34
18#	3.14	98.4	1.07	65	2.2	1.8	0.579	13.85
19#	3.14	98.4	1.3	70	1.9	1.6	0.629	14.36
20#	3.23	104.8	1.3	70	1.9	1.6	0.629	17.65
21#	3.23	104.8	1.76	75	1.6	1.4	0.679	14.5

The data sets 1 to 12 in the orthogonal test table of the input data are used as the training sample set of the model, 13–15 as the validation set of the model and the data sets 16 to 21 and other 6 groups are also used as the test sample set. The experimental factor is used as the input factor of the network, and the result is used as the output factor.

4.2.1. Determine the Number of Hidden Layer Nodes

A three-layer neural network with infinite hidden layer nodes can realize the selection of the number of hidden layer nodes when constructing any neural network. If the number of hidden layer nodes is too small, the network cannot establish a complex mapping relationship, and the network prediction error is large. However, if there are too many nodes, the network learning time system will increase, and the phenomenon of "over-fitting" may occur, that is, the prediction of training samples is accurate, and the prediction error of other samples is larger. At present, there is no theoretical equation for calculating hidden layer neurons, but in the process of practical use, it is generally based on empirical Equation (19):

$$l < \sqrt{(m+n)} + a \tag{19}$$

where, n—number of nodes of input layer;

l—number of hidden layer nodes;

m—number of output layer nodes;

a—the constant of 0–10, where n is 7 and m is 1,

From Equation (19), the interval of the number of neurons in the hidden layer BP network can be determined as [2,13]. Analyze the effect of different number of neurons in the hidden layer on the prediction performance of the BP model. As can be seen from Figure 11, with the increasing number of hidden nodes, the standard deviation of the training error of the BP model gradually decreases, and the standard deviation of the simulation error also shows an overall decreasing trend; while the standard deviation of the average simulation error shows a trend of first decreasing and then increasing. This indicates that the network gradually transitions from the "under-learning" state to the "over-learning" state as the number of implicit neurons increases. When the number of hidden nodes is 10, the training standard deviation, the simulation standard deviation, and the average error all achieve the best results.

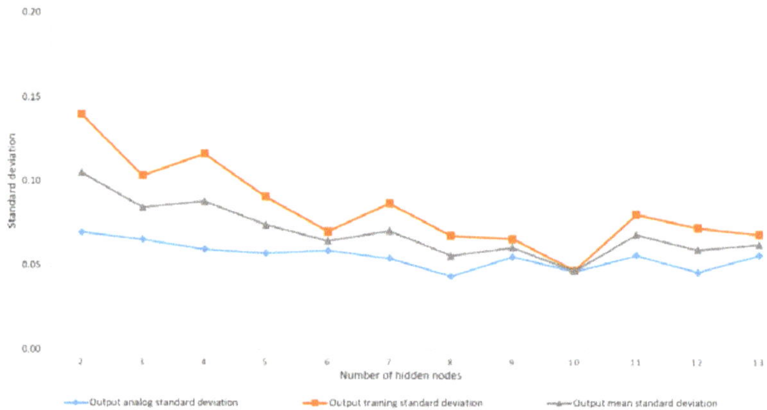

Figure 11. The effect of hidden layer nodes on the prediction effect.

4.2.2. Determining Inertia Weights ω

It is shown [34] that the particle swarm algorithm performs best and has the highest probability of finding the global optimal solution when $0.8 < \omega < 1.2$. Take ω as 0.8, 0.9, 1.0, 1.1, 1.2, run ten times, and calculate the average value of the fitness, and the impact results are shown in Figure 12. When $\omega = 0.9$, the global best fitness value calculated by the debugging parameters is the smallest and the model has the best solution performance.

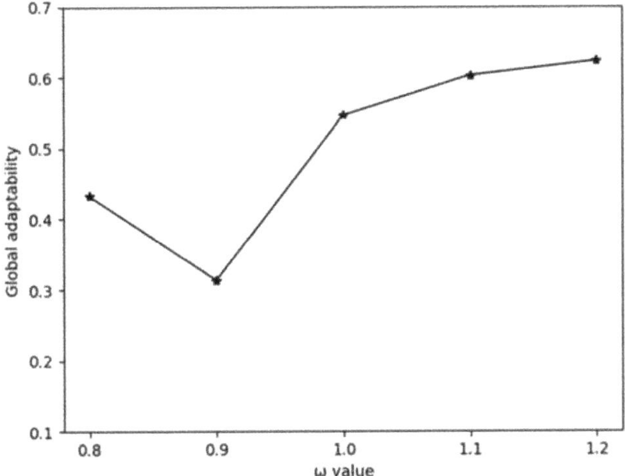

Figure 12. Effect of inertia weights on model performance.

4.2.3. Determining the Size of the Population

The population size M is generally taken from 20 to 90, and can be increased appropriately for complex or class-specific problems. However, the study shows that too large M values have no significant effect on improving the convergence accuracy of the model, while the computational complexity increases and the effect of finding the best becomes worse. M was divided into 15 groups and each group was run ten times to calculate the average value of the fitness, and the impact results were obtained as shown in Figure 13. When M = 70, the model had the best solution performance.

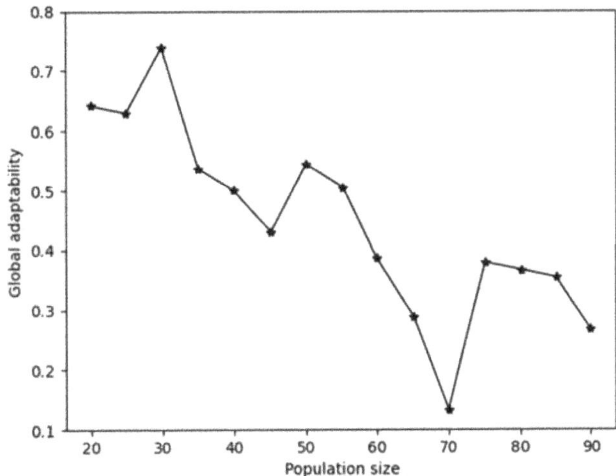

Figure 13. Effect of population size on model performance.

4.2.4. Determine the Maximum Velocity of the Particle

To speed up the model computation and prevent the model from diverging or falling into local optimum in the iteration, the maximum velocity $V_{max} = kX_{max}$ $(0.1 \leq k \leq 1)$ is generally set, X_{max} is the maximum search space, and X_{max} is taken as 7. The k values are divided into ten groups and run ten times to calculate the average value of the fitness, and the impact is derived as shown in Figure 14, when $k = 0.7$ and $V_{max} = 4.9$ the model has the best solution performance.

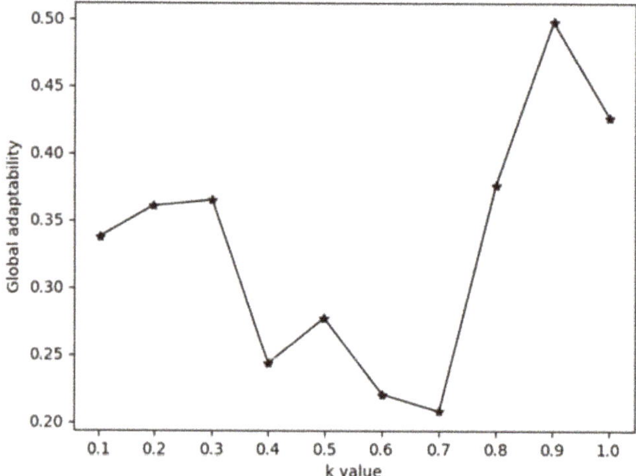

Figure 14. Effect of maximum particle velocity on model performance.

4.2.5. Number of Iterations

The PSO-BP algorithm is used to track the change of the fitness value of the objective function, and the change curve of the optimal individual fitness value of the particle swarm algorithm with the number of iterations is obtained, when the number of iterations reaches 15, the model gradually stabilizes, as shown in Figure 15.

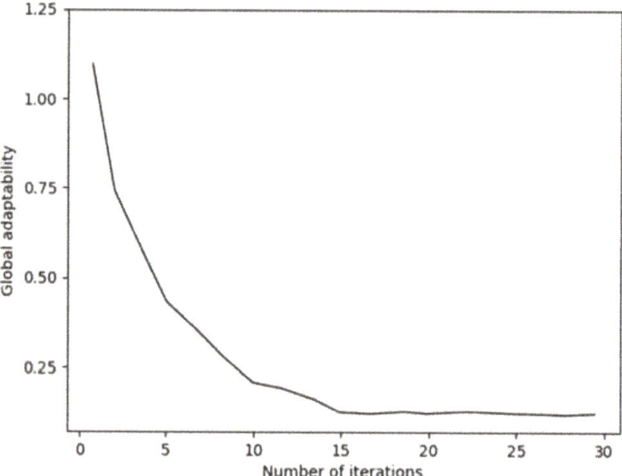

Figure 15. Effect of number of iterations on model performance.

The acceleration factors c_1, c_2 are dynamically adjusted with time [34] In summary, the model parameters are set as follows: number of nodes in the implicit layer l = 10, inertia weight $\omega = 0.9$, population size $M = 70$, maximum velocity $V_{max} = 4.9$, $c_1, c_2 \in [0.5, 2.5]$. The structure of the PSO-BP neural network model is as Figure 16.

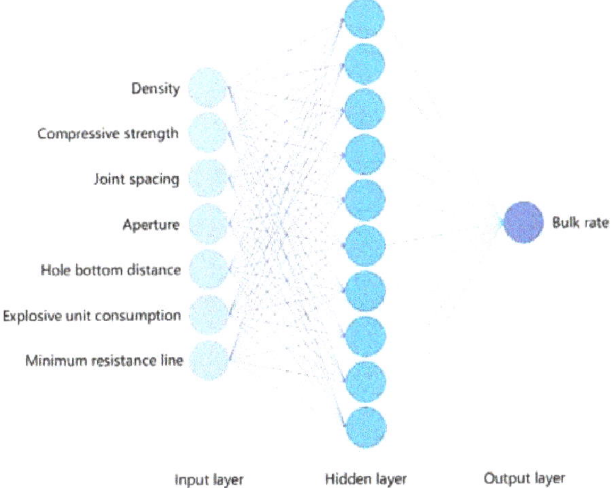

Figure 16. BP neural network model structure.

4.3. PSO-BP Deep Hole Blasting Block Prediction

The test sample Table 6 was input into the optimized PSO-BP neural network model, and the output values were compared as shown in the figure, trends almost overlap and the difference between the actual and predicted values was between 10^{-2} and 10^{-4}.

The calculation results of error are shown in Figures 17 and 18. It can be seen from Figure 17 that the error between the real value and the predicted value obtained by PSO-BP neural network algorithm optimized by particle swarm optimization is very small. The high accuracy of parameters obtained by learning sample inversion shows that the algorithm is feasible and can be used to predict the fragmentation of deep hole blasting. As we can see from Figure 18, the test data one of the model regression plot has reached 0.88316. This shows the high accuracy of the prediction, which can be used for practical applications.

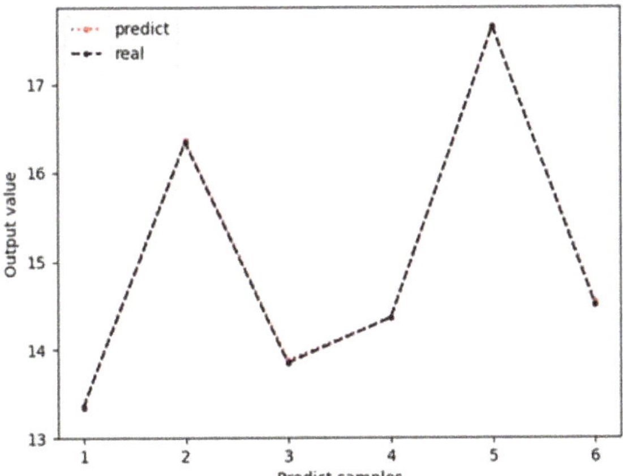

Figure 17. Comparison of actual and predicted values.

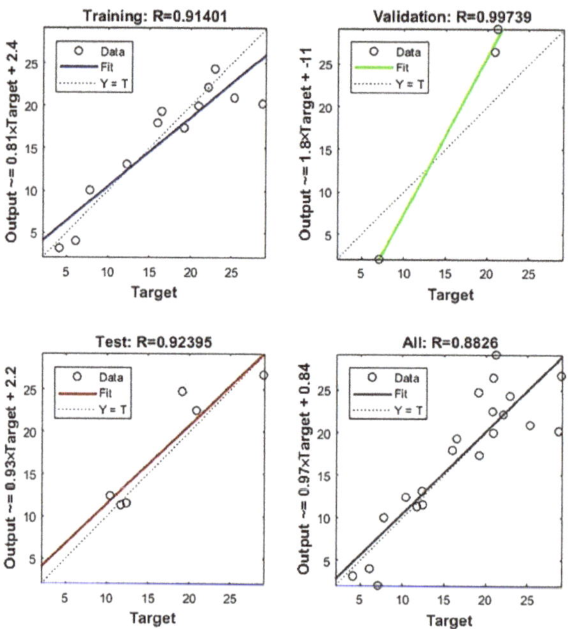

Figure 18. PSO-BP plot regression.

Further subdivision of hole bottom distance and minimum resistance line to predict blast block rate based on production accuracy requirements. Input the design samples for predicting the fragmentation of deep-hole blasting, and output the PSO-BP network prediction results of the fragmentation of deep-hole blasting after meeting the requirements of maximum iteration times or errors. Use PSO-BP network to calculate the data in Table 7. When the rock density is 3.14 t·m^{-3}, the compressive strength of the rock is 98.4 Mpa, and the Joint crack is 1.76 strip m^{-1}, the optimal blasting block rate of 6.83% can be achieved by arranging 75 mm drill holes, setting the bottom distance of the holes at 2.2 m, the minimum resistance line at 1.4 m, and the primary explosive unit consumption is taken as 0.629 kg·m^{-3}. This parameter has been applied to Sanxin Gold and Copper Mine Company.

Table 7. Predictive design data.

No.	Density/(t·m^{-3})	Compressive Strength/Mpa	Joint Crack/ (Strip m^{-1})	Aperture/mm	Hole Bottom Distance/m	Minimum Line of Resistance/m	Unit Consumption of Explosives/(kg·m^{-3})	Predicted Block Rate/%
1#	2.94	92.7	1.07	65	1.6	1.4	0.579	17.85
2#	2.94	92.7	1.3	70	1.75	1.5	0.679	14.05
3#	2.94	92.7	1.76	75	1.9	1.6	0.629	15.78
4#	2.94	92.7	1.07	70	2.05	1.7	0.604	11.50
5#	2.94	92.7	1.3	75	2.2	1.8	0.654	12.31
6#	3.14	98.4	1.07	65	1.6	1.8	0.579	14.30
7#	3.14	98.4	1.3	70	1.75	1.7	0.679	12.41
8#	3.14	98.4	1.76	75	1.9	1.6	0.659	15.71
9#	3.14	98.4	1.3	70	2.05	1.5	0.604	11.35
10#	3.14	98.4	1.76	75	2.2	1.4	0.629	6.83
11#	3.23	104.8	1.07	65	1.6	1.4	0.579	13.74
12#	3.23	104.8	1.3	70	1.75	1.5	0.604	13.01
13#	3.23	104.8	1.76	75	1.9	1.6	0.654	12.09
14#	3.23	104.8	1.3	70	2.05	1.7	0.629	17.78
15#	3.23	104.8	1.76	75	2.2	1.8	0.679	19.52

5. Conclusions

In this paper, the blasting parameters were optimized using blast funnel tests to address the problem of high blast block rate in the actual production of Sanxin gold and copper mine, and the blast parameters were optimized using PSO-BP neuron algorithm to predict the blast bulk output rate of sector deep hole blasting after the optimization of blast parameters, and the following conclusions were obtained in this study:

(1) Blast funnel tests were carried out in the 609 level to determine deep hole blasting parameters. In the field application of ammonium explosives, blasting funnel characteristics curve, blasting funnel volume and explosives burial depth relationship, funnel radius and explosives burial depth relationship, the best burial depth and critical burial depth of the blasting funnel, bursting pile block degree, and other content analysis and research were carried out, and then the deep hole blasting hole bottom distance, row spacing (minimum resistance line), and explosives unit consumption and other parameters were estimated. The final determination of the bottom distance of the hole is 2.2 m, the unit consumption of explosives is 0.629 kg/m^3, and the row spacing is 1.4 m.

(2) Establishment of PSO-BP deep hole blasting block prediction model, the rock properties and blast design parameters are selected as the input layer of the neural network, and the block rate as the output layer, the optimal number of hidden layer nodes is determined as 10, block prediction for deep sector blasting after optimization of hole network parameters, predicted block rate of 6.83% after optimization of orifice network parameters. It is predicted that its prediction accuracy is high and that blasting parameters can be optimized to effectively reduce the block rate.

(3) For underground mines with high blasting block yield, the results of the thesis can be applied with slight modifications based on the combination of the mine's own geological conditions, which can effectively reduce the generation of blasting block rate at the site, improving the mine's efficiency, and providing guidance to underground mining enterprises to reduce the blasting block rate in blasting construction.

Author Contributions: Conceptualization, B.K.; methodology, B.K.; software, Y.Y.; validation, R.P. and Y.Y.; formal analysis, B.K. and X.C.; investigation, R.P. and Y.Y.; resources, Y.H., W.W. and G.L.; data curation, R.P.; writing—original draft preparation, Y.Y. and R.P.; writing—review and editing, B.K. and J.Z.; visualization, Y.Y.; supervision, X.C. and G.R.; project administration, X.C.; funding acquisition, X.C. All authors have read and agreed to the published version of the manuscript.

Funding: This study was supported by the Fundamental Research Funds for the Central Universities (WUT: 2021IVA040).

Data Availability Statement: Not applicable.

Conflicts of Interest: The authors declare no conflict of interest.

References

1. Li, Q.; Luo, Z.; Huang, M.; Pan, J.; Wang, G.; Cheng, Y. Control of Rock Block Fragmentation Based on the Optimization of Shaft Blasting Parameters. *Geofluids* **2020**, *10*, 6687685. [CrossRef]
2. Xiao, S. Analysis of the impact of large diameter deep hole mining blasting on the filling body. *Nonferrous Met. Eng.* **2015**, *5* (Suppl. S1), 113–115. (In Chinese)
3. Xian, W. Optimization of blasting parameters to reduce the vibration to the filling body. *Chin. J. Saf. Sci.* **1998**, *8* (Suppl. S1), 99–102+123. (In Chinese)
4. Akande, J.M.; Lawal, A.I. Optimization of blasting parameters using regression models in Ratcon and NSCE granite quarries, Ibadan, Oyo State. *Nigeria* **2013**, *3*, 28–37. [CrossRef]
5. Rezaeineshat, A.; Monjezi, M.; Mehrdanesh, A.; Khandelwal, M. Optimization of blasting design in open pit limestone mines with the aim of reducing ground vibration using robust techniques. *Géoméch. Geophys. Geo-Energy Geo-Resour.* **2020**, *6*, 14. [CrossRef]
6. Shad, H.I.A.; Ahangari, K. An empirical relation to calculate the proper burden in blast design of open pit mines based on modification of the Konya relation. *Int. J. Rock Mech. Min. Sci.* **2012**, *56*, 121–126. [CrossRef]
7. Wang, Z.; Konietzky, H.; Shen, R. Coupled finite element and discrete element method for underground blast in faulted rock masses. *Soil Dyn. Earthq. Eng.* **2009**, *29*, 939–945. [CrossRef]
8. Ben Messaoud, S.; Hamdi, E.; Gaied, M. Coupled geomechanical classification and multivariate statistical analysis approach for the optimization of blasting rock boulders. *Arab. J. Geosci.* **2020**, *13*, 19. [CrossRef]

9. Monjezi, M.; Dehghani, H. Evaluation of effect of blasting pattern parameters on back break using neural networks. *Int. J. Rock Mech. Min. Sci.* **2008**, *45*, 1446–1453. [CrossRef]
10. Sadollah, A.; Bahreininejad, A.; Eskandar, H.; Hamdi, M. Mine blast algorithm: A new population based algorithm for solving constrained engineering optimization problems. *Appl. Soft Comput.* **2013**, *13*, 2592–2612. [CrossRef]
11. Liu, K.; Liu, B. Optimization of smooth blasting parameters for mountain tunnel construction with specified control indices based on a GA and ISVR coupling algorithm. *Tunn. Undergr. Space Technol.* **2017**, *70*, 363–374. [CrossRef]
12. Remli, S.; Benselhoub, A.; Rouaiguia, I. Optimization of Blasting Parameters in Open Cast Quarries of el Has-sa-Bouira (Northern Algeria). *GeoScience Eng.* **2019**, *65*, 53–62. [CrossRef]
13. Zhou, J.; Koopialipoor, M.; Murlidhar, B.R.; Fatemi, S.A.; Tahir, M.M.; Armaghani, D.J.; Li, C. Use of Intelligent Methods to Design Effective Pattern Parameters of Mine Blasting to Minimize Flyrock Distance. *Nonrenew. Resour.* **2019**, *29*, 625–639. [CrossRef]
14. Monjezi, M.; Ahmadi, Z.; Khandelwal, M. Application of neural networks for the prediction of rock fragmentation in Chadormalu iron mine. *Arch. Min. Sci.* **2012**, *57*, 787–798. [CrossRef]
15. Ghiasi, M.; Askarnejad, N.; Dindarloo, S.R.; Shamsoddini, H. Prediction of blast boulders in open pit mines via multiple regression and artificial neural networks. *Int. J. Min. Sci. Technol.* **2016**, *26*, 183–186. [CrossRef]
16. Ghaeini, N.; Mousakhani, M.; Amnieh, H.B.; Jafari, A. Prediction of blasting-induced fragmentation in Meydook copper mine using empirical, statistical, and mutual information models. *Arab. J. Geosci.* **2017**, *10*, 409. [CrossRef]
17. Hesarouieh, N.G.; Mousakhani, M.; Amnieh, H.B.; Jafari, A. Prediction of fragmentation due to blasting using mutual information and rock engineering system; case study: Meydook copper mine. *Int. J. Min. Geo-Eng.* **2017**, *51*, 23–28. [CrossRef]
18. Xie, C.; Nguyen, H.; Bui, X.-N.; Choi, Y.; Zhou, J.; Nguyen-Trang, T. Predicting rock size distribution in mine blasting using various novel soft computing models based on meta-heuristics and machine learning algorithms. *Geosci. Front.* **2021**, *12*, 101108. [CrossRef]
19. Bahrami, A.; Monjezi, M.; Goshtasbi, K.; Ghazvinian, A. Prediction of rock fragmentation due to blasting using artificial neural network. *Eng. Comput.* **2010**, *27*, 177–181. [CrossRef]
20. Murlidhar, B.R.; Armaghani, D.J.; Mohamad, E.T.; Changthan, S. Rock fragmentation prediction through a new hybrid model based on imperial competitive algorithm and neural network. *Smart Constr. Res.* **2018**, *2*, 38–49.
21. Ouchterlony, F.; Sanchidrian, J.A. A review of development of better prediction equations for blast fragmentation. *J. Rock Mech. Geotech. Eng.* **2019**, *11*, 1094–1109. [CrossRef]
22. Cunningham, C. The Kuz-Ram Model for production of fragmentation from blasting. In Proceedings of the 1st Symposium on Rock Fragmentation by Blasting, Lulea, Sweden, 23–26 August 1983.
23. Djordjevic, N. A two-component model of blast fragmentation. *AusIMM Proc.* **1999**, *304*, 9–13.
24. Kanchibotla, S.S.; Valery, W.; Morrell, S. Modelling fines in blast fragmentation and its impact on crushing and grinding. In *Explo '99–A Conference on Rock Breaking, Kalgoorlie, Australia, 7–11 November 1999*; The Australasian Institute of Mining and Metallurgy: Carlton, Australia, 1999; pp. 137–144.
25. Thornton, D.; Kanchibotla, S.; Esterle, J. A fragmentation model to estimate ROM size distribution of soft rock types. In Proceedings of the Twenty-Seventh Annual Conference on Explosives and Blasting Technique, Orlando, FL, USA, 28–31 January 2001; Volume I, pp. 41–53.
26. Cunningham, C. The Kuz-Ram fragmentation model–20 years on. In *European Federation of Explosives Engineers Brighton Conference Proceedings, Brighton, UK, 13–16 September 2005*; European Federation of Explosives Engineers: Brighton, UK, 2005; pp. 201–210.
27. Ouchterlony, F. What does the fragment size distribution of blasted rock look like? In Proceedings of the EFEE World Conference on Ex-Plosives and Blasting, Brighton, UK, 14–16 September 2005; pp. 189–199.
28. Dhekne, P.Y.; Pradhan, M.; Jade, R.K.; Mishra, R. Prediction Of Boulder Count In Limestone Quarry Blasting: Statistical Modeling Approach. *J. Min. Sci.* **2020**, *56*, 771–783. [CrossRef]
29. Miao, Y.; Zhang, Y.; Wu, D.; Li, K.; Yan, X.; Lin, J. Rock Fragmentation Size Distribution Prediction and Blasting Parameter Optimization Based on the Muck-Pile Model. *Min. Met. Explor.* **2021**, *38*, 1071–1080. [CrossRef]
30. Mutinda, E.; Alunda, B.; Maina, D.; Kasomo, R. Prediction of rock fragmentation using the Kuz-netsov-Cunningham-Ouchterlony model. *J. South. Afr. Inst. Min. Metall.* **2021**, *121*, 107–112. [CrossRef]
31. Bamford, T.; Esmaeili, K.; Schoellig, A.P. A deep learning approach for rock fragmentation analysis. *Int. J. Rock Mech. Min. Sci.* **2021**, *145*, 104839. [CrossRef]
32. Yan, Z.; Yan, C.; Mei, L.; Dong, Z. Analysis of the plugging length of the blast hole in drilling and blasting. *Blasting* **2021**, *38*, 45–49+129. (In Chinese)
33. Yun, D.; De, Z.; Gao, Y.; Qin, Z.; Xie, L. Optimal blasting of stope blasting drilling parameters based on BP network. *Blasting* **2014**, *31*, 57–62. (In Chinese)
34. Ratnaweera, A.; Halgamuge, S.K.; Watson, H.C. Self-Organizing Hierarchical Particle Swarm Optimizer With Time-Varying Acceleration Coefficients. *IEEE Trans. Evol. Comput.* **2004**, *8*, 240–255. [CrossRef]

Article

Variable Weight-Projection Gray Target Evaluation Model of Degree of Protection of Protective Layer Mining

Bing Qin [1,*], Zhanshan Shi [2,3], Jianfeng Hao [2], Bing Liang [1], Weiji Sun [1] and Feng He [1]

1 Institute of Mechanics and Engineering, Liaoning Technical University, Fuxin 123000, China; liangbing@lntu.edu.cn (B.L.); sunweiji@lntu.edu.cn (W.S.); hefeng@lntu.edu.cn (F.H.)
2 Institute of Mining, Liaoning Technical University, Fuxin 123000, China; shizhanshan@lntu.edu.cn (Z.S.); 471610041@stu.lntu.edu.cn (J.H.)
3 Liaoning Academy of Mineral Resources Development, Utilization Technical and Equipment Research Institute, Liaoning Technical University, Fuxin 123000, China
* Correspondence: qinbing@lntu.edu.cn; Tel.: +86-137-9508-4408

Abstract: In order to quantitatively evaluate the degree of protection in protective layer mining and provide guidance for the design of a secondary outburst elimination scheme, a variable weight-projection gray target dynamic evaluation model for the effectiveness of protective layer mining is established. The improved order relation analysis method was used to determine the subjective weight of each index toward the decision-making goal based on numerical diversity characteristics, and the initial fixed-weight calculation for mixed multi-attribute metrics was processed through the degree of index action. The variable weight function was used to dynamically adjust the fixed weight through the penalty and incentive index methods. Four indexes (gas content, gas pressure, coal seam permeability coefficient, and expansion deformation) were selected, the outburst elimination and anti-reflection were taken as the guide, and the critical value of each index for eliminating burst and the critical value of pressure relief were taken as the positive and negative bullseyes. Based on the variable weight-projection gray target decision model, the distance between the two target centers of each scheme was calculated; at the same time, the variable weight vector changed dynamically with the evaluation scheme to achieve the dynamic quantitative evaluation of the degree of protection. Additionally, compared with the calculation results of fixed weights, it was found that the variable weight-projection bullseye distance can more accurately reflect the dynamic control effect of differences in numerical combinations of multi-attribute indexes in different decision schemes based on the degree of protection of protective layer mining. Taking a mine in PingMei as the engineering background, Ding protected the Wu area, and the degree of protection in the Wu group coal seam reached 116.29%, eliminating the outburst risk of the coal seam. The Wu protected the Ji group coal seam, with the degree of outburst risk in the Ji group being reduced by 14.27, and the Ding + Wu group protected the Ji group coal seam, with the degree of outburst risk of the Ji group being reduced by 20.71%, but not eliminated. The evaluation model quantifies the degree of protection of protective layer mining, and provides a theoretical basis for further assessing whether the working face should strengthen the enhancing permeability or whether it needs to be used in tandem with high-strength outburst elimination methods.

Keywords: protective layer mining; degree of protection; variable weight theory; mixed multi-attribute index; projection gray target; dynamic evaluation

Citation: Qin, B.; Shi, Z.; Hao, J.; Liang, B.; Sun, W.; He, F. Variable Weight-Projection Gray Target Evaluation Model of Degree of Protection of Protective Layer Mining. *Energies* **2022**, *15*, 4654. https://doi.org/10.3390/en15134654

Academic Editors: Sergey Zhironkin and Krzysztof Tajduś

Received: 22 May 2022
Accepted: 22 June 2022
Published: 25 June 2022

1. Introduction

The outburst risk must be eliminated before the mining of a dangerous coal seam. However, protective layer mining is the most economical and effective regional measure to prevent coal and gas outburst [1,2] (pp. 21–36). The purpose of protective layer mining is to relieve the pressure in the coal seam and increase the permeability, and the gas extraction method is used to control the gas and prevent dangerous outbursts with a protective layer [3,4]. At present, the basis for judging the mining effect of the protective layer is mainly the provisions of the (AQ 1050–2008) Technical Specification for Protective Layer Mining [5] (pp. 2–3). The specification states that the index for judging the protection effect of the protected layer includes gas pressure, gas content, and gas extraction volume. The specification gives the specific value of the inspection index for determining whether to eliminate the burst, but does not specify the degree of pressure relief protection derived from the protective layer mining. At present, most of the research on evaluating the protection effect of protective layer mining only examines whether the outburst is eliminated [6]; however, there is currently no in-depth analysis of the situation in which the protected layer does not eliminate the outburst, but has a certain protective effect, quantifying that the degree of protection can accurately guide the formulation of secondary outburst elimination plans, improve outburst coal seam governance efficiency, and save production costs.

In terms of research on the evaluation of protective effects, Chen [7] and Kang [8] analyzed the effectiveness of upper protective layer mining by monitoring the gas pressure, the coal seam relative expansion and deformation, the gas flow, and the gas permeability changes in the protected coal seam. Yuan [9] proposed a technology to quickly and accurately measure the gas content in coal seams on site, and based on the gas content of the protected coal seam to determine the coal seam outburst elimination range of the protective layer. Liu [10] used the analytic hierarchy process to establish a reliability evaluation system for protective layer mining and proposed 28 evaluation indexes. Du [11] calculated the reliability of the gas content, expansion deformation, and gas pressure in the protected layer, obtained the index weight by fuzzy AHP algorithm, put forward a comprehensive quantitative index of the mining pressure relief effect, and judged the pressure relief effects and boundaries.

At the same time, when the protection effect evaluation index value for the protection layer is obtained, the evaluation index value is no longer a single exact number due to the incompleteness of coal mine site information acquisition and the particularity of the index; the index value may be a mixed number of exact numbers, interval numbers, or triangular fuzzy numbers. In order to solve the scheme evaluation of various data form indexes, SUN [12] realized the measurement of exact numbers, interval numbers, and triangular fuzzy numbers by constructing an interval attribute framework. Ma [13] proposed a mixed-attribute generalized gray target decision-making method, and used the vector method to deal with the mixed multi-attribute problem of deterministic and indeterminate numbers. Ma [14,15] used the two-element connection number to form a vector to deal with attributes and weights, in which both attributes contribute to the decision problem of mixed-interval grey numbers. The above methods can provide a reference for the treatment of index values in evaluating the degree of protection.

Due to differences in geological conditions, gas content and gas pressure are different in the same coal seam. When testing the same working face or the same coal seam, the data present a certain discreteness. This feature must be taken into account in the analysis of the outburst elimination range of protective layer mining [4]. When assigning index weights, traditional evaluation methods do not consider the impact of different index values on evaluation goals. In fact, the index value takes different values in different schemes and cannot be calculated according to the fixed weight [16]. Aiming to address the problem that a change in the index value will cause a dynamic change in the evaluation results, and that the fixed weight is not applicable in the evaluation, Wang [17] proposed the variable weight theory. Based on the variable weight theory, Wu [18] perfected the method for determining the threshold value of variable weight interval and weight adjustment parameters, and used the variable weight function to solve the evaluation problem of water inrush from the coal seam floor. Xu [19] proposed dynamic evaluation from the perspective of time, and proposed the dynamic evaluation method of grey target theory. This method can compare the evaluation values and ranking results of each scheme at each moment and overall, in a certain period of time. In addition, there are methods that combine subjective weights and objective weights for decision evaluation [20], and that can evaluate the synergy of coal and gas co-mining through fuzzy mathematics (objective) and analytic hierarchy process (subjective) -combined weighting. Chen [21] fused AHP (subjective) and the entropy weight method (objective) and built an optimization model based on the Lagrange function. The weight of the index combination was obtained, and the coal mine rock burst evaluation model was constructed. The above methods can provide a reference for the treatment of weights when evaluating the degree of protection.

The existing quantitative evaluation for the degree of protection in the protected coal seam that has not been eliminated remains to be solved, and the subjective fixed weights of the index cannot take into account the control effect of the various changes in the mixed multi-attribute index value on the protective effect. To address these issues, this paper proposes a dynamic evaluation model of protective layer mining effectiveness based on a variable weight-projection gray target. Aimed at the different types of indexes, the calculation of the mixed multi-attribute index is realized by using the index action degree, the improved order relation analysis method is used to calculate the subjective constant weight of the index, and the variable weight function is used to dynamically correct the subjective constant weight of the index. Two evaluation objectives of outburst elimination and increased permeability are designed, and the variable weight theory and the gray target theory are integrated, in order to eliminate the critical value of outburst risk and the critical value of increased permeability as the bullseye. Field data were selected before and after the protective layer mining of a mine in Pingdingshan, and the measurement for the degree of protection in protective layer mining was conducted according to the established variable weight-projection gray target model.

2. Quantitative Evaluation Index for Protection Degree of Protective Layer Mining

The aim of protective layer mining is to eliminate the danger of a coal seam outburst, relieve the pressure and increase the permeability of a coal seam, and promote gas drainage. Therefore, the four indexes of gas content, gas pressure, coal seam permeability coefficient, and expansion deformation were selected to establish a systematic evaluation index system.

2.1. Coal and Gas Outburst Risk Evaluation Index

The "Technical Specifications for Mining of Protective Layers" [5] (pp. 2–3) requires that the gas content or gas pressure in the protected layer be reduced to below the value of the initial outburst depth. If there are no data, the gas content of the coal seam must be reduced to below 8 m^3/t, or the gas pressure reduced to below 0.74 MPa, in order for it to be judged that the purpose of eliminating bursts has been achieved.

Coal seam gas content and pressure are important indexes for evaluating the risk of coal seam outburst. After the protected layer is depressurized, the gas content and pressure decrease significantly. According to the "Detailed Rules for Prevention and Control of Coal and Gas Outburst" [22] (pp. 22–23), for the elimination of the outburst risk, the coal seam gas content/pressure needs to be less than 8 m^3/t or 0.74 MPa.

2.2. Evaluation Index of Coal Seam Pressure Relief and Permeability Enhancement Effect

2.2.1. Coal Seam Expansion Value

During the protective layer mining, a goaf is generated, and during the movement of the roof and floor coal strata, due to the change in stress, the adjacent upper and lower coal seams will undergo vertical deformation. In the pressure relief area, the coal seam will expand and deform, and the deformation also reflects the stress change in the coal seam. When the coal seam expands, it indicates that pressure in the coal seam is relieved, and the larger the expansion value, the better the pressure relief. According to the "Regulations on the Prevention and Control of Coal and Gas Outbursts", the area where the maximum expansion value in the protected layer reaches 3‰ is the effective pressure relief range.

2.2.2. Coal Seam Permeability Coefficient

The coal seam permeability coefficient is a sign of the difficulty in coal seam gas flow, and it is also one of the important indexes in the degree of pressure relief. The classification of the coal seam permeability coefficient is shown in Table 1 [1]. The permeability of the coal seam is closely related to the stress state and fracture development characteristics in the coal seam. Under the action of the protective layer, the gas pressure in the protected layer reduction, expansion deformation and fracture development in the protected layer jointly promote a significant increase in the permeability; the permeability coefficient of the coal seam can increase from 100 to 1000-times. According to statistical analysis, when the relative value in the protected layer can reach more than 3‰, the permeability coefficient in the protected layer can increase by more than 300-times.

Table 1. Coal seam permeability coefficient classification.

Classification	Coal Seam Permeability Coefficient ($m^2 \cdot MPa^{-2} \cdot d^{-1}$)
easy draw	>10
normal draw	0.1~10
hard draw	<0.1

3. Dynamic Performance Evaluation Method of Protective Layer Mining Based on Variable Weight-Projection Gray Target

In order to solve the problem of inconsistency in the form of index evaluation values, and the difficulty in comparing the relative importance, the improved order relation analysis method (GI method) was used to determine the initial constant weight of the index, and then the index's advantage information and disadvantage information were synthesized. Referring to the grey target theory, taking the relative distance between the index evaluation value and the positive and negative bullseye as a variable, a new dominance function was defined. Combined with variable weight theory, a local state variable weight vector was constructed based on the punishment–reward mechanism, and the initial constant weight of the index was dynamically revised to solve the variable weight of the index. Finally, the projection gray target method was used, and the variable weight projection bullseye distance was used as the metric to determine the comprehensive efficiency and ranking of the evaluation objects [7].

3.1. Data Normalization

It was assumed that set $E = (e_1, e_2, \ldots, e_m)$ consists of m objects, which are waiting to be evaluated, and that m refers to the quantity. The set $X = (x_1, x_2, \ldots, x_n)$ consists of n indexes, which are under the same criterion layer; n refers to the quantity. The values of each index form an evaluation matrix $V = (v_{ij})_{mxn}$. Among them, v_{ij} can be given in three forms: exact number, interval number, and triangular fuzzy number. The data form of the same index in different evaluation schemes is the same. Because the dimensions of each index are different, they need to be standardized [16]. The index normalization matrix is $G = (g_{ij})_{mxn}$, g_{ij} represents the standardized evaluation value of the index x_j under the evaluation scheme e_i, and X^b and X^C represent the subscript sets of benefit-type and cost-type index in X, respectively.

The exact number $v_{ij} = v_{ij}$ can be normalized to:

$$g_{ij} = \begin{cases} v_{ij}/v_{\max j}, j \in X^b \\ 1 - v_{ij}/v_{\max j}, j \in X^c \end{cases} \quad (1)$$

where $v_{\max j} = \max\{v_{ij} | i = 1, 2, \cdots, m\}$.

The interval number $v_{ij} = \left[v_{ij}^L, v_{ij}^U\right]$ can be normalized to:

$$g_{ij} = \begin{cases} \left[v_{ij}^L / v_{\max j}^U, v_{ij}^U / v_{\max j}^U\right], j \in X^b \\ \left[1 - v_{ij}^U / v_{\max j}^U, 1 - v_{ij}^L / v_{\max j}^U\right], j \in X^c \end{cases} \quad (2)$$

where $v_{\max j}^U = \max\left\{v_{ij}^U | i = 1, 2, \cdots, m\right\}$.

The triangular fuzzy number $v_{ij} = \left(v_{ij}^L, v_{ij}^M, v_{ij}^U\right)$ can be normalized to:

$$g_{ij} = \begin{cases} \left[\frac{v_{ij}^L}{v_{\max j}^U}, \frac{v_{ij}^M}{v_{\max j}^U}, \frac{v_{ij}^U}{v_{\max j}^U}\right], j \in X^b \\ \left[1 - \frac{v_{ij}^L}{v_{\max j}^U}, 1 - \frac{v_{ij}^M}{v_{\max j}^U}, 1 - \frac{v_{ij}^U}{v_{\max j}^U}\right], j \in X^c \end{cases} \quad (3)$$

where $v_{\max j}^U = \max\left\{v_{ij} | i = 1, 2, \cdots, m\right\}$.

3.2. Measurement of Index Relative Distance

Definition 1. *Assuming that $\alpha = \left[\alpha^L, \alpha^U\right]$, $\beta = \left[\beta^L, \beta^U\right]$ is two interval numbers, the $D(\alpha, \beta)$ distance between α and β is calculated as*

$$D(\alpha, \beta) = \frac{1}{\sqrt{2}}\sqrt{\left(\alpha^L - \beta^L\right)^2 + \left(\alpha^U - \beta^U\right)^2} \quad (4)$$

When $\alpha^L = \alpha^U$, $\beta^L = \beta^U$, α and β are exact numbers, the distance is

$$D(\alpha, \beta) = |\alpha - \beta| \quad (5)$$

Definition 2. *Assuming that $\alpha = \left[\alpha^L, \alpha^M, \alpha^U\right]$ and $\beta = \left[\beta^L, \beta^M, \beta^U\right]$ is two triangular fuzzy numbers, the $D(\alpha, \beta)$ distance between α and β is calculated as*

$$D(\alpha, \beta) = \frac{1}{\sqrt{3}}\sqrt{\left(\alpha^L - \beta^L\right)^2 + \left(\alpha^M - \beta^M\right)^2 + \left(\alpha^U - \beta^U\right)^2} \quad (6)$$

3.3. Subjective Constant Weight Determined

The GI method (improved order relation analysis method) is a subjective weighting assignment method, which is based on the ranking of the degree of influence in the index on decision-making goals given by experts, and which calculates the relative importance of the index. The index action degree is introduced to solve the fixed subjective weight determination of different index forms [16].

A total of l experts are invited to rank the importance of n index affecting the evaluation target.

The Spearman coefficients of ranking given by the ith and vth experts are:

$$\rho_{iv} = 1 - \left[6\sum_{j=1}^{n}\left(\hat{x}_{ij} - \hat{x}_{vj}\right)^2\right] / \left[n\left(n^2 - 1\right)\right] \quad (7)$$

where $\hat{x}_{ij}$ and $\hat{x}_{vj}$ represent the ranking of the jth index given by the ith and vth experts, respectively. When the coefficient is greater than or equal to 0.5, the consistency test is passed; otherwise, the expert is eliminated.

The qualified ranking is converted into the corresponding score G_{ij}, and the average score $\overline{G}_j$ is calculated as follows:

$$\overline{G}_j = \frac{1}{l}\sum_{i=1}^{l} G_{ij} = \frac{1}{l}\sum_{i=1}^{l}(n - \hat{x}_{ij} + 1) \tag{8}$$

Ranking the index secondary based on average score, the relative distance of the index is calculated according to Section 3.2, and the sum of the distances used to represent the degree of action s_j; that is,

$$s_j = \sum_{i=1}^{m}\sum_{k=1}^{m} D\left(g_{ij}, g_{kj}\right) \tag{9}$$

If the means of the ranking scores of the two indexes are the same, the index with the largest effect will be ranked first. The ratio for the importance of adjacent index x_{j-1} and x_j is

$$r_j = \begin{cases} s_{j-1}/s_j, (j > 1, s_{j-1} \geq s_j) \\ 1, (j > 1, s_{j-1} < s_j) \end{cases} \tag{10}$$

The combined weight ω_j^0 of the nth index is calculated according to r_j, and ω_j^0 is a fixed subjective weight, expressed as

$$\omega_j^0 = \left(1 + \sum_{j=2}^{n}\prod_{i=j}^{n} r_i\right)^{-1} \tag{11}$$

3.4. Establish Variable Weight Function to Correct Fixed Subjective Weight

The variable weight function is used to dynamically adjust the fixed subjective weight of the index according to the specific value of the evaluation program index. When the dominance of an index is very low, even if the subjective weight of the index is relatively large, the overall effect of the scheme will be correspondingly reduced, and the index weight will be penalized. When an index has a high degree of dominance, even if its subjective weight is small, the overall effect of the scheme will be correspondingly improved, and the index weight will be encouraged. The penalty and incentive of the dynamic variable weight should correspond to the size of the constant weight, and the incentive amplitude of the variable weight function is smaller than the penalty amplitude.

3.4.1. Positive and Negative Bullseye

g_{ij} is the standardized evaluation value of the index. The mixed grey targets of different data form indexes can be expressed as follows:

Positive bullseye:

$$\begin{cases} \max\{g_{ij} | j \in C_1, i = 1, 2, \cdots, m\} \\ \max\left\{\frac{g_{ij}^L + g_{ij}^U}{2} | j \in C_2, i = 1, 2, \cdots, m\right\} \\ \max\left\{g_{ij}^M | j \in C_3, i = 1, 2, \cdots, m\right\} \end{cases} \tag{12}$$

Negative bullseye:

$$\begin{cases} \min\{g_{ij} | j \in C_1, i = 1, 2, \cdots, m\} \\ \min\left\{\frac{g_{ij}^L + g_{ij}^U}{2} | j \in C_2, i = 1, 2, \cdots, m\right\} \\ \min\left\{g_{ij}^M | j \in C_3, i = 1, 2, \cdots, m\right\} \end{cases} \tag{13}$$

C_1, C_2 and C_3 are the subscript sets of exact number, interval number, and triangular fuzzy number. Gas pressure and gas content are cost-type indexes; that is, the smaller the index value, the better. The coal seam permeability coefficient and expansion value are benefit-type indexes, and the larger the index value, the better. Among them, the gas pressure and gas content select the critical value of the outburst index in the "Detailed Rules for Prevention and Control of Coal and Gas Outburst" as the negative bullseye, and the critical value of the index formulated by each coal mine is the positive bullseye. The expansion value of the coal seam is 0 before the mining of the protective layer, which is used as the negative bullseye in this index, and the critical expansion value of 3‰ that needs to be reached in the protective layer mining is taken as the positive bullseye. The permeability coefficient takes the initial value of 0.1 for normal draw as the negative bullseye, and the intermediate value of 5 for normal draw as the positive bullseye. The selection of positive and negative bullseyes is shown in Table 2.

Table 2. Positive and negative bullseye values.

Category	Gas Pressure (MPa)	Gas Content ($m^3 \cdot t^{-1}$)	Coal Seam Permeability Coefficient ($m^2 \cdot MPa^{-1} \cdot d^{-1}$)	Expansion Value (‰)
Positive bullseye	critical value	critical value	5	3
Negative bullseye	0.74	8	0.1	0

3.4.2. Index Dominance Calculation

The dominance of x_j under the evaluation object e_i is R_{ij}, which is expressed as

$$R_{ij} = 1 - \exp\left(-\beta \frac{d_{ij}^{+}\left(g_{ij}, a_j^{+}\right) - d_{ij}^{-}\left(g_{ij}, a_j^{-}\right)}{d_{ij}^{\pm}}\right) \tag{14}$$

where $d_{ij}^{+}\left(g_{ij}, a_j^{+}\right)$, $d_{ij}^{-}\left(g_{ij}, a_j^{-}\right)$ and $d_{ij}^{\pm}$ represent the distance between the index evaluation value and the positive bullseye, the distance between the index evaluation value and the negative bullseye, and the distance between the positive and negative bullseye, respectively, calculated according to the indexes' relative distance in Section 3.2. β is the dominance adjustment coefficient, which takes a positive value; the larger the value, the greater the adjustment degree of the index dominance.

The matrix for the dominance of each index under each evaluation scheme can be given as $R = \left[R(g_{ij})\right]_{m\times n}$.

3.4.3. Variable Weight Function

The index weights are dynamically adjusted with the index status of each scheme. The penalty–incentive local state variable weight function based on index dominance is as follows:

$$S_{ij}\left(R_{ij}\right) = \begin{cases} (1/e)^{mv_j^0\left(R_{ij}-R^L\right)}, R_{ij} < R^L \\ 1, R^L \le R_{ij} \le R^U \\ e^{\delta mv_j^0\left(R_{ij}-R^U\right)}, R_{ij} > R^U \end{cases} \tag{15}$$

where, according to the index dominance, the indexes are clustered into three categories. R^L and R^U denote the critical values of punishment and incentive, respectively. The value is determined by the K-means algorithm [18].

Combining the constant weight calculation, Equation (11), and the dominance degree calculation, Equation (14), the variable weight in the jth index of the ith evaluation scheme is obtained, which is expressed as

$$\omega_{ij} = \frac{\omega_j^0 \cdot S_{ij}(R_{ij})}{\sum\limits_{j=1}^{n}\left[\omega_j^0 \cdot S_{ij}(R_{ij})\right]} \tag{16}$$

3.5. Weighted Projection Grey Target Decision Method to Determine the Advantages and Disadvantages of the Scheme

From the variable weight ω_{ij} and the positive and negative bullseye distances, the weighted positive bullseye distance $\lambda_i{}^+$, the weighted negative bullseye distance $\lambda_i{}^-$, and the average distance between the positive bullseye and the negative bullseye λ_0 are calculated for each evaluation object [16]; that is:

$$\begin{cases} \lambda_i^+ = \sum\limits_{j=1}^{n} d_{ij}^+ \omega_{ij} \\ \lambda_i^- = \sum\limits_{j=1}^{n} d_{ij}^- \omega_{ij} \\ \lambda^0 = \frac{1}{n}\sum\limits_{j=1}^{n} d_{ij}^{\pm} \end{cases} \tag{17}$$

The projected bullseye distance in the ith evaluation scheme is:

$$\eta_i = \lambda_i^- - \lambda_i^+ = \frac{(\lambda_i^-)^2 - (\lambda_i^+)^2}{\lambda^0} \tag{18}$$

The calculation flow chart for the variable weight projection target center distance is shown in Figure 1.

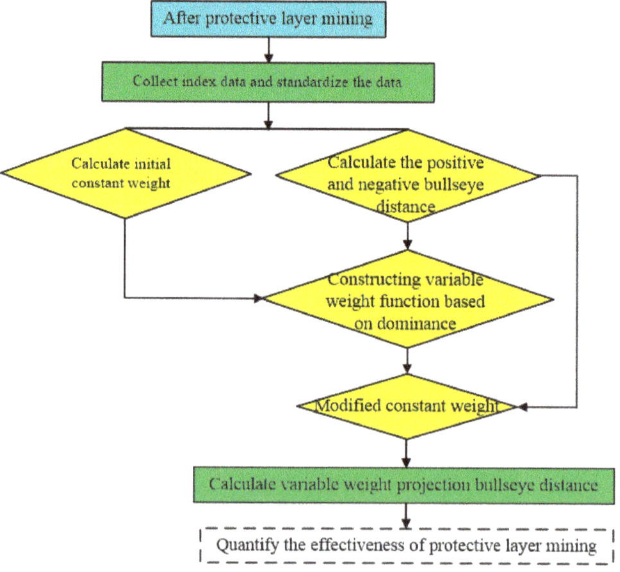

Figure 1. Calculation process.

4. Dynamic Evaluation of Degree of Protection by the Protective Layer

4.1. Evaluation Area Overview

4.1.1. Spatial Distribution Relationship of Working Face

The main coal seams in a mine of the Pingmei Group are the coal seams in group Ding, group Wu, and group Ji, in ascending order of buried depth. Protective layer mining is the preferred method for outburst prevention measures in this mining area. The location of the studied working face in the evaluation area is shown in Figure 2. Due to the staggered layout of the working face, there are mainly three types of protective seam mining in the current stage of mining: group Ding coal seam mining to protect the group Wu coal seam, group Wu coal seam mining to protect the group Ji coal seam, and group Ding + Wu coal seam mining to protect the group Ji coal seam.

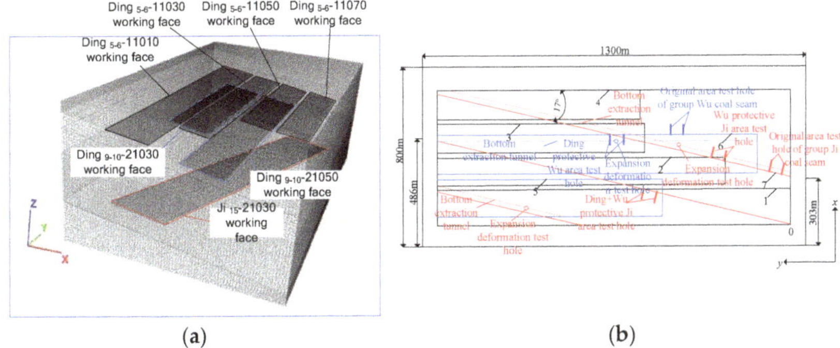

Figure 2. Spatial position relationship of working faces in the evaluation area: (**a**) working face space position; (**b**) working face plane projection. 1. Ding5-6-11010 working face; 2. Ding5-6-11030 working face; 3. Ding5-6-11050 working face; 4. Ding5-6-11070 working face; 5. Wu9-10-21030 working face; 6. Wu9-10-21050 working face; 7. Ji15-21030 working face. The black lines show the working face and measuring point located in coal seam Ding5-6, the blue lines show the working face and measuring point located in coal seam Wu9-10, and the red line shows the working face and measuring point located in coal seam Ji15.

4.1.2. Parameter Test Scheme of Evaluation Area

According to various protection forms in the layout of the working face, the test points are selected, as shown in Figure 2b. Among them, the gas content W, gas pressure P, and coal seam permeability coefficient λ are measured through the short-line measuring holes shown in the figure. During the measurement, the coal seam measuring holes are drilled along the short-line direction. First, the coal powder is taken to test the gas content, and then the hole is enlarged to observe the gas pressure. After the pressure observation, the valve is opened to release the gas. The gas permeability coefficient is calculated by testing the emission law. The expansion deformation is measured through the circle measuring hole in the figure. During the measurement, the deep base point displacement meter is constructed from the bottom extraction roadway in the working face of the coal seam Wu and Ji, and the expansion deformation is obtained by observing the displacement changes before and after mining.

4.1.3. Experimental Observation Data

1. Gas content

Two test holes are designed for each protection condition and the maximum test result is taken based on safety considerations. The gas content test results are shown in Table 3.

Table 3. Gas content test results.

Investigation Contents	Borehole Number	Coal Sample Quality (g)	Loss Amount W_1 (m^3/t)	Natural Desorption Amount W_2 (m^3/t)	Crushing Desorption Amount W_3 (m^3/t)	Non-Desorbable Amount W_c (m^3/t)	Gas Content W (m^3/t)
Original area of group Wu	NO. 1	693.4	1.06	1.59	1.53	2.23	6.41
	NO. 2	588.0	2.74	2.08	2.85	2.23	9.90
Original area of group Ji	NO. 1	575.5	1.51	3.09	3.25	1.28	9.13
	NO. 2	586.2	1.67	3.35	3.41	1.28	9.71
Group Wu protect group Ji area	NO. 1	494.2	0.78	2.19	2.13	1.28	6.38
	NO. 2	498.9	1.27	3.00	3.36	1.28	8.91
Group Ding+Wu protect group Ji area	NO. 1	614.0	0.57	2.22	2.15	1.28	6.22
	NO. 2	519.7	0.72	2.60	1.74	1.28	6.34
Group Ding protect group Wu area	NO. 1	531.7	0.62	2.31	1.67	1.28	5.88
	NO. 2	522.3	0.66	2.28	1.30	1.28	5.52

2. Gas pressure and coal seam permeability coefficient

Two test holes are designed for each protection condition and the maximum test result is taken based on safety considerations. The permeability coefficient is obtained according to the gas emission law of the borehole and the interval number is taken. The test results are shown in Figures 3–7.

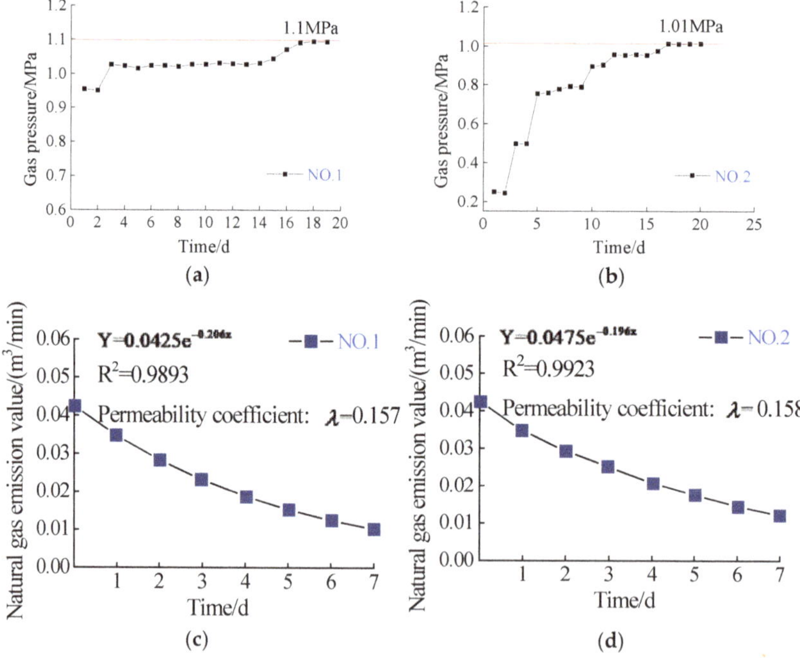

Figure 3. Original area of group Wu coal seam: (**a**) NO. 1 measuring point gas pressure test data; (**b**) NO. 2 measuring gas pressure point test data; (**c**) NO. 1 measuring point permeability coefficient test data; (**d**) NO. 2 measuring point permeability coefficient test data.

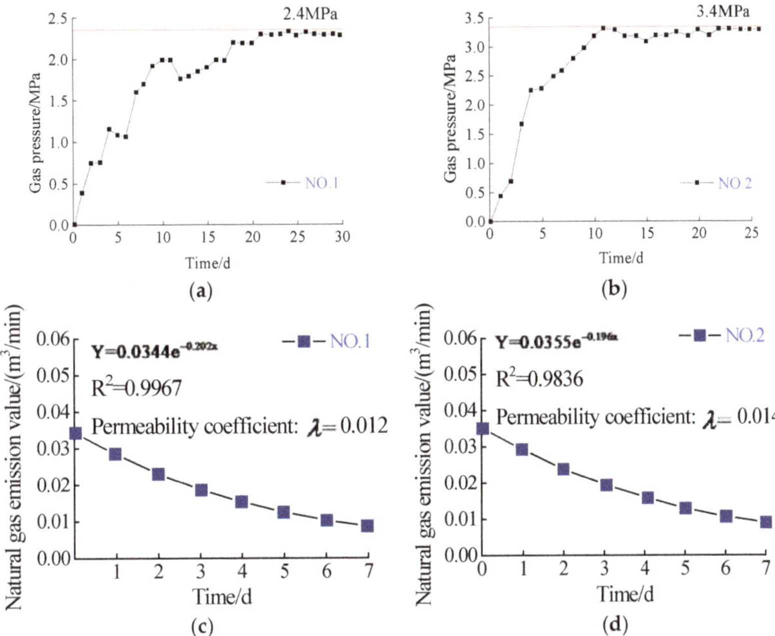

Figure 4. Original area of group Ji coal seam: (**a**) NO. 1 measuring point gas pressure test data; (**b**) NO. 2 measuring gas pressure point test data; (**c**) NO. 1 measuring point permeability coefficient test data; (**d**) NO. 2 measuring point permeability coefficient test data.

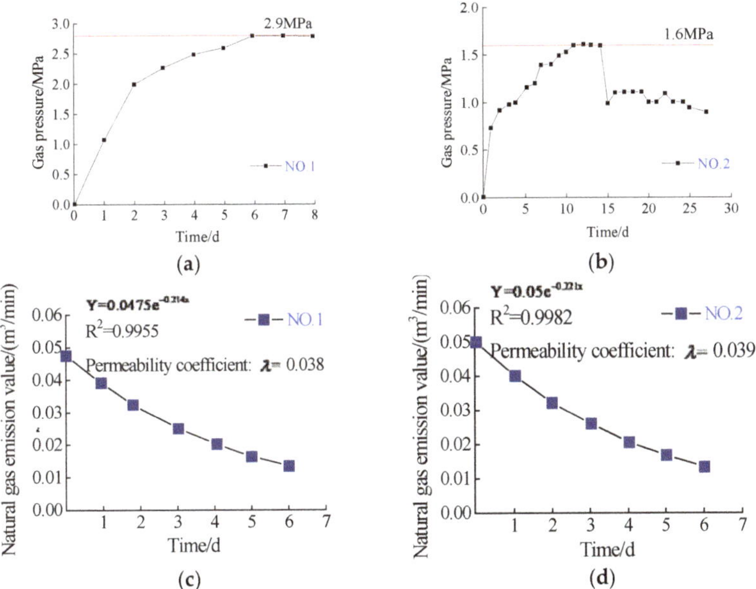

Figure 5. Wu protects Ji area: (**a**) NO. 1 measuring point gas pressure test data; (**b**) NO. 2 measuring gas pressure point test data; (**c**) NO. 1 measuring point permeability coefficient test data; (**d**) NO. 2 measuring point permeability coefficient test data.

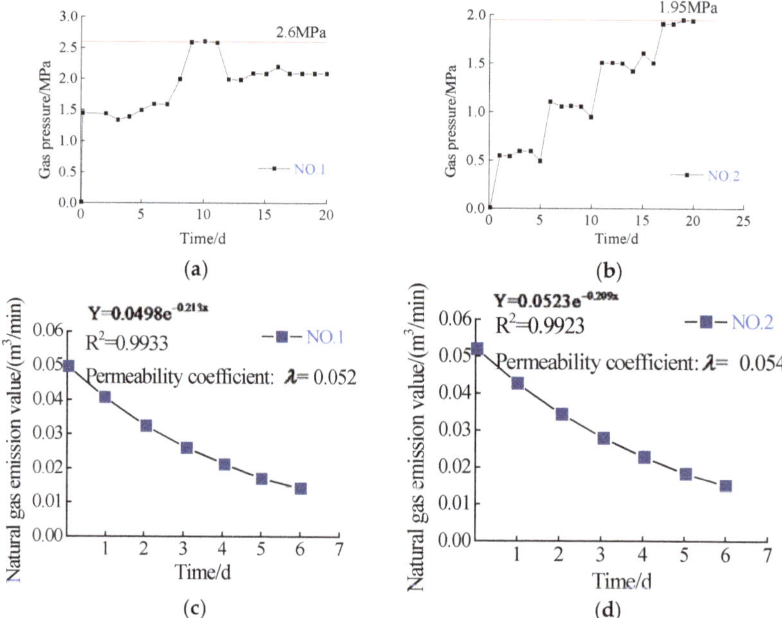

Figure 6. Ding + Wu protects Ji area: (**a**) NO. 1 measuring point gas pressure test data; (**b**) NO. 2 measuring gas pressure point test data; (**c**) NO. 1 measuring point permeability coefficient test data; (**d**) NO. 2 measuring point permeability coefficient test data.

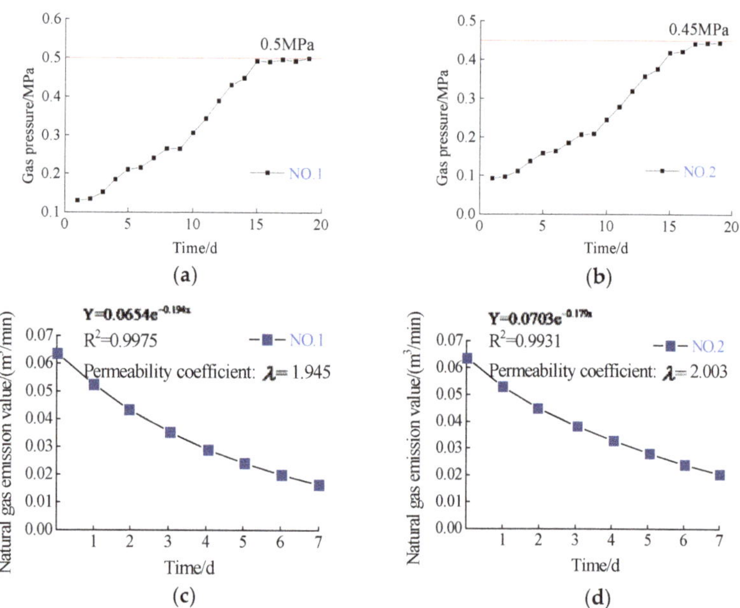

Figure 7. Ding protects Wu area: (**a**) NO. 1 measuring point gas pressure test data; (**b**) NO. 2 measuring gas pressure point test data; (**c**) NO. 1 measuring point permeability coefficient test data; (**d**) NO. 2 measuring point permeability coefficient test data.

3. Expansion deformation

The expansion deformation is tested with the device, as shown in Figure 8. Fixed points are set on the roof of the coal seam. The expansion deformation of the coal seam is measured by the change in the length of the steel wire in the hole.

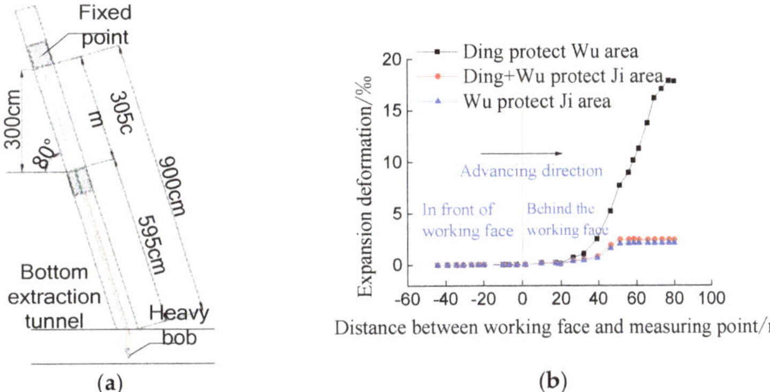

Figure 8. Expansion deformation test data: (**a**) test device; (**b**) test data.

4.2. Model Calculation

The field test obtained the original index parameters in the Wu group coal seam and Ji group coal seam in a mine of the Pingmei Group and the index parameters in the protective layer area after it was mined, as expressed in Table 4.

Table 4. Original test data of evaluation area.

Evaluation Scheme	Gas Content $W/m^3 \cdot t^{-1}$	Gas Pressure P/MPa	Permeability Coefficient $\lambda/m^2 \cdot MPa^{-1} \cdot d^{-1}$	Expansion Value/‰
Group Wu coal seam original area	9.90	1.1	0.157~0.158	0.000
Group Ji coal seam original area	9.71	3.4	0.012~0.014	0.000
The area where the group Ji coal seam is protected by the Wu group coal seam	8.91	2.9	0.038~0.039	2.140
The area where the group Ji coal seam is protected by the Ding + Wu group coal seam	6.34	2.6	0.052~0.054	2.130
The area where the group Wu coal seam is protected by the Ding group coal seam	5.88	0.5	1.945~2.003	17.80

The four indexes are represented by x_1, x_2, x_3, and x_4, respectively. The gas content and gas pressure in the collected index are exact numbers. The coal seam permeability coefficient and borehole natural flow attenuation coefficient belong to interval numbers. The index values are standardized according to Equations (1) and (3), respectively, and are dimensionless. Due to space reasons, this paper does not show the standardized results. Three experts were invited to provide their assessments of the order of importance of the indexes, and the Spearman coefficients for calculating the scores of the three experts are $\rho_1 = 0.967$, $\rho_2 = 0.967$, and $\rho_3 = 0.933$, all of which are greater than 0.5, thus, passing the consistency check. The average scores given in Table 5 are calculated according to Equation (8) in Section 3.3, and then the indexes are sorted according to the index action calculated by Equation (9). The relative importance of adjacent indexes is calculated according to Equation (10), and the subjective constant weight of the effect of each index on the protective layer mining is calculated according to Equation (11).

Table 5. Determining the subjective constant weight of index by the improved order relation analysis method.

Index	Expert 1	Expert 2	Expert 3	Average Score	Ideal Ordering	Index Action	Relative Importance	Subjective Constant Weight
x_1	3	4	4	1.333	x_2	4	\	0.333
x_2	1	1	2	3.667	x_4	2	1	0.250
x_3	4	3	3	1.667	x_3	4	0.5	0.217
x_4	2	2	1	3.333	x_1	2	2	0.200

According to Section 3.3, to correct the index subjective weights, the dominance degree R_{ij} of the four indexes is calculated by Equation (14) and formed into a matrix R:

$$R = \begin{bmatrix} -0.0049 & 0.0957 & 0.0098 & 0.1648 \\ 0.0099 & 0.0206 & 0.0101 & 0.1648 \\ 0.0038 & 0.0058 & 0.0101 & 0.1483 \\ -0.0072 & 0.0110 & 0.0101 & 0.1484 \\ -0.0097 & 0.1636 & 0.0025 & 0.0344 \end{bmatrix}$$

According to Equation (15), the independent variable in the variable weight function is divided into three regions. Therefore, the number of clusters is three, and the cluster centers are $f_1 = -0.0020, f_2 = 0.0527$, and $f_3 = 0.1566$, respectively. The penalty threshold and incentive threshold are obtained as follows:

$$\begin{cases} R^L = \frac{f_1+f_2}{2} = 0.0254 \\ R^U = \frac{f_2+f_3}{2} = 0.1047 \end{cases}$$

According to Equation (16), the variable weight is calculated based on the subjective weight ω_{ij}, as shown in Table 6.

Table 6. Index variable weight.

Scheme \ Index	x_1	x_2	x_3	x_4
1	0.1949	0.3229	0.2081	0.2742
2	0.1920	0.3195	0.2099	0.2767
3	0.1946	0.3236	0.2095	0.2724
4	0.1966	0.3218	0.2094	0.2722
5	0.1933	0.3606	0.2069	0.2391

The evaluation target in this paper is Henan Province, China, and the critical value [23] set by the locality is selected as the positive bullseye. The weighted positive bullseye distance, negative bullseye distance, and positive and negative bullseye distances of each scheme are calculated by Equation (17). The results can be substituted into Equation (18) to calculate the bullseye distance of variable weight projection, as shown in Table 7.

Table 7. Quantification of the coal seams' degree of protection before and after protective layer mining.

Evaluation Scheme	**The Original Area of the Ji Group Coal Seam**	**The Area Where the Group Ji Coal Seam Is Protected by the Wu Group Coal Seam**	**The Area Where the Group Ji Coal Seam Is Protected by the Ding + Wu Group Coal Seam**
Variable weight—projected bullseye distance	0.3119	0.2674	0.2473
Evaluation scheme	The original area of Wu group coal seam	The area where the group Wu coal seam is protected by the Ding group coal seam	\
Variable weight—projected bullseye distance	0.2321	−0.0378	\

4.3. Results Analysis

The degree of deviation in each scheme from the "optimal combination" can be determined according to the projected distance between the calculation result and the bullseye. If the bullseye distance is between (0, 1), the degree of protection is not significant, and mining is still dangerous. Corresponding to the danger intensity, the larger the value, the more serious the risk, and the smaller the value, the less dangerous; if the bullseye distance is between (−1, 0), there is effective protection, and the larger the absolute value is, that is, the farther the distance from 0, the higher the degree of protection, and vice versa.

From the calculation in Table 7, it can be seen that the distance between the test results and the bullseye in the original coal seam area, and the distance between the test results and the bullseye in the protected area, can be calculated by using this model, quantifying the degree of danger and the degree of protection, which can intuitively and accurately represent the protective layer mining effect. The variable weight-projected bullseye distance and the protective degree are reflected in Figure 9.

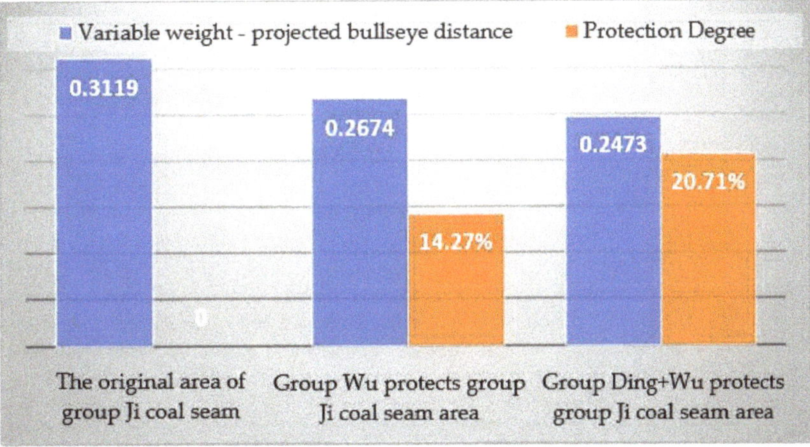

Figure 9. The degree of protection in the evaluation area.

- The projected bullseye distance of the original area in the group Wu coal seam calculated by the model is 0.2321, which is dangerous (greater than zero). After mining the Ding group coal seam of the upper protective layer, the bullseye distance is reduced to −0.0378, which is effective for protection (less than zero). The degree of protection in the original area in the group Wu coal seam reaches 116.29%, which has a good protection effect. In terms of individual indexes, by comparing the gas parameters of the original area in group Wu and the area of Ding protected by group Wu, as shown in Table 4, it can be seen that the safety of coal seam mining is improved, the gas content is reduced by 40.6%, the gas pressure is reduced from 1.1 MPa to 0.5 MPa, the gas pressure is reduced by 54.5%, the permeability coefficient increased by 12.38-times, and the expansion value reaches 17.8‰. The gas content and gas pressure decrease significantly, and the values decrease to within the safe range of the outburst prevention index. At the same time, the permeability of the coal seam increases by 12.4-times. The effect of group Wu being protected by group Ding alone is more obvious.

- The projected bullseye distance of the original area in the group Ji coal seam is 0.3119, which is dangerous (greater than zero). After mining the upper protective layer in the group Wu coal seam, the bullseye distance is reduced to 0.2674, which is still dangerous (greater than zero). The degree of protection in the original area in the group Ji coal seam reached 14.27%, which reduced the degree of danger of the original area of the group Ji coal seam by 14.27%. Comparing the gas parameters of the original area in the group Ji coal seam and the area where the group Ji coal seam is protected by the Wu group coal seam, as shown in Table 4, it can be seen that the gas content decreased from 9.71 m^3/t to 8.91 m^3/t, the gas content decreased by 8.2%, the gas pressure decreased from 3.4 MPa to 2.9 MPa, and the gas pressure decreased by 14.7%. The permeability of the coal seam increased by 3.2-times, and the expansion value reached 2.14‰, but it was still lower than the critical value of 3‰, as shown in Table 2. Only mining the protective layer in group Wu has a certain protective effect on group Ji.
- After mining the upper protective layer in the Ding + Wu coal seam, the bullseye distance was reduced to 0.2473 (greater than zero), and the degree of protection in the original coal seam area reached 20.71%, which reduced the danger level in the original coal seam area by 20.71%. Comparing the gas parameters of the original area in the group Ji coal seam and the area where the group Ji coal seam is protected by the Ding + Wu group coal seam, as shown in Table 4, it can be seen that the gas content decreased from 9.71 m^3/t to 6.34 m^3/t, the gas content decreased by 34.7%, the gas pressure decreased from 3.4 MPa to 2.6 MPa, and the gas pressure decreased by 23.5%. The permeability of the coal seam increased by 4.3-times, and the expansion value reached 2.13‰, which was also lower than 3‰, as shown in Table 2. From the results, it can be seen that the effect of the Ding + Wu protective layer on the group Ji coal seam is better than the effect of the separate mining in the group Wu coal seam to protect the group Ji coal seam.

The evaluation results for the working face are shown in Figure 10, showing the protective effect of the working face with outstanding danger before and after protective layer mining is carried out. In Figure 10, red indicates that there is a prominent danger that needs to be eliminated. The evaluation value is a positive value. The darker the color, the higher the risk. The larger the value, the lower the degree of protection, and the stronger the secondary means of eliminating the outburst degree must be. Green means that the outburst degree can be eliminated without secondary outburst elimination. The evaluation value is negative. The greater the absolute value, the higher the safety and the higher the degree of protection.

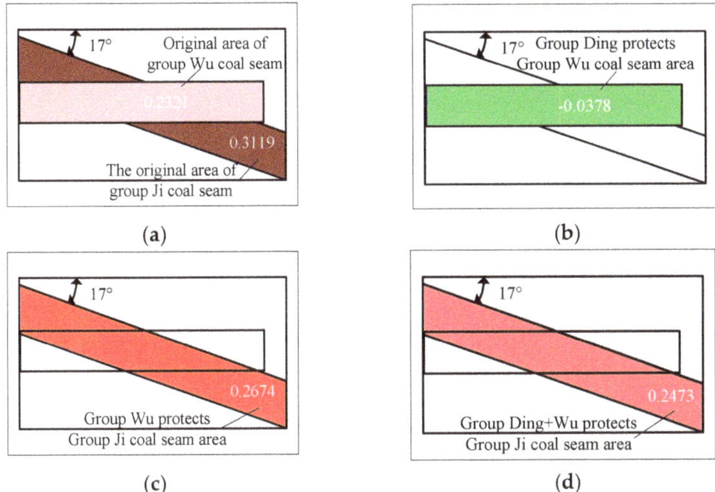

Figure 10. Quantitative results before and after mining of the protective layer in the evaluation area: (**a**) before the protective layer mining; (**b**) group Ding protects group Wu coal seam region; (**c**) group Wu protects group Ji coal seam region; (**d**) group Ding + Wu protects group Ji coal seam region.

4.4. Gas Control Data Analysis

After the protection layer was mined in this area of the Pingmei Group, the gas drainage volume of the working face where the protected layer was located increased significantly. The test data for the gas drainage parameters in the protected layer were statistically analyzed, and the changes in the gas drainage parameters in the protected layer working face were compared before and after the mining of the protective layer:

- After group Wu protective layer mining, the gas drainage rate for the group Ji protected layer increased from 19% to 50.1% and increased by 163.68%. There were still outstanding dangers.
- After group Ding + Wu upper protective layer mining, the gas drainage rate for the group Ji protected layer increased from 25.4% to 68.8% and increased by 170.87%. There were still outstanding dangers.
- After group Ding upper protective layer mining, the index for the group Wu protected layer was below the outburst critical value; the wind flow gas concentration in the protected layer working face increased from 0.15% to 0.41% and increased by 173.33%; and the gas extraction rate increased from 25.4% to 78.3% and increased by 208.27%.

5. Discussion

5.1. Comparative Analysis with the Calculation Results of the Constant Weight-Projection Gray Target

The projection bullseye distance obtained by using constant weight is shown in the first row in Tables 8 and 9, and the bullseye distance obtained by variable weight is shown in the second row in Tables 8 and 9. Through the evaluation of variable weight projection bullseye distance, the calculation results changed, the degree of protection of group Ding + Wu, protecting the group Ji area, decreased to a small extent, and other decision-making schemes improved compared with the constant-weight results. In the process of variable weight, the incentive advantage index penalizes the inferior index, so that the overall evaluation results can be balanced.

Table 8. Comparison of variable weight and constant weight projection bullseye distance for the group Wu protection effect.

Calculation Results	Group Wu Coal Seam Original Area	The Area Where the Group Wu Coal Seam is Protected by the Group Ding Coal Seam
Constant weight bullseye distance	0.2182	−0.0015
Variable weight bullseye distance	0.2321	−0.0378

Table 9. Comparison of variable weight and constant weight projection bullseye distance for the group Ji protection effect.

Calculation Results	Group Ji Coal Seam Original Area	The Area Where the Group Ji Coal Seam Is Protected by the Wu Group Coal Seam	The Area Where the Group Ji Coal Seam Is Protected by the Ding + Wu Group Coal Seam
Constant weight bullseye distance	0.2948	0.2630	0.2559
Variable weight bullseye distance	0.3119	0.2673	0.2473

The variable weight function is used to dynamically adjust the weight according to the index dominance of each scheme. In the scheme of group Ding protecting the group Wu coal seam area, the index dominance of x_1 is high; however, the constant weight of this index is low, and the overall evaluation value is significantly improved by variable weight. The index dominance of x_3 is low in this scheme; however, the constant weight of this index is high, and the overall evaluation value is reduced by variable weight. In the scheme of group Wu protecting the group Ji coal seam, the index dominance of x_2 is low; however, the constant weight of this index is high, and the overall evaluation value is reduced by variable weight. The index dominance of x_3 is high, and the constant weight of this index is high, but the improvement in the overall evaluation value is not obvious, satisfying the principle that the incentive range should be smaller than the penalty range. The variable weight calculation can better reflect the weight dynamic change caused by the test value's discreteness in the same coal seam or the same working face.

5.2. Comparison with Related Literature

Ref. [10] established a reliability evaluation method for protective layer mining from a macro perspective, and calculated the index weights by the analytic hierarchy process. From a macro perspective, the evaluation indexes were more comprehensive, but only the subjective constant weights were considered in the weight calculation, and the main basis was the expert-scoring method. In this paper, a variable weight function is established to correct the subjective weight, and the combination of subjective and objective measures is more reasonable. According to the gray target projection model, the bullseye distance is calculated for the targets of eliminated outbursts and the permeability is increased in each scheme. Given the quantitative results concerning the degree of protection in protective layer mining, it is possible to judge whether secondary outburst elimination is necessary and to provide a reference for formulating the most appropriate outburst elimination scheme, so that the quantification of the degree of protection is more precise and targeted.

Ref. [11] proposed an evaluation model for the reliability of protective layer mining, which predicts the pressure relief boundary, quantitatively examines the pressure relief effect, uses the fuzzy-analytic hierarchy process to calculate the weight, and only calculates the subjective constant weight. The index value only considers exact numbers and is not compared with the data for the original coal seam and, thus, does not quantify the degree of protection. This paper solves the multi-attribute decision-making problem of exact numbers, interval numbers, and triangular fuzzy numbers, revises the subjective constant weights based on the variable weight function, and quantitatively analyzes the degree of protection in the protective layer mining. The method in this paper can better reflect the multi-attribute characteristics of the evaluation indexes; there are differences in the combination of multi-attribute index values in each decision-making scheme. The method can also reflect the control effect of this difference on the degree of protection.

There is no other report in the literature on the quantification of the degree of protection in protective layer mining. The variable weight-projection gray target model for the effectiveness of protective layer mining quantifies the degree of protection. It provides a theoretical basis for further judging whether the working face needs to add high-strength outburst elimination means, and guides the formulation of a secondary outburst elimination plan for outburst coal seams.

6. Conclusions

- In the dynamic evaluation model of the variable weight projection bullseye, the subjective constant weight is corrected by the variable weight function to adjust the evaluation result, and the evaluation model reflects the dynamic changes in the evaluation index. Based on the punishment–incentive mechanism, the local state variable weight function is constructed, and the weight dynamic correction model is established. A weak index will affect the overall evaluation results. The penalty mechanism is used to reduce the weight of these indexes, so that the overall evaluation value is reduced after the weight adjustment. For the advantage index, the reward mechanism is used to increase the index weight and the overall evaluation value is increased.
- The research object is the outburst elimination effect in the protected layer in the engineering of protective layer mining. Taking the critical value of the evaluation index for outburst elimination and increasing permeability as the bullseye, the bullseye distance was calculated for each scheme, and the variable weight function was used to balance the evaluation results, quantifying the degree of protection. If the bullseye distance is a negative value, this indicates that there is no outburst risk, and the larger the absolute value, the higher the degree of protection. If the bullseye distance is a positive value, this indicates that there is an outburst risk, and the larger the value, the lower the degree of protection.
- Through the quantitative results, the protective layer mining of this mine was found to have a certain protective effect. Group Ding protects the group Wu coal seam area, with the degree of protection in the original coal seam reaching 116.29%, a good protective effect. After the mining of the protective layer, all indexes are within the safe value. Inside, the coal seam collapses. The group Wu coal seams protecting the group Ji coal seams area, and the group Ding + Wu coal seams protecting the group Ji coal seams, did not eliminate outbursts. In the area where the group Ji coal seam is protected by the Wu group coal seam, compared with the original area, the degree of risk in the group Ji coal seam was reduced by 14.27%. In the area where the group Ji coal seam is protected by the Ding + Wu group coal seam, the degree of risk in the group Ji coal seam was reduced by 20.71%. The degree of protection in group Ding + Wu to protect the group Ji coal seam scheme was slightly greater than that in group Wu to protect the group Ji coal seam scheme, and it can be increased by a small increase in permeability measures or by an improvement in gas drainage drilling design.
- This study obtains the variable weight-projection gray target model for evaluating the effectiveness of protective layer mining, and quantifies the degree of protection, provides a theoretical basis for judging whether the working face should be strengthened with enhanced permeability, and guides the formulation of a secondary outburst elimination scheme for an incompletely eliminated outburst coal seam.

Author Contributions: Conceptualization, B.Q. and Z.S.; Methodology, B.Q. and B.L.; Validation, W.S. and J.H.; Investigation, Z.S.; Data curation, B.Q. and F.H.; Writing—original draft preparation, B.Q.; Writing—review and editing, B.Q.; Visualization, J.H.; Supervision, W.S.; Project administration, B.Q.; Funding acquisition, Z.S.; Investigation, J.H.; Resources, F.H. All authors have read and agreed to the published version of the manuscript.

Funding: This research was funded by the National Natural Science Foundation of China, grant numbers 52004118 and 51874166; Liaoning Technical University, grant number LNTU20TD-11; and the Department of Education of Liaoning Province, grant number LJ2020QNL009.

Institutional Review Board Statement: Not applicable.

Informed Consent Statement: Not applicable.

Data Availability Statement: Data are contained within the article.

Conflicts of Interest: The authors declare no conflict of interest.

References

1. Cheng, Y.G. *Theory and Engineering Application of Coal Mine Gas Prevention and Control*, 1st ed.; China University of Mining and Technology Press: Beijing, China, 2010; pp. 21–36.
2. Zhang, H.T.; Wen, Z.H.; Yao, B.H.; Chen, X.Q. Numerical simulation on stress evolution and deformation of overlying coal seam in lower protective layer mining. *Alexandria Eng. J.* **2020**, *59*, 3623–3633. [CrossRef]
3. He, A.P.; Fu, H.; Huo, B.J.; Fan, C.J. Permeability Enhancement of Coal Seam by Lower Protective Layer Mining for Gas Outburst Prevention. *Shock Vib.* **2020**, *2020*, 8878873. [CrossRef]
4. Zou, Q.L.; Liu, H.; Zhang, Y.J.; Li, Q.M.; Fu, J.W.; Hu, Q.T. Rationality evaluation of production deployment of outburst-prone coal mines: A case study of nantong coal mine in Chongqing. *Saf. Sci.* **2020**, *122*, 104515. [CrossRef]
5. China University of Mining and Technology, National Engineering Research Center for Coal Mine Gas Control. *Technical Criterion of Protective Coal Seam Exploitation*, 1st ed.; China State Administration of Work Safety: Beijing, China, 2008; pp. 2–3.
6. Liang, B.; Qin, B.; Sun, W.J.; Wang, S.Y.; Shi, Y.S. The application of intelligent weighting grey target decision model in the assessment of coal-gas outburst. *J. China. Coal. Soc.* **2013**, *38*, 1611–1615.
7. Chen, C.X.; Su, J.; Li, W.B. Investigation on regional outburst prevention effect with mining of upper protective seam extraction. *Saf. Coal. Min.* **2017**, *43*, 110–112.
8. Kang, J.D. Study on the Influence of Coal Pillar on the Pressure Relief and Outburst Elimination of the Upper Protective Layer Mining. *Min. Saf. Environ. Prot.* **2018**, *45*, 17–21.
9. Yuan, L.; Xue, S. Defining outburst-free zones in protective mining with seam gas content-method and application. *J. China Coal Soc.* **2014**, *39*, 1786–1791.
10. Liu, Y.W.; Li, G.F. Reliability research of protective layer mining and pressure-relief gas drainage. *J. Min. Saf. Eng.* **2013**, *30*, 426–431.
11. Du, Z.S.; Qin, B.T.; Fan, Y.C. A reliability evaluation model of exploitation efficiency of the protective layer and its application. *J. Min. Saf.* **2017**, *34*, 186–191.
12. Sun, G.D.; Guan, X. Research on hybrid multi-attribute decision-making. In Proceedings of the 2016 International Conference on Cyber-Enabled Distributed Computing and Knowledge Discovery (CyberC), Chengdu, China, 13–15 October 2016; Computer Society: Washington, DC, USA; pp. 272–277.
13. Ma, J.S. Generalised grey target decision method for mixed attributes with index weights containing uncertain numbers. *J. Intell. Fuzzy Syst.* **2018**, *34*, 625–632. [CrossRef]
14. Ma, J.S. A generalized grey target decision method with mixes data type of index and weight. *Stat. Decis.* **2018**, *7*, 58–61.
15. Ma, J.S. Generalised grey target decision method with Index and weight both containing mixed attribute values based on improved gini-simpson index. *J. Stat. Inf.* **2019**, *34*, 28–34.
16. Zhang, Z.; Li, L.L.; Wei, Z.H.; Yu, H.F. Dynamic effectiveness evaluation of command control system based on variable weight-projection gray target. *Syst. Eng. Electron.* **2019**, *4*, 801–809.
17. Zou, R.H.; Wang, Y.C.; Deng, Y.J. Condition assessment method for transmission line with multiple outputs based on variable weight principle and fuzzy comprehensive evaluation. *High Volt. Eng.* **2017**, *43*, 1289–1295.
18. Li, S.Q.; Wu, Q.; Li, Z.; Zeng, Y.F.; Yuan, Q.D.; Yu, Y.L. Vulnerability evaluation and application of floor water inrush in mining area with multiple coal seams and single aquifer based on variable weight. *J. China Univ. Min. Technol.* **2021**, *50*, 587–597.
19. Xu, L.M.; Li, M.J.; Dai, Q.Z. Dynamic evaluation method based on grey target. *J. Syst. Sci. Math. Sci.* **2017**, *37*, 112–124.
20. Bing, L.; Qin, B.; Sun, F.Y.; Wang, Y.; Sun, Y.N.; Phuclo, K. Application of evaluation index system of coal and gas co-extraction and evaluation model. *J. China Coal Soc.* **2015**, *40*, 728–735.
21. Chen, G.B.; Teng, P.C.; Li, T.; Wang, C.Y.; Chen, S.J.; Zhang, G.H. Evaluation model of rock burst in coal mine and its application. *J. Taiyuan Univ. Technol.* **2021**, *52*, 966–973.

22. China State Administration of Coal Mine Safety. *Detailed Rules for Prevention and Control of Coal and Gas Outburst*, 1st ed.; Coal Industry Press: Beijing, China, 2019; pp. 22–23.
23. Henan Provincial People's Government. Available online: http://yjglt.henan.gov.cn/2014/10-25/987526.html (accessed on 25 October 2014).

Article

Stability Evaluation and Structural Parameters Optimization of Stope Based on Area Bearing Theory

Hao-Yu Qiu, Ming-Qing Huang * and Ya-Jie Weng

Zijin School of Geology and Mining, Fuzhou University, Fuzhou 350108, China; 161902104@fzu.edu.cn (H.-Y.Q.); wyj_569358641@163.com (Y.-J.W.)
* Correspondence: seango@fzu.edu.cn; Tel.: +86-188-5010-1619

Abstract: A reasonable and stable stope structure is the premise of realizing safe mining of underground metal ore. To safely mine the gently inclined medium-thick ore body, stope stability in Bainiuchang Mine was analyzed based on the pillar area bearing theory, and stope stability with regard to nine groups of structural parameters was numerically simulated. The results show that the existing stope structural parameters failed to maintain stability requirements and tended to be exposed to the risk of stope collapse. The middle section of the pillar as well as stope roofs and floors are vulnerable due to tensile stress when mining by open stoping, and the compressive stress concentration is prone to occur at the junction of the pillars, stope side walls, roofs and floors. Shear stress contributes little to pillar failure. The reasonable stope structural parameters of open stoping for the gently inclined medium-thick ore body in Bainiuchang mine are optimized using ANSYS numerical simulation: stope height 4.5 m, pillar diameter 4 m, pillar spacing 7 m and pillar row spacing 8 m. The onsite trial shows that the ore recovery rate reaches 82% under these parameters, which also realizes the equilibrium of safety and economy.

Keywords: gently inclined medium-thick ore body; structural parameter optimization; stope stability; area bearing theory; numerical simulation

Citation: Qiu, H.-Y.; Huang, M.-Q.; Weng, Y.-J. Stability Evaluation and Structural Parameters Optimization of Stope Based on Area Bearing Theory. *Minerals* **2022**, *12*, 808. https://doi.org/10.3390/min12070808

Academic Editor: Yosoon Choi

Received: 27 April 2022
Accepted: 21 June 2022
Published: 25 June 2022

1. Introduction

Gently inclined medium-thick ore body is one of the most widely distributed types of underground metal deposits worldwide. The inclination angle of this type of ore body is mostly 5~30°, which is not conducive to the normal operation of trackless mining equipment such as intelligent scrapers, nor to ore-drawing under gravity of blasted ore in the stope. It is internationally recognized as the most difficult type of ore body to mine [1–3]. Open stoping is commonly used in the world to mine gently inclined medium-thick ore bodies [4], and its mining structural parameters are the most important indicators for balancing safe, economic and efficient recovery. If pillars are too thin to resist ground pressure, it usually leads to safety accidents; in contrast, it will cause ore loss and reduce ore recovery rate and production profit. Structural parameters are the most influential factors for maintaining stope stability with open stoping. In 1973, Bieniawski proposed the RMR (Rock Mass Rating) rock mass classification method to evaluate the stability of various rock masses [5]. Bazaluk et al. [6] studied the stope stability of underground mining of iron ore by taking yuzhno-belozerskyi deposit as the research object. Wu et al. [7] used ANSYS and Flac3D to analyze the influence of mine room and pillar configuration on stope stability in highly fractured areas and preferred the stope structure configuration scheme based on the objective function and constraint function. Mikaeil et al. [8] used Flac2D to calculate the load cases such as axial forces, bending moments and shear forces in the vault, bottom and side walls of the underground roadway and analyzed the stability characteristics of the roadway support. Zhao et al. [9] used Flac3D to simulate the stope structure and design the safest stope exposure size based on the Mathews stability graphical solution method.

Qin et al. [10] analyzed the stability characteristics of ore pillars with different thicknesses by numerical simulation in an iron mine, which provided a reference for the optimization of structural parameters. Lan et al. [11] constructed a second-order response surface model and realized the comprehensive optimization of the stope structural parameters through multi-objective optimization and multi-attribute decision making. Liu et al. [12] used Flac3D to establish a numerical calculation model of the fractured ore body and studied the relationship between the stope span parameters, roof displacement and plastic zone volume. However, the aforementioned studies were based on safety and technology for optimization, and less consideration was given to the best economic efficiency during mine recovery.

Bainiuchang Mine has a typical gently inclined medium-thick ore body, which is mainly mined using the room and pillar method. The width of ore block is generally 20~25 m, the length is 30~40 m, and a circular pillar with a side length of 2~4 m is reserved. With the increasing mining depth, the ground pressure management is becoming more and more complicated, and there is a serious risk of collapse. It is urgent to optimize the original stope structural parameters. Based on the area-bearing theory, the mining process before optimization of the Bainiuchang Mine was simulated through ANSYS, and the stope stability was evaluated through stress analysis. Nine different stope structural parameters optimization schemes were proposed through orthogonal design, and the stress conditions of pillar, roof and floor were analyzed based on numerical simulation, and safe and economic stope structural parameters were further optimized, so as to guide the mining of gently inclined medium-thick ore body.

2. Area Bearing Theory

The area-bearing theory holds that the load on the pillar is the gravity of the overlying rock layer from the mining space it supports to the surface, and the area supported by the pillar is the sum of the mining area and the cross-sectional area of the pillar itself, from which the axial compressive stress of the pillar is approximated. It reveals the internal causes of pillar instability and failure (as shown in Figures 1 and 2). Its equilibrium equation is as follows Equation (1).

$$\sigma_p W_p = (W_o + W_p) p_z \tag{1}$$

where σ_p is the axial compressive stress of the mine pillar; p_z is the vertical positive stress component of the ground stress field before mining; W_o is the width of the mine room; W_p is the width of the mine pillar.

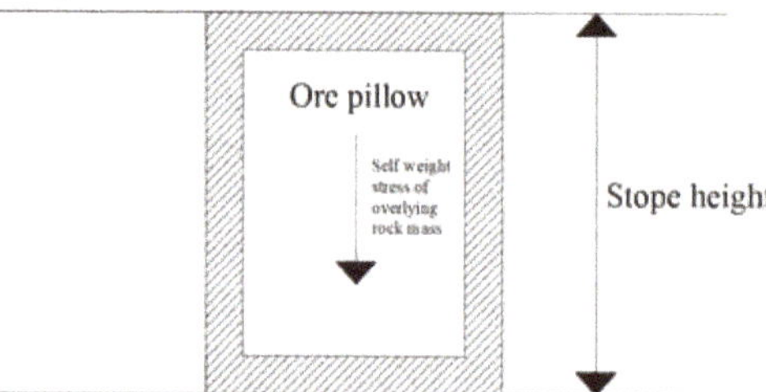

Figure 1. Schematic diagram of pillar stress.

The strip mine pillar and circular mine pillar are mainly arranged in Bainiuchang Mine. The calculation Equations (2) and (3) for the axial compressive stress of the two types of ore pillars can be derived from Equation (1).

$$\text{The strip mine pillar}: \ \sigma_p = \gamma H (1 + W_o / W_p) \tag{2}$$

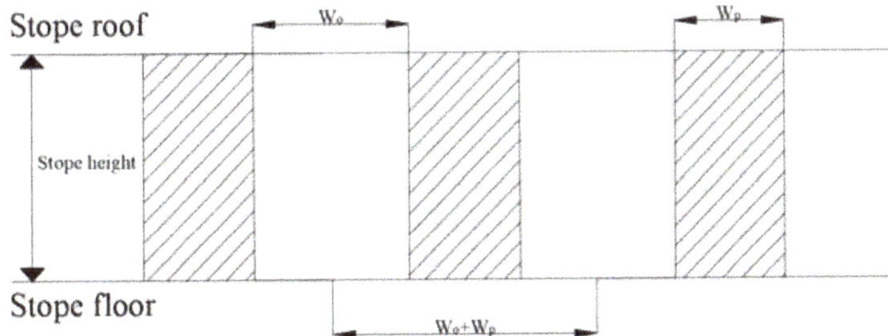

Figure 2. Schematic diagram of pillar bearing calculation.

$$\text{The circular mine pillar}: \ \sigma_p = \frac{4\gamma H}{\pi}(1 + W_o/W_p)^2 \quad (3)$$

where γ is the capacity of the overlying rock layer; H is the mining depth [13,14].

Based on the area bearing theory, the axial compressive stress of the ore pillar can be calculated. According to the area bearing theory, each pillar supports the ore rock in a certain area. The axial compressive stress of the ore pillar can be calculated by the gravity of the upper ore rock and the sectional area of the ore pillar. When the axial compressive stress exceeds the uniaxial compressive strength of the ore rock, the ore pillar will be crushed, resulting in fragmentation, deformation and even collapse. Then, cause the collapse of the stope roof. When the number of collapsed pillars reaches a certain amount, it will produce a chain effect, leading to the collapse of other pillars, and finally the overall collapse of the mine room. otherwise, it can be safely mined.

Combined with the area bearing theory (Equation (3)) and the Bieniawski pillar strength formula (Equation (4)), the safety factor calculation formula (Equation (5)) of the circular point pillar in Bainiuchang mine can be derived.

$$S_p = \sigma_c[0.64 + 0.36(W_p/h)] \quad (4)$$

where S_p is strength of ore pillar, MPa; σ_c is strength parameter of ore and rock, MPa; h is the room height, m; W_p is the width of the mine pillar.

$$K = \frac{\sigma_c[0.64 + 0.36(W_p/h)]}{\frac{4\gamma H}{\pi}(1 + W_o/W_P)^2} \quad (5)$$

where K is the safety factor of circular point column. Other physical quantities are described above.

3. Stope Stability Evaluation

The existing stope stability of Bainiuchang mine is evaluated. The geological structure of the mining area is well developed, and the weathering is strong. The hydrogeological conditions of the deposit are complex, with strong karst aquifers filling with water. The roof of the ore body is a carbonate rock mixed with clastic rock karst fissure aquifer, and the inflow is 0.0025~1.52 $L{\cdot}s^{-1}$; The floor of the ore body is a karst fissure aquifer of clastic rock mixed with carbonate rock, and the inflow is 0.0109~2.69 $L{\cdot}s^{-1}$. The ore body is stratiform and lenticular in shape, with siltstone and mudstone on the roof and argillaceous limestone on the floor. The surrounding rock mass of the upper and lower walls of the ore body is of good quality, and the engineering geological conditions are simple. The physical and mechanical parameters of the main ore rocks of the Bainiuchang mine are shown in Table 1. In Table 1, σ_c represents uniaxial compressive strength, σ_t represents uniaxial tensile strength, τ represents shear strength, c represents cohesion, φ represents internal friction angle, E represents elastic modulus, ρ represents density, and μ represents Poisson's ratio. The pillar of Bainiuchang mine are shown in Figure 3.

Table 1. Physical and mechanical parameters of Bainiuchang's ore rocks.

Ore Rock Type	σ_c /MPa	σ_t /MPa	τ /MPa	c /MPa	φ /(°)	E /GPa	ρ /g·cm^{-3}	μ
Argillaceous Limestone	58.30	3.76	36.81	3.47	28.38	9.37	2.74	0.25
Mudstone	35.42	1.99	30.64	1.14	17.79	4.51	2.79	0.19
Siltstone	77.14	2.78	37.57	4.08	35.97	14.88	2.64	0.27
Ore body	97.62	2.07	50.86	6.58	42.93	12.60	4.00	0.22

Figure 3. Picture of pillar in Bainiuchang mine.

The structural geology of site: The fault structures in the mining area are developed, mainly NW trending faults, with NE trending faults interspersed between them, forming a structural framework. F_2, F_3, F_6, F_7 and F_8 in the northwest direction are mainly exposed in the mining area. F_2 fault strike NW-SE, dip direction SW, dip of 65°. F_3 fault is the most important ore controlling and hosting structure in Bainiuchang mine, with a total length of 5000~6000 m, strike NW-SE, dip direction of 200~230° to SW, and dip of 20~35°. F_6 fault is about 1.7 km, strike NW-SE, dip direction SW, dip of 30~35°. F_7 fault is about 3.5 km, strike NW-SE, dip direction SW, dip of 34~50°. F_8 fault is about 3.4 km, strike NW-SE, dip direction SW, dip of 60~75°.

3.1. Simulation Conditions and Process

The onsite mining process of Bainiuchang mine (as shown in Figure 4): Strip rooms and strip pillars are arranged in intervals. Strip rooms are mined first. After strip rooms are mined, the strip pillars are mined, leaving round point pillars with a diameter of 3 m. The stope height is 3 m, stope length is 40 m and the width is 25 m. Next, a multi-step excavation finite element calculation model will be established according to the actual mining process by ANSYS 19.2.

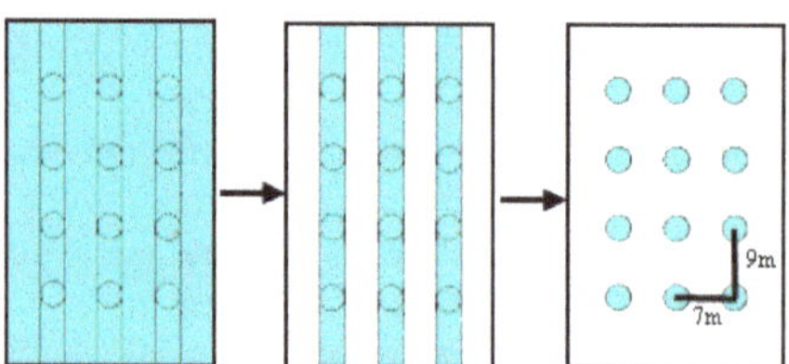

Figure 4. Arrangement and mining sequence of rooms and pillars.

Modeling: The finite element model is established according to the onsite geological conditions, and the mechanical model is established according to the influence range of 3~5 times the diameter of the excavation space. In order to balance the calculation speed and accuracy, a 500 m × 500 m × 625 m model is established (Figure 5). On the basis of not affecting the calculation results and considering the size coordination relationship of the model, the depth is taken to 1380 m above sea level. The model contains the ore body, and then contains the target ore block of numerical simulation calculation.

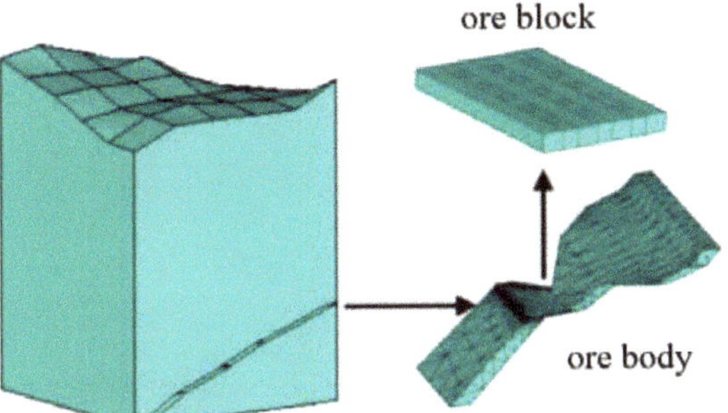

Figure 5. Finite element calculation model.

Mesh refinement analysis: If the mesh is sparse, the accuracy of calculation cannot meet the requirements. The finer the mesh, the higher the accuracy of calculation, but the amount of calculation will increase exponentially. In order to balance the calculation accuracy and efficiency, considering the regularity of the element, we established a material model based on the key points by using ANSYS. This work is according to the geological structure model and the distribution changes of the physical property parameters of each layer in the mining area. Using the tetrahedron (SOLID92) 10 node element type, we divided the finite element mesh of each volume element. The whole mesh was divided into 57,611 elements and 79,159 nodes (Figure 6).

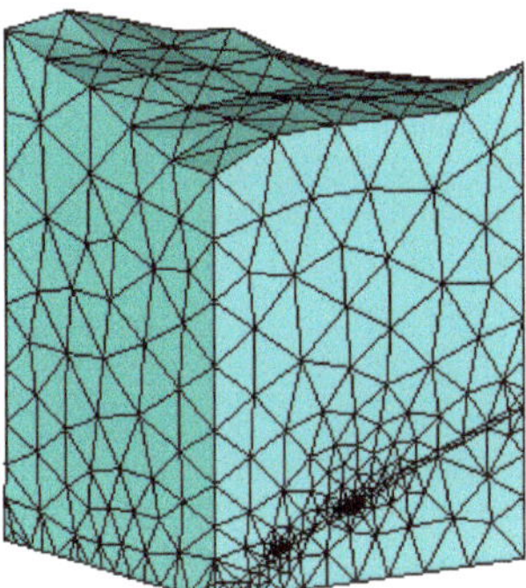

Figure 6. Finite element model meshing.

Initial stress conditions: Calculated the distribution of in situ stress according to gravity, as shown in Figure 7.

Boundary condition: The outer four sides of the model are fixed along the x-axis and y-axis, respectively, and the bottom of the model is fixed along the z-axis as the boundary condition.

Multi-step excavation numerical simulation process: the internal loading model that releases the load along the excavation perimeter is used. In the first step, the inversion of original rock stress is revealed according to the boundary conditions and the load conditions (dead weight stress field). In order to get close to the actual results, the second step is to carry out multi-step excavation simulation calculation according to the onsite mining process of Bainiuchang mine.

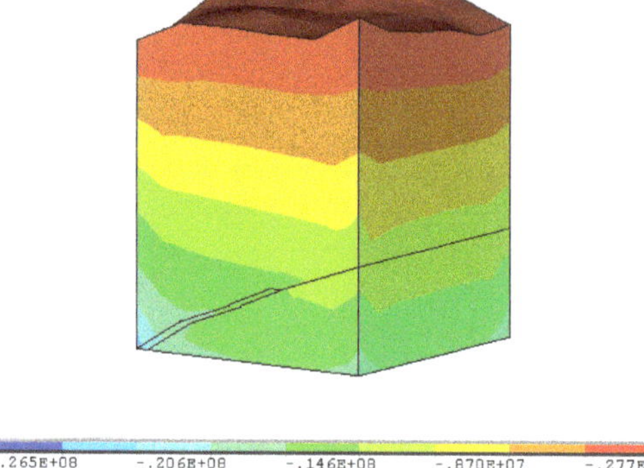

Figure 7. Dead-weight stress field calculation results.

Constitutive model: Three-dimensional elastic-plastic finite element calculation and analysis of the multi-step excavation was carried out using the Drucker–Prager plastic yielding criterion commonly used in geotechnical engineering [15,16]. This criterion is shown in Equation (6).

$$\alpha I_1 + J_2^{1/2} = k \tag{6}$$

where: $I_1 = \sigma_1 + \sigma_2 + \sigma_3$ is the first invariant of stress; $J_2 = \frac{1}{6}[(\sigma_1 - \sigma_2)^2 + (\sigma_2 - \sigma_3)^2 + (\sigma_3 - \sigma_1)^2]$ is the second invariant of stress deflection; $\alpha = \frac{2}{\sqrt{3}}\frac{\sin\phi}{(3\pm\phi)}$, $k = \frac{6}{\sqrt{3}}\frac{C\cdot\cos\phi}{(3\pm\sin\phi)}$ are experimental constants related to the internal friction angle and cohesion of rock; ϕ is the internal friction angle of rock; C is the cohesion of rock; σ_1, σ_2, σ_3 is the maximum, intermediate and minimum principal stress, respectively.

3.2. Simulation Results

Numerical simulations were performed according to the onsite mining process, and the numerical results of the maximum, intermediate and minimum principal stresses σ_1, σ_2, σ_3 and three axial shear stresses within the stope and pillars can be obtained (Table 2). The main stress distribution cloud images are shown in Figures 8–13.

Table 2. Numerical simulation results of the recovery process before optimization.

No.	Stress Parameter	Step 1 (Before Mining)	Step 2 (Mining Strip Rooms)	Step 3 (Mining Strip Pillars)
1	σ_1 (MPa)	−1.84~−0.49	−14.2~4.68	−26.0~4.42
2	σ_2 (MPa)	−2.07~−0.98	−16.3~−1.3	−36.4~1.41
3	σ_3 (MPa)	−16.4~−12.5	−68.5~−16.4	−142.0~−10.9
4	τ_{xy} (MPa)	−0.15~0.12	−1.70~−1.54	−9.48~9.63
5	τ_{yz} (MPa)	−0.67~0.70	−21.7~22.9	−25.3~25.8
6	τ_{zx} (MPa)	−0.007~0.177	2.16~2.83	−32.3~34.7

According to the area bearing theory, when the current stope structural parameters are adopted, the axial compressive stress borne by the ore pillar is 135.5 MPa, which exceeds the uniaxial compressive strength of the ore rock (97.62 MPa). At the same time, the simulation results show that compressive and tensile stresses exist simultaneously during the mining process. Compared with that before mining, all stresses have increased by tens of times.

The compressive stresses mainly exist in the junction part of the pillars, stope side walls, roofs and floors, while the tensile stresses mainly exist in stope side walls, roofs and floors. With the expansion of mining areas, the tensile stresses in the stope side walls and the roofs and floors gradually increase, and the compressive stresses in the ore pillars are also gradually concentrated, with the maximum compressive stress of 142.0 MPa in the pillar. This is much greater than the uniaxial compressive strength of the ore (97.62 MPa). The field observation also shows that under the existing stope structural parameters, the plastic yielding phenomenon of the ore pillars become more and more obvious as mining proceeds; after the end of the whole stope mining, all the pillars show yielding and the stope stability is poor. If the deep ore body continue to be mined according to the existing stope structural parameters, the risk of collapse is extremely high and safe and stable recovery cannot be guaranteed.

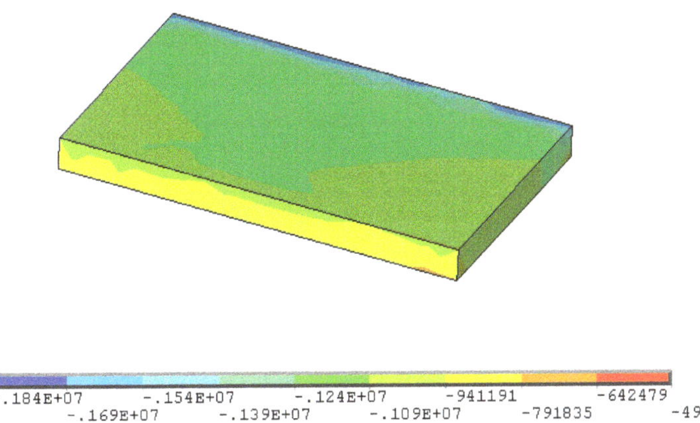

Figure 8. Maximum principal stress distribution in the ore block at 1480 mL before mining (parameters before optimization: stope height 3 m, pillar diameter 3 m, pillar spacing 7 m, pillar row spacing 9 m).

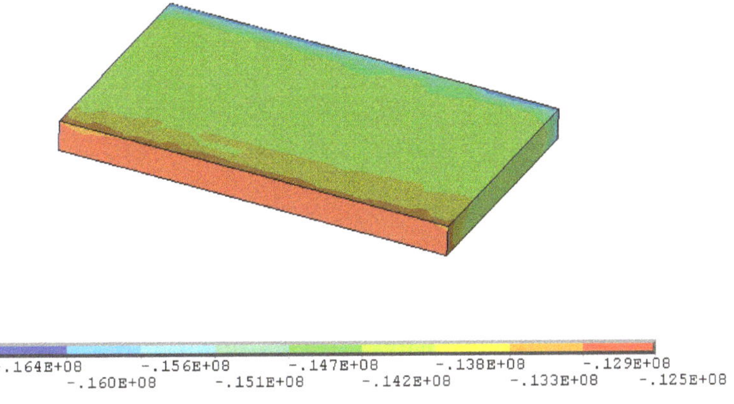

Figure 9. Minimum principal stress distribution in the block at 1480 mL before mining (parameters before optimization: stope height 3 m, pillar diameter 3 m, pillar spacing 7 m, pillar row spacing 9 m).

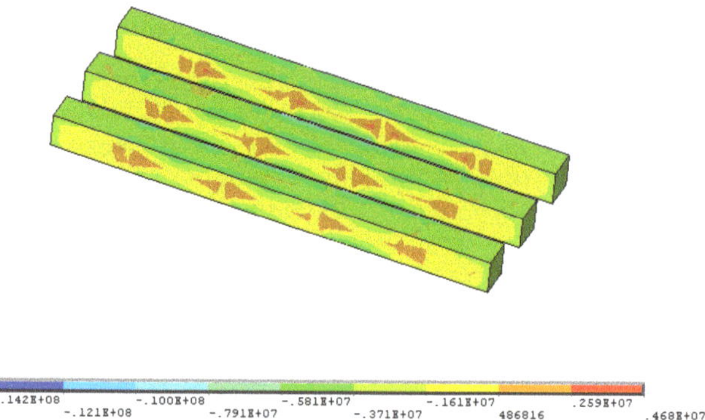

Figure 10. Maximum principal stress distribution in the ore block at 1480 mL after mining the strip room (parameters before optimization: stope height 3 m, pillar diameter 3 m, pillar spacing 7 m, pillar row spacing 9 m).

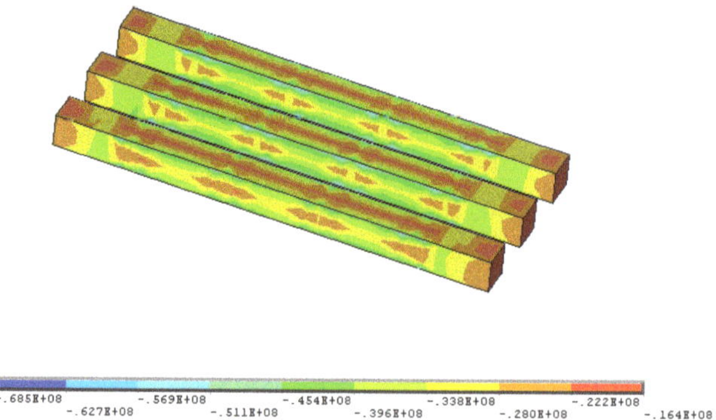

Figure 11. Minimum principal stress distribution in the ore block at 1480 mL after mining the strip room (parameters before optimization: stope height 3 m, pillar diameter 3 m, pillar spacing 7 m, pillar row spacing 9 m).

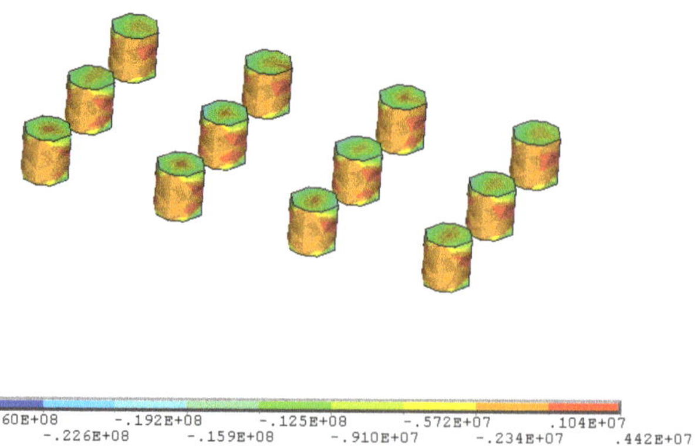

Figure 12. Maximum principal stress distribution in the ore block at 1480 mL after mining of strip pillars (parameters before optimization: stope height 3 m, pillar diameter 3 m, pillar spacing 7 m, pillar row spacing 9 m).

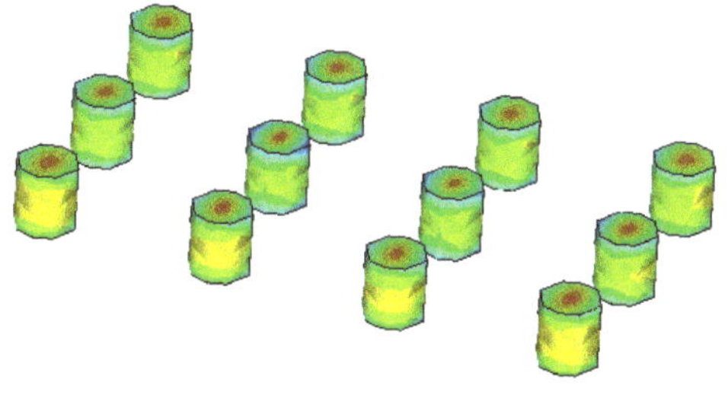

Figure 13. Minimum principal stress distribution in the ore block at 1480 mL after mining of strip pillars (parameters before optimization: stope height 3 m, pillar diameter 3 m, pillar spacing 7 m, pillar row spacing 9 m).

4. Optimization of Stope Structural Parameters

The stope height in Bainiuchang Mine is generally between 2.5 and 4.5 m. Therefore, the optimization of stope structural parameters was conducted for the three stope heights of 2.5 m, 3.5 m and 4.5 m. The nine optimization schemes after conducting orthogonal design are shown in Table 3. The finite element calculation continues the method in Section 3.1, and the ore pillar arrangement is shown in Figure 14.

Table 3. Nine optimization schemes for orthogonal design.

Scheme \ Factor	Stope Height/m	Pillar Diameter/m	Pillar Spacing/m	Pillar Row Spacing/m
1	2.5	2	4	6
2	2.5	3	6	8
3	2.5	4	8	10
4	3.5	2	5	8
5	3.5	3	7	7
6	3.5	4	6	9
7	4.5	2	6	7
8	4.5	3	5	9
9	4.5	4	7	8

4.1. Stope Height 2.5 m

The calculation results of schemes 1–3 are shown in Table 4, and the main stress distribution cloud images are shown in Figures 15–22. The results show that the maximum shear stress in the stope under the three schemes (36.3 MPa, 27.9 MPa, 28.0 MPa) does not exceed the shear strength of the ore rock (50.86 MPa). The tensile stresses mainly appear in the middle of the pillar, stope roof and floor, and the local tensile stresses of 5.83, 4.04 and 4.33 MPa is beyond the uniaxial tensile strength of ore by 2.07 MPa. It tends to result in a local rock fall in the stope roof. Pillars are mainly subjected to compressive stress, especially the junction area between the pillars and the roofs and floors is prone to stress concentration (Figures 17 and 21). In general, the maximum compressive stress values of 143.0, 118.0 and 123.0 MPa in the three schemes are much higher than the uniaxial compressive strength of the ore (97.62 MPa), and the ore pillar will be crushed by the overlying rock layer after the formation of mined-out areas. Hence, these three schemes are not advisable.

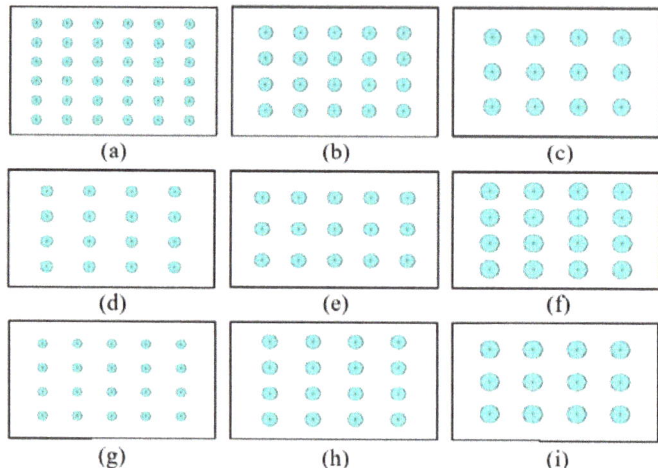

Figure 14. Nine schemes of room and pillar configurations (**a**) Scheme 1; (**b**) Scheme 2; (**c**) Scheme 3; (**d**) Scheme 4; (**e**) Scheme 5; (**f**) Scheme 6; (**g**) Scheme 7; (**h**) Scheme 8; (**i**) Scheme 9.

Table 4. Numerical calculation with stope height of 2.5 m.

No.	**Stress Parameter**	**Scheme 1** 2 m × 4 m × 6 m	**Scheme 2** 3 m × 6 m × 8 m	**Scheme 3** 4 m × 8 m × 10 m
1	σ_1/MPa	−25.6~5.83	−20.3~4.04	−19.6~4.33
2	σ_3/MPa	−143.0~−1.98	−118.0~−6.25	−123.0~−6.22
3	τ_{xz}/MPa	−33.7~36.3	−27.9~27.4	−28.0~25.0

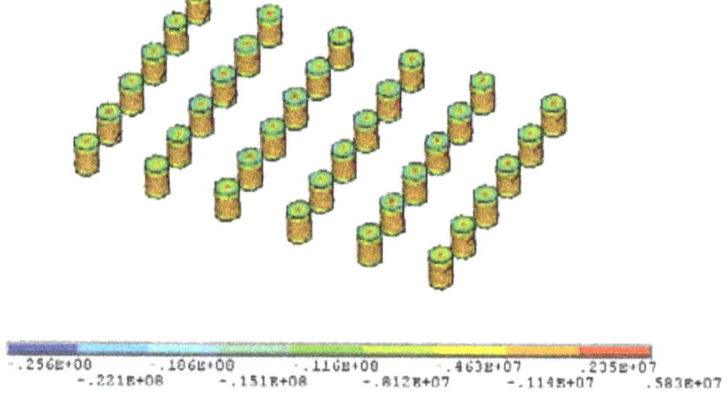

Figure 15. Maximum principal stress distribution of scheme 1 pillars at 1480 mL (parameters of scheme 1: stope height 2.5 m, pillar diameter 2 m, pillar spacing 4 m, pillar row spacing 6 m).

4.2. Stope Height 3.5 m

The calculation results of schemes 4–6 are shown in Table 5, and the main stress distribution cloud images are shown in Figures 23–30. The results show that the maximum shear stress in the stope under the three schemes (39.9 MPa, 28.0 MPa, 19.6 MPa) does not exceed the shear strength of the ore rock (50.86 MPa). The tensile stress distribution law as mentioned in Section 4.1, and the local rock body of the roof is prone to fall. From Figures 25 and 29, the compressive stresses are mainly concentrated in the pillars, and the maximum compressive stress values of 155.0 and 123.0 MPa in the pillars of scheme 4 and scheme 5 are already much higher than the uniaxial compressive strength of the ore (97.62 MPa). The maximum compressive stress of 86.8 MPa in scheme 6 is relatively low, and all stress indexes are greater than those of schemes 4 and 5.

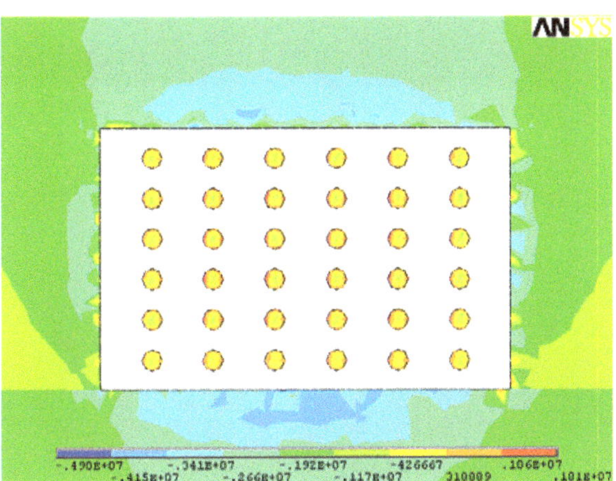

Figure 16. Maximum principal stress distribution in the middle of the pillar at 1480 mL of scheme 1 (parameters of scheme 1: stope height 2.5 m, pillar diameter 2 m, pillar spacing 4 m, pillar row spacing 6 m).

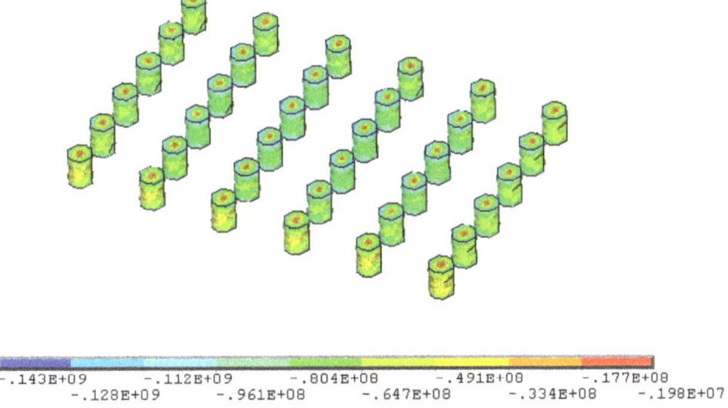

Figure 17. Minimum principal stress distribution of scheme 1 pillars at 1480 mL (parameters of scheme 1: stope height 2.5 m, pillar diameter 2 m, pillar spacing 4 m, pillar row spacing 6 m).

4.3. Stope Height 4.5 m

The calculation results of schemes 7–9 are shown in Table 6, and the main stress distribution cloud images are shown in Figures 31–38. The results show that the maximum shear stress in the stope under the three schemes (37.8 MPa, 28.3 MPa, 21.0 MPa) does not exceed the shear strength of the ore rock (50.86 MPa). The tensile stress distribution law as mentioned in Section 4.1, and the falling rock mass is prone to cause safety accidents. From Figures 33 and 37, the axes of the ore pillars suffer high compressive stresses, and the maximum compressive stress of 155.0 and 113.0 MPa in the pillars of scheme 7 and scheme 8 are much higher than the uniaxial compressive strength of the ore (97.62 MPa), and the pillars are susceptible to collapse by the pressure of the overlying rock after the formation of mined-out areas. In the perspective of safety, scheme 9 is greater than scheme 7 and 8.

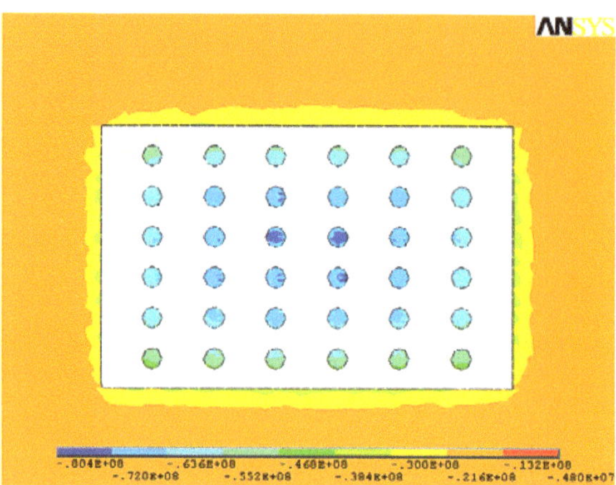

Figure 18. Minimum principal stress distribution in the middle of the pillar at 1480 mL of scheme 1 (parameters of scheme 1: stope height 2.5 m, pillar diameter 2 m, pillar spacing 4 m, pillar row spacing 6 m).

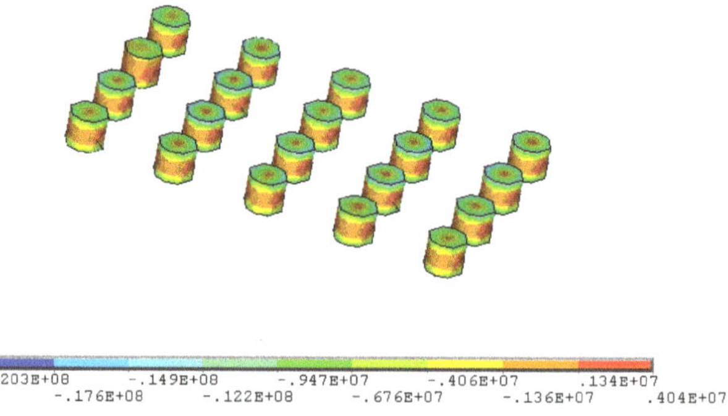

Figure 19. Maximum principal stress distribution of scheme 2 pillars at 1480 mL (parameters of scheme 2: stope height 2.5 m, pillar diameter 3 m, pillar spacing 6 m, pillar row spacing 8 m).

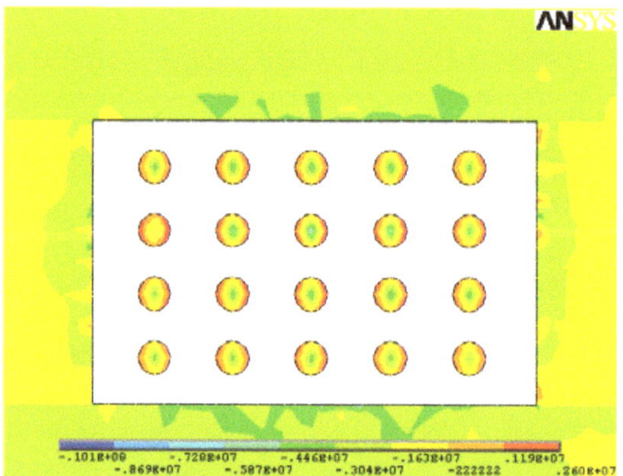

Figure 20. Maximum principal stress distribution in the middle of the pillar at 1480 mL of scheme 2 (parameters of scheme 2: stope height 2.5 m, pillar diameter 3 m, pillar spacing 6 m, pillar row spacing 8 m).

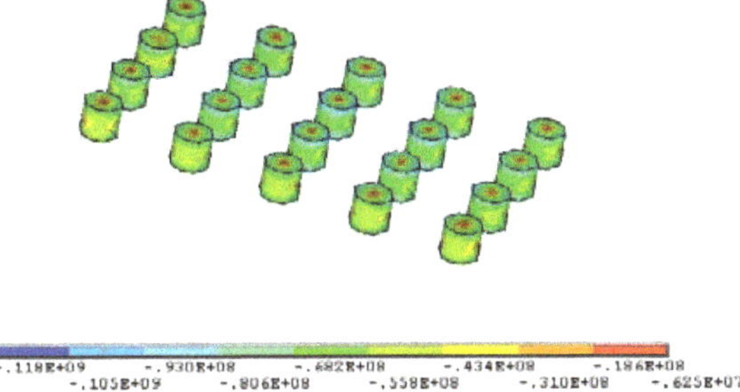

Figure 21. Minimum principal stress distribution of scheme 2 pillars at 1480 mL (parameters of scheme 2: stope height 2.5 m, pillar diameter 3 m, pillar spacing 6 m, pillar row spacing 8 m).

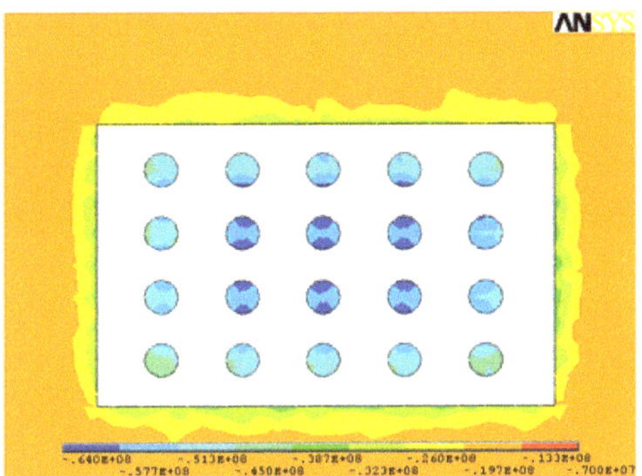

Figure 22. Minimum principal stress distribution in the middle of the pillar at 1480 mL of scheme 2 (parameters of scheme 2: stope height 2.5 m, pillar diameter 3 m, pillar spacing 6 m, pillar row spacing 8 m).

Table 5. Numerical calculation with stope height of 3.5 m.

No.	Stress Parameter	Scheme 4 2 m × 5 m × 8 m	Scheme 5 3 m × 7 m × 7 m	Scheme 6 4 m × 6 m × 9 m
1	σ_1/MPa	−24.6~4.96	−17.9~4.34	−14.4~3.18
2	σ_3/MPa	−155.0~−1.08	−123.0~−3.91	−86.8~−6.46
3	τ_{xz}/MPa	−39.9~37.3	−28.0~27.2	−19.6~19.6

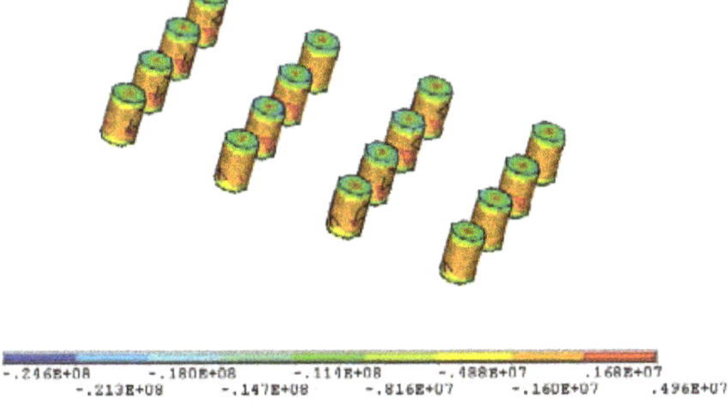

Figure 23. Maximum principal stress distribution of scheme 4 pillars at 1480 mL (parameters of scheme 4: stope height 3.5 m, pillar diameter 2 m, pillar spacing 5 m, pillar row spacing 8 m).

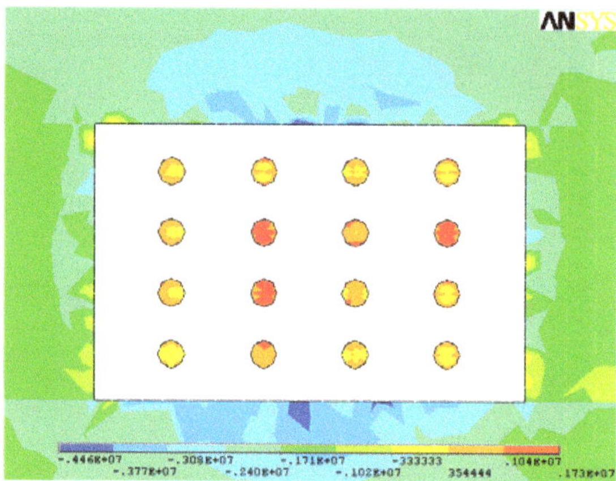

Figure 24. Maximum principal stress distribution in the middle of the pillar at 1480 mL of scheme 4 (parameters of scheme 4: stope height 3.5 m, pillar diameter 2 m, pillar spacing 5 m, pillar row spacing 8 m).

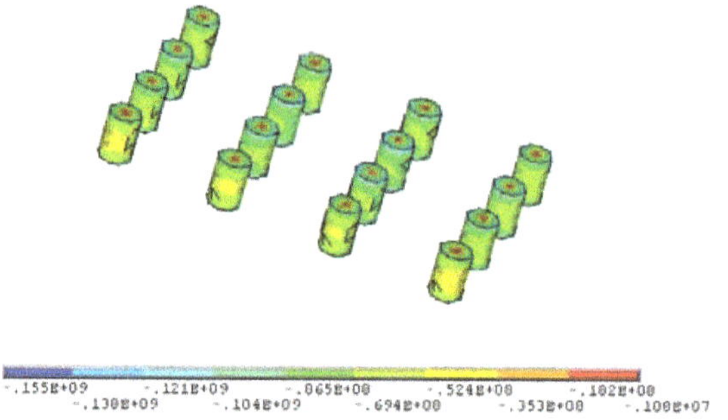

Figure 25. Minimum principal stress distribution of scheme 4 pillars at 1480 mL (parameters of scheme 4: stope height 3.5 m, pillar diameter 2 m, pillar spacing 5 m, pillar row spacing 8 m).

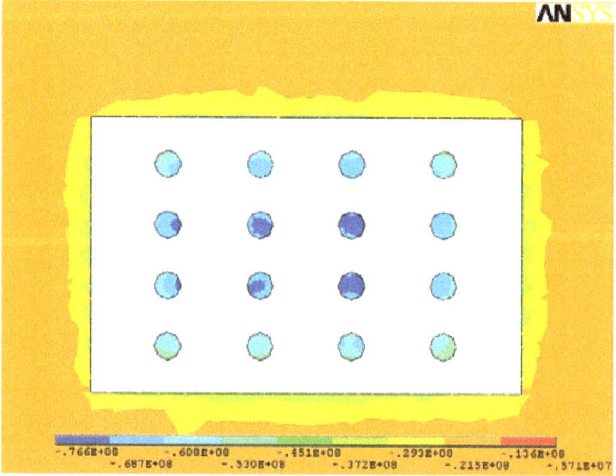

Figure 26. Minimum principal stress distribution in the middle of the pillar at 1480 mL of scheme 4 (parameters of scheme 4: stope height 3.5 m, pillar diameter 2 m, pillar spacing 5 m, pillar row spacing 8 m).

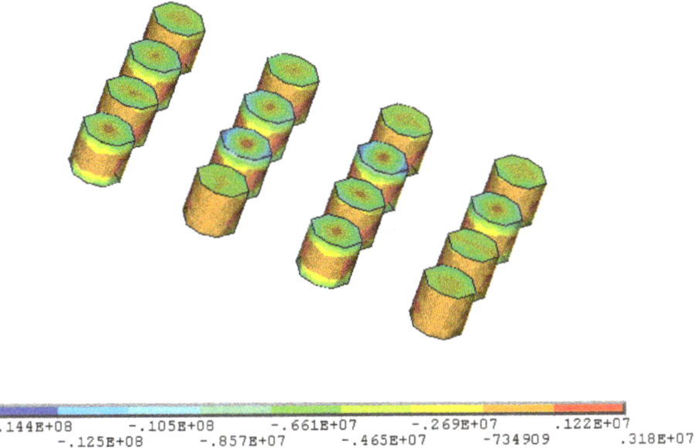

Figure 27. Maximum principal stress distribution of scheme 6 pillars at 1480 mL (parameters of scheme 6: stope height 3.5 m, pillar diameter 4 m, pillar spacing 6 m, pillar row spacing 9 m).

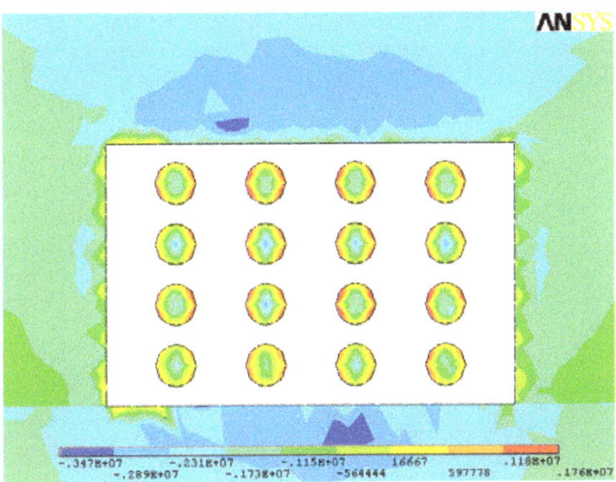

Figure 28. Maximum principal stress distribution in the middle of the pillar at 1480 mL of scheme 6 (parameters of scheme 6: stope height 3.5 m, pillar diameter 4 m, pillar spacing 6 m, pillar row spacing 9 m).

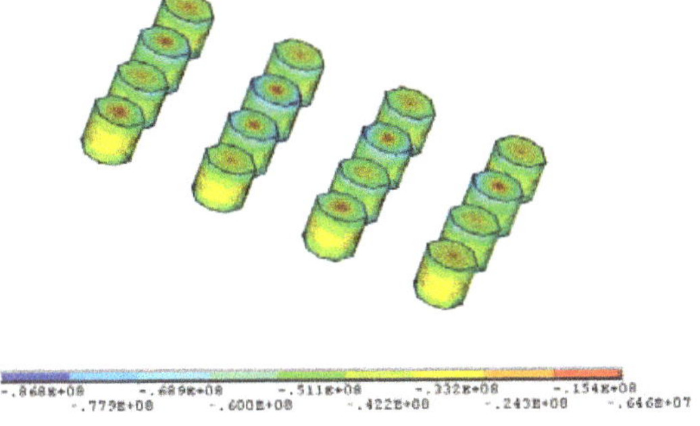

Figure 29. Minimum principal stress distribution of scheme 6 pillars at 1480 mL (parameters of scheme 6: stope height 3.5 m, pillar diameter 4 m, pillar spacing 6 m, pillar row spacing 9 m).

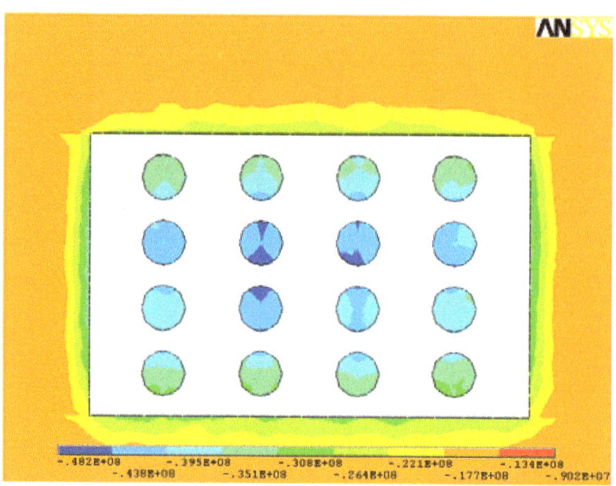

Figure 30. Minimum principal stress distribution in the middle of the pillar at 1480 mL of scheme 6 (parameters of scheme 6: stope height 3.5 m, pillar diameter 4 m, pillar spacing 6 m, pillar row spacing 9 m).

Table 6. Numerical calculation with stope height of 4.5 m.

No.	Stress Parameter	Scheme 7 2 m × 6 m × 7 m	Scheme 8 3 m × 5 m × 9 m	Scheme 9 4 m × 7 m × 8 m
1	σ_1/MPa	−25.3~6.01	−19.3~4.61	−16.3~3.89
2	σ_3/MPa	−155.0~−1.1	−113.0~−2.68	−92.5~−4.78
3	τ_{xz}/MPa	−35.3~37.8	−27.6~28.3	−20.6~21.0

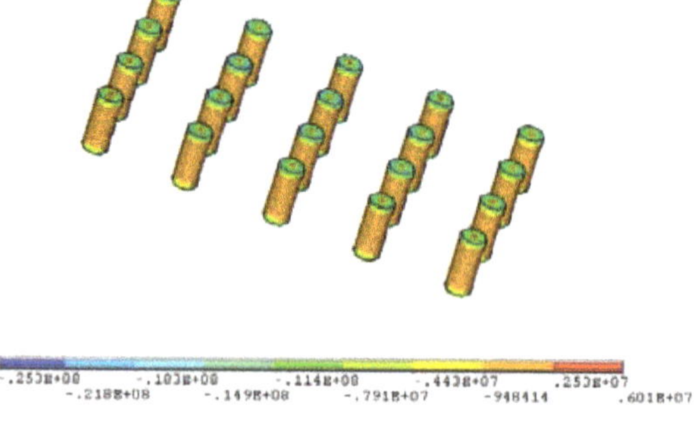

Figure 31. Maximum principal stress distribution of scheme 7 pillars at 1480 mL (parameters of scheme 7: stope height 4.5 m, pillar diameter 2 m, pillar spacing 6 m, pillar row spacing 7 m).

4.4. Optimization Results

In summary, the maximum shear stress in the ore pillar under the nine schemes is all lower than the shear strength of the ore rock of 50.86 MPa, and in the actual mining process, the occurrence state of the structural planes in the pillar can be disregarded, so the possibility of complete shear failure of the pillar is low. The maximum compressive stress of 86.8 and 92.5 MPa for schemes 6 and 9 are both lower than the uniaxial compressive strength of the ore (97.62 MPa) (Figure 39), which prevail over other schemes. Although stress values of scheme 6 are all low, scheme 9 has a lower ore loss than scheme 6 due to retained ore pillars, and the ore recovery rate of 84.92% is higher than that of 79.89% in scheme 6 (Figure 40). After comprehensive consideration, the stope structural parameters of scheme 9 are recommended.

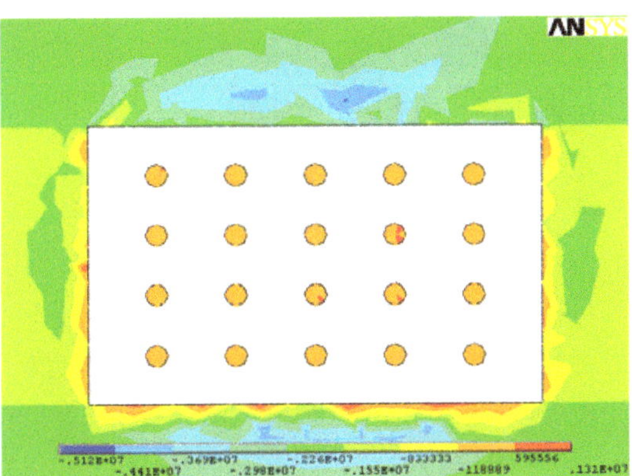

Figure 32. Maximum principal stress distribution in the middle of the pillar at 1480 mL of scheme 7 (parameters of scheme 7: stope height 4.5 m, pillar diameter 2 m, pillar spacing 6 m, pillar row spacing 7 m).

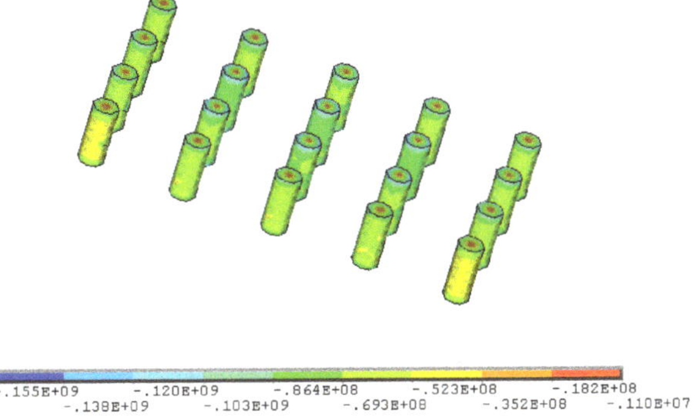

Figure 33. Minimum principal stress distribution of scheme 7 pillars at 1480 mL (parameters of scheme 7: stope height 4.5 m, pillar diameter 2 m, pillar spacing 6 m, pillar row spacing 7 m).

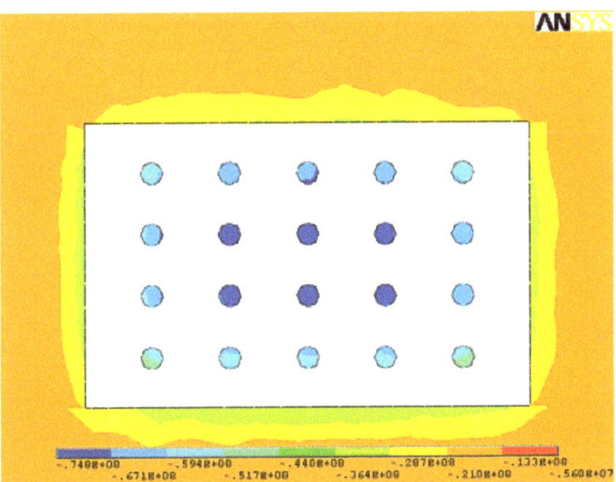

Figure 34. Minimum principal stress distribution in the middle of the pillar at 1480 mL of scheme 7 (parameters of scheme 7: stope height 4.5 m, pillar diameter 2 m, pillar spacing 6 m, pillar row spacing 7 m).

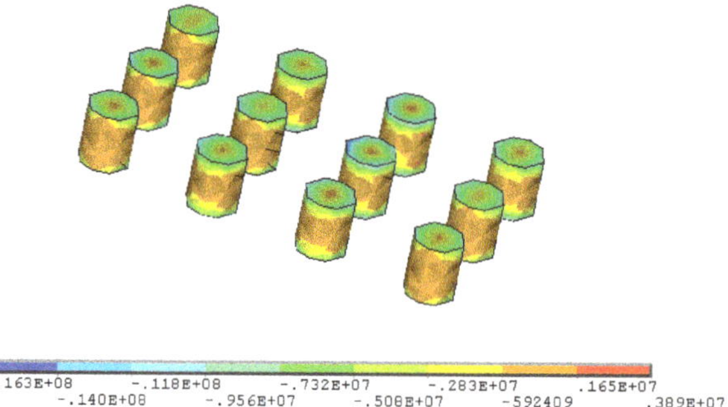

Figure 35. Maximum principal stress distribution of scheme 9 pillars at 1480 mL (parameters of scheme 9: stope height 4.5 m, pillar diameter 4 m, pillar spacing 7 m, pillar row spacing 8 m).

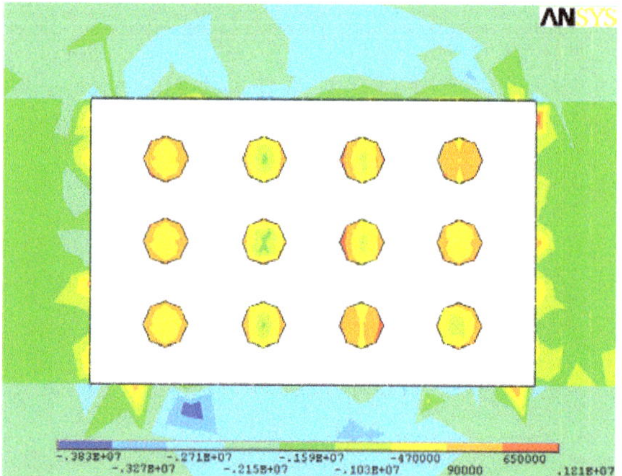

Figure 36. Maximum principal stress distribution in the middle of the pillar at 1480 mL of scheme 9 (parameters of scheme 9: stope height 4.5 m, pillar diameter 4 m, pillar spacing 7 m, pillar row spacing 8 m).

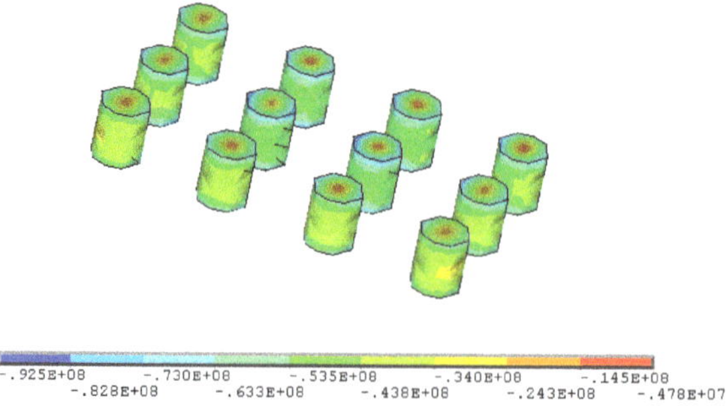

Figure 37. Minimum principal stress distribution of scheme 9 pillars at 1480 mL (parameters of scheme 9: stope height 4.5 m, pillar diameter 4 m, pillar spacing 7 m, pillar row spacing 8 m).

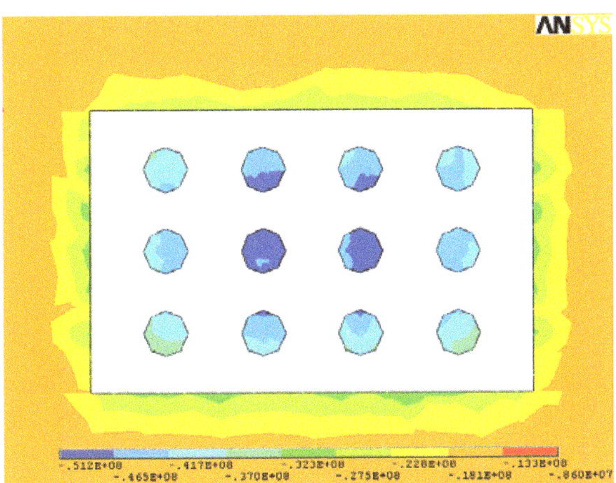

Figure 38. Minimum principal stress distribution in the middle of the pillar at 1480 mL of scheme 9 (parameters of scheme 9: stope height 4.5 m, pillar diameter 4 m, pillar spacing 7 m, pillar row spacing 8 m).

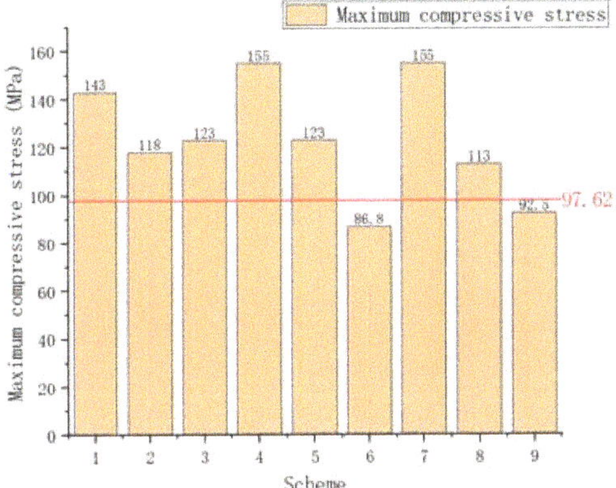

Figure 39. Maximum compressive stress in the ore pillars of schemes 1 to 9 (compared with the uniaxial compressive strength of the ore (97.62 MPa)).

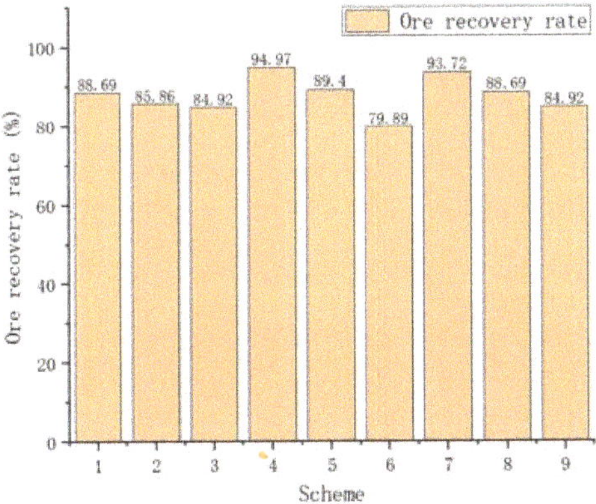

Figure 40. Designed ore recovery rate of schemes 1 to 9.

The optimal structural parameters of the room and pillar method are: stope height 4.5 m, ore pillar diameter 4 m, ore pillar spacing 7 m, and ore pillar row spacing 8 m. Based on the area-bearing theory, the axial compressive stress of the pillar under this parameter is 67.75 MPa, which is lower than the uniaxial compressive strength of the ore rock (97.62 MPa). The onsite trial is carried out in S3 stope at 1480 m using the recommended stope structural parameters, where all the stresses of the ore pillars meet the strength requirements of the ore rock, and the pillars ensures safety and stability after the ore room is mined. Additionally, the ore recovery rate of optimized stope reaches 82%, which also realizes the equilibrium of safety and economy.

5. Conclusions

1. Based on the area bearing theory, when the axial compressive stress is higher than the uniaxial compressive strength of the ore rock, the stope room will collapse. When the existing stope structural parameters are adopted, the maximum compressive stress in the ore pillar (142.0 MPa) is much higher than the uniaxial compressive strength of the ore (97.62 MPa). The existing stope structural parameters failed to maintain stability requirements and tended to be exposed to the risk of stope collapse.
2. When open stoping is applied to gently inclined medium-thick ore body, pillars are less affected by shear stress, and shear stress contributes little to pillar failure; the tensile stresses mainly occur in the middle of the pillars and within stope roofs and floors. Pillars mainly suffer compressive stress, especially the compressive stress concentrations that are prone to occur to the junction of the pillars, stope side walls, as well as roofs and floors.
3. The optimal structural parameters of the room and pillar method for the gently inclined medium-thick ore body are: stope height 4.5 m, pillar diameter 4 m, pillar spacing 7 m, and pillar row spacing 8 m. The onsite mining trial show that the optimized parameters are beneficial to realize the equilibrium of safety and economy, while the ore recovery rate reaches 82%.

Author Contributions: Writing—original draft preparation and editing, investigation, software, H.-Y.Q.; funding acquisition, writing—review, M.-Q.H.; translation, Y.-J.W. All authors have read and agreed to the published version of the manuscript.

Funding: This research was funded by the National Key Research and Development Program of China (Grant No. 2017YFC0602900), The National Natural Science Foundation of China (Grant No. 51804079) and National College Students' innovation and entrepreneurship training program (Grant No. 202210386022).

Data Availability Statement: The data presented in this study are available on request from the corresponding author.

Conflicts of Interest: The authors declare no conflict of interest.

References

1. Guo, Q.Q.; Yu, H.X.; Dan, Z.Y.; Li, S. Mining method optimization of gently inclined and soft broken complex ore body based on ahp and topsis: Taking miao-ling gold mine of china as an example. *Sustainability* **2021**, *13*, 12503. [CrossRef]
2. Wu, J. Research on sublevel open stoping recovery processes of inclined medium-thick orebody on the basis of physical simulation experiments. *PLoS ONE* **2020**, *15*, e0232640. [CrossRef] [PubMed]
3. Li, X.S.; Wang, Y.M.; Yang, S.; Xiong, J.; Zhao, K. Research progress in the mining technology of the slowly inclined, thin to medium thick phosphate rock transition from open-pit to underground mine. *Appl. Math. Nonlinear Sci.* **2021**, *6*, 319–334. [CrossRef]
4. Ge, Q.F.; Sun, X.S.; Zhu, W.G.; Chen, Q.G. Research of mining method for difficult-to-mine ore bodies in deep mine. *Adv. Mater. Res.* **2014**, *962–965*, 1041–1046. [CrossRef]
5. Serrano, A.; Olalla, C. Ultimate bearing capacity of rock masses. *Int. J. Rock Mech. Min. Sci. Geomech. Abstr.* **1994**, *31*, 93–106. [CrossRef]
6. Bazaluk, O.; Petlovanyi, M.; Zubko, S.; Lozynskyi, V.; Sai, K. Instability Assessment of Hanging Wall Rocks during Underground Mining of Iron Ores. *Minerals* **2021**, *11*, 858. [CrossRef]

7. Wu, A.X.; Huang, M.Q.; Han, B.; Wang, Y.M.; Yu, S.F.; Miao, X.X. Orthogonal design and numerical simulation of room and pillar configurations in fractured stopes. *J. Cent. South Univ.* **2014**, *21*, 3338–3344. [CrossRef]
8. Mikaeil, R.; Bakhshinezhad, H.; Haghshenas, S.S.; Ataei, M. Stability Analysis of Tunnel Support Systems Using Numerical and Intelligent Simulations (Case Study: Kouhin Tunnel of Qazvin-Rasht Railway). *Rud.-Geološko-Naft. Zb.* **2019**, *34*, 1–10. [CrossRef]
9. Zhao, K.; Wang, Q.; Li, Q.; Yan, Y.J.; Yu, X.; Wang, J.Q.; Cao, S. Optimization calculation of stope structure parameters based on mathews stabilization graph method. *J. Vibroeng.* **2019**, *21*, 1227–1239.
10. Qin, X.S.; Cao, H.; Wang, Z.X.; Zheng, Z.J. Analysis on safe thickness of the horizontal separation pillar in the upward horizontal slicing and filling method. *IOP Conf. Ser. Earth Environ. Sci.* **2021**, *861*, 052051. [CrossRef]
11. Lan, M.; Liu, Z.; Li, X. Multi-objective optimization and multi-attribute decision making on structural parameters of stage backfilling stope. *J. Cent. South Univ.* **2019**, *50*, 375–383.
12. Liu, D.; Shao, A.; Jin, C.; Ding, C.; Fan, F. Numerical model building for broken ore body and optimization of stope structural parameters. *J. Cent. South Univ.* **2019**, *50*, 437–444.
13. Qin, X.S.; Cao, H.; Guo, L.J. Sensitivity analysis of factors influencing pillar stability in the deep stope of underground salt mine. *IOP Conf. Ser. Earth Environ. Sci.* **2020**, *570*, 022002. [CrossRef]
14. Yao, G.H.; Wu, A.X.; Wang, Y.M. Stability analysis of stope retention pillars in broken rock conditions. *J. Univ. Sci. Technol. Beijing* **2011**, *33*, 400–405.
15. Ma, M.H.; Guo, Q.F.; Pan, J.L.; Ma, C.; Cai, M.F. Optimal Support Solution for a Soft Rock Roadway Based on the Drucker–Prager Yield Criteria. *Minerals* **2022**, *12*, 1. [CrossRef]
16. Drucker, D.C.; Prager, W. Soil mechanics and plastic analysis or limit design. *Quart. Appl. Math.* **1952**, *10*, 157–165. [CrossRef]

Article

Potential Uses of Artisanal Gold Mine Tailings, with an Emphasis on the Role of Centrifugal Separation Technique

Jeanne Pauline Munganyinka [1,2,*], Jean Baptiste Habinshuti [1,2], Jean Claude Ndayishimiye [3,4], Levie Mweene [5], Grace Ofori-Sarpong [1,6], Brajendra Mishra [2,*], Adelana R. Adetunji [1,7] and Himanshu Tanvar [2]

1 Department of Materials Science and Engineering, African University of Science and Technology, Abuja 900100, Nigeria; hbaptiste@aust.edu.ng (J.B.H.); gofori-sarpong@umat.edu.gh (G.O.-S.); aderade2004@yahoo.com (A.R.A.)
2 Mechanical Engineering, Worcester Polytechnic Institute, Worcester, MA 01609, USA; htanvar@wpi.edu
3 Department of Biology, Shenzhen MSU-BIT University, Shenzhen 518172, China; ndayiclaude2006@yahoo.fr
4 The Center for Earth and Natural Resource Sciences, Kigali P.O. Box 4285, Rwanda
5 Department of Materials Engineering, Indian Institute of Science, Bangalore 560012, India; leviemweene@alum.iisc.ac.in
6 Department of Minerals Engineering, University of Mines and Technology, Tarkwa P.O. Box 237, Ghana
7 Department Materials Science and Engineering, Obafemi Awolowo University, Ile-Ife 220282, Nigeria
* Correspondence: jmunganyinka@aust.edu.ng (J.P.M.); bmishra@wpi.edu (B.M.); Tel.: +1-570-664-4338 (J.P.M)

Abstract: Few investigations have focused on the potential uses of artisanal gold (Au) mine tailings, despite the fact that artisanal gold mining activity contributes to environmental issues such as greenhouse gas. Mineralogical characterizations of artisanal gold mine tailings in Miyove gold mine (Baradega and Masogwe) in Rwanda were investigated for potential utilization as a source of valuable gold, using the centrifugal separation technique. Results of X-ray diffraction analysis, energy dispersive X-ray spectroscopy, inductively coupled plasma mass spectrometry, inductively coupled plasma–optical emission spectroscopy, and X-ray fluorescence showed that artisanal gold mine tailings samples have significant amounts of gold to justify economical gold extraction opportunity. The gold grades in the ores and artisanal gold mine tailings were in the ranges of 37–152 and 2–7 g t^{-1}, respectively. Quartz was a major phase, with minor impurities in two different types of gold ores and their respective tailings. The beneficiation carried out using centrifugal separation, regarded as an extension of gravity separation, showed gold grades in the range of 535–1515 g t^{-1} for gold ores and 36–302 g t^{-1} for artisanal gold mine tailings. The gold recoveries for ores and artisanal gold mine tailings were in the range of 21.8–47.3% and 46.9–63.8%, respectively. The results showed that the centrifugal separation technique was more efficient in boosting gold recovery compared to the present panning approach employed at the site, which sometimes recover as low as 10%. The results suggest that mineralogical characterization of artisanal gold mine tailings allows for the development and design of a suitable methods for improving gold ore beneficiation and artisanal gold mine tailings reprocessing.

Keywords: gold mining; artisanal gold mine tailings; centrifugal separation; mineralogical characterization; artisanal and small-scale mining

Citation: Munganyinka, J.P.; Habinshuti, J.B.; Ndayishimiye, J.C.; Mweene, L.; Ofori-Sarpong, G.; Mishra, B.; Adetunji, A.R.; Tanvar, H. Potential Uses of Artisanal Gold Mine Tailings, with an Emphasis on the Role of Centrifugal Separation Technique. *Sustainability* **2022**, *14*, 8130. https://doi.org/10.3390/su14138130

Academic Editor: Rajesh Kumar Jyothi

Received: 10 May 2022
Accepted: 1 July 2022
Published: 3 July 2022

1. Introduction

Gold has been prized for its beauty and durability since ancient times [1]. Rwanda is a major producer of tin, tantalum, and tungsten (3Ts), as well as a gold and gemstone exporter [2]. Mineral exports brought in approximately USD 67 million in 2010, accounting for around 15% of total exports and making mining the country's main source of foreign earnings [3,4]. Due to the high global gold demand, the gold mining industry has recently emerged to have great potential, but most of this is carried out by individuals or cooperatives (artisanal small-scale miners) [5,6]. The gold mine tailings produced by small-scale miners are often dumped in designated tailing ponds with minimal treatment [5,7,8]. An increased artisanal gold mining in Rwanda has caused a rapid increase in artisanal gold mine tailings dumped yearly, and this can pose substantial risks to the health of people living around the mine sites as well as the neighboring ecosystems [9]. The mismanagement of the disposed tailings results in sliding, deforestation, and disturbance in the water table channel. In addition to decontaminating these mine tailings to prevent the release of harmful components, potential uses for these large quantities of mine tailings must be consistently evaluated for metal extraction [10,11]. Thus, re-processing of artisanal gold mine tailings is a sustainable way of minimizing the consequences of these wastes on the environment.

In nature, most gold occurs as nuggets and free grains in rocks, veins, and alluvial deposits [12]. Each gold deposit has unique mineralogical characteristics, thus requiring a specific type of processing to obtain pure gold [12–17]. The choice of beneficiation process depends on the nature of the gangue present in the ore and its association with the mineral of economic value [13,18–20]. For that reason, it is important to carry out a geometallurgical and diverse characterization analysis, which allows for a better observation of the different characteristics of the ore and its impact on the treatment of generated tailings [17,21,22]. The steps commonly employed in the processing of raw ore include crushing and grinding, screening, classification, concentration, dewatering, and disposal of tailings [23–25]. For the concentration and treatment of tailings, the steps commonly employed are classified using spirals, hydrocyclones, magnetic concentration, and flotation [14,26,27]. Chemical and mineralogical characterization allow for a better observation of the different mineralogical characteristics of gold ore and their relationship with the processing steps [28,29]. Grayson (2007) [30] identified about 70 different gold processing methods, except those based on cyanide or mercury, to recover gold lost by artisanal miners without side effects of cyanide and mercury. With the focus of avoiding the side effects caused by the mercury and cyanide, various gravity separation techniques have been tested to increase their suitability in artisanal gold mine [31]. These approaches allow the suggestion of different processing techniques, leading to changes in the processing steps already utilized. The inclusion of the information obtained by the mineral characterization converges to the optimization of mine development and, therefore, to the reduction of gold lost in the tailings, proving a better use of the mineral resources of a deposit. Thus, mineral characterization is of great importance for the future of mining since the comprehensive use of artisanal gold mine tailings is vital for the conservation of natural resources and, ensuring the sustainable development of the mining industry.

This research is therefore focused on the characterization and beneficiation of a gold ore, sourced from artisanal gold mine tailings. In light of this, a study on the mineralogical characteristics of both gold ores and resultant artisanal gold mine tailings was performed using different analytical tools. To achieve the best grade and recovery, gold ore was beneficiated, and artisanal gold mine tailings were treated using gravity concentration. This study is motivated by the search for alternatives that boost the re-treatment of tailings. The goal of this research was to perform a mineralogical characterization of artisanal gold mine tailings from a mine located in Rwanda and evaluate the technical feasibility of reprocessing this material and using it as a potential secondary source of gold.

2. Materials and Methods

2.1. Site Description and Sampling

The gold ores and artisanal gold mine tailings used in this work were collected from two zones (Baradega and Masogwe) of a mine in the region of Miyove in northern Rwanda (Figure 1). The topography of the mine area is defined by the steep hills and flat, marshy ground in the valleys. The elevation ranges from 1700 to 2200 m above sea level. Details of the local processing method (Figure S1) and regional geology of Miyove and the history of artisanal gold mining are shown in the Supplementary information. The samples were the representative fresh quartz vein of the gold ore deposit and disposed artisanal gold mine tailings that were mixed using cone and quartering methods. Two gold ore samples ($n = 2$, mass = 10 kg) and disposed artisanal gold mine tailings ($n = 2$, mass = 10 kg) were packed in labeled polyethylene plastic bags and immediately returned to the laboratory for further analysis. Each ore sample collected from the mine was crushed with a hammer to generate particles with a diameter of <40 mm. To produce particles with a diameter of 0.5 mm from those with <40 mm, the RETSCH 200520035 jaw crusher BB 50 Tungsten, which allows the user to set up a desired particle size, was used. A cone and quartering sampling were utilized to obtain an evenly dispersed sample. A ball mill was used to grind evenly dispersed samples to finer particles of a diameter ≤75 μm. After grinding, the sample was mixed thoroughly with the aid of the riffle sample splitter, and a representative sample was obtained for physico–chemical analysis and mineralogical characterization. For the process of beneficiation, a sieve shaker was used to produce particles with a diameter in the range of 45 to 600 μm. The particles with a diameter above 200 μm and the collected tailings were also ground by using a ball mill for the beneficiation process.

Figure 1. Artisanal gold mining. (**a**) The location of the Miyove gold mine and landscape of two sampling sites (red points). The photos of three artisanal gold mining methods at Miyove: (**b**) gold ore breaking with Hammer; (**c**) African stone grinding mill; (**d**) old-style table shaking system.

2.2. Mineralogy and Analysis of Concentrate

The particle size and shape analyses were carried out using a sieve shaker with meshes in the range of 45–600 μm and a Microtrac FlowSync (Microtrac Retsch GmbH,

Haan/Duesseldorf, Germany). The mineralogical characterization of the ore samples (*n* = 2) and tailings samples (*n* = 2) was carried out using an X-ray diffractometer (Malvern PANalytical, Westborough, MA, USA) with a Cu-Kα (λ = 0.1540 nm) radiation source operated at 45 kV and 40 mA. X'Pert HighScore Plus version 4.6a was used to identify the phases present in the samples. The chemical characterization of samples was performed using scanning electron microscopy in combination with energy-dispersive X-ray spectroscopy (JEOL JSM-700F, Hollingsworth & Vose, East Walpole, MA, USA). The solid samples were digested in an aqua regia solution (AR, 1:3 ratio of HNO_3 and HCl) to analyze the gold content with inductively coupled plasma mass spectrometry (ICP-MS, PerkinElmer, NEXION350x) [32]. For the quantitative elemental analysis, an inductively coupled plasma–optical emission spectroscopy (ICP-OES, Perkin Elmer Optima 8000) and X-ray fluorescence analyzer (XRF, Olympus Vanta M-Series) were also used. The samples for the ICP-OES were prepared by fusing the sample ore (0.1 g) with lithium tetraborate (1.0 g) as fluxing material in the graphite crucible at 1000 °C for 1 h. Then, 25% nitric acid was used to dissolve the melt and 2% of nitric acid was used to dilute the solution and also as a blank for ICP-OES analyses [33]. All analyses were carried out in triplicate, and the average values were reported.

2.3. Centrifugal Separation Technique

The procedures followed in the centrifugal separation of gold ores and tailings are depicted in Figure 2. In brief, beneficiation tests were carried out using a centrifugal concentrator blue bowl customized with kit w/pump and leg levelers in the presence of water as a fluid (Figure 2b). The riffle sample splitter was used to obtain homogeneous samples. Three beneficiation trials of 10 kg of gold ores and 10 kg of artisanal gold mine tailings of size <200 µm were processed. After cleaning the bowl and ensuring that all connected devices are in place and well controlled, a 12-volt pump was used to pump water into the bowl. The separation process uses centrifugal forces created by the vortex movement of water as it moves around the inner cone. Carefully ground samples were dropped into the bowl and the control valve was opened slowly to avoid overflow. As the water moved around, the inner cone carried the lighter materials to the surface, dropped them into a container through the discharge cone, and collected them as tailings. At the same time, the heavier particles remained and settled at the bottom of the bowl as concentrate. Beneficiation was carried out continuously with a solid to liquid ratio of 30% *w*/*w*. The obtained gravity concentrates were combined and analyzed for gold content and elemental composition. Thereafter, the recovery percentage was calculated. Results obtained using the centrifugal separation technique were further compared with the current panning approach used at the site (10%) in order to quantify the gold recovery.

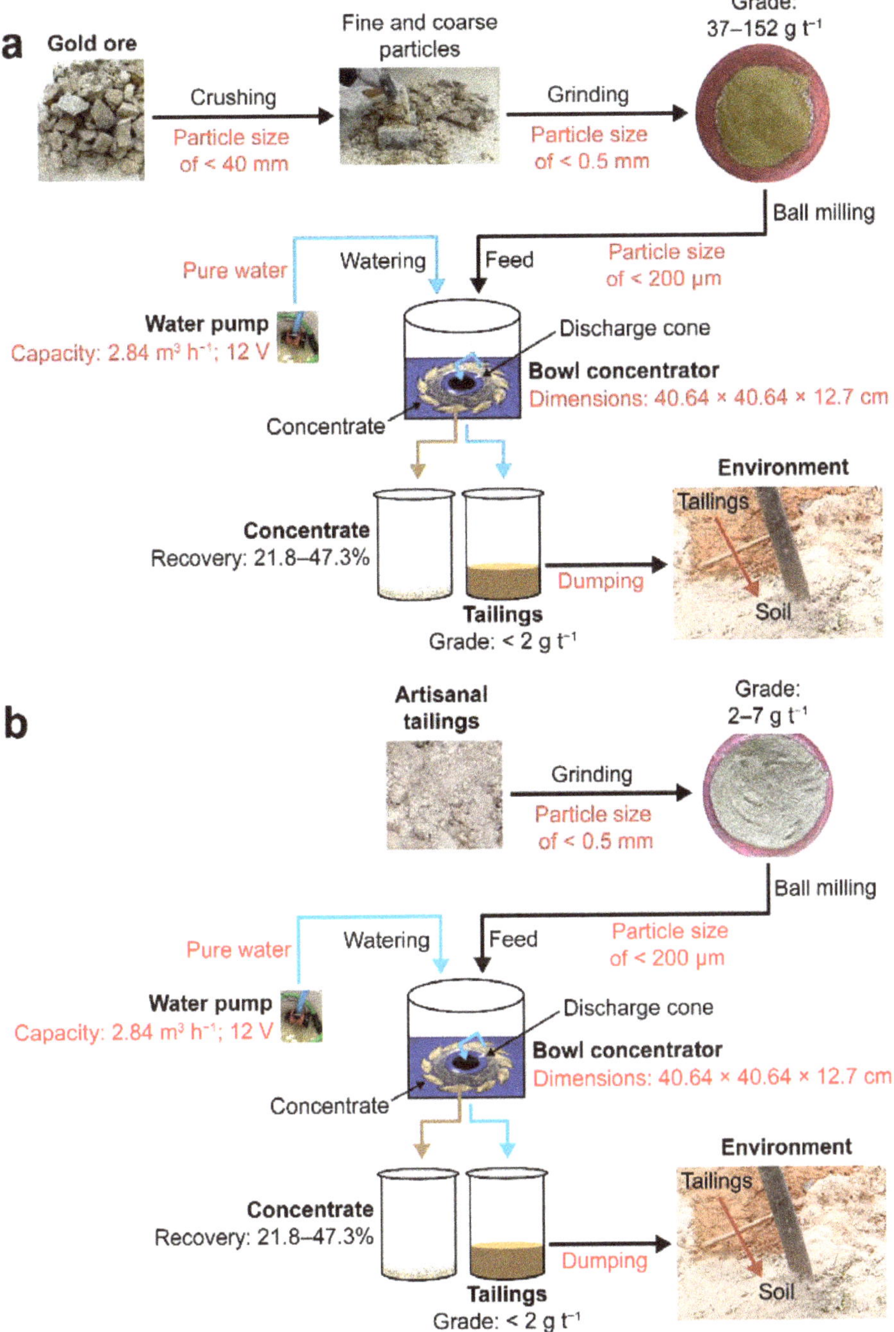

Figure 2. The workflow of the study indicates the importance of the centrifugal separation method in the remining of artisanal gold mine tailings and compares mineral grade and recovery percentage before and after beneficiation. (**a**) Gold ore; (**b**) artisanal tailings.

3. Results

3.1. Mineralogy of Raw Gold Ore and Resultant Artisanal Gold Mine Tailings

The gold ore particles were coarser than artisanal gold mine tailing particles (Figure 3). The diameters D10, D50, and D90 were respectively in the range of 4.8–6.8, 20.4–65.2, and 117.4–231.1 μm for gold ores, and 2.9–5.9, 18.0–56.3, and 88.7–117.2 μm for artisanal gold mine tailings. The gold grades for ores and artisanal gold mine tailings were in the range of 37–152 and 2–7g t^{-1}, respectively (Figure 2).

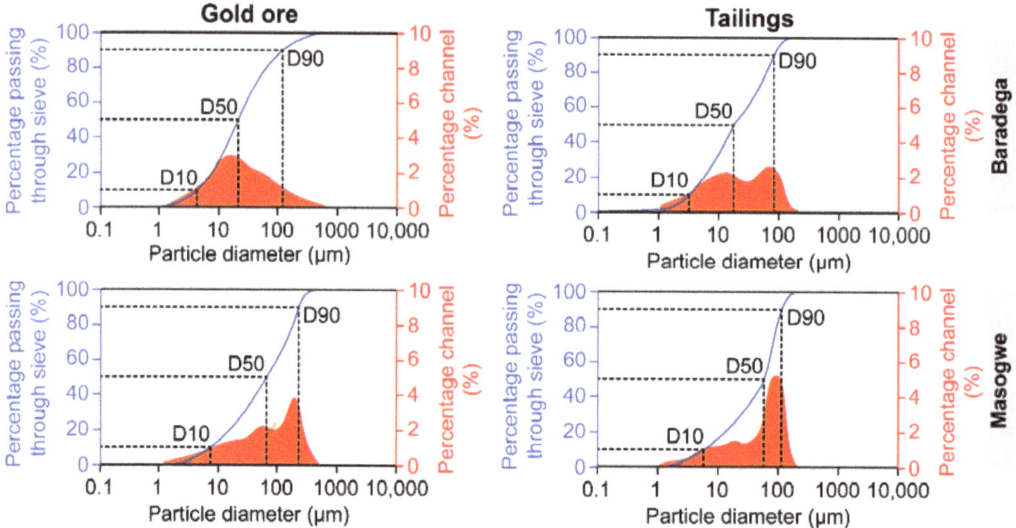

Figure 3. Particle size distribution for collected gold ores and resultant artisanal gold mine tailings.

Quartz (SiO_2) was the dominant phase in both gold ores and artisanal gold mine tailings. In addition, K-feldspar and aluminum oxide were evident in gold ores and artisanal gold mine tailings, respectively (Figure 4).

The elemental compositions revealed by the four studied samples were significantly different (Figure 5 and Table 1). Quantitative analysis of EDS results represented by the pie chart in Figure 5 showed that silicon was 34.70–44.50% in gold ores and 29.90–38.30% in tailings. Aluminium was 9.60–10.20% and 2.10–9.0%, respectively. Iron was 0.00–6.60% and 0.00–3.50%, respectively (Figure 5). Rare earth elements found in both gold ores and artisanal gold mine tailings were cerium, europium, lanthanum, neodymium, praseodymium, terbium, and yttrium. ICP-OES showed that silicon was 17.81–20.09% in gold ores and 32.42–33.81% in artisanal gold mine tailings. XRF showed that silicon was 41.06–42.71% and 38.41–39.19%, respectively (Table 1).

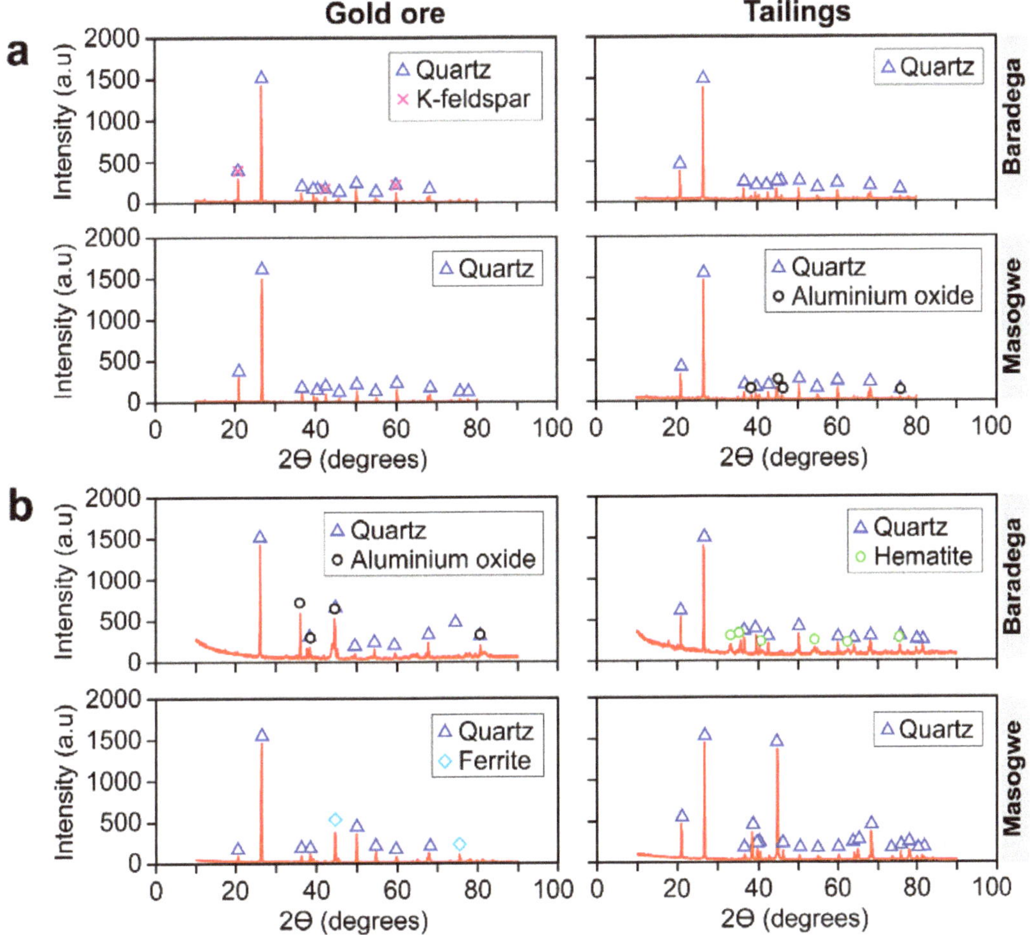

Figure 4. Crystallography of gold ore and resulting tailings, (**a**) before beneficiation, and (**b**) after beneficiation.

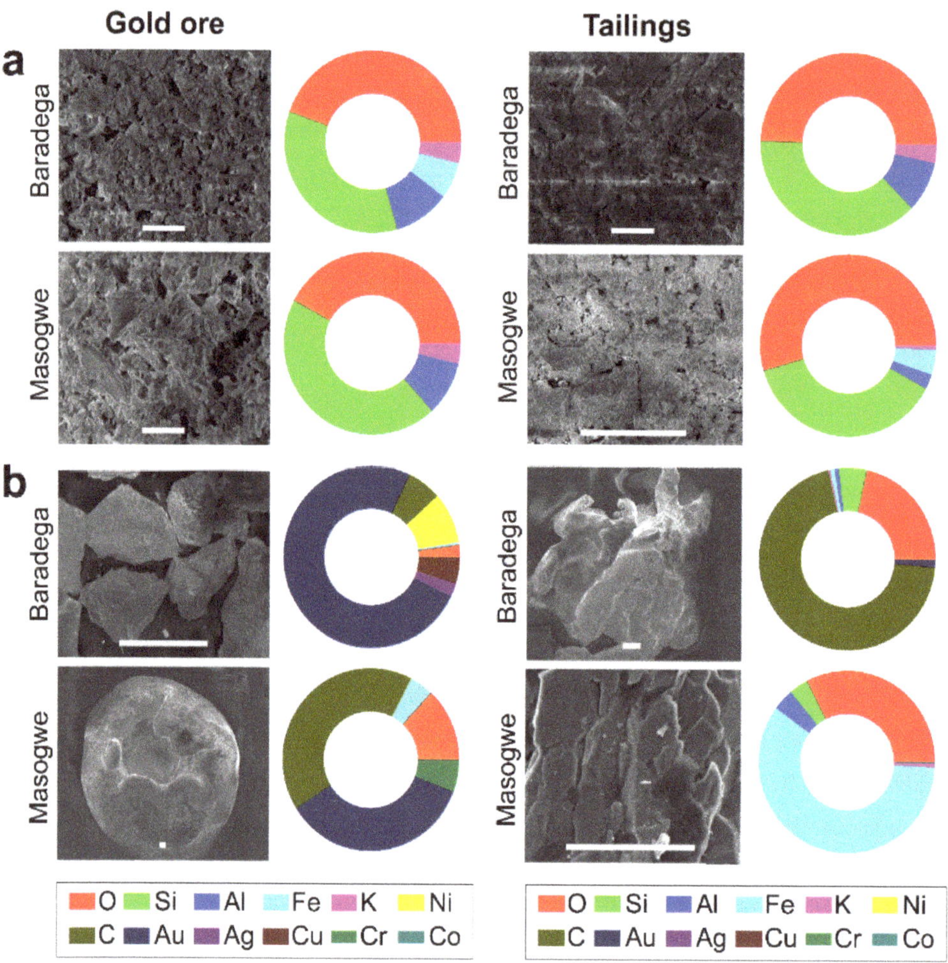

Figure 5. Elemental composition as measured in full scale X-ray on gold ore and resultant artisanal gold mine tailings, (**a**) before beneficiation, and (**b**) after beneficiation. The scale bars are 25 μm. The pie chart shows the proportions of 12 detect elements.

Table 1. Elemental composition of raw gold ore and resultant artisanal gold mine tailings.

Element	ICP-OES		XRF		ICP-MS	
	Gold Ore	Tailings	Gold Ore	Tailings	Gold Ore	Tailings
Trace element (%)						
Silicon	17.81–20.09	32.42–33.81	41.06–42.71	38.41–39.19		
Potassium	2.38–4.43	1.63–1.83	0.87–1.42	1.05–1.49		
Aluminium	2.06–3.70	3.23–4.70	4.73–6.49	6.36–6.67		
Iron	1.11–1.49	1.74–2.39	1.18–1.92	2.22–2.46		
Sodium	0.36–0.57	0.44–0.46	*	*		
Magnesium	0.10–0.11	0.06–0.08	*	*		
Titanium	0.09–0.14	0.18–0.19	0.16–0.21	0.18–0.25		
Copper	0.04–1.92	0.03–0.04	0.00–0.00	0.00–0.01		
Calcium	0.02–0.10	0.02–0.03	0.02–0.14	0.00–0.02		
Rare earth element (ppm)						
Lanthanum					2.98–14.50	4.90–7.90
Praseodymium					0.60–3.00	1.00–1.70
Yttrium					1.00–3.30	0.90–1.20
Europium					0.14–0.44	0.19–0.40
Terbium					0.03–0.12	0.05–0.05
Neodymium					2.38–11.00	0.11–6.30
Cerium					6.08–29.00	9.90–15.90

ICP-OES, inductively coupled plasma–optical emission spectroscopy; XRF, X-ray fluorescence. "–" indicates a range; "*" shows below detection limit.

3.2. Mineralogy of Gold Ore and Resultant Artisanal Gold Mine Tailings Concentrate

Quartz (SiO_2) was the dominant phase in both gold ore and artisanal gold mine tailings concentrate. Aluminum oxide and ferrite were evident in gold ores, while hematite was observable in artisanal gold mine tailings (Figure 4). The compositions of the samples were significantly different (Figure 5 and Table 2). Based on EDS, gold was 35.30–74.90% in gold ores and 0.00–1.40% in tailings. Silicon was 0.00% and 3.50–5.10%, respectively. Aluminium was 0.00% and 0.80–3.80%, respectively. Figure 5 shows that iron was 0.50–4.30% and 0.80–59.20%, respectively. ICP-OES indicated that silicon was 44.15–41.31% for gold ores and 41.74–45.50% for artisanal gold mine tailings. XRF indicated 40.89–46.92% and 41.56–41.86%, respectively (Table 2). The gold recoveries for ores were 21.8–47.3%, while those for artisanal gold mine tailings were 46.9–63.8% (Figure 2).

Table 2. Elemental composition of god ore and resultant artisanal gold mine tailings concentrates.

Element (%)	ICP-OES		XRF	
	Gold Ore	Tailings	Gold Ore	Tailings
Silicon	44.15–41.31	41.74–45.50	40.89–46.92	41.56–41.86
Iron	4.90–5.61	1.76–3.93	2.25–4.86	1.57–3.97
Potassium	0.46–1.13	0.67–1.36	0.06–0.61	0.35–0.77
Copper	0.13–0.17	0.14–0.18	0.00–0.03	0.00–0.00
Aluminium	0.05–0.07	0.21–0.28	2.72–2.72	1.82–3.19
Calcium	*	*	0.01–0.03	0.02–0.03

ICP-OES, inductively coupled plasma–optical emission spectroscopy; XRF, X-ray fluorescence. "–" indicates a range; "*" shows below detection limit.

4. Discussion

4.1. Characteristics of Artisanal Gold Mine Tailings and Potential Features as a New Source of Valued Gold

This study shows that gold associated rocks in Miyove Gold Mine allow for the extraction of gold resources efficiently (Figure 2). Manual breaking of gold ores with hammer, African stone grinding mill and old-style table shaking system are largely employed in the gold mining of Miyove (Figure 1b–d) possibly due to financial constraints on investment, lack of technology, and skilled labor [34,35]. For example, modern shaking tables that are very effective and can concentrate sizeable amounts of ore at a time, providing high-grade concentrates and liberated gold, are relatively expensive and require some skill to operate [30]. The use of primitive processing methods results in the generation of solid waste with high gold content [36]. This confirms our observation showing (1) the utilization of hammers in breaking gold ores as poor liberation of gold from the host rock [29], and as the main source of dispersed gold particles on the ground and in tailings (Figure 1b); (2) establishment of gold particle sizes in undesired ranges due to ineffectiveness in crushing and grinding gold ores (Figure 1b,c); and (3) inadequate separation of gold particles from tailings due to ineffectiveness in screening (Figure 1d).

The application of mineralogical information allows for a better understanding and solution of problems encountered during exploration and mining, and during the processing of ores, concentrates, smelter products, and related materials [13,16]. The geochemistry of artisanal gold mine tailings was conducted using various methods (XRD, EDS, ICP-MS, ICP-OES, and XRF) because gold particles and impurities in the form of rocks present in mine tailings are very tiny, and their smaller size derives amazing features, such as their chemical properties [28]. As the optimization depends on the mineralogical information of gold ore to identify the non-valuable minerals associated with gold as well as the gold host rock, quantitative analysis was performed and the information obtained provided an image of the sampled artisanal gold mine tailings behavior. Gold was hosted by quartz in the form of silicon dioxide (Figures 4 and 5 and Tables 1 and 2). The major phase identified was quartz, accompanied by minor impurities. Ferrite was identified as the phase associated with quartz, indicating the presence of iron in gold ores with small proportions of few metallic elements (Table 2). The presence of a hematite phase in the form of Fe_2O_3 (Figure 4) implies that gold ore and tailings samples show a paramagnetic property. Some phases detected in the studied tailings holds hazardous elements that could be considered as conveyor of human health and environment effects [37]. However, this iron-related phase was not peaked in the XRD of the gold ore sample, and possibly was masked by the other gangues presented in the ore [32,38]. The diameter of individual grains of rocks associated with gold play a major role in the processing of gold, especially when it is associated with the coarse gangue minerals [31]. Gold grades obtained after crushing and grinding all samples were high; gangues were rich in silicates (Figure 2). This suggests that processing gold ore requires specialized machinery that can be suited for the type of minerals being milled [39]. The processing of the tailings shows that about 2–7g t^{-1} was lost after using traditional techniques on the sites (Figure 2), suggesting the inadequacy of methods utilized by artisanal gold miners [40]. For gold ores, the enrichment ratio (ER) ranged from 9.9 to 14.4, and for tailings, it ranged from 18.0 to 43.1.

$$ER = \frac{\text{Grade of gold in the concentrate}}{\text{Grade of gold in the feed}} \quad (1)$$

4.2. Importance of Centrifugal Separation in Artisanal Gold Mine Tailings Remining

Different beneficiation techniques to extract gold from ore are chosen based on various factors [41–43]. The poor processing methods posed gold losses and health risks to the miners due to the contaminants exposure. In this study, each beneficiation step was designed to increase the concentration (grade) of the valuable components of the original ore (Figure 2). Mined ore and resulting artisanal gold mine tailings underwent comminution by crushing and/or grinding, feeding, and gravity concentration by centrifugal separation to remove the bulk of the rocks and gangue minerals. The utilization of a centrifuge was extremely beneficial as it eliminated large volumes of waste rocks from the gold. As a result, the gold in artisanal gold mine tailings was extracted at a reduced operating cost with respect to energy. The effectiveness of centrifugal separation is closely related to the gold particle size [31,39]. Additionally, particles in a centrifuge were segregated depending on their size, shape, density, and rotor speed; hence, the reported gold concentrates have enhanced grades and recoveries [15,40]. This shows the gold loss and ineffectiveness of traditional methods described in Figure 1b–d, and that mineral specific gravity facilitates pleasing separation of gold from its associated gangue minerals [28]. For example, a recovery of less than 10% of gold from solids of diameters of 74–150 μm and 40% of gold from solids of diameters of >212 μm [44].

4.3. Environmental and Management Implications

Artisanal and small-scale gold mining often occurs in locations where there is no large-scale mining presence [8]. Yet, artisanal small-scale mining is associated with a number of negative impacts [7,45,46]. For example, data and results indicate that poor handling of artisanal gold mine tailings can contribute to not only environmental degradation, but also abandoned pits and shafts, as well as a loss of gold resources. In addition, there is also a possibility of heavy metal contamination in the area due to poor handling processes. Previous study has examined the historical and current situation and issues of ASGM in connection to political, social, and environmental repercussions (such as population displacement, loss of livelihoods, migration of people, cost of living, water scarcity, and health implications) [47]. Thus, in this study, mineralogical characteristics of artisanal gold mine tailings using Miyove's samples in Rwanda and potential features as a new source of valued gold are shown (Figures 2–5, and Tables 1 and 2). In comparison to the current panning method used at the site (10%), the gold grades (Figure 2) reveal that the centrifugal separation technique is more effective in increasing gold recovery from artisanal gold mine tailings. This further demonstrates the need for efficiency, the use of environmentally friendly methods for gold recovery in mining (e.g., gravity methods by centrifugal separation), and artisanal gold mine tailings reuse. The priority areas for any further investigation must take into consideration more samples to sustain a successful scale-up of our findings across a wide array of mining sites and mineral separation techniques. This will thereby help to keep the artisanal and small-scale mining sector economically more productive rather than destructive.

5. Conclusions

This work presents mineralogical characteristics of artisanal gold mine tailings in two key mining areas of Miyove in Rwanda for potential use as a source of valuable gold using the centrifugal separation system. The results showed that artisanal gold mine tailings samples have significant amounts of gold, that can be mined using centrifugal separation practice. Quartz was a major phase, with minor impurities in both gold ores and resultant artisanal gold mine tailings. The centrifugal separation technique as applied to artisanal gold mine tailings was more beneficial in terms of gold recovery. The mineralogical characterization of artisanal gold mine tailings makes it possible to propose alternatives that improve gold ore beneficiation and the re-processing of artisanal gold mine tailings.

Supplementary Materials: The following supporting information can be downloaded at: https://www.mdpi.com/article/10.3390/su14138130/s1, Figure S1: Photos showing local gold processing method during samling (left) and pits excaveated by artisanal miners (right). References [48–55] are cited in the supplementary materials

Author Contributions: J.P.M.: investigation, formal analysis, writing—original draft, validation, writing—review and editing. J.B.H. and J.C.N.: methodology, formal analysis, writing—review and editing. L.M. and H.T.: data curation, validation, writing—review and editing. G.O.-S. and B.M.: conceptualization, methodology, resources, validation, writing—review and editing. A.R.A.: writing-review and editing. All authors have read and agreed to the published version of the manuscript.

Funding: This research was supported by Regional Scholarship and Innovation Fund (RSIF), a flagship program of the Partnership for Skills in Applied Sciences, Engineering and Technology (PASET), an Africa-led, World Bank-affiliated initiative.

Institutional Review Board Statement: Not applicable.

Informed Consent Statement: Not applicable.

Data Availability Statement: Raw data from the study is available on request.

Acknowledgments: The authors thank Rwanda Mines, Petroleum and Gas Board and Ngali Mining Ltd. authorizing sample collection and exportation and Miyove gold mining project for facilitating the sample collection and field investigation. The authors also thank Abdulhakeem Bello, Uwamungu Placide, and Pascaline Nyirabuhoro for their comments.

Conflicts of Interest: The authors declare that they have no known competing financial interests or personal relationships that could have appeared to influence the work reported in this paper.

References

1. Gold Statistics and Information | U.S. Geological Survey. Available online: https://www.usgs.gov/centers/national-minerals-information-center/gold-statistics-and-information (accessed on 30 March 2022).
2. Africa and the Middle East | U.S. Geological Survey. Available online: https://www.usgs.gov/centers/national-minerals-information-center/africa-and-middle-east#rw (accessed on 15 June 2022).
3. Lezhnev, S.; Swamy, M. Understanding Money Laundering Risks in the Conflict Gold Trade from East and Central Africal to Dubai and Onward. *Sentry* **2020**, 1–22. Available online: https://cdn.thesentry.org/wp-content/uploads/2020/11/ConflictGoldAdvisory-TheSentry-Nov2020.pdf (accessed on 9 May 2022).
4. WorldAtlas What Are the Major Natural Resources of Rwanda? Available online: https://www.worldatlas.com/articles/what-are-the-major-natural-resources-of-rwanda.html (accessed on 10 June 2022).
5. Crawford, A.; Bliss, M. IGF Mining Policy Framework Assessment. *Intergov. Forum Min. Min. Met. Sustain. Dev.* **2017**, 1–45.
6. Baptiste, J.; Pauline, J.; Adetunji, A.R. Mineralogical and Physical Studies of Low-Grade Tantalum-Tin Ores from Selected Areas of Rwanda. *Results Eng.* **2021**, *11*, 100248. [CrossRef]
7. Bansah, K.J.; Dumakor-Dupey, N.K.; Kansake, B.A.; Assan, E.; Bekui, P. Socioeconomic and Environmental Assessment of Informal Artisanal and Small-Scale Mining in Ghana. *J. Clean. Prod.* **2018**, *202*, 465–475. [CrossRef]
8. Hilson, G.; Amankwah, R.; Ofori-Sarpong, G. Going for Gold: Transitional Livelihoods in Northern Ghana. *J. Mod. Afr. Stud.* **2013**, *51*, 109–137. [CrossRef]
9. Dewaele, S.; Henjes-Kunst, F.; Melcher, F.; Sitnikova, M.; Burgess, R.; Gerdes, A.; Fernandez, M.A.; De Clercq, F.; Muchez, P.; Lehmann, B. Late Neoproterozoic Overprinting of the Cassiterite and Columbite-Tantalite Bearing Pegmatites of the Gatumba Area, Rwanda (Central Africa). *J. Afr. Earth Sci.* **2011**, *61*, 10–26. [CrossRef]
10. Mining, S.G.; Municipality, A. Waste Management and the Elimination of Mercury in Tailings from Artisanal Waste Management and the Elimination of Mercury in Tailings from Artisanal and Small-Scale Gold Mining in the Andes Municipality of Antioquia, Colombia. *Mine Water Environ.* **2020**, *40*, 250–256. [CrossRef]
11. Martinez, G.; Restrepo-baena, O.J.; Veiga, M.M. The Extractive Industries and Society The Myth of Gravity Concentration to Eliminate Mercury Use in Artisanal Gold Mining. *Extr. Ind. Soc.* **2021**, *8*, 477–485. [CrossRef]
12. Yalcin, E.; Kelebek, S. Flotation Kinetics of a Pyritic Gold Ore. *Int. J. Miner. Process.* **2011**, *98*, 48–54. [CrossRef]
13. Marsden, J.O.; House, C.I. The Chemistry of Gold Extraction. *Ellis Horwood Maylands Ave. UK* **2006**, *651*, 619.
14. Hylander, L.D.; Plath, D.; Miranda, C.R.; Lücke, S.; Öhlander, J.; Rivera, A.T.F. Comparison of Different Gold Recovery Methods with Regard to Pollution Control and Efficiency. *Clean-Soil Air Water* **2007**, *35*, 52–61. [CrossRef]
15. Coetzee, L.L.; Theron, S.J.; Martin, G.J.; van der Merwe, J.D.; Stanek, T.A. Modern Gold Deportments and Its Application to Industry. *Miner. Eng.* **2011**, *24*, 565–575. [CrossRef]
16. Gökelma, M.; Birich, A.; Stopic, S.; Friedrich, B. A Review on Alternative Gold Recovery Re-Agents to Cyanide. *J. Mater. Sci. Chem. Eng.* **2016**, *4*, 8–17. [CrossRef]
17. Napier-Munn, T.; Wills, B.A. *Wills' Mineral Processing Technology*; Elsevier Ltd: Oxford, UK, 2005; ISBN 9780750644501.
18. Venter, D.; Chryssoulis, S.L.; Mulpeter, T. Using Mineralogy to Optimize Gold Recovery by Direct Cyanidation. *JOM* **2004**, *56*, 53–56. [CrossRef]

19. Chryssoulis, S.L.; McMullen, J. *Mineralogical Investigation of Gold Ores*; Elsevier: Amsterdam, The Netherlands, 2016; ISBN 9780444636584.
20. Wang, X.; Qin, W.; Jiao, F.; Yang, C.; Cui, Y.; Li, W. Mineralogy and Pretreatment of a Refractory Gold. *Minerals* **2019**, *9*, 406. [CrossRef]
21. Fullam, M.; Watson, B.; Laplante, A.; Gray, S. *Advances in Gravity Gold Technology*; Elsevier: Amsterdam, The Netherlands, 2016; ISBN 9780444636584.
22. Melcher, F.; Graupner, T.; Gäbler, H.E.; Sitnikova, M.; Henjes-Kunst, F.; Oberthür, T.; Gerdes, A.; Dewaele, S. Tantalum-(Niobium-Tin) Mineralisation in African Pegmatites and Rare Metal Granites: Constraints from Ta-Nb Oxide Mineralogy, Geochemistry and U-Pb Geochronology. *Ore Geol. Rev.* **2015**, *64*, 667–719. [CrossRef]
23. Maponga, O.; Ngorima, C.F. Overcoming Environmental Problems in the Gold Panning Sector through Legislation and Education: The Zimbabwean Experience. *J. Clean. Prod.* **2003**, *11*, 147–157. [CrossRef]
24. Jena, M.S.; Mohanty, J.K.; Sahu, P.; Venugopal, R.; Mandre, N.R. Characterization and Pre-Concentration of Low Grade PGE Ores of Boula Area, Odisha Using Gravity Concentration Methods. *Trans. Indian Inst. Met.* **2017**, *70*, 287–302. [CrossRef]
25. Ernawati, R.; Idrus, A.; Petrus, H.T.B.M. Study of the Optimization of Gold Ore Concentration Using Gravity Separator (Shaking Table): Case Study for LS Epithermal Gold Deposit in Artisanal Small Scale Gold Mining (ASGM) Paningkaban, Banyumas, Central Java. *IOP Conf. Ser. Earth Environ. Sci.* **2018**, *212*, 012019. [CrossRef]
26. Bouchard, J.; Desbiens, A.; Nunez, E. Column Flotation Simulation and Control: An Overview. *Miner. Eng.* **2009**, *22*, 519–529. [CrossRef]
27. Nambaje, C.; Mweeneb, L.; Subramaniana, S.; Sajeev, K.; Santosh, M. Xanthan Gum Based Investigations into the Surface Chemistry of Cassiterite and Beneficiation of Cassiterite Tailings. *Miner. Procesing Extr. Metall. Rev.* **2020**, *2*, 150–164. [CrossRef]
28. Abdel-Khalek, N.A.; El-Shatoury, E.H.; Abdel-Motelib, A.; Hassan, M.S.; Abdel-Khalek, M.A.; El-Sayed, S. Mineralogical Study and Enhanced Gravity Separation of Gold-Bearing Mineral, South Eastern Desert, Egypt. *Physicochem. Probl. Miner. Process.* **2020**, *56*, 839–848. [CrossRef]
29. Bode, P.; McGrath, T.D.H.; Eksteen, J.J. Characterising the Effect of Different Modes of Particle Breakage on Coarse Gangue Rejection for an Orogenic Gold Ore. *Miner. Process. Extr. Metall. Trans. Inst. Min. Metall.* **2020**, *129*, 35–48. [CrossRef]
30. Grayson, R. Fine Gold Recovery-Alternatives to Mercury and Cyanide. *World Placer J.* **2007**, *7*, 66–161.
31. Veiga, M.M.; Gunson, A.J. Gravity Concentration in Artisanal Gold Mining. *Minerals* **2020**, *10*, 1026. [CrossRef]
32. Balaram, V.; Vummiti, D.; Roy, P.; Taylor, C.; Kar, P.; Raju, A.K.; Abburi, K. Determination of Precious Metals in Rocks and Ores by Microwave Plasma-Atomic Emission Spectrometry for Geochemical Prospecting Studies. *Curr. Sci.* **2011**, *104*, 1207–1215.
33. Habinshuti, J.B.; Munganyinka, J.P.; Adetunji, A.R.; Mishra, B.; Himanshu, T.; Mukiza, J.; Ofori-Sarpong, G.; Onwualu, A.P. Caustic Potash Assisted Roasting of the Nigerian Ferro-Columbite Concentrate and Guanidine Carbonate-Induced Precipitation: A Novel Technique for Extraction of Nb–Ta Mixed-Oxides. *Results Eng.* **2022**, *14*, 100415. [CrossRef]
34. Government of Rwanda (NISR, RMB). *Natural Capital Accounts for Mineral Resource Flows*; Version 1.0; National Institute of Statistics of Rwanda: Kigali, Rwanda, 2019.
35. Heizmann, J.; Liebetrau, M. *Efficiency of Mineral Processing in Rwanda' s Artisanal and Small-Scale Mining Sector: Quantitative comparison of Traditional Techniques and Basic Mechanized Procedures*; Bundesanstalt für Geowissenschaften und Rohstoffe: Hannover, Germany, 2017; ISBN 9783943566864.
36. Rintala, L.; Leikola, M.; Sauer, C.; Aromaa, J.; Roth-berghofer, T.; Forsén, O.; Lundström, M. Designing Gold Extraction Processes: Performance Study of a Case-Based Reasoning System. *Miner. Eng.* **2017**, *109*, 42–53. [CrossRef]
37. Akinyemi, S.A.; Oliveira, M.L.S.; Nyakuma, B.B.; Dotto, G.L. Geochemical and Morphological Evaluations of Organic and Mineral Aerosols in Coal Mining Areas: A Case Study of Santa Catarina, Brazil. *Sustainability* **2022**, *14*, 3847. [CrossRef]
38. Ogundare, O.D.; Adeoye, M.O.; Adetunji, A.R.; Adewoye, O.O. Beneficiation and Characterization of Gold from Itagunmodi Gold Ore by Cyanidation. *J. Miner. Mater. Charact. Eng.* **2014**, *2*, 300–307. [CrossRef]
39. Gül, A.; Kangal, O.; Sirkeci, A.A.; Önal, G. Beneficiation of the Gold Bearing Ore by Gravity and Flotation. *Int. J. Miner. Metall. Mater.* **2012**, *19*, 106–110. [CrossRef]
40. Hinton, J.J.; Veiga, M.M.; Veiga, A.T.C. Clean Artisanal Gold Mining: A Utopian Approach? *J. Clean. Prod.* **2003**, *11*, 99–115. [CrossRef]
41. Poloko, N. Physical Separation Methods, Part 1: A Review. In *The IOP Conference Series: Materials Science and Engineering*; IOP Publishing: Bristol, UK, 2019; Volumn 641, p. 12023.
42. Farrokhpay, S. Editorial for the Special Issue:"Physical Separation and Enrichment". *Minerals* **2020**, *10*, 173. [CrossRef]
43. Olyaei, Y.; Aghazadeh, S.; Gharabaghi, M.; Mamghaderi, H.; Mansouri, J. Gold, Mercury, and Silver Extraction by Chemical and Physical Separation Methods. *Rare Met. Mater. Eng.* **2016**, *45*, 2784–2789. [CrossRef]
44. Silva, M. *Placer Gold Recovery Methods*; California Department of Conservation, Division of Mines and Geology: Sacramento, CA, USA, 1986.
45. Zolnikov, T.R. Effects of the Government's Ban in Ghana on Women in Artisanal and Small-Scale Gold Mining. *Resour. Policy* **2020**, *65*, 101561. [CrossRef]
46. Eduful, M.; Alsharif, K.; Acheampong, M.; Nkhoma, P. Management of Catchment for the Protection of Source Water in the Densu River Basin, Ghana: Implications for Rural Communities. *Int. J. River Basin Manag.* **2020**, *20*, 167–183. [CrossRef]

47. Mestanza-Ramón, C.; Cuenca-Cumbicus, J.; D'orio, G.; Flores-Toala, J.; Segovia-Cáceres, S.; Bonilla-Bonilla, A.; Straface, S. Gold Mining in the Amazon Region of Ecuador: History and a Review of Its Socio-Environmental Impacts. *Land* **2022**, *11*, 221. [CrossRef]
48. Theunissen, K.; Hanon, M.; Fernandez, M. *Carte Géologique Du Rwanda Au 1/250.000*; Musée Royal de l'Afrique Centrale: Tervuren, Belgique, 1991.
49. Fernandez-Alonso, M.; Cutten, H.; De Waele, B.; Tack, L.; Tahon, A.; Baudet, D.; Barritt, S.D. The Mesoproterozoic Karagwe-Ankole Belt (Formerly the NE Kibara Belt): The Result of Prolonged Extensional Intracratonic Basin Development Punctuated by Two Short-Lived Far-Field Compressional Events. *Precambrian Res.* **2012**, *216–219*, 63–86. [CrossRef]
50. Baudet, D.; Hanon, M.; Lemonne, E.; Theunissen, K.; Buyagu, S.; Dehandschutter, J.; Ngizimana, W.; Nsengiyumva, P.; Rusanganwa, J.B.; Tahon, A. Lithostratigraphie Du Domaine Sédimentaire de La Chaine Kibarienne Au Rwanda. *Ann. la Société géologique Belgique*. 1989. Available online: https://popups.uliege.be/0037-9395/index.php?id=873 (accessed on 10 May 2022).
51. Uwiduhaye, J. d'Amour; Ngaruye, J.C.; Saibi, H. Defining Potential Mineral Exploration Targets from the Interpretation of Aeromagnetic Data in Western Rwanda. *Ore Geol. Rev.* **2021**, *128*, 103927. [CrossRef]
52. Cahen, L.; Snelling, N.J.; Delhal, J.; Vail, J.R.; Bonhomme, M.; Ledent, D. *The Geochronology and Evolution of Africa*; Oxford University Press: Oxford, UK, 1984.
53. Baudin, B.; Zigirababili, J.; Ziserman, A.; Petricec, V. *Carte Des Gîtes Minéraux Du Rwanda Ministère Des Ressources Naturelles (MIRENA)*; Ministere des Ressources Naturelles: Kigali, Rwanda, 1982.
54. SRK Exploration Services. *An Independent Technical Review of the Miyove Gold Project, Rwanda*; SRK Exploration Services: Cardiff, UK, 2014.
55. Wouters, S.; Hulsbosch, N.; Kaskes, P.; Claeys, P.; Dewaele, S.; Melcher, F.; Onuk, P.; Muchez, P. Late Orogenic Gold Mineralization in the Western Domain of the Karagwe-Ankole Belt (Central Africa): Auriferous Quartz Veins from the Byumba Deposit (Rwanda). *Ore Geol. Rev.* **2020**, *125*, 103666. [CrossRef]

Article

Experimental Study on the Performance of a Novel Unidirectional Explosive Element and an Explosive Logic Network

Fei Wang [1,2,*], Honghao Ma [3] and Zhaowu Shen [3]

[1] State Key Laboratory of Mining Response and Disaster Prevention and Control in Deep Coal Mines, Anhui University of Science and Technology, Huainan 232001, China
[2] Institute of Energy, Hefei Comprehensive National Science Center, Hefei 230031, China
[3] CAS Key Laboratory of Mechanical Behavior and Design of Materials, University of Science and Technology of China, Hefei 230026, China; hhma@ustc.edu.cn (H.M.); zwshen@ustc.edu.cn (Z.S.)
* Correspondence: kingfly@mail.ustc.edu.cn

Abstract: Considering the unsafety of the present blasting network used in the blasting mining of coalfield fires, a unidirectional explosive element (named explosive diode) is proposed according to explosive logic element principles. Through theoretical and experimental analysis, the internal structure and mechanism of the unidirectional transmission of the detonation signal were studied. For an explosive diode, the length of the quenching channel was defined to be the key parameter. The explosive diode was implemented in the traditional blasting network, obtaining an explosive logic network. To evaluate the safety and reliability of the explosive diode and explosive logic network, detonation propagation and explosion-proof experiments were conducted in the lab. The optimum length of the quenching channel to obtain unidirectional detonation transmission was established. The results showed that the explosive diode could reliably control the propagation direction of the detonation signal when the length of the quenching channel was between 15 mm and 25 mm. The explosive logic network achieved a reliable detonation propagation and was explosion-proof. In comparison with traditional networks, the explosive logic network showed increased safety and reliability as the number of subnets increased. This is a significant improvement to mining safety and demonstrates great promise for engineering applications.

Keywords: safety of blasting network; unidirectional explosive element; detonation propagation; blasting mining of coalfield fires

Citation: Wang, F.; Ma, H.; Shen, Z. Experimental Study on the Performance of a Novel Unidirectional Explosive Element and an Explosive Logic Network. *Energies* **2022**, *15*, 5141. https://doi.org/10.3390/en15145141

Academic Editor: Maxim Tyulenev

Received: 31 May 2022
Accepted: 13 July 2022
Published: 15 July 2022

1. Introduction

Worldwide, coal fires with a large area, a high temperature, and a long duration occur frequently in exposed or underground coal seams [1]. In many countries, such as China [2,3], South Africa [4], India [5], USA [6] and Australia [7], coal fires cause major disaster during opencast working. Domestic and international research on coal fires primarily include the following aspects: the distribution and development of coal fires, the detection and monitoring of coal fires [8–10], the modelling of underground coal fires [11–13], assessments of the impact on the environment and human health [14] and fire-fighting engineering [15,16]. Due to different geological conditions in China compared to other countries, there are fewer studies regarding blasting mining in coalfield fires outside of China. The major coal fire areas qualified for large-scale open-air mining are located in northwest China, where the climate is generally arid and rainless, vegetation on the ground is deficient, wild species are rare, the resident population is sparse, and public infrastructure near coal mines are underdeveloped. Open mining here adopted physical cooling construction procedures before blasting mining. Safety regulations for blasting stipulate that the temperature of the blast holes must be reduced to below 80 °C before blasting operations are conducted [17]. Blasting mining is distinct from traditional blasting because of the contact time and general lack of experience with it. The blasting site has low visibility, contains high-temperature blast holes and is an unsafe working environment. A field photograph is shown in Figure 1. Recently, in Ningxia province, two serious safety accidents occurred in coalfield fires, which resulted in the death of more than 10 people and in injuries to at least 20 more. These accidents were due to the premature explosion of the subnet, resulting in the entire network being detonated [18,19]. These accidents seriously affected the mineral mining efficiency and threatened the workers at the blasting site. Therefore, it is important to improve safety techniques in the blasting network.

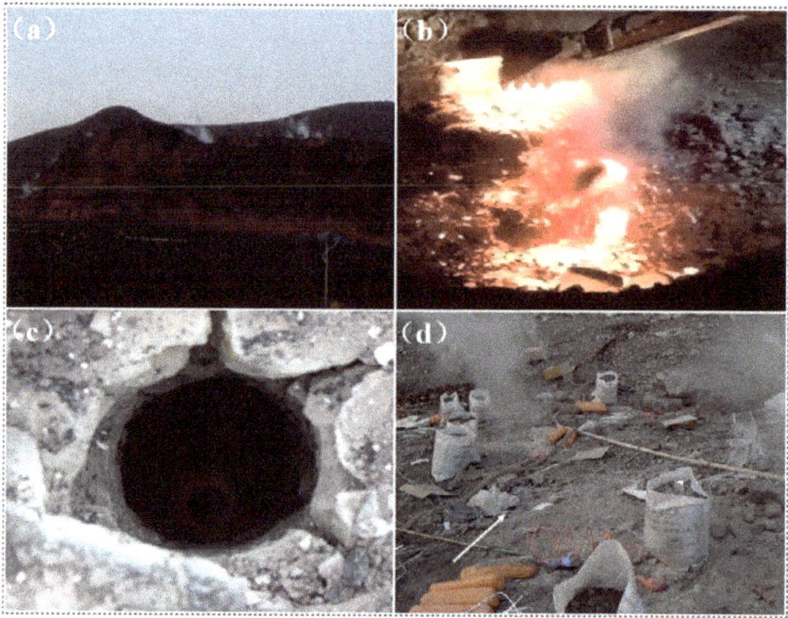

Figure 1. Blasting mining site, field photograph, December 2021, in Dashitou coalfield fires approximately 10 km northwest of Ruqigou in Ningxia province. (**a**) A long-time heated open pit coal mine (the color of the mountain is dark red), (**b**) spontaneous combustion coal seam, (**c**) high-temperature blast hole (the temperature can reach 400 °C) and (**d**) high-temperature blasting site.

It is significant that an explosive logic network was applied to field blasting to improve the safety of the blasting network. First proposed by D. A. Silvia [20,21] in the 1960s, an explosive logic network is composed of an explosive logic element and an initiating device. This network has a gate function, can make a logical judgement about the detonation signals coming from the different input ends and decides whether to output a signal and in which way to output it. A traditional blasting network is divided into the following categories: the detonating tube network, the detonating cord network, the electric detonator network and the hybrid network. In the design of a blasting network for coalfield fires, the blast holes are located in different high-temperature environments. After the implementation of physical cooling or resistant heat protection measures, the electric detonator and detonating tube networks are still not suitable for usage. By comparison, when adopting an RDX as the explosive, the detonating cord exhibited reliable detonation propagation and high-temperature endurance [22,23]. It is the optimal initiating equipment in the network design. In a traditional blasting network, the detonation signal can propagate bidirectionally. This is to say that the detonation signal can propagate from the main network to the subnet under conventional blasting. Similarly, it is able to detonate the main network when the detonation signal error is inputted from the subnet. Therefore, a traditional blasting network poses some potential hazard. In this paper, an element named explosive diode, which has a gate function that controls the propagation direction of the detonation signal, was designed. An explosive diode is a connector within the detonating cord network. Previously, some researchers studied the connector applied to the detonating fuse blasting network. Bartholomew S W et al. proposed a blasting signal transmission tube delay unit related to a time delay assembly [24], and Zakheim H put forward bidirectional delay connectors related to the time delay in the detonating cord network [25]. This connector mainly plays a time delay role, and research on the size of the detonation elimination chamber is limited. The author adopted the detonation elimination chamber structure and the shock initiation structure of a flying plate type in the detonation sequence [26]. Since conventional high explosives inside the explosive diode were not used, this structure was highly safe. In the experiments, the explosive diode can be applied to a traditional blasting network to form an explosive logic network, which performs well in field blasting. To evaluate the safety and reliable performance of the explosive diode and the explosive logic network, detonation propagation and explosion-proof experiments were performed by simulating a field-blasting network in the lab. Moreover, we conducted a theoretical calculation comparing the reliability of the new and the traditional networks. From the results of these experiments and the above calculation, the application of explosive logic networks would greatly increase the safety and efficiency of blasting mining in coal fire areas.

2. Composition of the Unidirectional Explosive Element

As shown in Figure 2, the unidirectional explosive element (named explosive diode) is composed of a metal shell (2), sealing elements (1 and 8) at both ends of the element, the input end of the detonation signal (14, End A), a hollow rubber structure (13), sealing wax paper (4), an excitation setting (5), an explosion-proof structure (6), and the output end of the detonation signal (7, End B). The relevant parameters of the initiating equipment (End A and End B) used in the experiment mentioned in this paper are listed in Table 1. The different densities of the layered charge in the explosion-proof structure 6 are listed in Table 2 [27]. The metal shell should possess sufficient strength to resist internal detonating and deflagrating reaction forces, as well as the longitudinal forces that may be applied during the detonation signal propagation. The preferred material is steel tubing. When the detonating cord (7 or 14) was inserted into the sealing element (1 or 8), a locking phenomenon could be produced. The tight fit between the detonating cord and the sealing element is waterproof and performs a fixed connecting function. The interior of the hollow rubber structure 13 forms an air channel called quenching channel (3). This quenching channel functions to control the shock wave attenuation from one side of the hollow rubber structure to the other. The sealing wax paper (4) prevented the excitation powder (11) from spilling. The excitation setting (5) was positioned directly abutting the quenching channel (3) and the second charge (10) to receive and transmit the blasting initiation signal. The explosion-proof structure (6) functioned to transmit the shock wave signal from the quenching channel (3) to the detonating cord of End B. Similar to the function of a diode in an electronic circuit, this element only allows a one-way transmission of the detonation signal and can also be called explosive diode. When the elements are connected to the blasting network, two connections are possible: forward connection and reverse connection. In the case of a forward connection to the blasting network, the detonation signal is input from end A and output from end B; this can realize a forward detonation transmission. In the case of a connection to the blasting network in reverse, the detonation signal is successfully blocked after being input from end B to realize reverse detonation suppression.

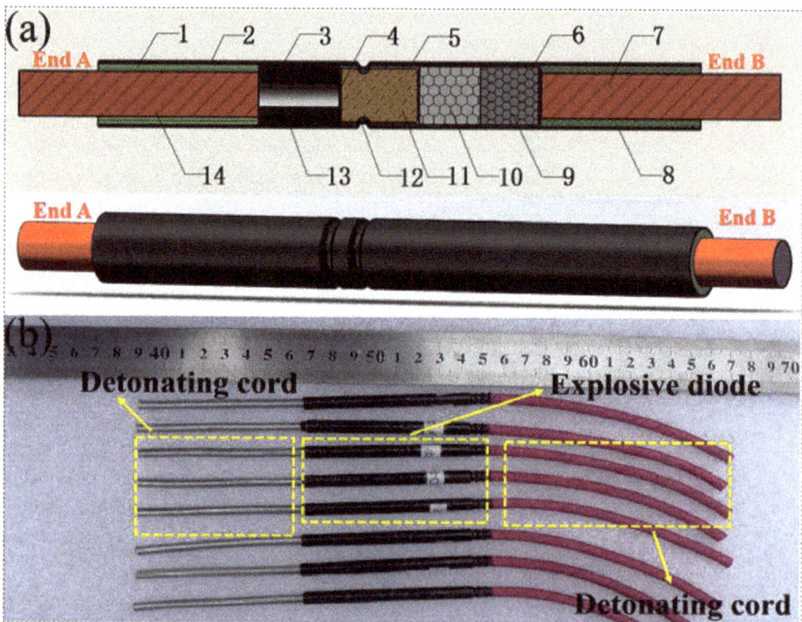

Figure 2. The unidirectional explosive element (explosive diode), (**a**) schematic, 1 and 8—Sealing element, 2—Metal shell, 3—Quenching channel, 4—Sealing wax paper, 5—Excitation setting, 6—Explosion-proof structure, 7—Output end of detonation signal, 9—1st Charge, 10—2nd Charge, 11—Excitation powder, 12—Bayonet fixing, 13—Hollow rubber structure, 14—Input end of detonation signal, (**b**) sample.

Table 1. Initiating equipment (End A and End B).

Types of Initiating Equipment	The Core Load of Detonating Cord (g/m)
Detonating cord	2.4; 4.0; 6.7; 9.7; 12.0; 14.0

Table 2. Layered structure's charge densities.

Serial Number	Layered Structure	Charge	Density (g/cm^3)
11	Excitation Powder	PETN/Graphite/Al/Oxidant	0.8
9	1st Charge	Granulation PETN	1.2–1.3
10	2nd Charge	Granulation PETN	0.9–1.0

3. Experimental Methods and Results

In this paper, reliable explosion propagation and explosion-proof experiments were carried out by verifying the iron plate explosion propagation method (judging the propagation of the detonation signal by the explosion trace on the iron plate), and the application range of the key parameters of the unidirectional explosion element was obtained. Based on the unidirectional explosion propagation model, the explosive logic network was designed. The logic function of the unidirectional explosive element in the explosion network was also confirmed by reliable explosion propagation and explosion-proof experiments. At the same time, the reliability of the traditional explosive network and the explosive logic network was analyzed. The flow diagram of the experimental methods is shown in Figure 3.

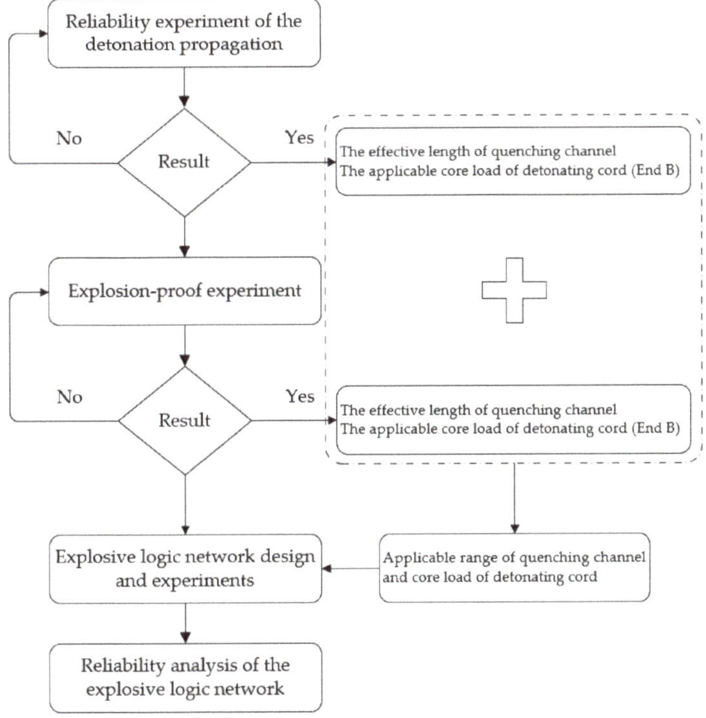

Figure 3. Flow diagram of the experimental methods.

3.1. Experiment to Evaluate the Reliability of the Detonation Propagation

Through the detonation propagation experiment, the explosive diode accessed the network through a forward connection. In Figure 4, the propagation direction of the detonation signal is indicated by an arrow (i.e., from End A to End B), and the configuration is shown.

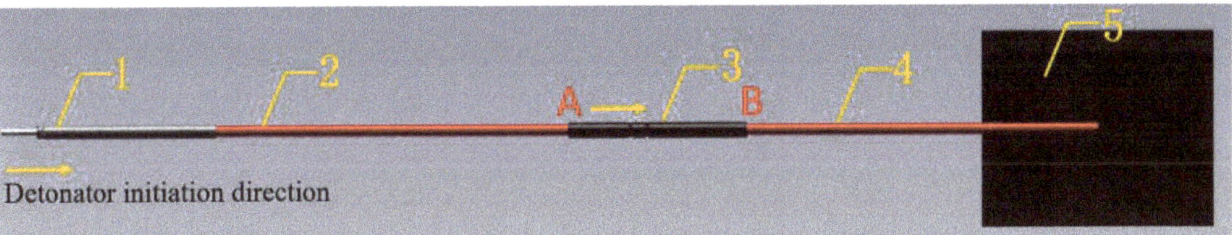

Figure 4. Configuration of the detonation propagation experiment, 1—detonator, 2—detonating cord of end A, 3—explosive diode, 4—detonating cord of end B, 5—the thickness of 0.2 mm iron-plate.

In the process of detonation propagation, a reliable detonation wave propagation is the desired goal. The key parameters within this element contain the detonating cord of End B with different core loads and quenching channel lengths. As shown in Figure 4, an industrial detonating cord weighing 14 g per meter was inserted into End A and End B. The element selected a 10 mm quenching channel, and the End B detonating cord was connected to a 0.2 mm experimental iron plate. An 8# industrial detonator ignited the detonating cord of End A to test whether the detonation signal could propagate successfully from End A to End B.

Based on the experiments shown in Figure 5, multi-group experiments were conducted to test the effect on the detonation propagation by adjusting the length of the quenching channel and by using different liner densities of the detonating cord at End B. The results of these experiments are listed in Table 3.

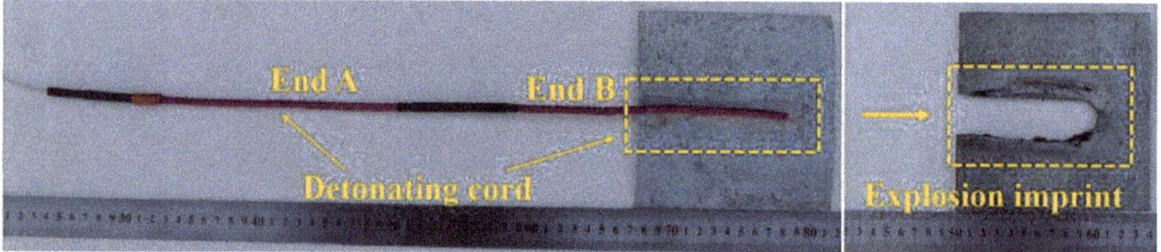

Figure 5. Successful detonation propagation experiments.

Table 3. Detonation propagation results using different design parameters.

Core Load of the Detonating Cord (End A) (g/m)	Length of the Quenching Channel (mm)	Core Load of the Detonating Cord (End B) (g/m)	Results
14.0	10	2.4; 4.0; 6.7; 9.7; 12.0; 14.0;	Y Y
	15	2.4; 4.0; 6.7; 9.7; 12.0; 14.0;	Y Y
	20	2.4; 4.0; 6.7; 9.7; 12.0; 14.0;	Y Y
	25	2.4; 4.0; 6.7; 9.7; 12.0; 14.0;	Y Y

Y—Reliable explosion, N—Failed, mm—millimeter, g/m—gram per meter.

Considering Figure 5 and Table 3, we concluded that the explosive diode could reliably allow detonation propagation from End A to End B to achieve a gate function, and the experimental iron plate was successfully destroyed, as shown in Figure 5. The propagation process of the detonation signal was as follows. First, a stable detonation signal was inputted from the detonating cord of End A; then, it was transformed to a shock wave after passing through the quenching channel. Finally, this shock wave triggered the excitation setting of the sublayer charge structure to ignite the End B detonating cord, whose core load ranged from 2.4 g per meter to 14 g per meter.

3.2. *Explosion-Proof Experiment*

With the experimental method described above, the element accessed the experimental networks following a reverse connection. The propagation direction of the detonation signal is indicated by an arrow (i.e., from End B to End A) in the structure diagram shown in Figure 6.

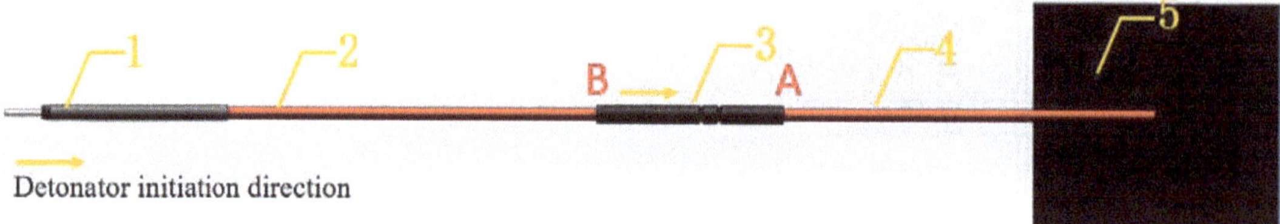

Figure 6. Structure diagram of the explosion-proof experiment, 1—detonator, 2—detonating cord of end B, 3—explosive diode, 4—detonating cord of end A, 5—the thickness of 0.2 mm iron-plate.

In the explosion-proof experiment, the desired goal was for the element to prevent the propagation of a detonation signal reliably from End B for different core loads of the End B detonating cord and quenching channel lengths. The explosion-proof experiment for the explosive diode was performed as shown in Figures 7 and 8. An 8# industrial detonator was used to ignite the End B detonating cord and test whether the detonation signal could be prevented during the propagation from End B to End A.

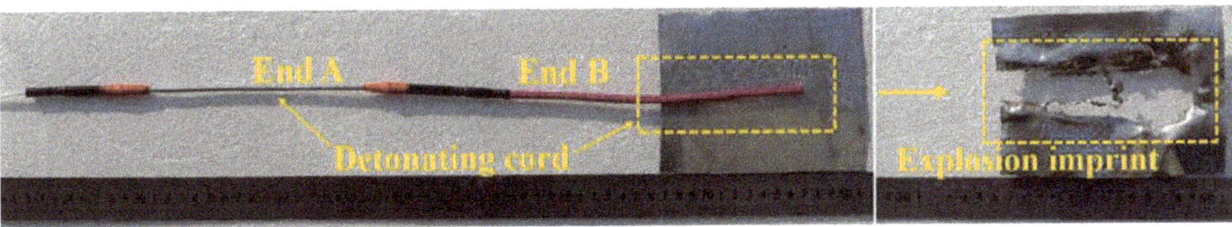

Figure 7. Failed explosion-proof experiments.

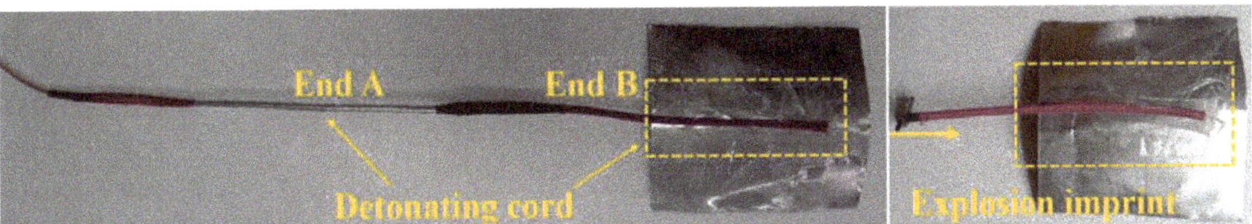

Figure 8. Successful explosion-proof experiments.

Based on the experiments shown in Figures 7 and 8, multi-group experiments to test the explosion-proof effect by adjusting the length of the quenching channel and by varying the core load of the End B detonating cord were performed. The results of these experiments are listed in Table 4.

Table 4. Explosion-proof results using different design parameters.

Core Load of the Detonating Cord (End A) (g/m)	Length of the Quenching Channel (mm)	Core Load of the Detonating Cord (End B) (g/m)	Results
14.0	10	2.4; 4.0; 6.7;	Y
		9.7; 12.0; 14.0;	N
	15	2.4; 4.0; 6.7;	Y
		9.7; 12.0; 14.0;	Y
	20	2.4; 4.0; 6.7;	Y
		9.7; 12.0; 14.0;	Y
	25	2.4; 4.0; 6.7;	Y
		9.7; 12.0; 14.0;	Y

Y—Reliable explosion-proof, N—Failed, mm—millimeter, g/m—gram per meter.

From Figures 7 and 8 and Table 4, when the length of the quenching channel was 10 mm and the core load of the End B detonating cord was 9.7 g per meter or above, it was concluded that the element could not reliably prevent the reverse propagation of the detonation signal. The experimental iron plate was destroyed, as shown in Figure 7. When the length of the quenching channel was between 15 mm to 25 mm, the element could reliably prevent the reverse propagation of the detonation signal. The detonation signal failed to ignite the End A detonating cord, and the element realized its explosion-proof function, as shown in Figure 8.

3.3. Explosive Logic Network Design and Experiments

The explosive logic network (new network) consisted of an ignition end, a main network (detonating cord), a subnet (detonating cord) and an explosive diode, as shown in Figure 9. In comparison with a traditional blasting network, the explosive logic network had a higher level of safety in the gate function.

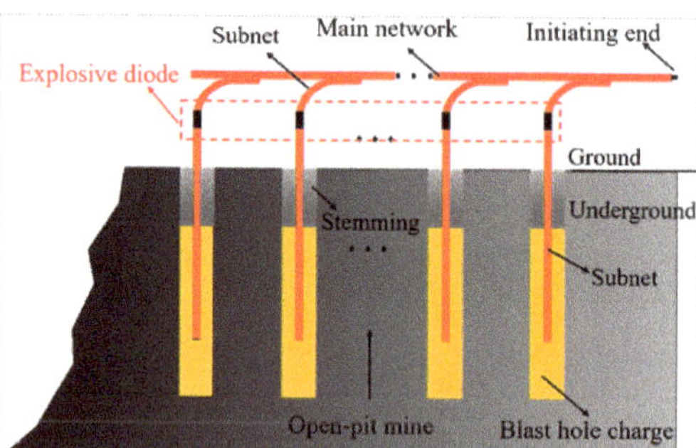

Figure 9. Schematic diagram of the application of the new explosive logic network.

In the explosive logic network, the objective was to control the unidirectional propagation of the detonation signal in the blasting network. An explosive diode element, which played an important role as a logic control switch, connected the main network to the subnet according to a forward connection. The propagation of the detonation signal included the following processes: inputting from the ignition end, passing through the main network and the explosive diode and propagating into the subnet of the blast hole. If a premature explosion occurred in the subnet, the detonation signal would be halted when it passed through the element in according to a reverse propagation to protect the safety of the main network. The explosive logic network was utilized to simulate a field-blasting network. The main network was similar to the ground-blasting network, and each subnet was similar to the charge holes. The experimental iron plate represented the blasting effect of a blast hole. In the detonation propagation experiments of this new network, when the detonation signal was inputted from the ignition end, it could pass through the main network smoothly, and the explosive diode detonated each subnet reliably. Five obvious explosion imprints in the iron plate indicated that the reliable explosion met the network design for field blasting. The experimental apparatus and results are shown in Figure 10.

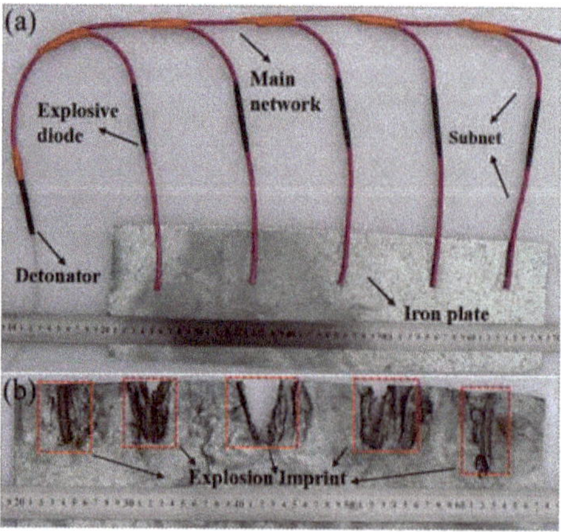

Figure 10. Network of the detonation propagation experiment (**a**) and results (**b**).

In the explosion-proof experiments, the sudden spontaneous firing or premature explosion of the blast subnet in coalfield fires due to accidents (such as high temperature, electro discharge, or stray current) caused by the subnet was simulated. From the explosion imprint on the experimental iron plate and the explosive diode, the propagation of the detonation signal was halted and did not arrive at the main network. The integrity of the rest of the network was protected, and the explosion of the entire blasting network was avoided. The experimental apparatus and results are shown in Figure 11. This experiment demonstrated that the blast logic network has a high safety in engineering blasting and is able to protect the staff at the blasting site.

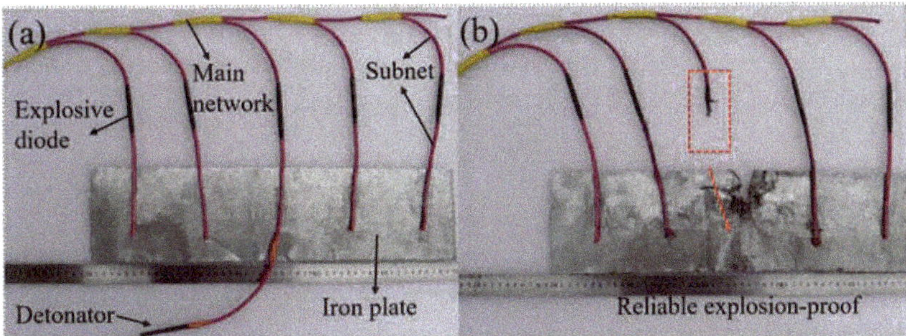

Figure 11. Network of the explosion-proof experiment (**a**) and results (**b**).

3.4. Reliability Analysis of the Explosive Logic Network

A reliability analysis [28] that compared the traditional blasting network with the new network was made. In a traditional network, connection reliability between the subnet and the main network is equal to that of a serial network. The reliability of a subnet is defined as P_i (i.e., the reliability of every subnet is $P_1, P_2, P_3, \ldots, P_n$). The reliability of the main network is P and is indicated in the formula (1).

$$P = P_1 \times P_2 \times \cdots P_n = \prod_1^n P_i \tag{1}$$

In the explosive logic network, the connection reliability between the subnet and the main network equals that of a parallel network. The reliability of the main network is P and is shown in the formula (2).

$$P = 1 - (1 - P_1) \times (1 - P_2) \times \cdots (1 - P_n) = 1 - \prod_1^n (1 - P_i) \tag{2}$$

In the above formulas, n is the number of subnets. The reliability of a subnet was assumed to be 0.9990, and the reliability of the two networks with a different number of subnets was compared and is shown in Table 5.

Table 5. Comparison of the reliability of two types of blasting networks.

Subnet Numbers	Traditional Network Reliability	Explosive Logic Network Reliability
1	0.9999	0.9999
10	0.9900	0.9999
20	0.9802	0.9999
30	0.9704	0.9999
40	0.9608	0.9999
50	0.9512	0.9999

As shown in Table 5, the reliability of a traditional network decreased as the number of subnets increased. When the number of subnets reached 50, the reliability of a traditional network was 0.9512, whereas the reliability of the new network was 0.9999. In a traditional network, the propagation of the detonation signal was bidirectional (i.e., two-way propagation), and the error input of the detonation signal from any subnet could trigger a premature explosion in the main network. In the explosive logic network, the propagation of the detonation signal was unidirectional (i.e., one-way network), and every explosive diode played the role of a controlling switch and stopped the input of errors from the subnet to protect the main network, as shown in Figure 11b. This new type of network, whose safety and reliability were demonstrated to be higher than those of a traditional one, is more suitable for blasting mining in coalfield fires.

4. Analysis of the Propagation Mechanism of the Unidirectional Explosive Element

The detonation propagation mechanism is illustrated as follows. First, the stable detonation signal was inputted from the detonating cord of End A, and then the detonation signal was transformed to a shock wave when it passed through the quenching channel. Finally, the shock wave and high-temperature gas products ignited the excitation powder in the established excitation setting. As the heat accumulated, local hot spots formed and developed into combustion in a short time. Large amounts of gas were produced and resulted in a high temperature and high pressure in the examined excitation setting [29]. Afterwards, a slapper with a high velocity formed from the bottom in the considered excitation setting at a high temperature and a high pressure. Finally, the slapper struck the layered charge with different densities (2nd charge and 1st charge) in the examined excitation setting, allowing the propagation of the detonation. In the process of detonation propagation, the sensitivity of the charge in the explosion-proof structure was reduced from a high level to a low level, and the detonation energy increased from a low level to a high level. Therefore, a stable detonation energy ignited the initiating device, and detonation propagation was achieved.

The explosion-proof mechanism is illustrated as follows. If the detonation signal was inputted from the detonating cord of End B, it passed through the explosion-proof structure from the reverse direction, and the detonation energy was greatly reduced under the conditions where there was a lack of hot spots and a low detonation sensitivity for the first insensitive charge. Therefore, the detonation energy of End B could not ignite the initiating device at End A because the sympathetic detonation was attenuated in the quenching channel, and the system was therefore explosion-proof in the reverse direction.

Analysis of the quenching channel length was carried out. If the quenching channel was short, the detonation signal from End B attempted to detonate the initiating device at End A with a sympathetic detonation. The element would directly ignite the detonating cord of End A when there was a great amount of energy in the sympathetic detonation and therefore lost the explosion-proof feature. On the contrary, if the quenching channel was long, the detonation signal would reduce excessively and could not ignite the next excitation setting needed for the detonation signal to propagate smoothly from End A to End B. The results of the multi-group experiments were analyzed. If the quenching channel was longer than 25 mm, the reliability of detonation propagation was low. If the quenching channel was shorter than 15 mm, the reliability of the explosion-proof system was low. Therefore, a length of the quenching channel between 15 mm and 25 mm should be selected for the optimal design of the element.

5. Conclusions

In this paper, an explosive diode as an element of detonation propagation was designed and added to a traditional blasting network to form a new explosive logic network. To evaluate the performance of this new network, detonation propagation and explosion-proof experiments were performed. Moreover, the reliability of the two networks was analyzed and compared. Some key points are as follows:

(1) An explosive element (named explosive diode) with unidirectional detonation signal transmission function was designed. The internal charge structure and detonation sequence of the element were described in detail, and the unidirectional explosive propagation of the element was analyzed. The optimal length of the quenching channel was demonstrated to be between 15 mm and 25 mm. In the detonation propagation experiments, the explosive diode could reliably propagate the detonation signal in the forward direction and successfully prevent the reverse propagation of the detonation signal. When the length of the quenching channel was 10 mm and the core load of the End B detonating cord was 9.7 g per meter or higher, it was concluded that the element could not reliably prevent the reverse propagation of the detonation signal.

(2) In the detonation propagating experiments of the explosive logic network, the detonation signal could successfully pass through the explosive diode to detonate each subnet reliably. Moreover, in the explosion-proof experiments, the propagation of the detonation signal was halted by the explosive diode, the main network was not ignited, and therefore, the explosive diode protected the rest of the network and avoided the premature explosion of the entire blasting network. The reliability of explosive logic network and traditional explosive network was analyzed. With the increase of the number of subnets, the reliability of the explosive logic network appeared greater than that of the traditional explosive network. The explosive element and explosive logic network can be applied to engineering environments prone to premature explosion or sudden spontaneous firing, such as high-temperature coal mine blasting, blast furnace nodulation blasting, high-temperature tunnel blasting, and can improve the safety and productivity of blasting operations.

Author Contributions: Methodology and writing—original draft, F.W.; validation and data curation, H.M.; writing—review and editing, and Z.S. All authors have read and agreed to the published version of the manuscript.

Funding: This research was funded by the Natural Science Foundation of the Anhui Higher Education Institution (No. KJ2021A0461), the Independent subject of State Key Laboratory of Mining Response and Disaster Prevention and Control in Deep Coal Mines (No. SKLMRDPC20ZZ07), the Anhui Province Natural Science Foundation (No. 2108085QA40), University-level key projects of Anhui University of science and technology (No. xjzd2020-03), the Institute of Energy, Hefei Comprehensive National Science Center (Grant No. 21KZS216).

Data Availability Statement: Not applicable.

Acknowledgments: The authors gratefully acknowledge the CAS Key Laboratory of Mechanical Behavior and Design of Materials that contributed to the research results reported within this paper.

Conflicts of Interest: The authors declare no conflict of interest.

References

1. Kuenzer, C.; Stracher, G.B. Geomorphology of coal seam fires. *Geomorphology* **2012**, *138*, 209–222. [CrossRef]
2. Song, Z.; Kuenzer, C. Coal fires in China over the last decade: A comprehensive review. *Int. J. Coal Geol.* **2014**, *133*, 72–99. [CrossRef]
3. Su, H.T.; Zhou, F.B.; Shi, B.B.; Qi, H.N.; Deng, J.C. Causes and detection of coalfield fires, control techniques, and heat energy recovery: A review. *Int. J. Miner. Metall. Mater.* **2020**, *27*, 275–291. [CrossRef]
4. Bell, F.G.; Bullock, S.E.T.; Halbich, T.F.J.; Lindsay, P. Environmental impacts associated with an abandoned mine in the Witbank Coalfire, South Africa. *Int. J. Coal Geol.* **2001**, *45*, 195–216. [CrossRef]
5. Agarwal, R.; Singh, D.; Chauhan, D.S.; Singh, K.P. Detection of coal mine fires in the Jharia coal field using NOAA/AVHRR data. *J. Geophys. Eng.* **2006**, *3*, 212–218. [CrossRef]
6. Stracher, G.B.; Taylor, T.P. Coal fires burning out of control around the world: Thermodynamic recipe for environmental catastrophe. *Int. J. Coal Geol.* **2004**, *59*, 7–17. [CrossRef]
7. Ellyett, C.D.; Fleming, A.W. Thermal infrared imagery of the Burning Mountain coal fire. *Remote Sens. Environ.* **1974**, *3*, 79–86. [CrossRef]

8. Zhang, J.; Wagner, W.; Prakash, A.; Mehl, H.; Voigt, S. Detecting coal fires using remote sensing techniques. *Int. J. Remote Sens.* **2004**, *25*, 3193–3220. [CrossRef]
9. Guha, A.; Kumar, K.V.; Kamaraju, M.V.V. A satellite-based study of coal fires and open-cast mining activity in Raniganj coalfield, West Bengal. *Curr. Sci.* **2008**, *95*, 1603–1607.
10. Du, B.; Liang, Y.; Tian, F. Detecting concealed fire sources in coalfield fires: An application study. *Fire Saf. J.* **2021**, *121*, 103298. [CrossRef]
11. Huo, H.; Jiang, X.; Song, X.; Li, Z.L.; Ni, Z.; Gao, C. Detection of coal fire dynamics and propagation direction from multi-temporal nighttime Landsat SWIR and TIR data: A case study on the Rujigou coalfield, Northwest (NW) China. *Remote Sens.* **2014**, *6*, 1234–1259. [CrossRef]
12. Roy, P.; Guha, A.; Kumar, K.V. Structural control on occurrence and dynamics of Coalmine fires in Jharia Coalfield: A remote sensing based analysis. *J. Indian Soc. Remote Sens.* **2015**, *43*, 779–786. [CrossRef]
13. Fei, J.; Wen, H. Experimental research on temperature variation and crack development in coalfield fire. *Combustion* **2017**, *28*, 29. [CrossRef]
14. Cheng, X.J.; Wen, H.; Xu, Y.H.; Fan, S.X.; Ren, S.J. Environmental treatment technology for complex coalfield fire zone in a close distance coal seam—A case study. *J. Therm. Anal. Calorim.* **2021**, *144*, 563–574. [CrossRef]
15. Lu, X.; Wang, D.; Qin, B.; Tian, F.; Shi, G.; Dong, S. Novel approach for extinguishing large-scale coal fires using gas–liquid foams in open pit mines. *Environ. Sci. Pollut. Res.* **2015**, *22*, 18363–18371. [CrossRef] [PubMed]
16. Shao, Z.; Wang, D.; Wang, Y.; Zhong, X.; Tang, X.; Hu, X. Controlling coal fires using the three-phase foam and water mist techniques in the Anjialing Open Pit Mine, China. *Nat. Hazards* **2015**, *75*, 1833–1852. [CrossRef]
17. *GB 6722-2014*; Safety Regulations for Blasting. China National Standardization Management Committee: Beijing, China, 2014.
18. The Office of the Safety Committee of the State Council Reported the "10.16" Blasting Accident in Dafeng Open Pit Mine. Available online: http://www.gov.cn/jrzg/2008-10/21/content_1126202.htm (accessed on 21 October 2008).
19. Notification on the "10.14" Major Explosive Explosion Accident in Dafeng Open Pit Mine of Shenhua Ningmei Group. Available online: https://www.chinamine-safety.gov.cn/zfxxgk/fdzdgknr/sgcc/sgtb/202004/t20200401_350496.shtml (accessed on 12 October 2009).
20. Silvia, D.A.; Ramsey, R.T.; Spencer, J.H. Explosive Gate, Diode and Switch. U.S. Patent 3,430,564, 4 March 1969.
21. Silvia, D.A. Explosive Logic Network. U.S. Patent 5,311,819, 17 May 1994.
22. Liao, M.Q.; Sun, F.Q. Applications of Common Detonating Fuses in High-Temperature Blasting Jobs. *Explos. Mater.* **1991**, *20*, 19–21. (In Chinese)
23. Lee, J.S.; Hsu, C.K.; Chang, C.L. A study on the thermal decomposition behaviors of PETN, RDX, HNS and HMX. *Thermochim. Acta* **2002**, *392*, 173–176. [CrossRef]
24. Bartholomew, S.W.; Rontey, D.C.; Necker, W.J.; Adams, C.F. Blasting Signal Transmission Tube Delay Unit. U.S. Patent 4,742,773, 10 May 1988.
25. Zakheim, H. Bidirectional Delay Connector. U.S. Patent 3,727,552, 17 April 1973.
26. Du, J.G.; Ma, H.H.; Shen, Z.W. Laser Initiation of Non-Primary Explosive Detonators. *Propellants Explos. Pyrotech.* **2013**, *38*, 502–504. [CrossRef]
27. Mei, Q. Study on Key Technology and Application of Low Energy Detonating Fuses. Ph.D. Thesis, University of Science and Technology of China, Hefei, China, 2007.
28. Cui, X.; Li, Z.; Zhou, T.; Shen, Z. Reliability Analysis of Large-scale Priming Circuit Used in Blasting Demolition. *Blasting* **2012**, *2*, 030. (In Chinese)
29. Starkenberg, J. Ignition of solid high explosive by the rapid compression of an adjacent gas layer. In Proceedings of the Seventh Symposium (International) on Detonation, Annapolis, MD, USA, 16–19 June 1981; pp. 3–16.

 *sustainability*

Case Report

Mechanism Analysis of Roof Deformation in Pre-Driven Longwall Recovery Rooms Considering Main Roof Failure Form

Bonan Wang [1,*], Lin Mu [1], Mingming He [1] and Shuancheng Gu [2]

1 School of Civil Engineering and Architecture, Xi'an University of Technology, Xi'an 710048, China; muller@xaut.edu.cn (L.M.); hemingming@xaut.edu.cn (M.H.)
2 School of Architecture and Civil Engineering, Xi'an University of Science and Technology, Xi'an 710054, China; kjdxjgxy@163.com
* Correspondence: wangbonan@xaut.edu.cn

Abstract: Pre-driven recovery rooms are used extensively for the removal of mining equipment and hydraulic supports in longwall coal mining. Roof stability is a crucial factor influencing the speed and safety of the removal of operators in pre-driven recovery rooms. The characterization of roof deformation mechanisms in recovery rooms under front abutment pressures is significant for surrounding rock control and stability evaluation. In this study, three different roof subsidence evaluation models, considering different main roof failure forms, were established. It was noticed that the main roof break position had a significant effect on recovery room roof sag. The breaking of the main roof above the main recovery room and protective coal pillar was found to be the main driving force for large roof deformations. Furthermore, field monitoring data of roof sag and coal pillar stress in the 15205 and 15206 panels of Hongliulin Coal Mine were analyzed. According to evaluation models and field monitoring data, we propose determination methods for the evaluation of recovery room roof sag and main roof break position. During the study it was found that the inversion results of the main roof break position of the recovery room in 15205 and 15206 panels were 4.2 m and 9.1 m, respectively, which are basically consistent with the results calculated by periodic weighting. The research findings provide a reference for the quantitative evaluation of recovery room roof stability and the design of support parameters and yield mining.

Keywords: pre-driven recovery room; roof sag; coal pillar stress; periodic weighting

Citation: Wang, B.; Mu, L.; He, M.; Gu, S. Mechanism Analysis of Roof Deformation in Pre-Driven Longwall Recovery Rooms Considering Main Roof Failure Form. *Sustainability* **2022**, *14*, 9093. https://doi.org/10.3390/su14159093

Academic Editor: Adam Smoliński

Received: 24 May 2022
Accepted: 22 July 2022
Published: 25 July 2022

1. Introduction

Pre-driven recovery room technology is a commonly applied equipment removal method in longwall fully mechanized faces. It was first tested and popularized in the United States in the late 1980s [1–3]. In this method, one or two roadways are excavated parallel to the longwall face and approach the terminal line in advance to provide space for equipment removal and the use of trackless rubber-tired mine cars to transport equipment. In the process of the gradual popularization of this technology, some researchers have studied pre-driven recovery room support design methods and developed support technology that includes single hydraulic props or concrete pillars combined with bolts and cables [4–11]. After the initial support of the recovery room with bolts and cables, a reinforcement support is constructed as the longwall face approaches 200 m from the terminal line. Single hydraulic props or concrete pillars can be used to reinforce and support recovery rooms. This support method has been widely used in the recovery rooms of many coal mines since the 1990s. Since the beginning of the 21st century, high resistance hydraulic supports have been extensively employed for roof support in many coal mines, some of which have used chock hydraulic supports for reinforcement of recovery room supports and to replace single hydraulic props and concrete pillars. This support method has significantly improved control of the surrounding rock and the safety degree of recovery rooms [12–14].

Research on the support technology and design methods of pre-driven recovery rooms has formed a relatively mature system. Meanwhile, several achievements have been made in studies on the ground, strata and support responses of recovery rooms at the end of mining stage and the deformation laws of the surrounding rock of recovery rooms has been basically determined. Based on a large number of field monitoring and case studies, researchers have qualitatively determined the factors affecting the surrounding rock stability of recovery rooms [15–17]. The obtained results show that the surrounding rock deformation of pre-driven recovery rooms was large in the middle and small at opposite ends and the main roof failure mode is the key factor determining the roof deformation and stability. Further, support intensity, overburden depth, mining height, width of coal pillar, etc. has affected the surrounding rock stability in recovery rooms [18–25]. Furthermore, according to overburden theory, some researchers have studied the load and stability of the coal pillars of recovery rooms and have revealed the instability mechanism of unmined coal pillars in longwall faces and protective coal pillars between the main and secondary recovery rooms, developing corresponding coal pillar design methods [26–31]. However, although these studies have determined the surrounding rock deformation law of pre-driven recovery rooms and solved coal pillar design problems to a certain extent, it is obvious that our understanding of surrounding rock deformation and failure of recovery rooms is still qualitative rather than quantitative. For example, many field-monitoring projects and case studies have focused on support technology, most of which have suggested that the break position of the main roof was the main factor determining recovery room roof stability. However, there is still lack of comprehensive mechanical models to explain the quantitative relationship between main roof break position and roof sag of recovery rooms. In addition, as one of the main measures in controlling roof deformation, the quantitative relationship between support intensity and recovery room roof deformation also needs to be studied. So far, some researchers have established mechanical models for recovery room roof sag by using cantilever beam theory [32]; however, such a simple model cannot explain large roof deformations in some recovery rooms. Therefore, in many mining areas, due to the inability of quantitative calculation and evaluation of recovery room roof deformation, some coal mines have formulated inappropriate support and removal schemes, resulting in support crushing accidents. Such accidents frequently occur in many mining areas with favorable geological and mining conditions, which also indicates that the roof deformation and stability evaluation of recovery rooms is still a problem that needs to be solved. Shennan mining area is located in Shenfu-Dongsheng Coalfield at the junction of Inner Mongolia and Shaanxi province in China, which is an area with typical shallow coal seams. Shallow burial, small dip angle and simple geological structures are the typical features of coal seam occurrence in this mining area. Recently, high-strength chock hydraulic supports have been applied to surrounding rock supports in pre-driven recovery rooms. However, by using such advanced production equipment and simple geological mining conditions, large roof deformations of the recovery rooms frequently occur, resulting in serious support crushing accidents which threaten the safety of operators and equipment. Figure 1 shows large roof deformations and hydraulic support crushing accidents of a pre-driven recovery room in a typical shallow longwall panel in Shennan mining area. In the figure, shield hydraulic supports with maximum working resistance of 9000 kN were used in the longwall face and chock hydraulic supports with maximum working resistance of 12,000 kN were used in the recovery room. However, roof sag still exceeded 1.2 m and the main recovery room height was reduced from 3.5 m to just over 2.0 m. Such excessive roof deformations reduced the stroke of most shield and chock hydraulic support columns to below 100 mm and in some cases, even to 0; dozens of hydraulic supports were even crushed. In these hydraulic support crushing accidents at the end of longwall face mining stage, although the main recovery room roof sag exceeded 1.2 m, the roof sags of headgate and tailgate did not exceed 40 mm, indicating that surrounding rock deformation and failure mechanisms of recovery room were more complex than other mining roadways arranged along the face advancing direction. Therefore, it is necessary to determine roof deformation mechanisms in recovery rooms and estimate and identify main roof break positions.

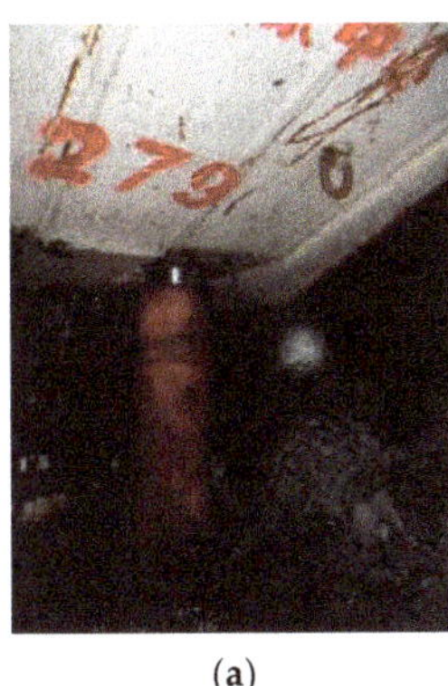

(**a**)

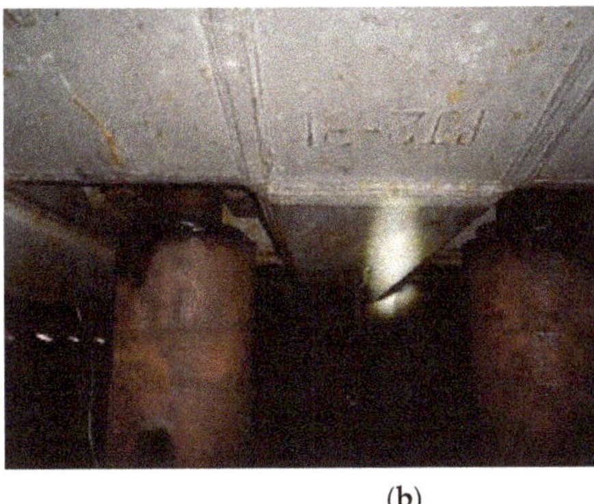

(**b**)

Figure 1. Large deformation of roof and hydraulic support crushing accident of a recovery room in Shennan mining area: (**a**) less than 100 mm stroke of hydraulic support column; and (**b**) completely crushed hydraulic support.

In this study, roof deformation and coal pillar stress analyses of recovery rooms in two adjacent longwall panels in Hongliulin Coal Mine were performed. Three different roof sag evaluation models for recovery rooms considering different main roof failure forms were established and the roof sag variation law of recovery rooms was analyzed. From the case studies and theoretical analysis results obtained for two adjacent panels of Hongliulin Coal Mine in Shennan mining area, the inversion discrimination method of the main roof break position in the recovery room was established based on theoretical and experimental data of roof sag and coal pillar stress. According to this inversion discrimination method, along with periodic weighting observations at the end of the mining stage, the main roof break position was identified, and a basis was provided for formulating surrounding rock control measures for recovery rooms during equipment removal stage under different roof failure forms.

2. Case Study

2.1. Roof and Coal Seam Conditions

Table 1 summarizes the roof lithology and occurrence conditions of 5^{-2} coal seam. The average thicknesses of 5^{-2} coal seam was 7.23 m and was therefore classified as a thick coal seam. The immediate and main roofs of the 5^{-2} coal seam were mainly composed of sandstone with high strength and thickness. The average depth of the 5^{-2} coal seam at the recovery room was about 150 m. The roof in Shennan mining area had good stability, was easy to maintain and did not usually cause large roof deformations and collapse.

Table 1. Lithology and roof condition of 5^{-2} coal seam.

Coal Seam Number	Coal Seam Average Thickness (m)	Immediate Roof		Main Roof		Average Depth (m)
		Lithology	Average Thickness (m)	Lithology	Average Thickness (m)	
5^{-2}	7.23	Sandstone	7.50	Medium sandstone	19.27	23–206

2.2. Support and Mining Conditions

Both panels 15205 and 15206 used pre-driven double recovery rooms. Support patterns in the main recovery rooms of the two panels are given in Table 2. The recovery room was supported by bolts, cables, and hydraulic supports. Initial supports included bolts and cables and two rows of chock hydraulic supports were installed as reinforcement support before the longwall face entrance into the main recovery room.

Table 2. Pattern of supports in the main recovery room of panels 15205 and 15206.

Longwall Panel	Arrangement	Support Pattern (Main Recovery Room)		
		Roof	Coal Pillar Rib	Mining Rib
15205 15206	Pre-driven double recovery room	Steel bolts, cables, and hydraulic supports	Steel bolts	FRP bolts

The support parameters and mining conditions of the two panels are illustrated in Table 3. The length of the recovery room in both panels was 300 m. The width of coal pillar between the main and secondary recovery rooms was 20 m. The maximum working resistances of shield and chock hydraulic supports in the two panels were 17,000 kN and 18,000 kN, respectively. Figure 2 shows a plan view of the support design for the main recovery room.

Table 3. Support parameters and mining conditions of panels 15205 and 15206.

Longwall Panel	Longwall Face		Main Recovery Room			
	Mining Height (m)	Shield Supports Resistance(kN)	Height (m)	Width (m)	Chock Supports Resistance (kN)	Coal Pillar Width (m)
15205 15206	6.7	17,000	4.5	6.0	18,000	20

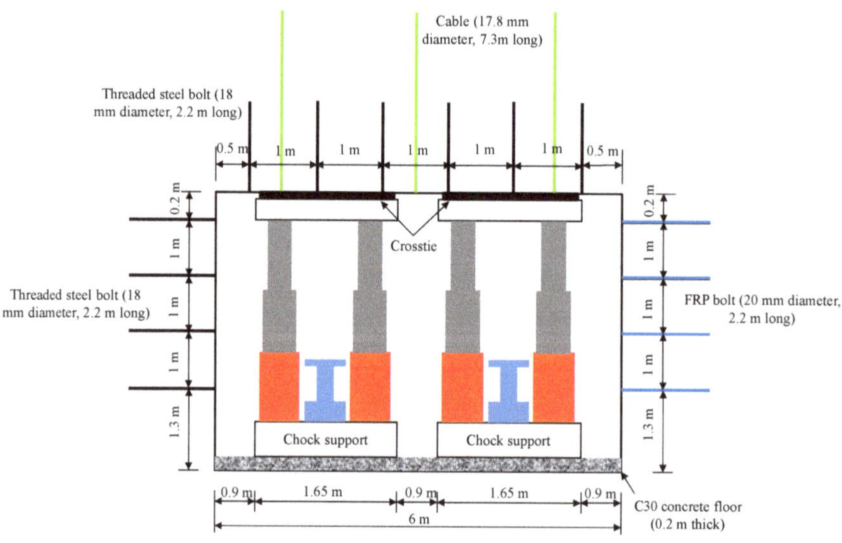

Figure 2. Plan view of support design for the main recovery room.

2.3. Field Monitoring

Previous studies have shown that recovery room roof deformation presented significant non-uniformity with a large deformation in the middle and small deformations at opposite ends. Therefore, according to the longwall face dip length of the 5^{-2} coal seam in Hongliulin coal mine, the main recovery room was divided into six roof deformation and six protective coal pillar stress monitoring areas, as shown in Figure 3. Recovery room roof sag was characterized by canopy-to-base convergence of chock supports. The average canopy-to-base convergence of the two chock support rows was taken and measured by a laser range finder. Coal pillar stress changes were monitored by HCZ-2 borehole pressure cells which were installed at depths of 2, 4, 6, and 8 m. The monitoring results of roof sag and coal pillar stress in panels 15205 and 15206 are illustrated in Figure 4.

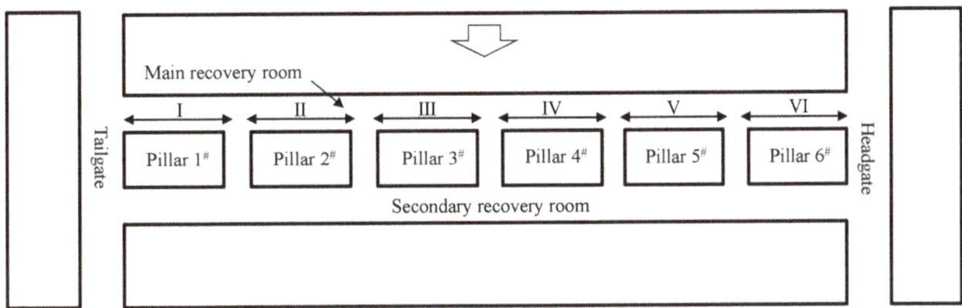

Figure 3. Layout of roof deformation and coal pillar stress monitoring areas in the main recovery room.

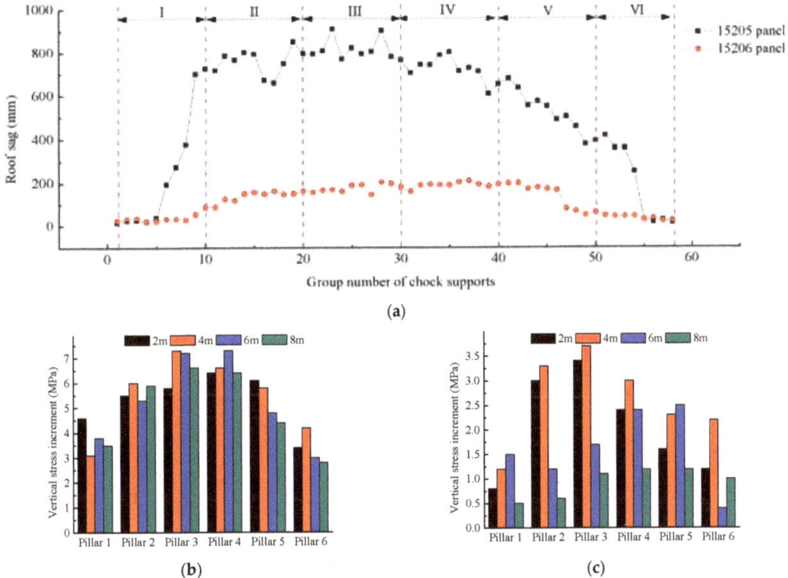

Figure 4. Roof deformation and coal pillar stress monitoring results of recovery room: (**a**) distribution and comparison of the roof sag of the main recovery rooms in panels 15205 and 15206; (**b**) field monitoring results of protective coal pillar stress in panel 15205; (**c**) field monitoring results of protective coal pillar stress in panel 15206.

The monitoring results presented in Figure 4 reveal that the deformations of surrounding rocks around the recovery room in panels 15205 and 15206 were quite different. The maximum recovery room roof sag in panel 15205 was four times higher than that in panel 15206 and was even larger than 900 mm in the third monitoring area. Coal pillar stress monitoring results also reveal that the overall increase of coal pillar stress in the recovery room of panel 15205 was large, and the increase of coal pillar stress in the middle was higher than 5 MPa. However, the overall increase of coal pillar stress in the recovery room of panel 15206 was small and the maximum increase of coal pillar stress in the middle was only 3 to 3.5 MPa with small stress changes in the deep part of coal pillar, which indicate little disturbance in the elastic zone of coal pillar.

Roof sag monitoring data obtained from the middle of the recovery rooms of panels 15205 and 15206 are presented in Figure 5. It was observed that recovery room roof sags of the two panels were greatly increased near the entrance of the longwall face. The difference was that, beyond the entrance of the longwall face in panel 15205, recovery room roof sag continued to increase by about 42 mm, while the corresponding value for panel 15206 was relatively smaller. This shows that the roof of the recovery room in panel 15205 was still in a state of dynamic change after mining was stopped, which is unfavorable for roof control in the process of hydraulic support removal.

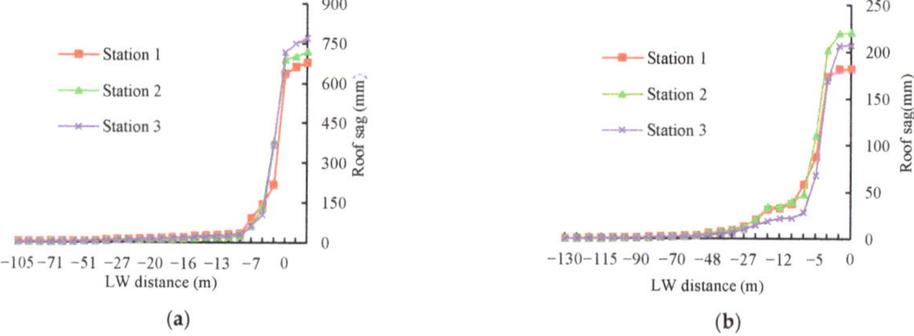

Figure 5. Variation curve of roof subsidence with mining progress: (**a**) panel 15205; (**b**) panel 15206.

Figure 6 shows coal pillar failure in the recovery rooms of panels 15205 and 15206. In panel 15206, only a spalling failure in the lower part of the coal pillar with dirt band and unsupported areas was observed while in panel 15205, the whole middle and lower parts of the coal pillar bulged out and some supporting structures such as bolts and joists were also damaged and failed.

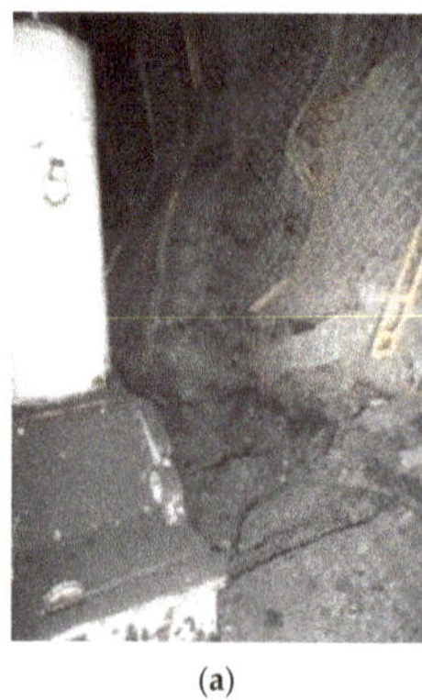

(a)

(b)

Figure 6. Spalling failure of protective coal pillar in the main recovery room: (**a**) panel 15205; (**b**) panel 15206.

Based on the above results, it was concluded that although there were no roof fall accidents in the two monitored panels, there was a great risk of excessive roof deformation in the recovery room of panel 15205. According to the classification of surrounding rock stability of coal mine roadways in China, mining roadways with roof sags exceeding 400 mm were in an unstable state [33,34], as presented in Table 4. From roof sag monitoring results, it was found that the recovery room in panel 15205 was basically in an extremely unstable state, which greatly increased safety risk during equipment removal process and made recovery room maintenance difficult; therefore, temporary reinforcement support was required. For the recovery room in panel 15206, with an overall roof sag of about 200 mm, surrounding rock was relatively stable. This meant there was no need to add extra temporary supports during the equipment removal process, which not only enhanced removal speed, but also reduced safety risk. Hence, for pre-driven longwall recovery room, if roof sag limit value could be controlled at 100–400 mm, equipment could be safely and quickly removed.

Table 4. Classification of surrounding rock stability of mining roadway.

Stability Category of Roadway Surrounding Rock	Stability of Roadway Surrounding Rock	Roof Sag of Mining Roadway (mm)	
		Average	Range
I	Extremely stable	30	10~50
II	Stable	75	50~100
III	Moderately stable	250	100~400
IV	Unstable	500	400~600
V	Extremely unstable	1200	600~1800

3. Mechanical Model of Roof Deformation of Pre-Driven Recovery Room

3.1. *Roof Failure Form in Recovery Room*

Based on strata behavior observations in Chinese mining practices, when a longwall face enters a recovery room, the main roof could continue to break and collapse. Depending on the main roof break position following the entrance of the longwall face into the main recovery room, three forms of main roof failure can occur [22,24,27,35–37], as shown in Figure 7.

From the three main roof failure forms, the second and third forms were not conducive to equipment removal. As shown in Figure 7a, when the main roof breaks behind the hydraulic supports of the longwall face, the roof forms a cantilever rock beam. It is obvious that the roof is in a relatively complete and stable state and only flexural deformations occur, which are conducive to hydraulic support removal. As shown in Figure 7b,c, when the main roof breaks above the main recovery room or protective coal pillar, the entire, or a part, of the recovery space becomes unstable. After primary removal of some chock or shield hydraulic supports, supporting intensity reduction leads to further rotation of the key block and roof sag continues to increase, making the removal of remaining hydraulic supports potentially difficult. On the other hand, some case studies have also shown that the breaking of the main roof behind the longwall face hydraulic supports does not necessarily ensure recovery room roof stability. When support strength was insufficient or cantilever rock beam was too long, the main roof broke again, changing from the failure form in Figure 7a to those presented in Figure 7b,c.

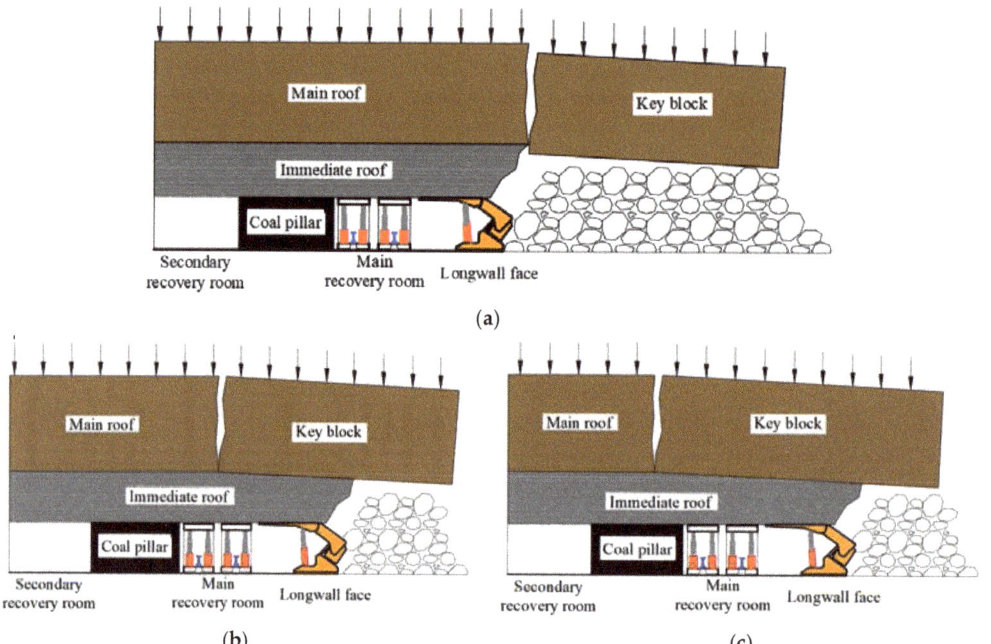

Figure 7. Main roof failure forms following the entrance of longwall face into main recovery room: (**a**) main roof breaking behind longwall face; (**b**) main roof breaking above main recovery room and longwall face; and (**c**) main roof breaking above protective coal pillar.

3.2. Establishing Mechanical Models

3.2.1. Main Roof Breaks behind Shield Hydraulic Supports

When the main roof breaks behind the shield hydraulic supports, the recovery room roof takes on a cantilever beam structure since the mining rib is completely mined. If the breaking of the cantilever rock beam does continue, immediate roof deflection could be considered as recovery room roof sag. The upper part of the immediate roof bears front abutment pressure and hydraulic supports directly supported the roof. The developed mechanical model is shown in Figure 8 where q_1 is front abutment pressure; f_1 and f_2 are supporting intensities of chock and shield hydraulic supports, respectively; w_1 and w_2 are the widths of main recovery room and longwall face, respectively; h_0 is the immediate roof thickness; h is the recovery room height; and d_1 is the break position when the main roof breaks behind shield hydraulic supports.

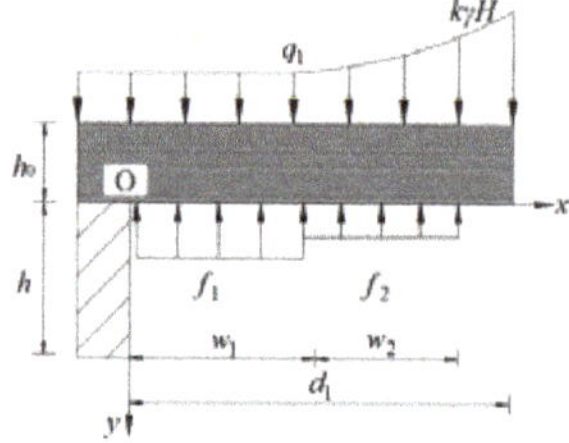

Figure 8. Mechanical model for roof sag when the main roof breaks behind shield hydraulic supports.

Main roof front abutment pressure in front of break position is stated as [38]:

$$q_1(x) = (k-1)\gamma H e^{-\frac{2f}{h\beta}(d_1-x)} + \gamma H \tag{1}$$

where k is stress concentration factor at main roof break position; f is friction coefficient and is defined as $f = \tan\varphi_1$, φ_1 is the internal friction angle of main roof; β is a coefficient and is defined as $\beta = 1/\lambda$, and λ is lateral pressure coefficient.

The deflection equation of cantilever rock beam under front abutment pressure is expressed as:

$$u_1(x) = \frac{(q_1 - f_1 - f_2)x^2}{24E_0 I_0}\left[x^2 - 4d_1 x + 6d_1^2\right] \tag{2}$$

where E_0 is immediate roof elastic modulus and I_0 is the section moment of immediate roof inertia.

In field monitoring, the roof sag monitoring point is generally located in the mid-span of the main recovery room, i.e., $x = W_1/2$. Substituting Equation (1) into Equation (2) results in the calculation equation of roof sag in the mid-span of the main recovery room as:

$$u_1 = \frac{w_1^2\left[(k-1)\gamma H\left(1 - e^{-\frac{2fd_1}{h\beta}}\right) - f_1 - f_2\right]\left[\frac{w_1^2}{4} - 2d_1 w_1 + 6d_1^2\right]}{8E_0 h_0^3} \tag{3}$$

3.2.2. Main Roof Breaks above the Main Recovery Room

When the main roof beaks above the main recovery room and longwall face, the immediate roof is affected by front abutment pressure q_1 as well as the pressure of key block and its overburden. The roof sag of the main recovery room is equal to the sum of deformation under the action of q_1 and q_2 and is written as:

$$u_2 = u_{2a} + u_{2b} \tag{4}$$

where u_{2a} and u_{2b} are roof sags under the action of q_1 and q_2, respectively.

u_{2a} was calculated according to immediate roof deflection and u_{2b} was determined by the principle of energy conservation in the process of roof rotary deformation. The calculation model is shown in Figure 9.

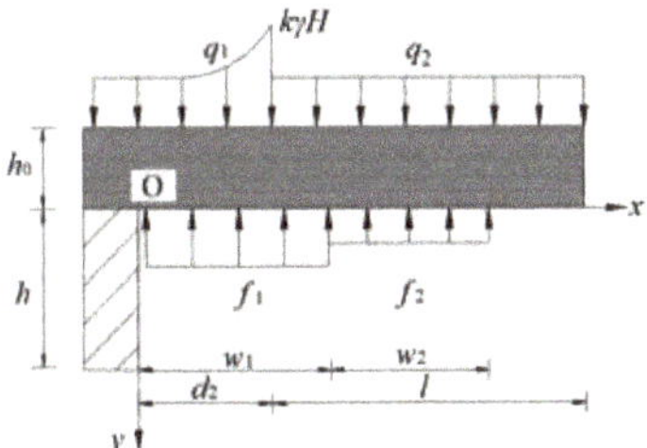

Figure 9. Mechanical model for roof sag when the main roof breaks above the main recovery room and longwall face.

Roof sag under the action of q_1 is given as:

$$u_{2a} = \frac{w_1^2\left[(k-1)\gamma H\left(1 - e^{-\frac{2fd_2}{h\beta}}\right) - f_1 - f_2\right]\left[\frac{w_1^2}{4} - 2d_2 w_1 + 6d_2^2\right]}{8E_0 h_0^3} \tag{5}$$

In Figure 9, the immediate roof could be considered as deformation body, its upper part was the key block along with its overburden, and its lower part was supported by shield and chock hydraulic supports. According to the principle of energy conservation, the work done by roof rotary deformation was equal to the sum of the strain energy stored in the immediate roof and the work done by hydraulic supports. The energy conservation equation is expressed as:

$$W_1 + W_2 = W_3 + W_4 + W_5 \tag{6}$$

where W_1 is the work done by the key block and its overburden rotary deformation; W_2 is the work done by the immediate roof rotary deformation; W_3 is the strain energy stored in the immediate roof; and W_4 and W_5 are the work done by chock and shield hydraulic supports to the roof, respectively.

In Equation (6), the work done by the hydraulic support and the strain energy stored in protective coal pillar were all related to recovery room roof sag. Therefore, after the determination of other influential parameters in Equation (6), roof sag calculation equation was derived. The line distribution stress q_2 of the key block and its overburden acting on the immediate roof was taken as unit length. When the rotation angle was α, the work done by key block is given by:

$$W_1 = \int_{d_2}^{w_1+w_2} q_2 \alpha x dx = \frac{1}{2}\gamma_1 h_1 \alpha \left(w_1^2 + 2w_1 w_2 + w_2^2 - d_2^2\right) \tag{7}$$

where γ_1 is the average volumetric weight of the main roof; h_1 is the height of the key block and its overburden; α is the rotation angle of the key block; and l is the step of the main roof's last weighting.

Under the action of the key block, the immediate roof deformations and the center of gravity were also changed. The deformation between the immediate roof and the key block was considered as coordinated, i.e., the rotation angle of the immediate roof was also equal to α. Therefore, according to the geometric relationship of immediate roof deformation shown in Figure 10, the displacement of the center of gravity along the vertical direction Δ_c is illustrated as [39]:

$$\begin{cases} \Delta_c = y_{OC_1} - y_{OC_2} = \frac{h_0 - \sin\theta\sqrt{h_0^2+(d_2+l)^2}}{2} \\ \sin(\alpha+\theta) = \frac{h_0}{\sqrt{h_0^2+(d_2+l)^2}} \end{cases} \tag{8}$$

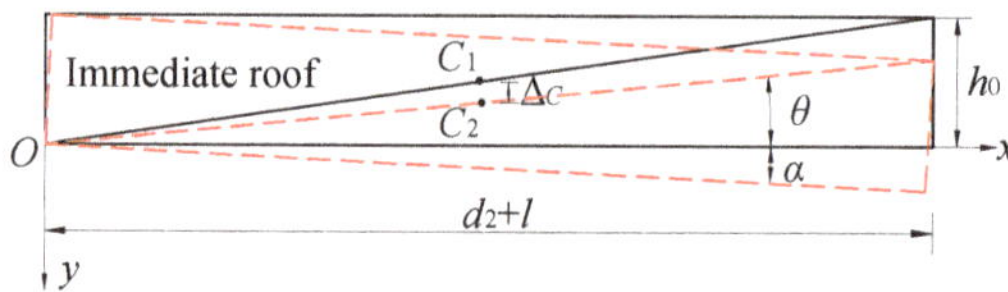

Figure 10. Geometric relationship of the rotary motion of immediate roof.

The simplified expression of Equation (8) is rearranged as:

$$\Delta_c = \frac{1}{2}[h_0(1 - \sec\alpha) - \tan\alpha(d_2 + l)] \tag{9}$$

Therefore, the work done by immediate roof rotary deformation is expressed as:

$$W_2 = \frac{1}{2}\gamma_0 h_0 l[h_0(1 - \sec\alpha) - \tan\alpha(d_2 + l)] \tag{10}$$

where γ_0 is the average volumetric weight of immediate roof.

The immediate roof stored energy under the action of the key block, and this part of strain energy is calculated as:

$$W_3 = \frac{F^2 l}{2EA} = \frac{\gamma_1^2 h_1^2 l^2 h_0}{2E_0(d_2 + l)} \tag{11}$$

By bearing roof deformation of the recovery room, the work done by the chock and the shield hydraulic supports are stated as:

$$W_4 = \int_0^{w_1} \frac{f_1 u_{2b} x}{w_1/2} dx = f_1 w_1 u_{2b} \tag{12}$$

$$W_5 = \int_{w_1}^{w_1+w_2} \frac{f_2 u_{2b} x}{w_1/2} dx = \frac{f_2 w_2 u_{2b}(w_2 + 2w_1)}{w_1} \tag{13}$$

By substituting Equations (7), (10), (11), (12) and (13) into Equation (6), u_{2b} is stated as:

$$u_{2b} = \frac{E_0 w_1 (d_2 + l)\{\gamma_1 h_1 \alpha (w_1^2 + 2w_1 w_2 + w_2^2 - d_2^2) + \gamma_0 h_0 l [h_0 (1 - \sec\alpha) - \tan\alpha (d_2 + l)]\} - 2\gamma_1^2 h_1^2 l^2 h_0 w_1}{2E_0 (d_2 + l)[f_1 w_1^2 + f_2 w_2 (w_2 + 2w_1)]} \tag{14}$$

Substituting Equations (5) and (14) into Equation (4) gives the main recovery room roof sag.

3.2.3. Main Roof Breaks above Coal Pillar

When the main roof breaks above the protective coal pillar, the immediate roof is completely controlled by the key block and its overburden. Similarly, roof sag under this main roof failure form could also be calculated by the principle of energy conservation.

According to the mechanical model presented in Figure 11, the calculation equation of each parameter in Equation (6) was derived. The work done by the key block is stated as:

$$\overline{W_1} = \int_{-d_3}^{w_1+w_2} q_2 \alpha x dx = \frac{1}{2}\gamma_1 h_1 \alpha \left(w_1^2 + 2w_1 w_2 + w_2^2 - d_3^2\right) \tag{15}$$

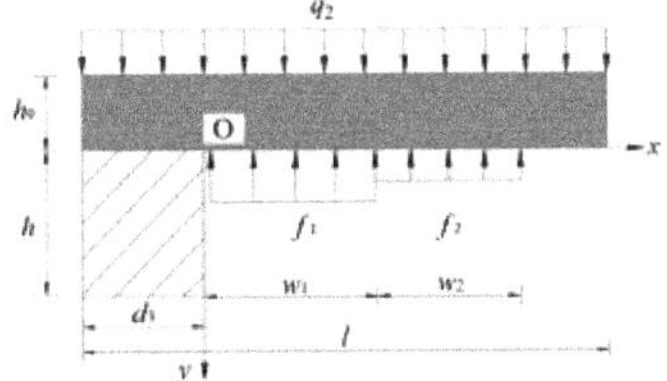

Figure 11. Mechanical model for roof sag when the main roof breaks above the main recovery room.

The work done by the immediate roof rotary deformation is expressed as:

$$\overline{W_2} = \frac{1}{2}\gamma_0 h_0 l [h_0 (1 - \sec\alpha) - l \tan\alpha] \tag{16}$$

The strain energy stored in the immediate roof is obtained as:

$$\overline{W_3} = \frac{\gamma_1^2 h_1^2 l^2 h_0}{2E_0 l} \tag{17}$$

The work done by the chock and the shield hydraulic supports are calculated as:

$$\overline{W_4} = \int_0^{w_1} \frac{f_1 u_3 x}{w_1/2} dx = f_1 w_1 u_3 \tag{18}$$

$$\overline{W_5} = \int_{w_1}^{w_1+w_2} \frac{f_2 u_3 x}{w_1/2} dx = \frac{f_2 w_2 u_3 (w_2 + 2w_1)}{w_1} \tag{19}$$

When the main roof breaks above the protective coal pillar, the roof rotary deformation also compresses the outside of the coal pillar and, therefore, the strain energy stored in it, which is stated as:

$$\overline{W_6} = \frac{d_3 h(\gamma_0 h_0 + \gamma_1 h_1)^2}{2E_c} \tag{20}$$

where E_c is coal elastic modulus.

Energy conservation equation is written as:

$$\overline{W_1} + \overline{W_2} = \overline{W_3} + \overline{W_4} + \overline{W_5} + \overline{W_6} \tag{21}$$

By substituting Equations (15) to (20) into Equation (21), the calculation equation of u_3 is expressed as:

$$u_3 = \frac{E_0 E_c l w_1 \{\gamma_1 h_1 \alpha (w_1^2 + 2w_1 w_2 + w_2^2 - d_3^2) + \gamma_0 h_0 l [h_0(1 - \sec\alpha) - l\tan\alpha]\} - E_c \gamma_1^2 h_1^2 l^2 h_0 w_1 - E_0 d_3 l h w_1 (\gamma_0 h_0 + \gamma_1 h_1)^2}{2E_0 E_c l [f_1 w_1^2 + f_2 w_2 (w_2 + 2w_1)]} \tag{22}$$

3.3. Parameter Analysis

This section can be divided into two subheadings. This should provide a concise and precise description of the experimental results, their interpretation, and experimental conclusions.

In Equations (3), (5), (14) and (22), some parameters could be directly determined by geological data and laboratory tests, while others, such as key block rotation angle, supporting intensity of shield, chock hydraulic supports and length of the key block (last weighting step) needed to be determined based on experimental and calculation data. By using the real-time mine pressure observation system of the Hongliulin Coal Mine, the last weighting steps of panels 15205 and 15206 were determined to be 19.7 and 18.3 m, respectively [31]. There were two layers of hard rock in the overburden of the 5^{-2} coal seam in Hongliulin Coal Mine. Based on key stratum theory, it was determined that when the main key stratum broke, h_1 was 49 m [40]. The maximum rotation angle of key block is calculated as [41]:

$$\alpha_{\max} = \arcsin\frac{M - h_0(k_c - 1)}{l} = \begin{cases} \arcsin\frac{7 - 4.02 \times 0.3}{19.7} = 17^\circ (15205 \text{ panel}) \\ \arcsin\frac{7 - 4.02 \times 0.3}{18.3} = 19^\circ (15206 \text{ panel}) \end{cases} \tag{23}$$

where M is mining height and k_c is crushing expansion which was 1.3.

Supporting intensities of shield and chock hydraulic supports needed to be calculated along with field monitoring data. The average column pressure and calculated support resistance of shield and chock hydraulic supports after the entrance of the longwall face into the main recovery room are given in Table 5. According to the data, the supporting intensities f_1 and f_2 could be calculated using Equations (24) and (25), respectively.

Table 5. Pressure calculation results of hydraulic supports in panels 15205 and 15206 after the entrance of the longwall face into the main recovery room.

Longwall Panel	Average Column Pressure of Hydraulic Supports (MPa)		Average Value of Support Resistance (kN)	
	Chock Supports	**Shield Supports**	**Chock Supports**	**Shield Supports**
15205	23.09	26.62	10,469	10,833
15206	22.15	30.56	10,041	12,436

The supporting intensity of chock hydraulic supports is calculated as:

$$f_1 = \frac{nF_c}{w_1 L_c} = \begin{cases} \frac{2 \times 10469}{6 \times 5.5} = 634\text{kPa (Panel 15205)} \\ \frac{2 \times 10041}{6 \times 5.5} = 608\text{kPa (Panel 15206)} \end{cases} \tag{24}$$

where F_c is the average value of chock supports resistance; L_c is the top beam length of chock support; and n is the number of chock supports for each supporting section.

The supporting intensity of shield hydraulic supports is calculated as:

$$f_2 = \frac{F_s}{w_2 L_{sw}} = \begin{cases} \frac{10833}{6\times1.75} = 1032\text{kPa (Panel 15205)} \\ \frac{12436}{6\times1.75} = 1184\text{kPa (Panel 15206)} \end{cases} \tag{25}$$

where F_s is the average value of shield support resistance and L_{sw} is the top beam width of shield support.

The parameters required for the calculation and analysis are summarized in Table 6. The change rule of recovery room roof sag with main roof break position was obtained by taking break position d as a variable. The value of the main roof break position d is obtained by Equation (26):

$$d = \begin{cases} d_1 > w_1 + w_2 \ \langle 12 < d_1 \leq 20 \rangle \\ 0 \leq d_2 \leq w_1 + w_2 \ \langle 0 \leq d_2 \leq 12 \rangle \\ d_3 < B \ \langle -10 \leq d_3 < 0 \rangle \end{cases} \tag{26}$$

where B is the width of protective coal pillar.

Table 6. Statistics of occurrence conditions, surrounding rock mechanical parameters and engineering factors of the 5^{-2} coal seam.

Longwall Panel	Surrounding Rock Occurrence Conditions			Surrounding Rock Mechanical Parameters						Mining and Engineering Parameters						
	H (m)	h_0 (m)	h_1 (m)	γ (kN/m^3)	γ_0 (kN/m^3)	γ_1 (kN/m^3)	φ_1 (°)	E_0 (GPa)	E_c (GPa)	α (°)	l (m)	f_1 (kPa)	f_2 (kPa)	h (m)	w_1 (m)	w_2 (m)
15205 15206	155 147	4.02	49	22.2	23.5	23.9	38	1.9	1.1	17 19	19.7 18.3	634 608	1032 1184	4.5	6.0	6.0

Roof sag deviation curve and main roof break position were plotted as presented in Figure 12 by substituting the above parameters and main roof break position d as variables into Equations (3), (5), (14) and (22). It was seen from the figure that when the main roof breaks behind the shield hydraulic supports, recovery room roof sag is relatively small (no more than 80 mm). When the main roof break position moves to goaf, rock beam cantilever length was enhanced, so roof sag was gradually increased. When the main roof breaks above the main recovery room, roof sag is sharply increased and reaches its maximum when the break position is located at the edge of protective coal pillar. At this time, roof sag is close to 900 mm, which is extremely unfavorable for controlling recovery room surrounding rock. When the main roof breaks above the protective coal pillar, roof sag is gradually decreased due to the coal pillar support. In this situation, recovery room roof sag is relatively small, but because the outside of the coal pillar is affected by the main roof rotary deformation, coal pillar stress is sharply increased. Once coal stress exceeds its ultimate strength, the coal pillar is damaged by spalling, which is also unfavorable for maintaining the stability of recovery room and coal pillar.

By adjusting the parameters of panel 15206 as shown in Table 5, the effect of the supporting intensity of hydraulic supports on recovery room roof sag was analyzed, as presented in Figure 13. It was concluded that the improvement of the supporting intensity of hydraulic supports reduced recovery room roof sag. When the main roof breaks above the recovery room or the shallow part of the coal pillar, the improvement of the supporting intensity of hydraulic supports significantly controlled roof sag. When the main roof breaks above or behind the longwall face, roof sag still decreases with the increase of supporting intensity, but the decline is small. However, in engineering practice, due to the upper limit of the supporting intensity provided by hydraulic supports, it is unrealistic to blindly improve the supporting intensity of hydraulic supports. For example, the maximum supporting intensity of shield supports installed in the longwall face of the 5^{-2} coal seam in Hongliulin Coal Mine was 1600 kPa and the maximum supporting intensity provided by the two rows of chock supports in the recovery room was 1100 kPa. When the main roof beaks at an unfavorable position (-5 m $\leq$ d $\leq$ 5 m), even if all hydraulic supports reach their maximum supporting intensity, recovery room roof sag still ranges from 500 to 600 mm.

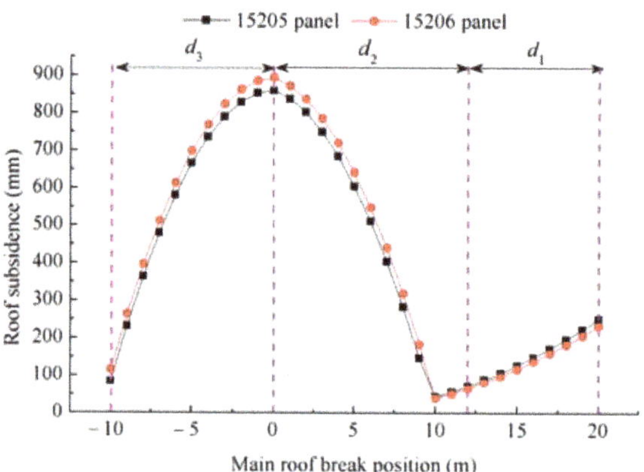

Figure 12. Influence of main roof break position on the roof sag of panels 15205 and 15206.

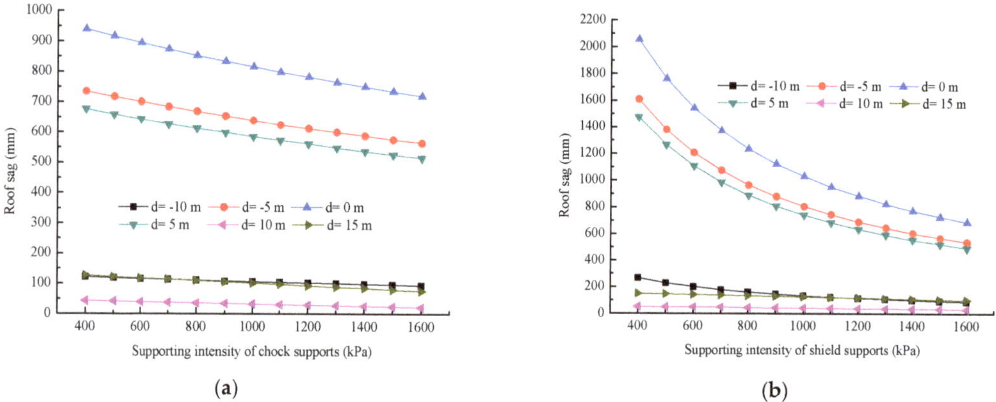

Figure 13. Influence of the supporting intensity of hydraulic supports on the roof sag of recovery room: (**a**) chock supports; (**b**) shield supports.

3.4. Inversion of Main Roof Break Position

Panels 15205 and 15206 were two adjacent longwall panels of the 5^{-2} coal seam in Hongliulin Coal Mine in Shennan mining area. The surrounding rock lithology and roof conditions of the recovery rooms in the two panels were the same with completely consistent support form and strength. However, roof sag modes in the two panels were quite different. In order to determine the deformation mechanism of the recovery room roof, the proposed mechanical model along with field monitoring data of panels 15205 and 15206 were applied to analyze roof deformation differences under similar conditions.

Based on previous analyses, after all relevant parameters were determined, the main roof break position of the recovery room could be calculated according to roof sag. However, as shown in Figure 12, two or three main roof break positions could be obtained based on the inversion calculation of roof sag. If other additional conditions were not considered, such inversion results would be meaningless. In field monitoring, roof sag and coal pillar stress are the most frequently monitored items and relevant data are also the easiest to obtain directly. Therefore, by combining field monitoring data for panels 15205 and 15206 in Hongliulin Coal Mine, we developed a method for the determination of the main roof break position for the recovery room in the 5^{-2} coal seam, as shown in Figure 14. The inversion discrimination of the main roof break position was divided into three steps:

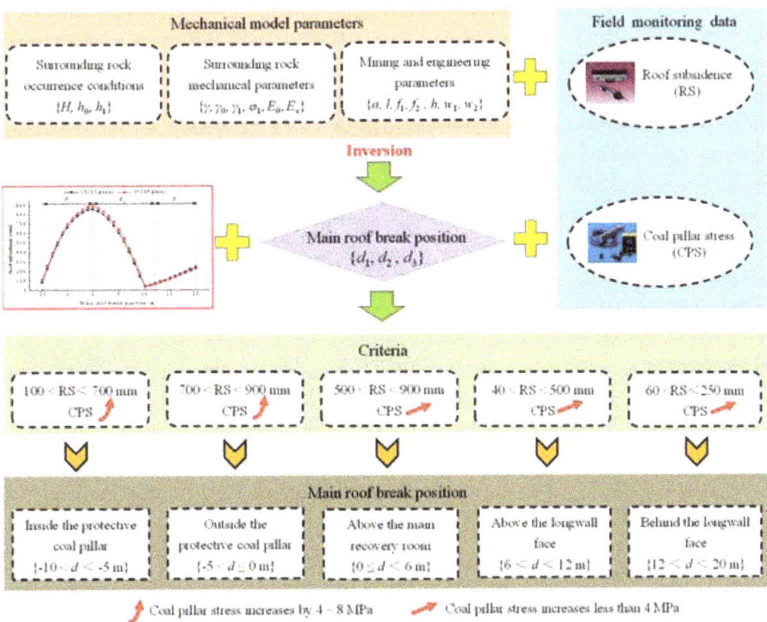

Figure 14. Inversion discrimination method of main roof break position in recovery room: taking the 5^{-2} coal seam as an example.

Step 1: Determining overburden occurrence conditions, surrounding rock mechanical characteristic parameters and mining and engineering parameters required for theoretical models, followed by substituting the roof sag value obtained from field monitoring into the developed calculation model to obtain two or three alternative values for main roof break positions after inversion.

Step 2: Determining roof sag and coal pillar stress discrimination criteria for various main roof break positions according to coal pillar stress monitoring data and roof sag variation curve with the main roof break position.

Step 3: Filtering the inverted main roof break position data to obtain the final value to meet the conditions according to the discrimination criteria of roof sag and coal pillar stress.

Based on field monitoring data, the main roof break position of each monitoring area in the recovery room for panels 15205 and 15206 was calculated, as presented in Table 7. According to the main roof break position data calculated by inversion, the schematic diagram of the main roof break lines of panels 15205 and 15206 after the entrance of the longwall face into the main recovery room were drawn, as shown in Figure 15. Obviously, the reason for the large deformations in the surrounding rock of the recovery room in panel 15205 was that the main roof break line was above the protective coal pillar, while that in panel 15206 was relatively small because the main roof break line was above the shield support, which was closer to the 12 m calculated by Equation (27).

Table 7. Calculation of main roof break position in each monitoring area.

Number of Monitoring Area	15205 Panel		15206 Panel	
	Average Roof Sag (mm)	Main Roof Break Position (m)	Average Roof Sag (mm)	Main Roof Break Position (m)
I	240	8.3 (*19.6, −8.9*)	38	10
II	760	−3.6 (*2.8*)	142	9.3 (*16.2, −9.8*)
III	816	−2.4 (*1.6*)	177	9.1 (*17.8, −9.6*)
IV	724	−4.2 (*3.4*)	193	8.9 (*18.4, −9.5*)
V	521	−6.6 (*5.8*)	153	9.2 (*16.7, −9.8*)
VI	184	8.7 (*17.5, −9.3*)	39	10

Note: the italicized numbers in brackets represent the data removed after filtering.

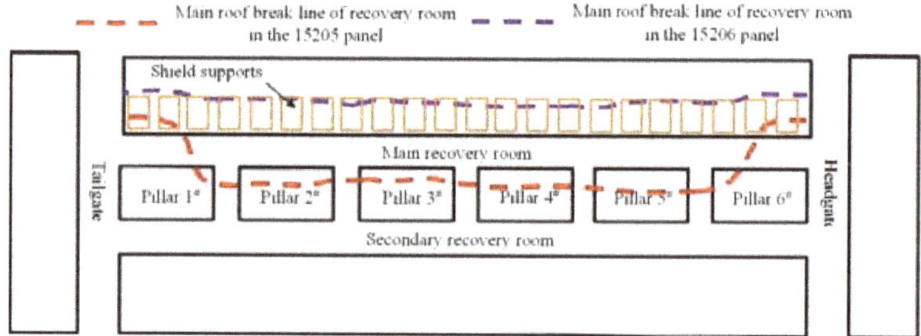

Figure 15. Distribution diagram of main roof break line of recovery room in panels 15205 and 15206.

For verification of the rationality of the developed theoretical models, the main roof break positions calculated by periodic weighting steps were compared with inversion results. As summarized in Table 8, periodic weighting steps in the middle of recovery room in panels 15205 and 15206 were analyzed. The main roof break positions of the last weighting of panels 15205 and 15206 were estimated to be −8.2 m and 4.0 m; that is, distance from the main roof break position to the terminal line. The main roof break positions converted to distance from coal pillar edge were −2.2 m and 10 m, which were considered as the coordinate system in the model applied in this study.

Table 8. Periodic weighting positions and steps of each panel.

Longwall Panel	15205		15206	
Number of Periodic Weighting	Position (m)	Step (m)	Position (m)	Step (m)
1	208	—	205	—
2	189.5	18.5	186.5	18.5
3	169.5	20	163.5	23
4	150	19.5	148	15.5
5	131	19	136.5	11.5
6	110	21	119	17.5
7	91.5	18.5	103	16
8	72	19.5	83	20
9	53	19	60.5	22.5
10	34	19	42.5	18
11	11.5	22.5	22	20.5
12	−8.2	19.7	4	18
Main roof break position of last weighting (m)	−2.2		10	

From inversion results, the average values of main roof break positions in II~V monitoring areas in the middle of the recovery room in panels 15205 and 15206 were −4.2 m and

9.1 m, respectively, which accord with the main roof break position obtained by periodic weighting steps.

4. Discussion

From the above analyses, it can be seen that under similar coal mining and geological conditions, the main reason for large roof deformations in recovery rooms is the main roof break position. It can also be seen that improving support intensity could also control roof deformation to a certain extent.

In practice, hydraulic supports can rarely achieve maximum support intensity due to their own working conditions and the premise they should not replace hydraulic support selection; therefore, support intensity cannot be increased indefinitely. Hence, monitoring the working resistance of hydraulic supports and evaluating the stability of recovery room roof deformation in time are the basic conditions for safe equipment removal [42,43]. When support strength reaches the upper limit and roof sag becomes larger, additional reinforcement support measures need to be taken.

Improving support intensity is not the best approach to control recovery room roof sag. The key to control recovery room roof sag is to manually control the main roof break at a favorable position. It can be seen from the above analysis that if surrounding rock stability and equipment removal process safety in the recovery room were to be ensured, the most reasonable breaking position of the main roof last weighting is calculated as:

$$d_0 = w_1 + w_2 \tag{27}$$

By pre-splitting the main roof to decrease the front abutment pressure at the position of d_0 before the entrance of the longwall face into recovery room, the main roof would not break after the entrance of the longwall face into the recovery room and excessive roof deformation of recovery room could be avoided. At the same time, it is necessary to ensure that recovery rooms have sufficient supporting intensity to avoid the formation of adverse failure forms due to secondary breaks in the roof.

The present study was carried out under simple, shallow and hard roof conditions. Hence, the established mechanical models could be mainly applied to coal mines with similar overburden conditions. Based on the mechanical models and corresponding inversion method, more factors such as the evolution of overburden structure, elastoplasticity and meso damage failure of coal and rock mass, and dynamic load of mining should be considered in future studies.

5. Conclusions

In this study, the effect of the main roof break position on pre-driven longwall recovery room stability was investigated. Three different mechanical models were developed based on the three main roof failure modes. Also, the main roof break position of last weighting in two adjacent panels of Hongliulin Coal Mine was inversed by combining them with field monitoring data. The following conclusions were drawn:

Main roof break position is a key factor influencing recovery room roof sag. In the 5^{-2} coal seam, when the main roof broke near the edge of the coal pillar, the maximum roof sag of the recovery room was close to 900 mm and when it was broken behind the hydraulic supports of longwall face, it was about 40 mm, indicating a difference of more than 20 times.

Based on the established theoretical models and field monitoring data and taking the longwall face of the 5^{-2} coal seam in Hongliulin Coal Mine as an example, a main roof break position discrimination method was developed for recovery rooms based on the inversion of the monitoring values of roof sag and the variation amplitude of coal pillar stress. The inversion results of the main roof break position of the recovery room in panels 15205 and 15206 of Hongliulin Coal Mine were −4.2 m and 9.1 m, respectively, which were basically consistent with the −2.2 m and 10.0 m obtained from ground pressure observation results.

When the main roof breaks in an unfavorable position, it is necessary to take temporary support and reinforcement measures in the recovery room, such as adding bolts and cables and installing single hydraulic props, to control roof deformation during equipment removal. Meanwhile, it is necessary to speed up equipment removal to avoid support crushing accidents due to the continuous rotation and deformation of the key block.

Author Contributions: Conceptualization, B.W. and S.G.; methodology, B.W.; software, B.W.; validation, B.W.; formal analysis, L.M.; investigation, B.W.; resources, B.W.; data curation, B.W.; writing—original draft preparation, B.W.; writing—review and editing, L.M. and M.H.; visualization, B.W.; supervision, S.G.; project administration, B.W.; funding acquisition, B.W., L.M. and M.H. All authors have read and agreed to the published version of the manuscript.

Funding: This research was funded by the Scientific Research Program Funded by Shaanxi Provincial Education Department, grant number No.21JK0782, Natural Science Basic Research Program of Shaanxi, grant number No.2022JQ-357, National Natural Science Foundation of China, grant number No.42177158, China Postdoctoral Science Foundation, grant number No.2019M653878XB. The APC was funded by Natural Science Basic Research Program of Shaanxi, grant number No.2022JQ-357.

Institutional Review Board Statement: Not applicable.

Informed Consent Statement: Not applicable.

Data Availability Statement: Not applicable.

Conflicts of Interest: The authors declare no conflict of interest.

References

1. Bauer, E.; Jeffery, M.; Listak, J.; Berdine, M.; Bookshar, W. Longwall recovery utilizing the open entry method and various cement-concrete supports. In Proceedings of the 7th International Conference on Ground in Mining, Morgantown, WV, USA, 3 August 1988; pp. 30–42.
2. Bauer, E.; Listak, J.; Berdine, M. *Assessment of Experimental Longwall Recovery Rooms for Increasing Productivity and Expediting Equipment Removal Operations*; Report of Investigation 9248; US Department of Interior, Bureau of Mines: Washington, DC, USA, 1989; p. 20.
3. Bauer, E.; Listak, J. Productivity and equipment removal enhancement using pre-driven longwall recovery rooms. In Proceedings of the 1989 Multinational Conference on Mine Planning and Design, Lexington, KY, USA, 1 January 1989; pp. 119–124.
4. Smyth, J.; Stankus, J.; Wang, Y.; Guo, S.; Blankenship, J. Mining through in-panel entries and full-face recovery room without standing support at U.S. Steel Mine 50. In Proceedings of the 17th Conference on Ground Control in Mining, Morgantown, WV, USA, 4–6 August 1998; pp. 21–30.
5. Tadolini, S.C.; Barczak, T.M.; Zhang, Y. The effect of standing support stiffness on primary and secondary bolting systems. In Proceedings of the 22nd International Conference on Ground Control Mining, Morgantown, WV, USA, 5–7 August 2003; pp. 300–307.
6. Tadolini, S.C.; Barczak, T.M. Design parameters of roof support systems for pre-driven longwall recovery rooms. *SME Annu. Meet. Exhibit.* **2006**, *318*, 87.
7. Tadolini, S.C.; Barczak, T.M. Rock mass behavior and support response in a longwall panel pre-driven recovery room. In Proceedings of the 6th International Symposium on Ground Support in Mining and Civil Engineering Construction, Cape Town, South Africa, 30 March–3 April 2008; pp. 167–182.
8. Barczak, T.M.; Tadolini, S.C.; Zhang, P. Evaluation of support and ground response as longwall face advances into and widens pre-driven recovery room. In Proceedings of the 26th International Conference Ground Control Mining, Morgantown, WV, USA, 31 July–2 August 2007; pp. 160–172.
9. Barczak, T.M.; Tadolini, S.C. Pumpable roof supports: An evolution in longwall roof support technology. *Trans. Soc. Min. Met. Explor.* **2008**, *324*, 19–31.
10. Barczak, T.M.; Esterhuizen, G.S.; Ellenberger, J.L. A first step in developing standing roof support design criteria based on ground reaction data for Pittsburgh seam longwall tailgate support. In Proceedings of the 27th International Conference on Ground Control Mining, Morgantown, WV, USA, 29–31 July 2008; pp. 349–359.
11. Thomas, R. Recent developments in pre-driven recovery road design. In Proceedings of the 27th International Conference on Ground Control in Mining, Morgantown, WV, USA, 29–31 July 2008; pp. 197–205.
12. Zorkov, D.; Renev, A.; Filimonov, K.; Zainulin, R. The roof support load analysis for pre-driven recovery room parameters design. In Proceedings of the 5th International Innovative Mining Symposium, Kemerovo, Russia, 19–21 October 2020.

13. Shu, C.X.; Jiang, F.X.; Han, Y.W.; Li, D. Long-distance multi-crosscut rapid-retracement technique in deep heavy fully mechanized face. *J. Min. Safe Eng.* **2018**, *35*, 473–480.
14. Pan, W.D.; Li, X.Y.; Li, Y.W.; Li, X.B.; Qiao, Q.; Gong, H. A supporting design method when longwall face strides across and passes through a roadway. *Adv. Mater. Sci. Eng.* **2020**, *2020*, 8891427. [CrossRef]
15. Listak, J.; Bauer, E. Front abutment effects on supplemental supports in pre-driven longwall equipment recovery rooms. In Proceedings of the 30th U.S. Symposium on Rock Mechanics, Morgantown, WV, USA, 19–22 June 1989.
16. Wynne, T.; John, S.; Guo, S.; Peng, S.S. Design, monitoring and evaluation of a pre-driven longwall recovery room. In Proceedings of the 12th International Conference on Ground Control in Mining, Morgantown, WV, USA, 3–5 August 1993; pp. 205–216.
17. Oyler, D.C.; Frith, R.C.; Dolinar, D.R.; Mark, C. International experience with longwall mining into pre-driven rooms. In Proceedings of the 17th International Conference on Ground in Mining, Morgantown, WV, USA, 4–6 August 1998; pp. 44–53.
18. Oyler, D.C.; Mark, C.; Dolinar, D.R.; Frith, R.C. A study of ground control effects of mining longwall faces into pre-driven longwall recovery room. *Geotech. Geol. Eng.* **2001**, *19*, 137–168. [CrossRef]
19. Tadolini, S.C.; Zhang, Y.; Peng, S.S. Pre-driven experimental longwall recovery room under weak roof conditions design, implementation, and evaluation. In Proceedings of the 21st International Conference on Ground Control Mining, Morgantown, WV, USA, 6–8 August 2002; pp. 1–10.
20. Wu, Z.G.; Li, W.Z. Strata response principle and stress distribution characters of pre-driven recovery room. In Proceedings of the 30th Annual International Pittsburgh Coal Conference, Beijing, China, 15–18 September 2013; pp. 4124–4130.
21. Wu, Z.G.; Li, W.Z. Surrounding rock convergence rule along the working face tendency of recovery room. In Proceedings of the 3rd International Conference on Energy and Environmental Protection, Xi'an, China, 26–28 April 2014; pp. 962–965.
22. Zhu, C.; Yuan, Y.; Chen, Z.; Meng, C.; Wang, S. Study of the stability control of rock surrounding longwall recovery roadways in shallow seams. *Shock Vib.* **2020**, *2020*, 2962819. [CrossRef]
23. Kang, H.P.; Lv, H.W.; Zhang, X.; Gao, F.Q.; Wu, Z.; Wang, Z. Evaluation of the ground response of a pre-driven longwall recovery room supported by concrete cribs. *Rock Mech. Rock Eng.* **2016**, *49*, 1025–1040. [CrossRef]
24. Wang, B.; Dang, F.; Chao, W.; Miao, Y.; Li, J.; Chen, F. Surrounding rock deformation and stress evolution in pre-driven longwall recovery rooms at the end of mining stage. *Int. J. Coal Sci. Technol.* **2019**, *6*, 536–546. [CrossRef]
25. Feng, G.; Li, S.; Wang, P.; Guo, J.; Qian, R.; Sun, Q.; Hao, C.; Wen, X.; Liu, J. Study on floor mechanical failure characteristics and stress evolution in double predriven recovery rooms. *Math. Probl. Eng.* **2020**, *2020*, 9391309. [CrossRef]
26. Gu, S.C.; Huang, R.B.; Li, J.H.; Su, P.L. Stability analysis of un-mined coal pillars during the pressure adjustment prior to working face transfixion. *J. Min. Safe Eng.* **2017**, *34*, 151–156.
27. Wang, F.T.; Shao, D.L.; Niu, T.C.; Dou, J.F. Progressive loading characteristics and accumulated damage mechanisms of shallow-buried coal pillars in withdrawal roadways with high-strength mining effect. *Chin. J. Rock Mech. Eng.* **2022**, *41*, 1148–1159.
28. He, Y.J.; Song, Y.X.; Shi, Z.S.; Li, J.Q.; Chen, K.; Li, Z. Study on failure mechanism of withdrawal channel's surrounding rock during last mining period. *China Safe Sci. J.* **2022**, *32*, 158–166.
29. Zhang, Y.; Li, H.; Zhang, X.; Zhao, Y.; Zhao, R. Study on impact mechanism of double withdrawal channels in fully mechanized mining face based on overburden theory. *Chin. J. Undergr. Space Eng.* **2022**, *18*, 305–312.
30. Lv, H.W. The mechanism of stability of pre-driven rooms and the practical techniques. *J. China Coal Soc.* **2014**, *39* (Suppl. S1), 50–56.
31. Wang, B.N.; Dang, F.N.; Gu, S.C.; Huang, R.B.; Miao, Y.; Chao, Y. Method for determining the width of protective coal pillar in the pre-driven longwall recovery room considering main roof failure form. *Int. J. Rock Mech. Min. Sci.* **2020**, *130*, 104340. [CrossRef]
32. Wang, S.S.; Wang, Z.Q.; Huang, X.; Su, Z.H. Calculation of direct roof subsidence of retracement channel and analysis of its influencing factors. *J. Min. Sci. Technol.* **2021**, *6*, 409–417.
33. Chen, Y.G.; Lu, S.L. *Strata Control around Coal Mine Roadways in China*; China university of mining and technology press: Xuzhou, China, 1994.
34. Qian, M.G.; Shi, P.W.; Xu, J.L. *Mine Pressure and Strata Control*; China university of mining and technology press: Xuzhou, China, 2010.
35. Wang, X.Z.; Ju, J.F.; Xu, J.L. Theory and applicable of yield mining at ending stage of fully-mechanized face in shallow seam at Shendong mine area. *J. Min. Safe Eng.* **2012**, *29*, 151–156.
36. Liu, C.; Yang, Z.; Gong, P.; Wang, K.; Zhang, X.; Zhang, J.; Li, Y. Accident analysis in relation to main roof structure when longwall face advances toward a roadway: A case study. *Adv. Civ. Eng.* **2018**, *2018*, 3810315. [CrossRef]
37. Chen, Z.S.; Yuan, Y.; Zhu, C.; Wang, W.M. Stability mechanism and control factors on equipment removal area under "goaf-roof-coal" structure. *Adv. Civ. Eng.* **2021**, *2021*, 6628272. [CrossRef]
38. Chen, Y.G.; Qian, M.G. *Strata Control around Coal Face in China*; China university of mining and technology press: Xuzhou, China, 1994.
39. Chen, Y. Study on Stability Mechanism of Rock Mass Structure Movement and Its Control in Gob-Side Entry Retaining. Doctor Thesis, China University of Mining and Technology, Xuzhou, China, 2012.
40. Xu, J.L.; Qian, M.G. Method to distinguish key strata in overburden. *J. China Univ. Min. Tech.* **2000**, *29*, 463–467.

41. Fu, Y.P.; Song, X.M.; Xing, P.W.; Yan, G.C. Stability analysis on main roof key block in large mining height workface. *J. China Coal Soc.* **2014**, *34*, 1027–1031.
42. Herezy, L.; Janik, D.; Skrzypkowski, K. Powered Roof Support-Rock Strata Interactions on the Example of an Automated Coal Plough System. *Studia Geo. Mech.* **2018**, *40*, 46–55. [CrossRef]
43. Skrzypkowski, K.; Korzeniowski, W.; Duc, T.N. Choice of powered roof support FAZOS-15/31-POz for Vang Danh hard coal mine. *Inz. Miner.-J Pol. Min. Eng. Soc.* **2019**, *2*, 174–181. [CrossRef]

Case Report

Interaction Mechanism of the Upper and Lower Main Roofs with Different Properties in Close Coal Seams: A Case Study

Shengrong Xie [1], Yiyi Wu [1], Fangfang Guo [1], Dongdong Chen [1,*], En Wang [1,2], Xiao Zhang [1], Hang Zou [1], Ruipeng Liu [1], Xiang Ma [1] and Shijun Li [3]

[1] School of Energy and Mining Engineering, China University of Mining & Technology, Beijing 100083, China; xsrxcq@163.com (S.X.); 15662045768@163.com (Y.W.); gffever@163.com (F.G.); wang_en1994@126.com (E.W.); zx_hlh@outlook.com (X.Z.); zh981113zh@163.com (H.Z.); lrp18839113025@163.com (R.L.); 15109544321@163.com (X.M.)
[2] Department of Civil Engineering, The University of British Columbia, Vancouver, BC V6T 1Z4, Canada
[3] State Key Laboratory of Hydroscience and Engineering, Tsinghua University, Beijing 100084, China; thulsj@163.com
* Correspondence: chendongbcg@163.com

Abstract: Close-distance coal seams are widely distributed in China, and the mining of overlying coal seams leads to floor damage. To grasp the properties and the fracture spans of the damaged main roof in the underlying coal seam, combining the calculation of the floor damage depth with rock damage theory and the formulas for calculating the first and periodic weighting intervals of the damaged main roof and the instability conditions of the damaged key blocks are obtained. Three interaction stability mechanics models are proposed for key blocks with different properties of the upper and lower main roof, and the instability conditions of the lower damaged key blocks are obtained when the fracture lines overlap. When combined with a specific example, the field monitoring verified the calculation results. The research results are as follows: (1) The first and periodic weighting intervals, horizontal thrust between blocks, and critical load of instability of the damaged main roof are significantly reduced. Still, there are differences in its reduction under different loads, rotation angles, and lumpiness. (2) When the fracture lines of the upper and lower main roofs overlap, the stability of the damaged key blocks is the lowest. There are three linkage stability regions in the critical load curves of the two key blocks. (3) In this case, the damage equivalent of the main roof is 0.397, which belongs to the local damage type. Its first and periodic weighting intervals are 40 m and 16 m, which is 22% and 24% less than when there is no damage. (4) A supporting load of 0.489 MPa is required to maintain the stability of the upper key block, and the lower damaged key block is prone to rotary and sliding instability during the first and periodic weighting, respectively. Thus, the supports need to bear a total of 0.988 MPa and 0.761 MPa to maintain the stability of the two key blocks simultaneously. The ground pressure data monitored on-site is in accord with the calculation results.

Keywords: multi-coal seam mining; depth of floor damage; ultimate span of fracture; instability condition; linkage of key blocks; support bearing capacity

Citation: Xie, S.; Wu, Y.; Guo, F.; Chen, D.; Wang, E.; Zhang, X.; Zou, H.; Liu, R.; Ma, X.; Li, S. Interaction Mechanism of the Upper and Lower Main Roofs with Different Properties in Close Coal Seams: A Case Study. *Energies* **2022**, *15*, 5533. https://doi.org/10.3390/en15155533

Academic Editors: Longjun Dong, Yanlin Zhao and Wenxue Chen

Received: 24 March 2022
Accepted: 22 July 2022
Published: 30 July 2022

1. Introduction

In underground coal mining in China, the mining of close-distance coal seam groups is common [1,2], such as Datong mining area [3], Bulianta mining area [4], Zhungeer mining area [5], and so on. For the two layers of coal in close distance, the overlying coal seam is generally mined first, and then the underlying coal seam is mined after the overlying rock collapse is stable [6,7]. As a result, the breaking and periodic pressure of the overlying strata during the mining of the underlying coal seam will be affected by the mining of the overlying coal seam, especially the two layers of coal at close distance and ultra-close distance [8–10].

In the mining of single-layer coal, regarding the breaking regularity of key strata in overlying strata, Academician Qian Minggao established a mechanical model according to the field practice and put forward the "voussoir beam" theory [11,12], which has been widely used in the field and become the guiding theory of underground coal mining in China at present. For the mining of the underlying coal seam in the close-distance coal seam group, because the mining of the overlying coal seam has been completed, the overlying strata of the underlying coal seam are no longer complete. Still, they are mined under the goaf of the overlying coal seam [13–15]. And the breaking of the main roof will be directly affected by the broken main roof of the overlying coal seam [16]; that is, the two main roofs of the upper and lower coal seams interact with each other, which is different from the composite key layer [17] and the double-layer hard and thick main roof [18].

The upper coal seam was mined in advance, the main roof and its overlying strata were broken, and the goaf was compacted after mining. The goaf floor is affected by the mining of the working face, resulting in rock failure and damage zone, which shows that the rock mass of the floor is in a plastic state [19]. Zhao et al. [20–22] analyzed the mechanical properties of the damaged rock from the experimental point of view and developed the damage model of the fractured rock mass. Zuo et al. [23] calculated the impact load of cantilever beam fracture on the goaf floor area in the longwall face. Li et al. [24] pointed out that the floor damage depth is affected by the inclined length of the working face, roof lithology combination, mining depth, floor lithology combination, coal seam thickness, and coal seam dip angle. Hu et al. [25] proposed the backpropagation neural network (BPNN) method to calculate the floor failure depth, and the result was close to the measured value. Li et al. [26] introduced the thin plate yield theory and obtained the low-stress area where the maximum displacement area and failure area of the floor are located above the inner boundary of the rock stress shell of the stope floor. To sum up, the mining of the overlying coal seam will inevitably cause destruction and damage to its floor strata, which is precisely the roof strata of the underlying coal seam. Therefore, when the underlying coal seam is mined, its roof strata have been partially destructed and damaged [27,28]. Because it is under the goaf of the overlying coal seam, the main roof of the overlying coal seam that has been broken directly acts on the roof strata of the underlying coal seam, forming the linkage action of the upper and lower double main roof with different properties.

As the first key layer of coal mining, the mechanical research between its key blocks, the main roof, has matured [29–31], and its mechanical principle has successfully guided the field practice [32,33]. As for the extension of the study of the single-layer main roof, He et al. [34] studied the instability and catastrophe conditions of the single-layer damaged main roof. Chai et al. [35] studied the structure of highly thick composite key strata in the stope. Huang et al. [36] studied the double key strata structure of the working face with large mining height in the shallow coal seam. Ning et al. [18] studied the instantaneous fracture effect of the double-layer hard and thick main roof. The above research has progressed in single-layer main roof property and multi-layer key strata joint action. However, the formula of the common instability condition of double key strata has not been obtained. The respective properties of double key strata have not been considered, which is insufficient for the widely existing mining of close-distance coal seams.

Therefore, a method for calculating the damage depth of the floor and four damage states of the main roof of the coal seam under the close-distance coal seam group are put forward. Meanwhile, the formulas for calculating the first and periodic weighting intervals of the damaged main roof and the instability conditions of the damaged key blocks are obtained by using the clamped beam and cantilever beam models. Therefore, the interaction mechanism of the two key blocks and discriminant formula of rotary and sliding instability are obtained by using the damage mechanics and elastic mechanics basic methods when the coal seam is mined under the gob area. Combined with an example, the linkage effect of the key blocks of the two main roofs is analyzed, and the instability types of damaged key blocks during first and periodic fracture and the support load required to maintain linkage stability are analyzed. In order to verify the calculation results, the ground pressure observation of the working face was carried out on the spot. This study can provide a theoretical basis for the calculation of the main roof weighting step and the bearing capacity of the support when the coal seam is mined at a close distance.

2. Property Analysis of Upper and Lower Main Roofs

In the close coal seam group, the mining of the overlying coal seam causes the floor strata to be damaged, resulting in the mining of the underlying coal seam under the condition of the damaged roof.

2.1. Study on Floor Damage due to Mining of Overlying Coal Seam

In the mining process of the working face, the abutment pressure in front of the coal wall will lead to the destruction of the floor. As shown in Figure 1, according to the plastic theory, the limit equilibrium zone generated by the front abutment pressure on the floor can be divided into three regions: active stress zone (zone I), transition zone (zone II), and passive stress zone (zone III) [37]. However, the mining of the working face is a continuous process. With the constant advancement of the working face, the front abutment pressure is constantly moved forward in the limit equilibrium zone. By connecting the peak point of its maximum depth $y_{\max}$, the damage zone in the floor generated by working mining can be obtained.

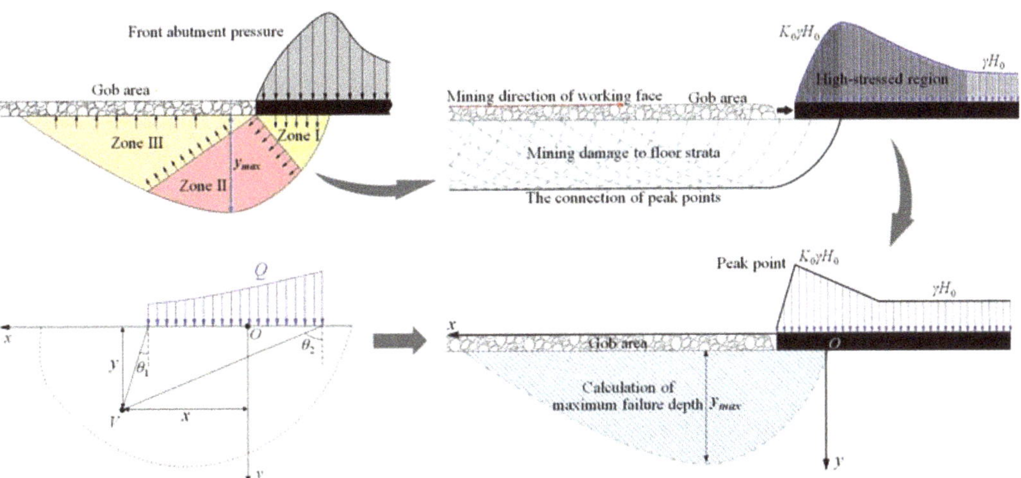

Figure 1. Mechanical model of floor damage caused by mining.

As shown in Figure 1, the original rock stress is γH_0, γ is the average unit weight, kN/m^3. H_0 is the buried depth of coal seam, m. The peak value of the front abutment pressure of the working face is set as $K_0\gamma H_0$; this linear load Q is simplified into a uniform load for calculation which is sufficient to meet the accuracy requirements of the project [38], as shown in Equation (1).

$$Q = \frac{(K_0 + 1)\gamma H_0}{2} \tag{1}$$

where K_0 is the maximum stress concentration factor. According to the stress solution of a half-plane body under uniformly distributed normal load on the boundary in elasticity, the three stress components of any point V can be obtained:

$$\begin{cases} \sigma_x = -\frac{Q}{2\pi}[2(\theta_2 - \theta_1) - (\sin 2\theta_2 - \sin 2\theta_1)] \\ \sigma_y = -\frac{Q}{2\pi}[2(\theta_2 - \theta_1) + (\sin 2\theta_2 - \sin 2\theta_1)] \\ \tau_{xy} = \tau_{yx} = -\frac{Q}{2\pi}(\cos 2\theta_1 - \cos 2\theta_2) \end{cases} \tag{2}$$

where θ_1 and θ_2 are the two included angles between point V and stress section, respectively. According to the elasticity, the principal stress at any point in the bottom plate is:

$$\left.\begin{matrix} \sigma_1 \\ \sigma_3 \end{matrix}\right\} = \frac{\sigma_x + \sigma_y}{2} \pm \sqrt{\left(\frac{\sigma_x - \sigma_y}{2}\right)^2 + \tau_{xy}^2} \tag{3}$$

where σ_1 and σ_2 are the maximum and minimum principal stress of the rock mass of the floor, MPa. From Equations (2) and (3), the principal stress at any point of the floor caused by mining is shown below [37].

$$\begin{cases} \sigma_1 - \frac{Q}{\pi}[(\theta_1 - \theta_2) + \sin(\theta_1 - \theta_2)] - \gamma y \\ \sigma_3 = \frac{Q}{\pi}[(\theta_1 - \theta_2) - \sin(\theta_1 - \theta_2)] - \gamma y \end{cases} \tag{4}$$

Mohr–Coulomb yield criterion in geotechnical engineering is:

$$\frac{1}{2}(\sigma_1 - \sigma_3) = c_0 \cos \varphi_0 + \frac{1}{2}(\sigma_1 + \sigma_3) \sin \varphi_0 \tag{5}$$

where c_0 is the average cohesion of floor rock mass, MPa, φ_0 is the average internal friction angle of floor rock mass, °. The mining damage depth y of the floor can be obtained from Equations (4) and (5):

$$y = \frac{Q}{1000\pi\gamma}\left(\theta - \frac{\sin\theta}{\sin\varphi_0}\right) + \frac{c_0 \cos\varphi_0}{1000\gamma \sin\varphi_0} \tag{6}$$

where $\theta = \theta_1 + \theta_2$. From Equation (6), θ = arccos (sin φ_0). Therefore, the maximum depth $y_{\max}$ is:

$$y_{\max} = \frac{Q}{1000\pi\gamma}\left(\arccos(\sin\varphi_0) - \frac{\sin[\arccos(\sin\varphi_0)]}{\sin\varphi_0}\right) + \frac{c_0\cos\varphi_0}{1000\gamma\sin\varphi_0} \tag{7}$$

From Equation (7), the relationship between the depth of floor damage caused by mining and the depth of burial, the stress concentration factor, the internal friction angle, and cohesion can be obtained as shown in Figure 2.

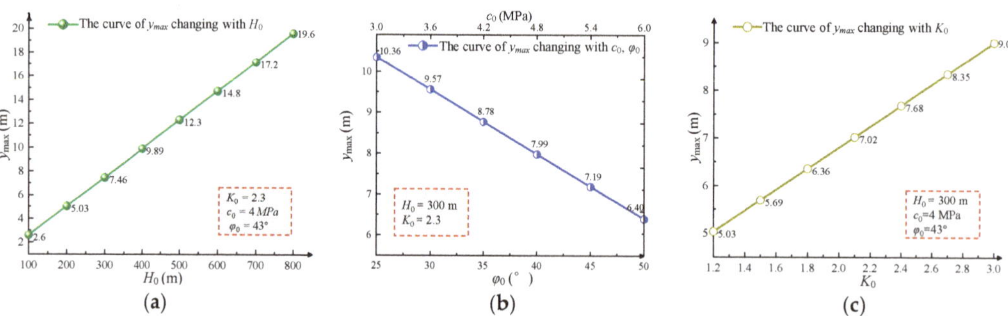

Figure 2. Influence of various parameters on the maximum damage depth of the floor. (**a**) The effect of H_0 on $y_{\max}$ (**b**) The effect of c_0 and φ_0 on $y_{\max}$ (**c**) The effect of K_0 on $y_{\max}$.

It can be seen from Figure 2a that the maximum damage depth of the bottom plate increases by 7.5 times when the buried depth is increased by eight times. This depth increases by 2.4 m for every 100 m increase in buried depth. As shown in Figure 2b, the maximum damage depth decreases with the rise of the cohesion and internal friction angle. The cohesion and internal friction angle increase two times, and the depth decreases by 0.62 times. It can be seen from Figure 2c that the higher the stress concentration coefficient, the greater the maximum damage depth gradually. The stress concentration factor increases two times, and the depth increases 1.53 times.

The above is a calculation method to obtain the damage depth of the bottom plate. If it can be measured in the field (such as seismic wave velocity tomography, AE, micro-seismic monitoring, etc.) [39–42], the actual measurement data can be used; if the upper coal seam has already been mined and the field data cannot be obtained, you can refer to the calculation method in this paper.

2.2. Types of the Damaged Main Roof in the Lower Coal Seam

The mining of the upper coal seam causes damage to the floor of the gob area, and the floor is the roof of the lower coal seam. As shown in Figure 3, the main roof is divided into four categories according to the degree of damage to the main roof of the lower coal seam. Figure 3a represents complete damage to the base roof. In Figure 3b, the main roof damage area accounts for more than 50% in the vertical direction and belongs to extensive damage. In Figure 3c, the percentage of main roof damage area in the vertical direction is ≤50%, which belongs to local damage. Figure 3d shows no damage to the main roof.

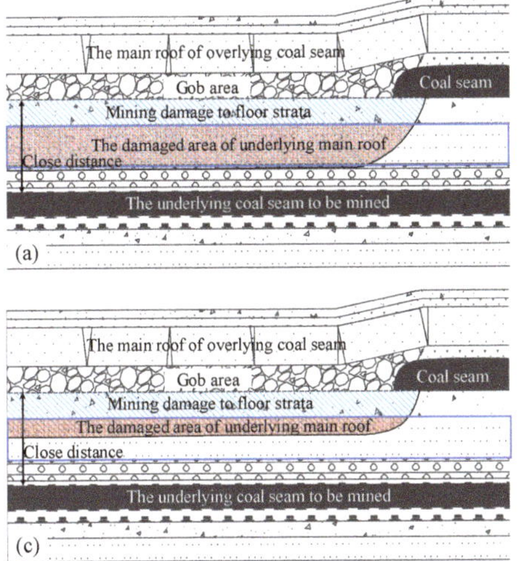

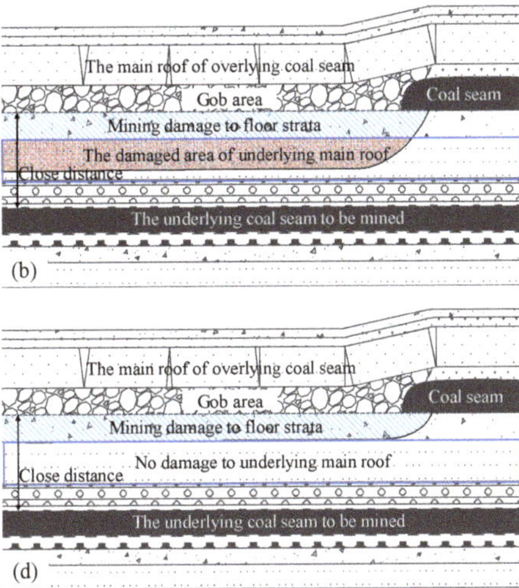

Figure 3. Four types of damaged main roof in the lower coal seam. (**a**) Complete damage to the lower main roof (**b**) Wide area damage of the lower main roof (≥50%) (**c**) Local damage to the lower main roof (<50%) (**d**) No damage to the lower main roof.

In this paper, the linkage between the upper broken main roof and the lower damaged main roof is mechanically analyzed and studied.

3. Stability Calculation and Analysis of Damaged Main Roof in Underlying Coal Seam

According to the basic principle of damage mechanics, the damaged main roof's mechanical properties and instability conditions are studied.

3.1. Analysis of First and Periodic Fracture of the Damaged Main Roof

As shown in Figure 4, for continuous materials, the uniaxial tensile specimen is subjected to tensile force P, the intact non-damaged micro section is subjected to uniformly distributed load, and its area is S. The actual stress area of the damaged rock block is S_a, the cross-sectional area of the lesion zone is S_d. According to the damage theory, the damage variable D is shown in Equation (8) below [34].

$$\begin{cases} \sigma_a = \frac{P}{S_a}, \sigma_d = 0 \\ S_a + S_d = S \\ D = \frac{S_d}{S} \end{cases} \quad (8)$$

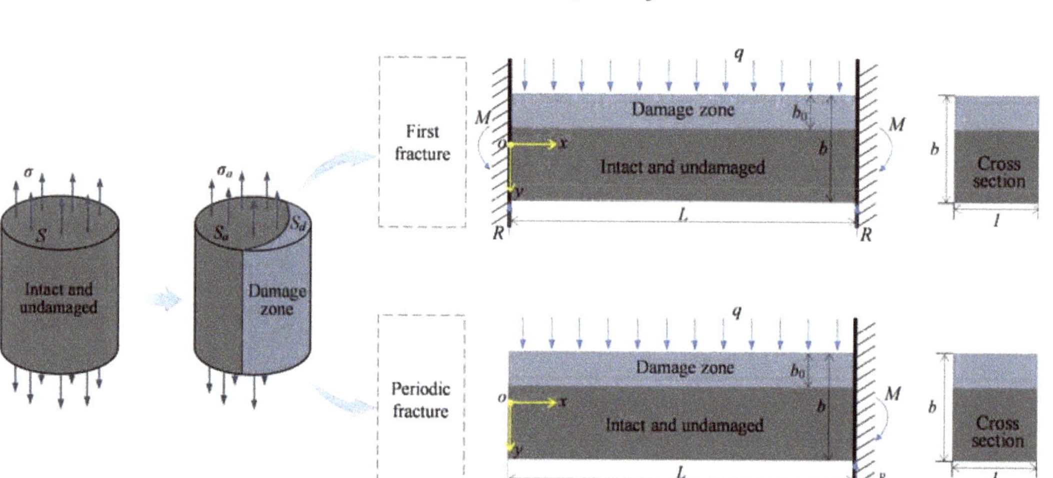

Figure 4. Mechanics model of first and periodic fracture of the damaged main roof.

(1) Clamped beam model for the first fracture of the damaged main roof

In general, the first fracture of the main roof is simplified as a two-dimensional clamped beam model [12]. As shown in Figure 4, the length of the beam is L, the height of the beam is b, the width is 1, the height of the damage band is b_0, the upper surface of the beam is subjected to a uniformly distributed load q, the exact solution of the horizontal stress is obtained [43]:

$$\sigma_x = -\frac{6}{b^3}qx(x-L)y + \frac{1}{b^3}q\left(4y^2 - L^2\right)y + \frac{3(2v-1)}{5b}qy - \frac{1}{2}vq \quad (9)$$

where v is Poisson's ratio. The maximum bending moment occurs at both ends of the beam, where the maximum tensile stress first exceeds the tensile strength limit of the material $[\sigma_t]$ [34]. For the damaged main roof beam model, its tensile strength limit is:

$$\frac{\sigma}{1-D} \leq [\sigma_t] \quad (10)$$

Then $(0, -b/2)$ is substituted into Equation (9) to obtain the limit span L_f of the damaged main roof beam during the first fracture. The lumpiness i_f of the fractured rock block (lumpiness is also called fracture degree [12], which refers to the height-length ratio of rock block, that is, $i = b/L$):

$$\begin{cases} L_f = b\sqrt{\frac{2[\sigma_t](1-D)}{q} + \frac{2+11\nu}{5}} \\ i_f = \frac{1}{\sqrt{\frac{2[\sigma_t](1-D)}{q} + \frac{2+11\nu}{5}}} \end{cases} \quad (11)$$

(2) Cantilever beam model for periodic fracture of the damaged main roof

The periodic weighting interval of the damaged main roof is usually determined according to the cantilever beam [12]. As shown in Figure 4, the exact solution for the horizontal stress of the damaged cantilever beam is obtained with the same mechanical parameters [43]:

$$\sigma_x = -\frac{6}{b^3}qx^2y + \frac{4}{b^3}qy^3 - \frac{3}{5b}qy \quad (12)$$

Substituting $(L, -b/2)$ into Equation (12), and the limit span L_p of periodic fracture of the damaged main roof beam and the lumpiness i_p are as follows:

$$\begin{cases} L_p = \frac{\sqrt{3}b}{3}\sqrt{\frac{[\sigma_t](1-D)}{q} + \frac{1}{5}} \\ i_p = \frac{\sqrt{3}}{\sqrt{\frac{[\sigma_t](1-D)}{q} + \frac{1}{5}}} \end{cases} \quad (13)$$

(3) The first and periodic weighting intervals of the damaged main roof

As shown in Figure 5, with the increase in damage equivalent D, the main roof's first and periodic fracture intervals decrease, and the decreasing rate increases gradually. At the same time, when the load q of the damaged main roof is small, the damage degree of the ultimate span with a fracture is more obvious. When q is small, with the increase in damage degree, the first fracture interval of the main roof can be reduced by 61.3%, while the periodic fracture interval can be reduced by 78.7%.

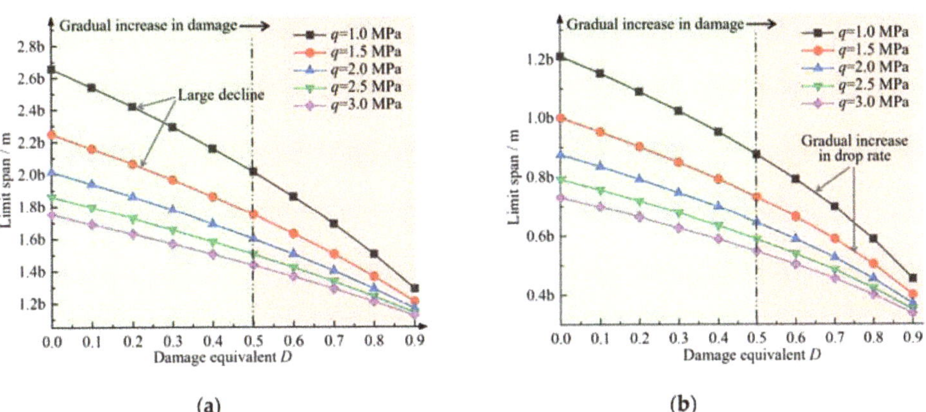

Figure 5. Relationship between damage degree and first and periodic fracture intervals of the main roof. (**a**) First fracture of the damaged main roof (**b**) Periodic fracture of the damaged main roof.

3.2. Stress Analysis of Key Block of the Damaged Main Roof

As shown in Figure 6, the damage degree will affect the contact length and rotation angle of the key block, and the stress analysis of the damaged main key block is carried out.

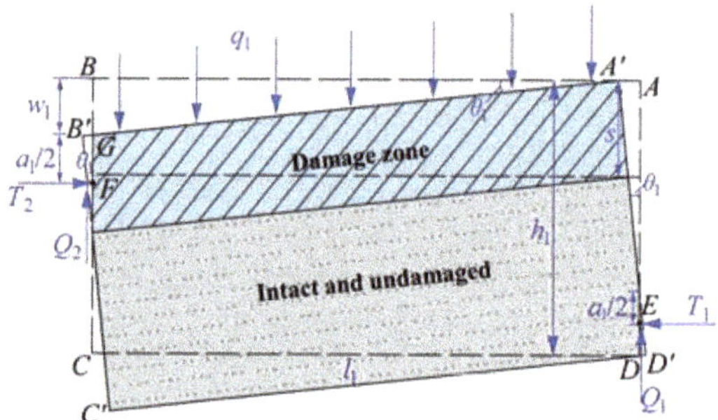

Figure 6. Stress of key block of the damaged main roof.

For the damaged key blocks, the contact length between the two ends is a_1 [12]:

$$a_1 = \frac{1}{2}(h_1 - l_1 \sin\theta_1) \tag{14}$$

where l_1 and h_1 are the length and height of key blocks. According to the stress analysis, the equilibrium relations between $\sum F_X = 0$ and $\sum F_Y = 0$ are listed:

$$\begin{cases} T_1 = T_2 \\ q_1 w_1 \csc\theta_1 = Q_1 + Q_2 \end{cases} \tag{15}$$

where T_1 and T_2 are horizontal thrust or extrusion forces between rock blocks, q_1 is the overlying load of the rock block, Q_1 and Q_2 are shear forces at the contact of rock blocks.

According to the "voussoir beam" theory, there is only horizontal compressive force at the hinge area F [12], $Q_2 = 0$. w_1 is the vertical subsidence of point B during rotation, and its formula is:

$$w_1 = l_1 \sin\theta_1 \tag{16}$$

Take the moment balance $\sum M_E = 0$ for the action point E of horizontal extrusion pressure:

$$q_1 w_1 \csc\theta_1 \cdot \left[\frac{1}{2}w_1 \cot\theta_1 + (h_1 - a_1)\tan\theta_1\right] = T_2 \cdot \left(h_1 - w_1 - \frac{1}{2}a_1 - \frac{1}{2}a_1\right) \tag{17}$$

The expression of horizontal thrust T can be obtained from Equation (17):

$$T = \frac{q_1 l_1 \cdot [l_1 \cos\theta_1 + (h_1 + l_1 \sin\theta_1)\tan\theta_1]}{h_1 - l_1 \sin\theta_1} \tag{18}$$

The expression of Equation (18), which is transformed into containing i is:

$$T = \frac{q_1 L \cdot [\cos\theta_1 + (i + \sin\theta_1)\tan\theta_1]}{i - \sin\theta_1} \tag{19}$$

Substituting L_f, L_p, i_f, and i_p can show that the horizontal thrust at the first and periodic fracture of the damaged main roof is affected by the damage equivalent D, as shown in Figure 7.

As shown in Figure 7, the horizontal thrust between key blocks decreases with the increase in damage equivalent when the damaged main roof is broken for the first and periodic fracture. The difference is that when the rotation angle is large, the decreasing trend of the horizontal thrust of the first fracture is fast at first and then slow, and the larger the rotation angle, the greater the decreasing degree (84%→91%), while the horizontal thrust of the periodic fracture has no apparent decreasing trend. The decreasing degree is mainly unchanged (94%) with the larger rotation angle.

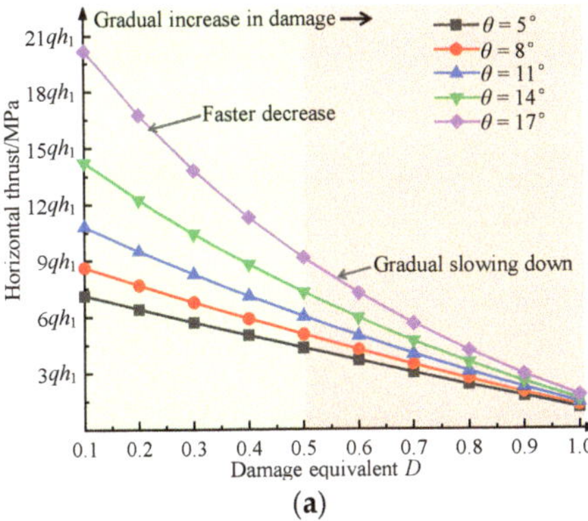

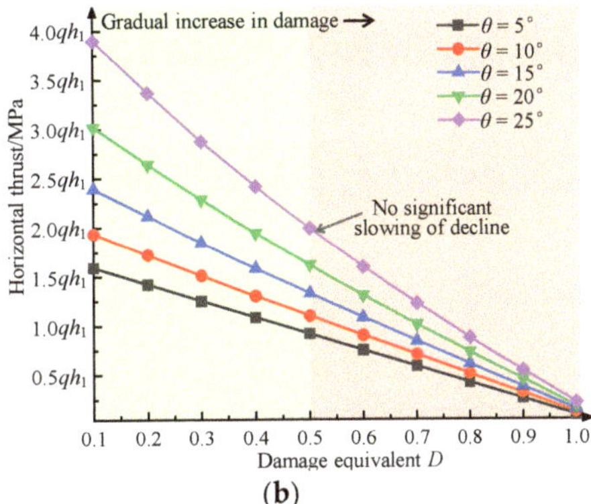

Figure 7. Influence of damage degree on horizontal thrust between key blocks. (**a**) First fracture of the damaged main roof (**b**) Periodic fracture of the damaged main roof.

3.3. Conditional Formula of Main Roof Instability

The sliding and rotary instability conditions of the damaged main roof during the first and periodic fracture are derived [44].

(1) Sliding instability of the damaged key block

The maximum shear stress of the damaged key block occurs at point E. To prevent the structure from slipping instability at point E [12], it should meet the following requirements:

$$T\tan\varphi \geq Q_1 \tag{20}$$

where $\tan\varphi$ is the friction coefficient between rocks, Q_1 is the shear force at point E, which can be obtained from Equation (15):

$$Q_1 = q_1 w_1 \csc\theta_1 \tag{21}$$

The critical lumpiness i_{sm} of sliding instability is obtained respectively when the damaged key block is broken:

$$i_{sm} = \frac{\sin\theta_1 + (\sin\theta_1 \cdot \tan\theta_1 + \cos\theta_1)\tan\varphi}{1 - \tan\theta_1 \tan\varphi} \tag{22}$$

As shown in Figure 8, the critical lumpiness i_{sm} of sliding instability increases with the increase in the rotary angle of the key block, and the larger the $\tan\varphi$, the larger the growth rate and the faster the growth rate, and the smaller the $\tan\varphi$, the constant growth rate of the critical lumpiness i_{sm}. Therefore, the smaller the friction coefficient is, the lower the critical lumpiness of the main roof sliding instability is, and the smaller the influence of the rotary angle is.

i_{sm} is substituted into Equations (11) and (13), respectively, and the relations between critical loads q_{fsm}, q_{psm} and D (damage variable) in two states are obtained:

$$\begin{cases} q_{fsm} = \frac{10 i_{sm}^2 [\sigma_t](1-D)}{5 - i_{sm}^2 (2+11\nu)} \\ q_{psm} = \frac{5 i_{sm}^2 [\sigma_t](1-D)}{15 - i_{sm}^2} \end{cases} \tag{23}$$

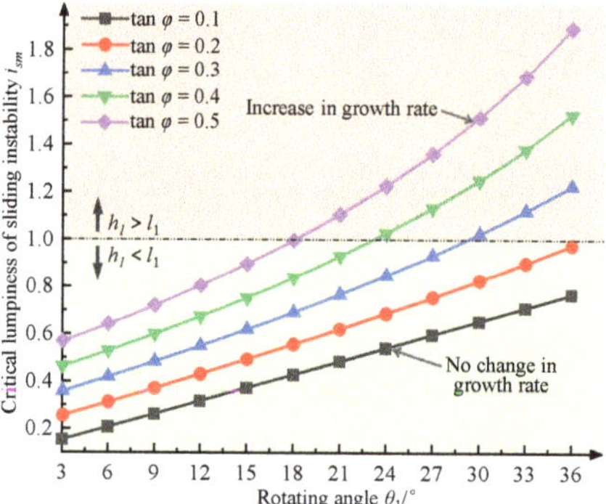

Figure 8. The critical lumpiness of sliding instability of the damaged key block.

As shown in Figure 9, the critical load of key block sliding instability when the main roof is broken for the first time is more significant than that of periodic fracture. Under the same rotation angle, the critical load of the first fracture is about 8.3 times that of the periodic fracture. In addition, the critical loads of both of them decrease with the increase in damage degree, and the reduction degree exceeds 80%.

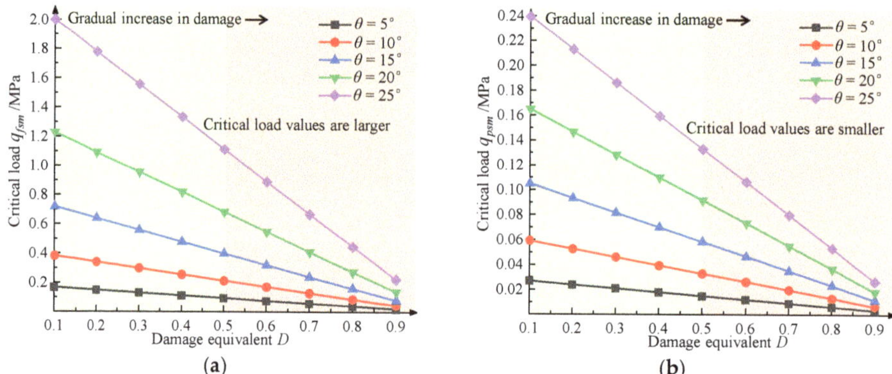

Figure 9. Critical load of sliding instability of the damaged key block. (**a**) First fracture of the damaged main roof (**b**) Periodic fracture of the damaged main roof.

(2) Rotation instability of the damaged key block

Excessive horizontal compressive force T between rock blocks leads to the crushing of rock blocks at the corner and the instability of the whole key block [12], and the conditional expression for the instability of rotation is:

$$T \leq a_1 \eta [\sigma_c]_D \tag{24}$$

where η is the coefficient; $[\sigma_c]_D$ is the compressive strength of damaged rock mass, and its value should be:

$$[\sigma_c]_D = [\sigma_c](1 - D) \tag{25}$$

where $[\sigma_c]$ is the compressive strength of intact rock mass. Therefore, the critical load q_{rm} of the damaged key block for rotary instability is:

$$q_{rm} = \frac{\eta [\sigma_c](1 - D)(i - \sin\theta_1)^2}{2[\cos\theta_1 + (i + \sin\theta_1)\tan\theta_1]} \tag{26}$$

As shown in Figure 10, the critical load of the damaged key block for rotary instability decreases with the increase in damage equivalent, and the larger the rotary angle of the key block, the smaller the critical load of rotary instability; while the larger the lumpiness of the key block, the larger the critical load of rotary instability. Therefore, no matter what kind of instability forms, the damage degree dramatically reduces the stability of key blocks.

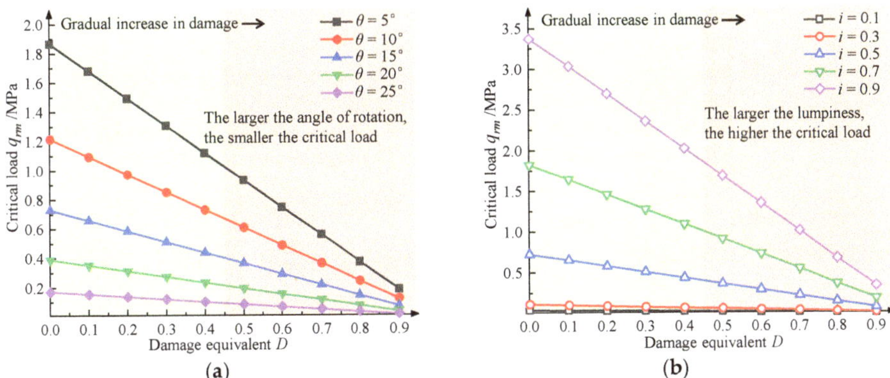

Figure 10. Rotary instability of the damaged key block. (**a**) Influence of the inclination factor (**b**) Influence of the lumpiness factor.

4. Calculation and Analysis of the Interaction Mechanism of the Upper and Lower Main Roofs

In the close coal seam group, when the underlying coal seam is mined under the gob area of the overlying coal seam, it is necessary to consider the interaction between the overlying broken main roof and the damaged main roof of the underlying coal seam.

4.1. Main Roof Mechanics Analysis of the Upper Coal Seam

The periodic fracture only needs to be considered to analyze the main roof of the overlying coal seam because the first limit span is minimal compared with the whole propulsion distance.

(1) Periodic fracture interval of the main roof of the overlying coal seam

As shown in Figure 11, cantilever beam model with periodic fracture of the main roof, the limit span of periodic fracture of main roof beam in overlying coal seam L_{up} and the lumpiness of fractured rock blocks i_{up} are as follows:

$$\begin{cases} L_{up} = \frac{\sqrt{3}b}{3}\sqrt{\frac{[\sigma_t]}{q} + \frac{1}{5}} \\ i_{up} = \frac{\sqrt{3}}{\sqrt{\frac{[\sigma_t]}{q} + \frac{1}{5}}} \end{cases} \quad (27)$$

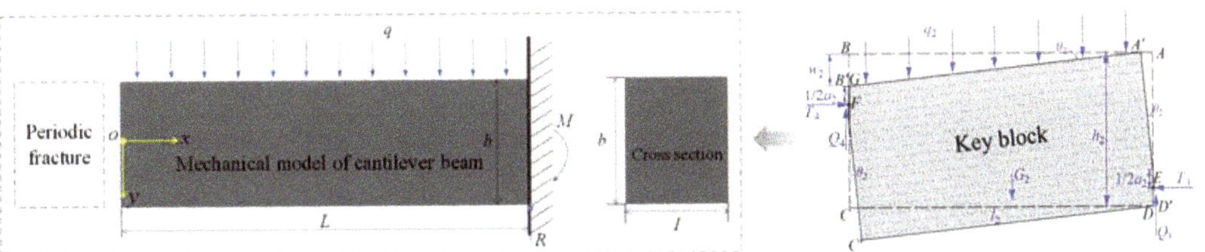

Figure 11. Critical load of rotary instability of the damaged key block.

The key block of the upper coal seam moves because it is unable to bear the load of its load layer, then its structure becomes unstable and turns around; the load value q_s transmitted downward by the key block is a specific value, the most fundamental is whether this load acts on the key block in the underlying main roof or other areas. From this analysis, it can be seen that when the fracture lines of the overlying and underlying main roof overlap, the load transmitted by the upper key block acts on the key block in the underlying main roof, which is the worst case of instability.

Therefore, this paper mainly studies the stability of the two key blocks in the case of deterioration; that is, the lower damaged key block bears the effect of $q + q_s$, first fracture interval L_{sf}, periodic fracture interval L_{sp}, and lumpiness i_{sf}, i_{sp}:

$$\begin{cases} L_{sf} = b\sqrt{\frac{2[\sigma_t](1-D)}{q+q_s} + \frac{2+11\nu}{5}}, i_{sf} = \frac{1}{\sqrt{\frac{2[\sigma_t](1-D)}{q+q_s} + \frac{2+11\nu}{5}}} \\ L_{sp} = \frac{\sqrt{3}b}{3}\sqrt{\frac{[\sigma_t](1-D)}{q+q_s} + \frac{1}{5}}, i_{sp} = \frac{\sqrt{3}}{\sqrt{\frac{[\sigma_t](1-D)}{q+q_s} + \frac{1}{5}}} \end{cases} \tag{31}$$

For the sliding instability, similarly, the horizontal thrust T_{st} of the damaged key block under the additional load of the key block in the overlying coal seam is obtained:

$$T_{sl} = \frac{(q_1 + q_s)l_1 \cdot [l_1 \cos\theta_1 + (h_1 + l_1 \sin\theta_1)\tan\theta_1]}{h_1 - l_1 \sin\theta_1} \tag{32}$$

From Equation (22), when the key block in the main roof is broken, the critical lumpiness is independent of the load value; thus, $i_{stm} = i_{sm}$. And substitute i_{stm} into Equation (22) to obtain the relationship between critical load q_{sfs}, q_{sps} and damage variable D of key block sliding instability when the underlying damaged main roof is broken for the first time and periodically:

$$\begin{cases} q_{sfs} = \frac{10i_{stm}^2[\sigma_t](1-D)}{5-i_{stm}^2(2+11\nu)} \\ q_{sps} = \frac{5i_{stm}^2[\sigma_t](1-D)}{15-i_{stm}^2} \end{cases} \tag{33}$$

For rotary instability, the critical load q_{srm} of rotation instability of the damaged key block under the additional load of key block in the upper coal seam can be calculated:

$$q_{srm} = \frac{\eta[\sigma_c](1-D)(i-\sin\theta_1)^2}{2[\cos\theta_1 + (i+\sin\theta_1)\tan\theta_1]} - q_s \tag{34}$$

Thus, the catastrophe condition formula for the instability of the damaged key block when the overlying broken main roof and the underlying damaged main roof overlap the fracture line is derived, the stability and the additional load transmitted of the key block in the overlying main roof are considered, the following examples will describe the application of this conditioning formula.

5. Analysis of Field Application Results

5.1. Geological Survey of Engineering Application Examples

As shown in Figure 14, the No. 4 coal seam of Swallow Mountain mine is firstly mined, and its main roof has collapsed and compacted the gob area, while the underlying No. 3 coal seam is the coal seam being mined, and the vertical distance between the two coal seams is 18.4 m, which belongs to the close-distance coal seams. The formulas in this paper are all applied to study the interaction between two main roofs in close-distance coal seams, so the theory is used to analyze the fracture step and block stability of the main roofs in the case of coal seam mining.

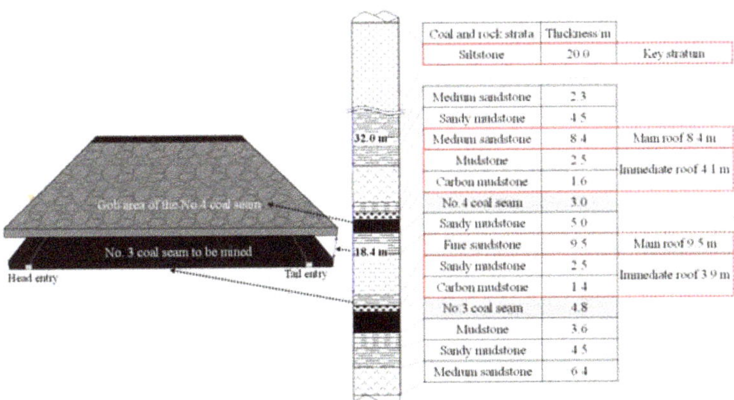

Figure 14. Generalized stratigraphic column of coal seam and strata.

The buried depth of the underlying coal seam is about 400 m, the main roof thickness is 9.5 m, and its distance from the overlying coal seam is 5.0 m. According to Equation (7), it can be seen that the damage depth of the floor caused by the mining of the overlying coal seam is 8.77 m, so the damage depth of the main roof of the underlying coal seam is 3.77 m, and the intact undamaged depth is 5.73 m. Therefore, the damage equivalent D of the main roof is 0.397, which belongs to the local damage type.

The calculation shows that the main roof of the overlying coal seam bears the load of 0.864 MPa, the limit span of periodic weighting L_{up} is 14 m, the broken expansion coefficient is 1.4 [12], and the maximum rotary angle is 5.7°. The damaged main roof of the underlying coal seam bears a load of 0.502 MPa. The ultimate span of the first weighting L_f is 40 m, which is 22% lower than that of 51 m without damage, and the maximum rotary angle is 4.6°; The ultimate span L_p of periodic weighting is 16 m, which is 24% lower than that of 21 m without damage, and the maximum rotary angle is 11.6°.

5.2. Stability Calculation of the Upper and Lower Main Roofs with Different Properties

(1) Instability analysis of the main roof of the overlying coal seam

When mining the working face of the underlying coal seam, it is necessary to analyze the linkage effect of the key blocks in the main roof of the two coal seams. To maintain stability, the key block in the overlying coal seam will transfer the load to the damaged key block in the underlying coal seam. For the damaged key block, this load is an additional load. Therefore, firstly, the stability of the key block in the main roof of the underlying coal seam is analyzed.

As shown in Figure 15a, for the sliding instability, the critical lumpiness of the key block can be calculated as 0.41. Its bearing capacity is 0.375 MPa, which is insufficient to bear the overburdened load. The sliding instability will occur, and the load value of 0.489 MPa will be transmitted downward. As shown in Figure 15b, for rotation instability, it has a bearing capacity of 2.889 MPa at the maximum angle of 5.7°, so rotation instability will not occur.

(2) Instability analysis of the damaged main roof of the underlying coal seam

① The first fracture of the damaged main roof

As shown in Figure 16a, when the damaged main roof is first fracture, for sliding instability, its critical lumpiness is 0.39, corresponding to the bearing capacity of 1.405 MPa, so the damaged key block will not occur sliding instability. As shown in Figure 16b, for rotary instability, when the maximum rotary angle is 4.6°, it has only a bearing capacity of 0.162 MPa, so the damaged key block will have rotary instability.

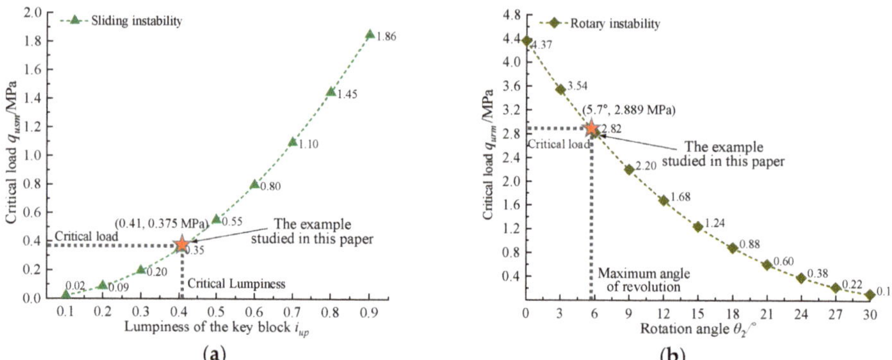

Figure 15. Stability analysis of the main roof of the upper coal seam. (**a**) Sliding instability of the upper main roof (**b**) Rotary instability of the upper main roof.

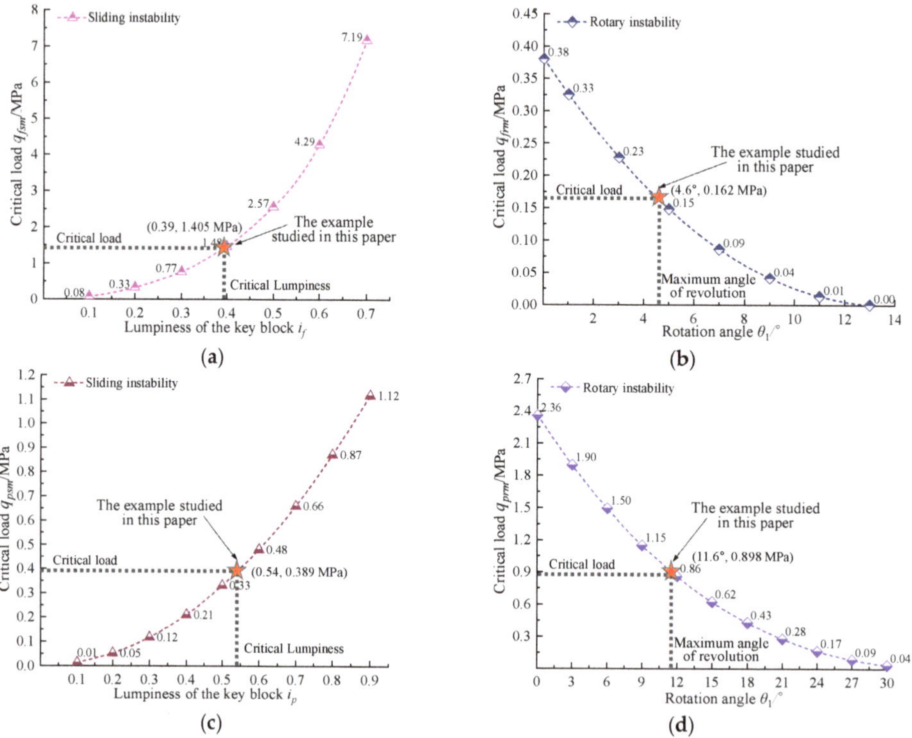

Figure 16. Stability analysis of the first and periodic fracture of the damaged main roof. (**a**) Sliding instability during the first fracture (**b**) Rotary instability during the first fracture (**c**) Sliding instability during the periodic fracture (**d**) Rotary instability during the periodic fracture.

② The periodic fracture of the damaged main roof

When the damaged main roof is fractured periodically, similarly, as shown in Figure 16c, for the sliding instability, its critical lumpiness is 0.54, and its bearing capacity is 0.389 MPa, so the sliding instability of the damaged key block will occur. As shown in Figure 16d, for rotary instability, when the maximum angle is 11.6°, it corresponds to the bearing capacity of 0.898 MPa, so rotary instability will not occur.

(3) Combined stability analysis of the upper and lower main roofs with different properties

According to the stability of the overlying and underlying main roofs, the critical load curve is divided into three regions. Region I: joint instability region, and the upper and lower key blocks are all unstable; Region II: stability judgment region, the lower damaged key block remains stable under its own load (the load of the rock layer between the two main roofs), but it bears the additional load, and its stability needs to be judged; Region III: combined stability region, the upper and lower key blocks interact with each other and remain stable.

① The first fracture of the damaged main roof under the additional load

The upper main roof cannot keep itself stable and will transmit additional load downward, so the damaged key block that is the first fracture will bear the additional load. As shown in Figure 17a, there are three regions in the critical load curve of sliding instability. When combined with this calculation example, the stability state of region II is judged. From the above, the value of the additional load is 0.489 MPa, and the bearing capacity of the damaged key block that is the first fracture is 0.903 MPa. Therefore, the sliding instability of the damaged key block will not occur under the action of the additional load. However, as shown in Figure 17b, the critical load curve of rotary instability only exists in region I. That is, the two key blocks cannot maintain their own balance, and the support needs to provide 0.829 MPa to ensure the balance.

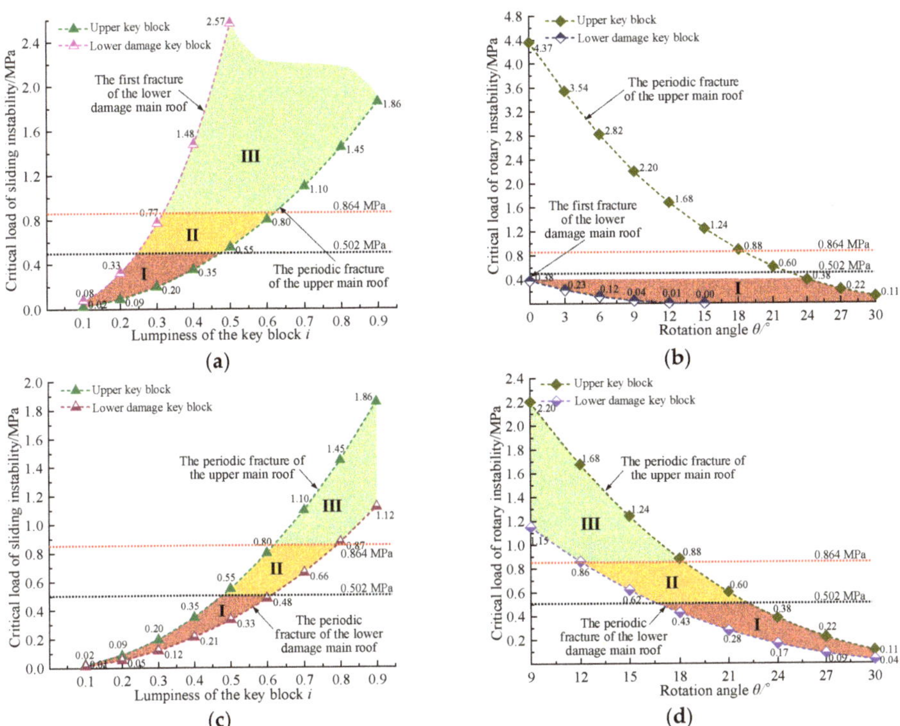

Figure 17. Stability analysis of the first fracture of the damaged main roof when the two main roofs interact. (**a**) Sliding instability during the first fracture (**b**) Rotary instability during the first fracture (**c**) Sliding instability during the periodic fracture (**d**) Rotary instability during the periodic fracture.

The support bears a load of top coal mass and immediate roof (0.159 MPa) and inhibits the rotary instability of the damaged main roof, with a total load of 0.988 MPa. ZF12000/23/35 top coal caving support is adopted on-site, with the roof-control distance of 5.18 m and the bearing capacity of the support reaching 1.3 MPa. Therefore, during the initial pressure, the support function of the support can keep the overlying and underlying main roofs with different properties.

② The periodic fracture of the damaged main roof under the additional load

Similarly, for the damaged key block that bears additional load and breaks periodically, as shown in Figure 17, there are three regions in the critical load curve of sliding and rotary instability. For the discrimination of region II, as shown in Figure 17c, it is known that the instability of the overlying key block transmits a critical load of 0.489 MPa, and the damaged key block cannot maintain its own stability and instability. Therefore, the support of the working face needs to bear a total load of 0.761 MPa to maintain the stability of the two key blocks; As shown in Figure 17d, it is known that the load is 0.898 MPa of the damaged main roof to inhibit the rotary instability during periodic fracture, and there is 0.396 MPa after bearing its own load layer. Therefore, the rotary instability will occur under the action of additional load, but the support of the support is sufficient to maintain its stability.

In this calculation example, the two main roof key blocks move jointly; the lower damaged key block is prone to rotation instability with the first fracture of the main roof and sliding instability when the main roof is broken periodically. Still, it can remain stable under the support of the working face support, which also verifies the rationality of the support selection.

5.3. Observation of Working Face Support Resistance

Observe the time-weighted mean resistance of the support in the 3215 working face when mining the lower coal seam, as shown in Figure 18.

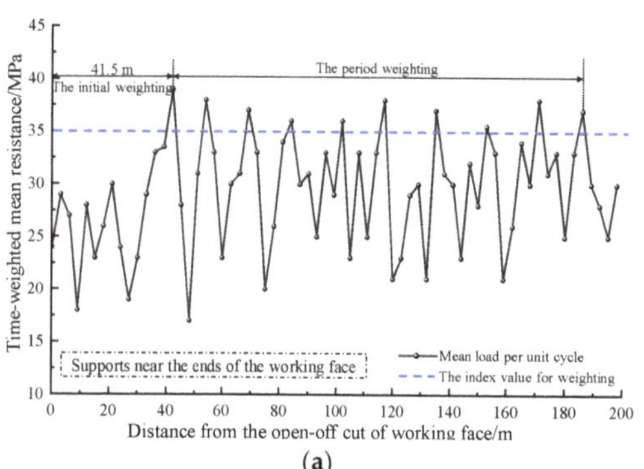

(**a**)

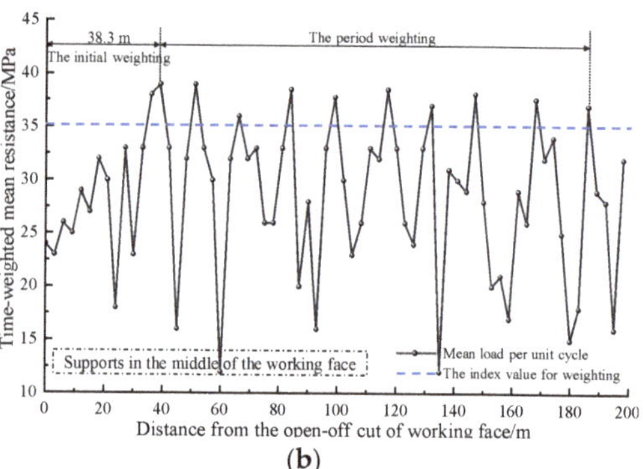

(**b**)

Figure 18. Observation of the time-weighted mean resistance of the support in 3215 working face. (**a**) Supports near the ends of the working face (**b**) Supports in the middle of the working face.

As shown in Figure 18a, the first pressure interval of the support near the end of the 3215 working face is 41.5 m, and the average periodic weighting interval of the first nine times is 16.5 m. As shown in Figure 18b, the first pressure interval of the support in the middle of the working face is 38.3 m, and the average periodic weighting interval of the first nine times is 15.8 m. The results are close to the theoretical calculation. At the same time, the variation range of ground pressure strength and pressure in the middle of the working face is more significant than that in the end area, and the supports are in good working condition. When the working face is weighted for the first time, the support bears the maximum pressure and is close to the ultimate bearing capacity; When the working face is weighted periodically, the support pressure is significant, but they are all within the controllable range, which shows that the support selection is reasonable and verifies the above calculation results.

6. Conclusions

(1) The mining of the upper coal seam in the group of close-distance coal seams will trigger damage to the floor. The formula for calculating the damage depth of the floor is proposed, and the damage depth increases by 2.4 m for every 100 m increase in burial depth; the cohesion and internal friction angle increase to double, and the damage depth decreases to 0.62 times; the stress concentration coefficient increases to double and the damage depth increases to 1.53 times.

(2) Four types of the damaged main roof of the lower coal seam are proposed, and the mechanical models of the first and periodic fractures of the damaged main roof are established. The limited span of the main roof during the first and periodic fractures, the horizontal thrust of the key block, and the critical load of instability are significantly reduced by the damage; however, there are differences in reducing regularity under different loads rotary angles, and lumpiness.

(3) The mechanical model of the interaction mechanism and stability of the upper and lower main roofs with different properties is established. When the fracture lines of the upper and lower main roofs overlap, the stability of the damaged key blocks is the lowest. There are three linkage stability regions in the critical load curves of the two key blocks: joint stability region, stability judgment region, and combined stability region.

(4) In the example of this paper, the damage equivalent D is 0.397 of the damaged main roof, belonging to the local damage type. The first and periodic pressure interval is 40 m and 16 m, decreasing by 22% and 24%, respectively, compared with the no damage.

(5) A supporting load of 0.489 MPa is required to maintain the stability of the upper key block, and the lower damaged key block is prone to rotary and sliding instability during the first and periodic weighting, respectively. Thus, the supports need to bear a total of 0.988 MPa and 0.761 MPa to maintain the stability of the two key blocks simultaneously. The ground pressure data monitored on-site accord with the calculation results.

Author Contributions: Conceptualization, S.X., D.C. and Y.W.; methodology, D.C., Y.W. and F.G.; software, Y.W., X.Z., H.Z., R.L., X.M. and S.L.; formal analysis, Y.W.; investigation, F.G., E.W., H.Z., R.L., X.M. and S.L.; data curation, E.W. and X.Z.; writing—original draft preparation, Y.W.; writing—review and editing, S.X. and Y.W.; supervision, S.X.; project administration, D.C.; funding acquisition, D.C. All authors have read and agreed to the published version of the manuscript.

Funding: This work was financially supported by the National Natural Science Foundation of China (Grant No. 52074296), the Fundamental Research Funds for the Central Universities (Grant No. 2022YJSNY18), the National Natural Science Foundation of China (Grant No. 52004286), the Fundamental Research Funds for the Central Universities (Grant No. 2022XJNY02), the China Postdoctoral Science Foundation (Grant No. 2020T130701, 2019M650895), all of which were gratefully acknowledged.

Informed Consent Statement: Informed consent was obtained from all subjects involved in the study.

Conflicts of Interest: The authors declare no conflict of interest.

References

1. Xie, S.R.; Sun, Y.J.; He, S.S.; Li, E.P.; Gong, S.; Li, S.J. Distinguishing and controlling the key block structure of close-spaced coal seams in China. *J. S. Afr. Inst. Min. Metall.* **2016**, *116*, 1119–1126. [CrossRef]
2. Wang, J.; Yang, S.; Wei, W.; Zhang, J.; Song, Z. Drawing mechanisms for top coal in longwall top coal caving (LTCC): A review of two decades of literature. *Int. J. Coal Sci. Technol.* **2021**, *8*, 1171–1196. [CrossRef]
3. Xia, H.C.; Gao, G.S.; Yu, B. Research on the overlying strata in full-mechanized coal mining of datong jurassic coal multi-layer goaf. *Adv. Mat. Res.* **2013**, *616–618*, 402–405. [CrossRef]
4. Yang, X.; Wen, G.; Dai, L.; Sun, H.; Li, X. Ground Subsidence and Surface Cracks Evolution from Shallow-Buried Close-Distance Multi-seam Mining: A Case Study in Bulianta Coal Mine. *Rock Mech. Rock Eng.* **2019**, *52*, 2835–2852. [CrossRef]
5. Ma, L.; Cao, X.; Liu, Q.; Zhou, T. Simulation Study on Water-Preserved Mining in Multiexcavation Disturbed Zone in Close-Distance Seams. *Environ. Eng. Manag. J.* **2013**, *12*, 1849–1853. [CrossRef]

6. Wei, X.; Bai, H.; Rong, H.; Jiao, Y.; Zhang, B. Research on Mining Fracture of Overburden in Close Distance Multi-seam. *Procedia Earth Planet. Sci.* **2011**, *2*, 20–27. [CrossRef]
7. Cui, F.; Jia, C.; Lai, X.; Yang, Y.; Dong, S. Study on the law of fracture evolution under repeated mining of close-distance coal seams. *Energies* **2020**, *13*, 6064. [CrossRef]
8. Gao, X.; Zhang, S.; Zi, Y.; Pathan, S.K. Study on Optimum Layout of Roadway in Close Coal Seam. *Arab. J. Geosci.* **2020**, *13*, 746. [CrossRef]
9. Liu, X.; Li, X.; Pan, W. Analysis on the floor stress distribution and roadway position in the close distance coal seams. *Arab. J. Geosci.* **2016**, *9*, 83.
10. Xiong, Y.; Kong, D.; Wen, Z.; Wu, G.; Liu, Q. Analysis of coal face stability of lower coal seam under repeated mining in close coal seams group. *Sci. Rep.* **2022**, *12*, 509. [CrossRef]
11. Qian, M. Review of the Theory and Practice of Strata Control Around Longwall Face in Recent 20 Years. *J. China Univ. Min. Technol.* **2000**, *1*, 1–4.
12. Qian, M.; Miao, X.; He, F. Analysis of Block in the Structrue of Voussoir Beam in Longwall Mining. *J. China Coal Soc.* **1994**, *19*, 557–563.
13. Pan, W.; Zhang, S.; Liu, Y. Safe and Effcient Coal Mining Below the Goaf: A Case Study. *Energies* **2020**, *13*, 864. [CrossRef]
14. Guo, G.; Yang, Y. The Study of Key Stratum Location and Characteristics on the Mining of Extremely Thick Coal Seam under Goaf. *Adv. Civ. Eng.* **2021**, *2021*, 8833822. [CrossRef]
15. Liu, Q.; Huang, J.; Guo, Y.; Zhong, T. Research on the temporal-spatial evolution of ground pressure at short-distance coal seams. *Arab. J. Geosci.* **2020**, *13*, 1014. [CrossRef]
16. Cai, G.S.; An, L.Q.; Zhu, X.X.; An, J.L.; Mao, L.T. Study on Close Coal Seam Pressure Behavior Law of Zhong Xing Mine by Physical Model. *Appl. Mech. Mater.* **2012**, *204–208*, 1439–1444. [CrossRef]
17. Miao, X.; Mao, X.; Sun, Z. Formation Conditions of Compound Key Strata in Mining Overlayer Strata and Its Discriminance. *J. China Univ. Min. Technol.* **2005**, *34*, 547–550.
18. Ning, J.; Wang, J.; Jiang, L.; Jiang, N.; Liu, X.; Jiang, J. Fracture analysis of double-layer hard and thick roof and the controlling effect on strata behavior: A case study. *Eng. Fail Anal.* **2017**, *81*, 117–134. [CrossRef]
19. Xinggao, L.; Yanfa, G. Damage Analysis of Floor Strata. *Chin. J. Rock Mech. Eng.* **2003**, *22*, 35–39.
20. Zhao, Y.; Wang, Y.; Wang, W.; Tang, L.; Liu, Q.; Cheng, G. Modeling of rheological fracture behavior of rock cracks subjected to hydraulic pressure and far field stresses. *Theor. Appl. Fract. Mech.* **2019**, *101*, 59–66. [CrossRef]
21. Zhao, Y.; Zhang, C.; Wang, Y.; Lin, H. Shear-related roughness classification and strength model of natural rock joint based on fuzzy comprehensive evaluation. *Int. J. Rock Mech. Min. Sci.* **2020**, *137*, 104550. [CrossRef]
22. Zhao, Y.; Liu, Q.; Zhang, C.; Liao, J.; Lin, H.; Wang, Y. Coupled seepage-damage effect in fractured rock masses: Model development and a case study. *Int. J. Rock Mech. Min. Sci.* **2021**, *144*, 104822. [CrossRef]
23. Li, C.; Zuo, J.; Shi, Y.; Wei, C.; Duan, Y.; Zhang, Y.; Yu, H. Deformation and Fracture at Floor Area and the Correlation with Main Roof Breakage in Deep Longwall Mining. *Nat. Hazards* **2021**, *107*, 1731–1755. [CrossRef]
24. Li, J.; Zhang, M.; Wang, X.; Xue, Y.; Wang, L. Prediction of destroyed coal floor depth based on improved vulnerability index method. *Arab. J. Geosci.* **2022**, *15*, 192. [CrossRef]
25. Hu, Y.; Li, W.; Wang, Q.; Liu, S.; Wang, Z. Study on Failure Depth of Coal Seam Floor in Deep Mining. *Environ. Earth Sci.* **2019**, *78*, 697. [CrossRef]
26. Li, J.; Guo, R.; Yuan, A.; Zhu, C.; Zhang, T.; Chen, D. Failure Characteristics Induced by Unloading Disturbance and Corresponding Mechanical Mechanism of the Coal-Seam Floor in Deep Mining. *Arab. J. Geosci.* **2021**, *14*, 1170. [CrossRef]
27. Hou, Y.; He, S.; Xie, S. Damage and rupture laws of main roof between coal seams with a close distance. *Rock Soil Mech.* **2017**, *38*, 2989–2999+3008.
28. Chen, B. Stress-induced trend: The clustering feature of coal mine disasters and earthquakes in China. *Int. J. Coal Sci. Technol.* **2020**, *7*, 676–692. [CrossRef]
29. Chen, D.; Wu, X.; Xie, S.; Sun, Y.; Zhang, Q.; Wang, E.; Sun, Y.; Wang, L.; Li, H.; Jiang, Z.; et al. Study on the Thin Plate Model with Elastic Foundation Boundary of Overlying Strata for Backfill Mining. *Math. Prob. Eng.* **2020**, *2020*, 8906091. [CrossRef]
30. Chi, X.; Yang, K.; Wei, Z. Breaking and mining-induced stress evolution of overlying strata in the working face of a steeply dipping coal seam. *Int. J. Coal Sci. Technol.* **2021**, *8*, 614–625. [CrossRef]
31. Qian, M.; Miao, X.; Xu, J. Theoretical Study of Key Sratum in Ground Conrol. *J. China Coal Soc.* **1996**, *21*, 2–7.
32. Wang, H.; Shuang, H.-Q.; Li, L.; Xiao, S.-S. The Stability Factors' Sensitivity Analysis of Key Rock B and Its Engineering Application of Gob-Side Entry Driving in Fully-Mechanized Caving Faces. *Adv. Civ. Eng.* **2021**, *2021*, 9963450. [CrossRef]
33. Xu, X.; He, F.; Li, X.; He, W. Research on Mechanism and Control of Asymmetric Deformation of Gob Side Coal Roadway with Fully Mechanized Caving Mining. *Eng. Fail Anal.* **2021**, *120*, 105097. [CrossRef]
34. He, S.; Xie, S.; Song, B.; Zhou, D.; Sun, Y.; Han, S. Breaking laws and stability analysis of damage main roof in close distance hypogynous seams. *J. China Coal Soc.* **2016**, *41*, 2596–2605.
35. Jing, C.; Wulin, L.; Wengang, D.; Dingding, Z.; Zhe, M.; Qiang, Y. Deformation of huge thick compound key layer in stope based on distributed optical fiber sensing monitoring. *J. China Coal Soc.* **2020**, *45*, 44–53.
36. Qingxiang, H.; Jinlong, Z.; Longtao, M.; Pengfei, T. Double key strata structure analysis of large mining height longwall face in nearly shallow coal seam. *J. China Coal Soc.* **2017**, *42*, 2504–2510.

37. Weitao, L.; Dianrui, M.; Li, Y.; Liuyang, L.; Chenhao, S. Calculation method and main factor sensitivity analysis of inclined coal floor damage depth. *J. China Coal Soc.* **2017**, *42*, 849–859.
38. Lu, H. Stress Distribution and Failure Depths of Layered Rock Mass of Mining Floor. *Chin. J. Rock Mech. Eng.* **2014**, *33*, 2030–2039.
39. Dong, L.; Tong, X.; Ma, J. Quantitative Investigation of Tomographic Effects in Abnormal Regions of Complex Structures. *Engineering* **2020**, *7*, 1011–1022. [CrossRef]
40. Zhang, Y.; Yao, X.; Liang, P.; Wang, K.; Sun, L.; Tian, B.; Liu, X.; Wang, S. Fracture evolution and localization effect of damage in rock based on wave velocity imaging technology. *J. Cent. South. Univ.* **2021**, *28*, 2752–2769. [CrossRef]
41. Dong, L.; Sun, D.; Li, X.; Ma, J.; Zhang, L.; Tong, X. Interval non-probabilistic reliability of surrounding jointed rockmass considering microseismic loads in mining tunnels. *Tunn. Undergr. Space Technol.* **2018**, *81*, 326–335. [CrossRef]
42. Dong, L.; Zou, W.; Li, X.; Shu, W.; Wang, Z. Collaborative localization method using analytical and iterative solutions for microseismic/acoustic emission sources in the rockmass structure for underground mining. *Eng. Fract. Mech.* **2019**, *210*, 95–112. [CrossRef]
43. Chunxiao, Z. *Theoretical Analysis of Beams with Fixed Ends*; Hefei University of Technology: Hefei, China, 2017.
44. Li, Z.; Xu, J.; Ju, J.; Zhu, W.; Xu, J. The effects of the rotational speed of voussoir beam structures formed by key strata on the ground pressure of stopes. *Int. J. Rock Mech. Min. Sci.* **2018**, *108*, 67–79. [CrossRef]

Article

Avoiding Buffer Tank Overflow in an Iron Ore Dewatering System with Integrated Control System

Ênio L. Junior [1,2], Moisés T. da Silva [3,*] and Thiago A. M. Euzébio [3]

1 Programa de Pós-Graduação em Instrumentação, Controle e Automação de Processos de Mineração, Universidade Federal de Ouro Preto e Instituto Tecnológico Vale, Ouro Preto 35400-000, MG, Brazil; enio.lopes.junior@vale.com
2 Vale S.A., Canaã dos Carajás 68537-000, PA, Brazil
3 Instituto Tecnológico Vale (ITV), Ouro Preto 35400-000, MG, Brazil; thiago.euzebio@itv.org
* Correspondence: moises.silva@pq.itv.org

Abstract: High water usage is necessary while ore passes through the many stages of a mineral processing plant. However, a dewatering system filters the final ore pulp product to remove the water, which is reutilized in the previous processes. This step is fundamental to reducing the fresh new water consumption. Usually, several tanks, pumps, and filters form a dewatering system—any failure or shutdowns from those components disbalance the pulp flow. The waste of many tons of water and ore products for a tailing dam is the worst consequence of a mass disbalance in a dewatering system. This paper proposes an advanced regulatory control strategy composed of cascade and override loops for a dewatering system. The main purpose is to increase the production period, even under filter failure and changes in the inlet pulp characteristics. This control strategy is evaluated using a digital model of a large-scale Brazilian iron ore processing plant. Two scenarios are investigated: the simultaneous failure of two filters and disturbances in the flow and density of the thickener. The simulation results show that the proposed control strategy could extend the period of operation of the dewatering plant under failures in the disc filters and reject significant disturbances. For the considered simulation period, the proposed solution increases the time to overflow by 72% when compared to the previous control strategy. Thus, it is possible to avoid the waste of approximately 2448.36 tons of ore pulp that would be sent to the tailings dam.

Keywords: advanced regulatory control; dewatering process; zero waste

Citation: Lopes, Ê.L.; da Silva, M.T.; Euzébio, T.A.M. Avoiding Buffer Tank Overflow in an Iron Ore Dewatering System with Integrated Control System. *Sustainability* **2022**, *14*, 9347. https://doi.org/10.3390/su14159347

Academic Editor: Glen Corder

Received: 9 May 2022
Accepted: 13 July 2022
Published: 30 July 2022

1. Introduction

Recycling water in the process industry is important from a sustainability point of view. Mineral processing plants widely use dewatering systems to recover water and partially dry the processed ore. This operation has two purposes: to ensure the simple transport of ore and to reduce freshwater consumption [1]. The early stages of iron ore processing consume large volumes of water. After the dewatering operation, approximately 90% of the process water is recovered for reuse [2].

Basically, three units of operation comprise a dewatering system. The first unit consists of pulp thickening, in which the inlet pulp is separated into two products: clarified water to be reused and thickened pulp containing a high level of solids (60% to 70% by weight). The second unit is made up of buffer tanks, whose purpose is to temporarily store the thickened pulp to minimize the flow difference between the previous stage and the next one, i.e., the filtration stage. Filters comprise the third and last unit and are used to remove part of the residual water so that the final product is approximately 90 wt% solids. For instance, in [3], a maximum water recovery of 97% was achieved by treating the slurry in a dewatering circuit with a combination of high rate thickener and press filter in a chromite ore beneficiation plant. Due to the importance of water recovery, this operation is fundamental, and it is monitored continuously to ensure maximum performance.

Even when applying the best maintenance practices, equipment failure is an inevitable event that can affect the productivity of an industry. In dewatering plants, filters frequently present malfunctions that inhibit their use for several minutes to hours. These well-known failures motivate the parallel positioning of filters to let them work simultaneously and independently. Despite the alternative routes, when one or more filters suddenly fail, the dewatering circuit needs to be adapted to the new temporary condition. In some cases, buffer tanks are not enough to support the flow difference, and the thickened pulp overflows into the tailings dam. This waste of valuable processed ore and water negatively impacts key production performance and sustainability indices [4].

To control the dewatering process, it is essential to consider all subprocesses. However, a significant part of the existing control approaches focus on just one specific part, such as thickening [5–9] and pulp tanks [10,11]. These control strategies do not consider the effects of a unit's operations on other stages of the process. Furthermore, restrictions must be respected for the safe operation of the process. For instance, tank levels must remain in safe ranges, the pulp density must remain in a range that does not compromise the operation of pumps and valves, and so on.

In this context, few academic papers address the control of mineral processing plants by taking into account the effects of unit operations on the other stages. In [12], a multivariable controller, which combines fuzzy control and other strategies, was applied to a thickener in an iron ore concentration plant. In that work, the effects of the underflow flow rate and density variations, which cause problems in a subsequent flotation step, were taken into account. The proposed control strategy seeks to reduce the variability of these process variables so that the flotation stage's efficiency is not impaired. Similarly, an intelligent control strategy was proposed by [13] for an iron ore pumping circuit. In that study, the control system attenuates flow and density disturbances at the entrance of a passage box. The strategy allows the box level to fluctuate within the operating limits, reducing the harmful effects of the disturbance on the efficiency of the hydrocyclone powered by the box. On the other hand, in [14], a fault-tolerant fuzzy controller was designed to manipulate the setpoints of the regulatory tank level in buffer tanks to control the storage tank level. Since this strategy uses a natural language, the controllers are easily understood by a nonspecialist. However, these controllers are difficult to tune and maintain.

In a practical way, advanced control solutions are still challenging to implement and maintain in the mineral industry. According to [15], regulatory control strategies are still dominant in this industry. Advanced control applications, such as predictive control, can be found, though in considerably fewer numbers. The main reasons for the still rare use of advanced control in the mineral industry are the lack of process models, the high variability of the process and points of operation and the lack of trained professionals to keep the system running. On the other hand, regulatory control strategies can be used to control more complex processes, including the interactions and restrictions between different stages. Techniques referred to as advanced regulatory control (ARC) can be used [16]. These techniques are often implemented in addition to basic process controls. Basic process controls are designed to meet the basic operating requirements of an industrial plant. In contrast, ARC control techniques are normally added later to achieve better performance and sustainability of the operation of a process. For instance, in [17], a cooling water circuit controlled by an ARC controller obtained a 30% reduction in its energy consumption. This type of application does not require the use of detailed process models or the acquisition of new assets, as it can be implemented in the control systems typically in use in the industry as a programmable logic controller (PLC). In [18], several advanced control schemes were applied to pressure buffering control in industrial gas headers. The goal was to reduce gas emissions and improve consumer stability. The application of an adaptive model-based predictive controller in different mineral processing operations was performed by [19]. The proposed controller enables significant performance improvements compared to conventional control strategies for processes with long-time delays and multivariable interactions.

The main scope of this paper is to present a novel control strategy to automatically adjust the dewatering system to minimize the loss of valuable materials into tailings dams. An advanced regulatory controller is designed and compared to a traditional and nonintegrated controller usually encountered on the shop floor. The primary purpose is to enlarge the production period, even under random filter failure and dynamic changes in the inlet pulp characteristics. This study is conducted via a dynamic simulator of a large iron ore processing plant in Brazil. To the best of our knowledge, no control strategy manipulates all dewatering stages, and our strategy is failure tolerant regarding faults in filters.

This paper is organized as follows: In Section 2, the dewatering process is described, as well as the current control strategy. In Section 3, the operational problems of the dewatering plant under study are discussed. The proposed control strategy applied to the dewatering process is presented in Section 4. The results attained using the proposed control strategy are presented in Section 5. Finally, in Section 6, the conclusions are discussed.

2. The Dewatering System

In this section, we describe the dewatering process and the existing regulatory control strategy.

2.1. The Dewatering Process Description

The proposed control strategy is developed for a mineral processing plant of Vale S.A., a mining company in Brazil. In this process, water is mixed with iron ore. This mixture produces a pulp, which goes through several unit operations. At the end of the process, iron ore must be dewatered to be transported as a final product. The dewatering process removes water from iron ore for reuse as process water [20].

The dewatering process is composed of three stages: (i) thickening, (ii) the transport and storage of pulp, and (iii) filtration. The schematic diagram of the dewatering process under study is shown in Figure 1. In this process, the pulp with low-grade solids (about 57.4 wt%) feeds a thickener (feed flow, q_A). The thickener has a diameter of 35 m and is designed to process up to 1957 m^3/h of pulp. This equipment continuously makes a solid–liquid separation, producing a clearer overflow and an underflow (thickened flow, q_U) with a higher grade of solids (about 65 wt%). The flocculant (anionic polyacrylamide) is used to accelerate the sedimentation of the solids and ensure the recovered water with turbidity around 200 NTU. The thickener is discharged using a centrifugal pump (BP-001), and part of the material is recirculated. The estimated period of effective operation per year is 7930 h.

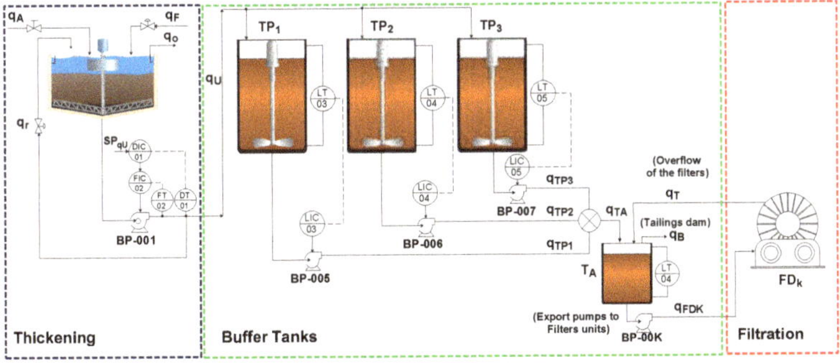

Figure 1. The dewatering process.

After the thickening process, the pulp flows into three buffer tanks (TP_1, TP_2, and TP_3), with volume of 250 m^3 each tank. Due to the high residence time, agitators are used to prevent the sedimentation of particles. Pumps BP-005, BP-006, and BP-007, at the outlet of each buffer tank, regulate the pulp flow that goes to the storage tank (T_A). All tanks have a level measurement, and in the case of overflow, all material is wasted on the tailings dam. In this case, the plant operation must be stopped.

Pulp from the storage tank feeds a set of vacuum disc filters (filters feed flow, q_{FDK}) through eight export pumps (BP-00K). The filtering unit provokes a negative differential pressure, which makes the pulp pass through a semipermeable membrane designed to hold almost all the solids and leave the clearer water to pass through [2]—the solid deposit on the membrane increases, forming a cake. After the cake is thick enough, the deposited solids are removed by applying a positive differential pressure on the disc membrane. At the end of this cycle, two products are yielded: a solid cake (93 wt% of solids) and the filtrate (water recovered).

The filtering units, named FD_k in Figure 1, have 12 discs with a diameter of 1.9 m and a volume of 5 m^3 in the pulp basins. These units operate independently and each unit can process up to 190 m^3/h of pulp. If a larger volume of pulp is pumped into the filters, any excess material will leak into the basins and return via overflow pipes to the storage tank (T_A).

2.2. The Current Control Strategy

The current control strategy consists of the nonintegrated operation of the different stages of the dewatering process. The closed-loop system is depicted in Figure 2. Signal r denotes the setpoint, u is the system inputs (manipulated variable), and y is the process outputs (controlled variable). $G(s)$ represents the system dynamics and $C(s)$ indicates the controller transfer function. The setpoints of the thickener underflow density, thickener outflow, and buffer tank level are modified over time by a human operator. In addition, the operator constantly monitors the storage tank level (H_{TA}) to prevent overflow into the tailings dam if one or more filters fail.

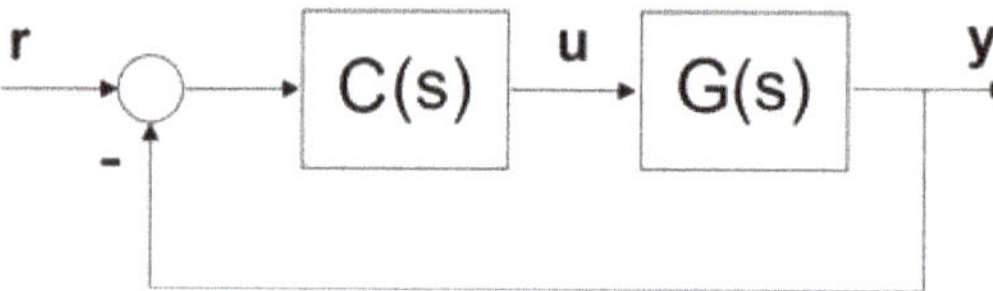

Figure 2. Closed-loop control system.

Among the several strategies applied to control loops, the most used in the industry for regulatory layer is the PID controller [21]. The main reasons for this are: it is easier to maintain; there are few tuning parameters; offers good performance and robustness regardless of operating point [22]. There are five regulatory control loops in the current control strategy: FIC02, DIC01, LIC03, LIC04, and LIC05. The FIC02 loop controls the thickener outflow (q_U), and the DIC01 controls the thickener underflow density (ρ_U). These loops make up a cascading structure, where the master loop is DIC01 and the slave is FIC02. The LIC03, LIC04, and LIC05 loops manipulate the respective frequency bump to control the levels of the buffer tanks TP_1, TP_2, and TP_3. For these loops, the regulatory controller setpoints are defined by the human operator. Table 1 shows the PI controller parameters for the loops used in the current control strategy with the respective filter time constant (λ). These controllers are defined using the method proposed by [23]. For all loops, the PI controller considered in this paper is formulated as

$$C(s) = K_p\left(1 + \frac{1}{T_i s}\right), \tag{1}$$

where K_p and T_i are the proportional gain and integral time, respectively.

Table 1. PI parameters of the current control strategy.

Loop	K_p	T_i	λ
FIC02	0.07	2.25	1.88
DIC01	−65.83	1382.30	30.00
LIC03/04/05	24.24	66.00	10.00

3. Problem Description

As described in Section 2, the dewatering plant is composed of subprocesses. The first subprocess consists of pulp thickening. Then, the pulp is transported and stored in tanks, which have volumes capable of absorbing variations in the process flow. These tanks feed the filtration stage, which removes residual water, delivering ore with low moisture content.

The current control strategy consists of a nonintegrated operation. Thus, it does not take into account the effects of unit operations on the other stages of the process. Beyond that, the control strategy does not consider a global objective. Thus, the operation of the process can be severely impaired. In addition, constraints must be respected. For instance, the pulp density must remain within a range that does not compromise the pump's operation.

The main problem in the dewatering process under study is the frequent failure of the disc filters. To illustrate this problem, consider data from 2520 h of operation of the filter units. During this period, for a total of 1498 h, i.e., 59.4% of the time, at least one filter was out of service due to a failure condition. For this period under failure, Figure 3 shows the percentage of the number of filters that had problems. Note that 64.6% of the time, failure occurred in only one disc filter; 27.4% of the time, the failure occurred in two filters simultaneously. The high frequency of failure has a direct impact on the productivity of the dewatering plant. In this way, the implementation of a control strategy capable of maintaining the operation of the process even in the event of disc filter failure becomes essential.

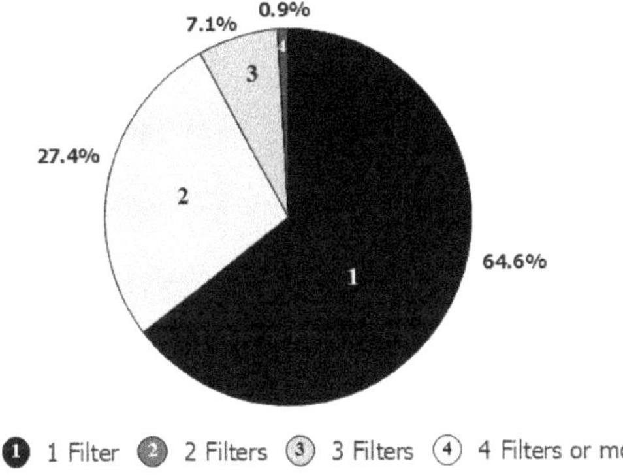

Figure 3. Percentage of filter failures.

Upon failure of the disc filters, several variables can simultaneously affect the dewatering process. Thus, the human operator cannot act according to the overall context of the process. In this context, the operational problem is as follows: In the case of disc filter failure, the filter feed pump stops, and it is not possible to empty the pulp into the filters, causing the level of the T_A tank to increase. This leads to overflow of tank T_A, which causes tons of concentrated pulp to be lost to tailings dams every year.

To overcome this operational problem, the control system should consider the several variables of the dewatering process and their interactions. The control solutions proposed in this paper consist of an ARC strategy used to reduce the loss of iron ore concentrate into the tailings dam by keeping the level of the T_A tank within its operational limits, especially under disc filter failure.

4. The Proposed Control Strategy

This section describes the proposed ARC strategy. Initially, the variables and loops used in the proposed control of the dewatering process are described. Then, the proposed control strategy is presented.

4.1. Variables and Loops of the Proposed Control Strategy

A simplified diagram of the dewatering process with the loops defined for the proposed control strategy is shown in Figure 4. For the sake of simplicity, the buffer tanks (TP_1, TP_2, and TP_3) are represented by only one block (TP_N). In the proposed control strategy, the outlet flow of the thickener (q_U) comprises a slave loop of a cascading structure. The master loop of this structure is the result of an override control composed of the average level of the buffer tanks ($\overline{H}_{TP}$) and the minimum and maximum restrictions on the thickener underflow density (ρ_U). In addition, according to Figure 4, the outlet flow of the buffer tanks is the result of the action of an override controller consisting of the storage tank level (H_{TA}) and minimum and maximum restrictions on the level of each buffer tank (H_{TP1}, H_{TP2} and H_{TP3}). For the sake of simplicity, in the simplified diagram shown in Figure 4, the level loops associated with the storage tank are represented by LIC0X, whereas the minimum and maximum restrictions on the level loops related to the buffer tanks are represented by LIC0Y and LIC0Z, respectively.

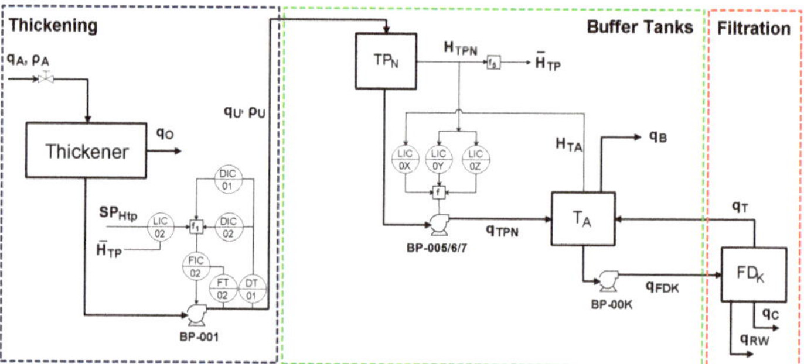

Figure 4. Simplified diagram of the dewatering process.

In this way, there are seven process variables that must be controlled: the thickener underflow density (ρ_U), the flow rate of the thickener underflow (q_U), the level of each buffer tank (H_{TP1}, H_{TP2} and H_{TP3}), the average level of the buffer tanks ($\overline{H}_{TP}$) and the storage tank level (H_{TA}). As manipulated variables (MVs), the following variables are initially considered: the thickener underflow pump frequency (BP-001), the setpoint of the thickener outlet flow, the frequency pump (BP-005/6/7) of each buffer tank, and the frequency of the export pumps (BP-00K). These MVs are analyzed below:

- Thickener underflow pump frequency: This variable is coupled to the density of the underflow (ρ_U), the level of the three buffer tanks and, consequently, the average level of these tanks ($\overline{H}_{TP}$).
- Setpoint of the thickener outlet flow: This variable is the setpoint of the slave loop of the cascade structure and is the result of the override control between the average level of the buffer tanks ($\overline{H}_{TP}$) and the minimum and maximum restrictions on the thickener underflow density (ρ_U).
- Frequency of the buffer tank pumps: Each of the pumps is coupled with the respective level of the buffer tank (H_{TP1}, H_{TP2} and H_{TP3}). In addition, the pumps are coupled to the storage tank level (H_{TA}).
- Frequency of the export pumps: Considering the increase in the storage tank level (H_{TA}), it is necessary to increase the export flow (q_{FDK}) and to pump more pulp into the filter. The filter has limited dewatering capacity, which means that pulp overfeed results in overflow of the filter; all overflowed material returns via overflow lines

(q_T), as illustrated in Figure 4. Therefore, the frequency of the export pumps is not considered an MV but rather a disturbance.

According to the analysis of the MVs, the degree of freedom of the system is reduced. Now, there are five MVs to control seven process variables (PVs). In this way, the correct assignment of MVs and PVs becomes essential for adequate operation of the process.

With the reduced degree of freedom, the proposed control approach uses a constraint-based strategy. This advanced regulatory controller is composed of PI controllers using override algorithms, connected in a cascade structure. In all cases, it is assumed that pumps are at the outlets of the buffer tanks.

Table 2 summarizes the input and output variables for the defined scope of the dewatering control system, as well as the respective loops associated with these variables. Note that the buffer tank loops have three MVs associated with the override controller. For example, for buffer tank TP_1, the loops are LIC03, LIC06 and LIC09. These loops are associated with the storage tank (u_5), the minimum (u_6) and the maximum (u_7) restrictions of the level of this tank, respectively. More details about the dewatering control system loops are presented below.

Table 2. Loops of the dewatering system.

Loop	Manipulated Variables		Controlled Variables	
	Variable	Name	Variable	Name
FIC02	u_1	BP-001 frequency	y_1	Thickened flow
DIC01	u_2	SP thickener outlet flow	y_2	Underflow density
DIC02	u_3		y_2	
LIC02	u_4		y_3	Average level TPs
LIC03 LIC06 LIC09	u_5 u_6 u_7	BP-005 frequency	y_7 y_4 y_4	T_A level TP_1 level
LIC04 LIC07 LIC10	u_8 u_9 u_{10}	BP-006 frequency	y_7 y_5 y_5	T_A level TP_2 level
LIC05 LIC08 LIC11	u_{11} u_{12} u_{13}	BP-007 frequency	y_7 y_6 y_6	T_A level TP_3 level

4.2. The Proposed Control Using Cascade and Override Structures

The storage tank level (H_{TA}) is controlled by manipulating the flow rate through pumps BP-005, BP-006 and BP-007, as illustrated in Figure 5. In nontypical operating conditions, such as the shutdown of one export pump, BP-00K, there is an imbalance between the flows q_{FDK} and q_{TA}. In this way, the pumps BP-005, BP-006 and BP-007 decrease the flow rate q_{TA}, keeping the storage tank level H_{TA} close to the setpoint. Thus, with the flow through the pumps reduced, there is an imbalance between the flows q_{TA} and q_U, raising the levels of the buffer tanks TP_1, TP_2 and TP_3. In this case, the proposed control strategy uses maximum and minimum level restriction controllers to prevent the operating limits from being exceeded in each of the buffer tanks. According to Figure 5, for each buffer tank pump (BP-005, BP-006 and BP-007), there are three controllers whose outputs are connected by means of the selectors f_2, f_3 and f_4, respectively.

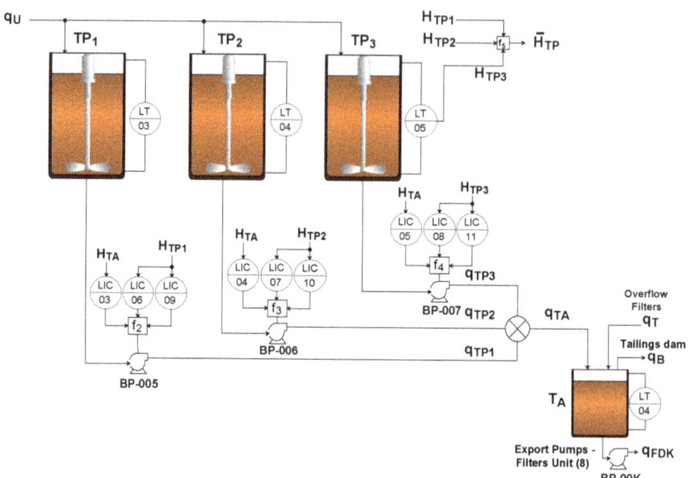

Figure 5. Proposed control strategy—pulp storage and transportation.

Algorithm 1 describes the operation of the selectors f_2, f_3 and f_4. To understand this algorithm, consider the control of pump BP-005: Under normal operation, the signal that controls the frequency of pump BP-005 is the output of controller LIC03 (u_5), which controls the H_{TA} level. If the buffer tank level, H_{TP1}, reaches a maximum or minimum threshold, LIC03 will no longer control H_{TA} and will be manipulated by controller LIC06 (u_6) or LIC09 (u_7) to keep the H_{TP1} level at its maximum or minimum value, respectively. A similar operation occurs for the BP-006 and BP-007 pump controllers. Table 3 summarizes the control loops and their respective variables of the buffer tanks.

Algorithm 1 Override selector—pumps BP-005/6/7.

```
if u_{7/10/13} > u_{5/8/11} then
    u_{selected} = u_{7/10/13};
else if u_{6/9/12} < u_{5/8/11} then
    u_{selected} = u_{6/9/12}
else
    u_{selected} = u_{5/8/11}
end if
```

Table 3. Control loops of the buffer tanks.

Loop	Description	Tank	MV	PV
LIC03	Storage tank level controllers	T_A	u_5	y_7
LIC04			u_8	y_7
LIC05			u_{11}	y_7
LIC06	Minimum level restriction controllers for each buffer tank	TP_1	u_6	y_4
LIC07		TP_2	u_9	y_5
LIC08		TP_3	u_{12}	y_6
LIC09	Maximum level restriction controllers for each buffer tank	TP_1	u_7	y_4
LIC10		TP_2	u_{10}	y_5
LIC11		TP_3	u_{13}	y_6

Note that the control structure described provides selectivity to the system since only the pumps connected to buffer tanks operating under normal conditions control the storage tank level. Note also that the level control of the buffer tanks is carried out using pumps BP-005, BP-006 and BP-007. These controllers are activated only when the operational limits of the buffer tanks are reached. In this way, under normal operating conditions, the proposed control strategy uses the flow rate of the underflow of the thickener as the MV to control the buffer tank levels. This strategy is illustrated in Figure 6, where LIC02 controls the average level of the buffer tanks, $\overline{H}_{TP}$. In addition, loops DIC01 and DIC02 are related to the minimum and maximum density restriction controllers, respectively.

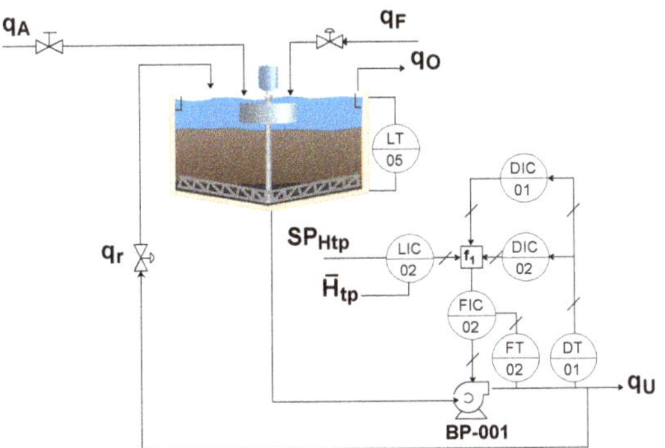

Figure 6. Proposed control strategy—thickening.

According to Figure 6, loops LIC02, DIC01, and DIC02 comprise an override control, which is the master loop of the cascade structure. Moreover, FIC02 is the slave loop of this cascade structure. The control action of the override control of loops LIC02, DIC01, and DIC02 is selected according to the selector f_1, which is described in Algorithm 2. In this algorithm, signals u_4 and u_3 represent the MV of the maximum (DIC02) and minimum (DIC01) density restriction controllers, respectively. These controllers have direct action, and signal u_4 tends to saturate at its minimum value, while signal u_3 tends to saturate at its maximum value. The signal of the medium level controller of the buffer tanks u_2 has a reverse action and is the predominant signal as long as all restrictions are met. The buffer tanks, despite being decoupled from each other, are kept at very similar levels, as the loop LIC02 controls the medium level of these tanks. Table 4 summarizes the control loops and their respective variables of the thickener.

Algorithm 2 Override selector—pump BP-001

if $u_4 > u_2$ **then**
    $u_{selected} = u_4$;
else if $u_3 < u_2$ **then**
    $u_{selected} = u_3$;
else
    $u_{selected} = u_2$
end if

Table 4. Control loops of the thickening.

Loop	Description	MV	B
LIC02	Buffer tank medium level controller (master)	u_2	y_3
DIC01	Minimum density restriction controller (master)	u_3	y_2
DIC02	Maximum density restriction controller (master)	u_4	y_2
FIC02	Underflow flow rate controller (slave)	u_1	y_1

Under normal operating conditions, the density value of the underflow stays within the operating limits; thus, signal u_4 (DIC02) saturates at the minimum value and u_3 (DIC01) saturates at the maximum value. At the same time, signal u_2 (LIC02) remains within these limits, defining the flow rate of pump BP-001. For example, according to the override selector presented in Algorithm 2, if the maximum density limit is reached, control signal u_4 increases, and when the control signal becomes larger than signal u_2, it becomes the signal effectively applied in FIC02.

Table 5 shows the PI controller parameters for all loops used in the proposed control strategy with the respective filter time constant (λ). All controllers were defined using the method proposed by [23]. Note that for density loops, the same PI controller is used for the minimum and maximum constraints of the override control. We use the same parameters for the controllers because the minimum and maximum density loops have similar dynamics and are independent of the operating point. The same is true for the buffer tank level and storage tank loops. Note also that for the proposed control strategy, the same parameters of the PI controllers of the current and nonintegrated strategy described in Section 2.2 (DIC01, FIC02, LIC03, LIC04 and LIC05) are used. As the control objective is the same, the same controller parameters are used in both strategies. Thus, the performance of the two control strategies presented are evaluated under the same controllers.

Table 5. PI parameters of the proposed control.

Loop	K_p	T_i	λ
FIC02	0.07	2.25	1.88
DIC01/02	−65.83	1382.30	30.00
LIC02	1372.60	86.00	15.00
LIC06-11	−29.55	9014.00	50.00
LIC03/04/05	24.24	66.00	10.00

5. Results and Discussion

In this section, the results obtained with the proposed control strategy through dynamic simulations are presented. The proposed control was implemented in Simulink®, and the simulation of the dewatering process was developed using the commercial simulator IDEAS® (Andritz Automation). All elements of the dewatering process under study were modeled using the principles of mass and energy conservation and population balance.

The framework used to evaluate the control strategy works via an OPC (open platform communication) connection between Simulink (OPC client) and the IDEAS dynamic process simulator. In this framework, the OPC works as a bridge between IDEAS and Simulink. The OPC client uses the OPC server to obtain data from or send commands to the simulator. Figure 7 illustrates the framework of the control system with the simulator.

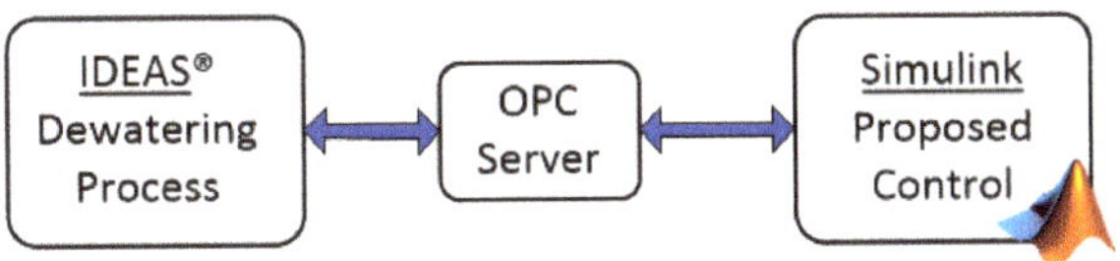

Figure 7. Schematic diagram of the simulation framework.

As presented in Section 3, data from 2520 h of operation of the filter units were analyzed and classified. According to Figure 3, for 92% of the time under failure, at most 2 filters were stopped. Due to its relevance, the scenario with the simultaneous failure of two filters was chosen to evaluate the proposed control. In addition, the scenario with disturbances in the flow rate and density of the thickener was evaluated.

5.1. Case 1: Two Filters Fail

At time $t = 0$ h, the stopping of two filters of the dewatering process is simulated. The aim is to analyze how long the plant could operate without a complete shutdown due to the overflow of the storage tank (T_A). Figure 8 shows the main process variables of the proposed and current control strategies. The MVs are illustrated in Figure 9. Because the outputs of the controllers are subject to selector override control, the variable u_s is used to highlight the signal actually selected and applied to the actuators. For the sake of simplicity, the loops for tanks TP_2 ($u_{8,9,10}$) and TP_3 ($u_{11,12,13}$) are omitted in all cases since they present similar characteristics to the loop of tank TP_1.

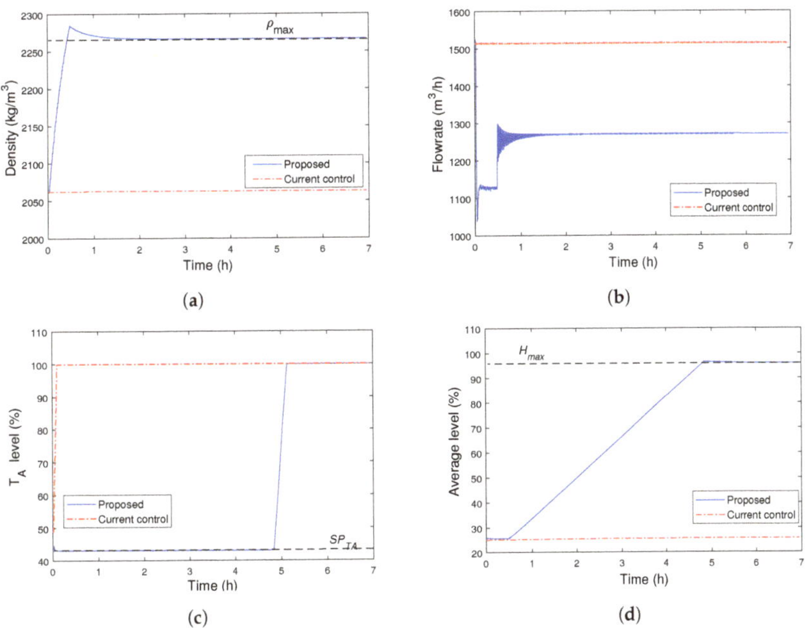

Figure 8. Process variables—filter failure case, (**a**) density of the underflow, (**b**) flow rate of the underflow, (**c**) level—storage tank, (**d**) average level—buffer tanks.

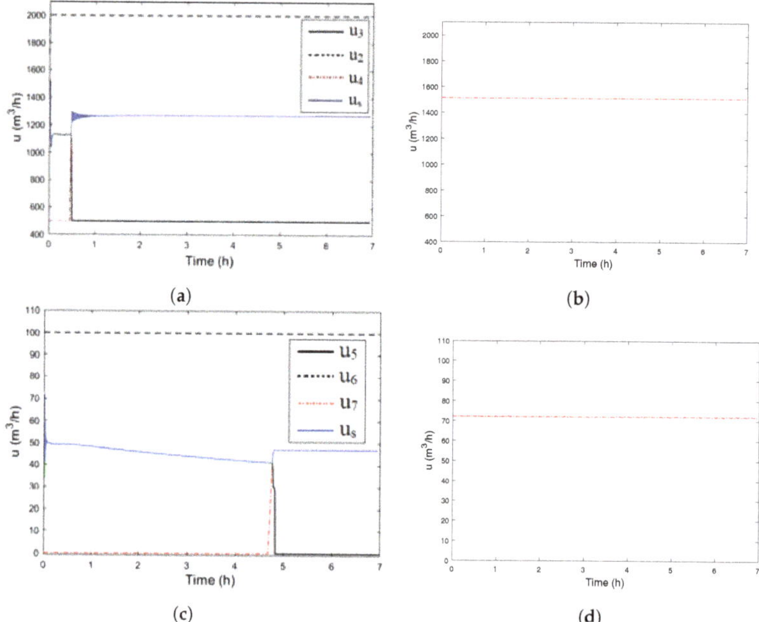

Figure 9. Manipulated variables—filter failure case, (**a**) setpoint of loop FIC02—proposed control, (**b**) Setpoint of loop FIC02—current control, (**c**) BP-005 frequency (TP_1)—proposed control, (**d**) BP-005 frequency (TP_1)—current control.

According to Figure 9c, under the failure of two filters, the LIC03/04/05 loop controllers of the proposed strategy quickly restrict the frequency of the pumps to keep the storage tank level at its setpoint. Because these pumps are also coupled to the buffer tanks, at the moment the pump frequency is reduced, the level of the buffer tanks tends to rise as shown in Figure 8d. In addition, the controller of loop LIC02 is tuned to maintain the average level of the buffer tanks at its setpoint; thus, the thickener underflow flow rate is quickly reduced, as shown in Figure 8b. Moreover, according to Figure 8a, with the reduction in the flow rate q_U, the underflow density starts to rise toward the upper limit of restriction. When this limit is reached at $t = 0.4$ h, the master controller becomes the maximum density restriction loop (DIC02), signal u_4. Thus, the DIC02 loop defines the setpoint for the underflow flow (slave) loop (FIC02), as shown in Figure 9a. In this case, the average level of the buffer tanks starts to rise.

When the level of the buffer tanks reaches the limit, at time $t = 4.7$ h, the maximum level restriction loops (LIC09/10/11) take control over the pumps (BP-005/6/7); see Figure 9c. At this moment, it can be seen that the density and level restrictions have been reached, and there is no MV capable of controlling the level of the storage tank, causing its level to rise until overflow occurs at time $t = 5.1$ h, as shown in Figure 9c.

Because the current control strategy consists of a nonintegrated operation, the storage tank level increases rapidly, as shown in Figure 8c. Note also that although the storage tank overflows, there is no action from the control system, as there is no integration to take into account the effects of unit operations in the other stages of the process. According to Figure 8a, there are no changes in the underflow density. However, because the overflow of storage tank T_A occurs due to the failure of the filters, tons of products are destined for the tailings dam. Thus, in this case and from the practical point of view, maintaining the underflow density at the setpoint means that ore pulp will be wasted, which directly impacts the plant's productivity. The nonalteration of the underflow flow rate and buffer tank levels, respectively, shown in Figure 8b,d, is also directly related to the waste of ore pulp into the tailings dam. Figure 9b shows the setpoint for the underflow flow (slave) loop

(FIC02) for current control. As expected, there is no change in this flow rate because the control strategy is not integrated. For the same reason, a similar result is obtained at the pump frequency of BP-005, shown in Figure 9d.

For the period of operation analyzed, Figure 10 shows a duration of filter failure corresponding to 92% of the time under the failure of up to two filters. Note that when two filters fail simultaneously, 87.1% of the stoppages last less than three hours and 95.5% of the failures last less than 5 h. Thus, the proposed control strategy is able to guarantee the continuity of operation in the majority of failure cases.

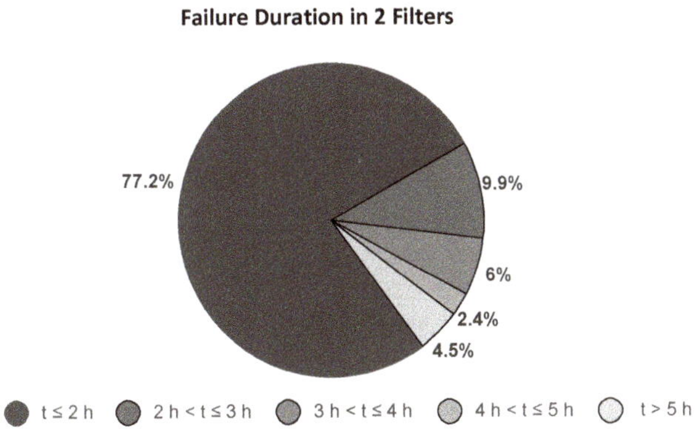

Figure 10. Failure time of two filters.

According to Figure 9d, in the current control strategy, the flow at the outlet of tank TP_1 is 72 m^3/h. The same is obtained for the other TP_2 and TP_3 tanks. Therefore, the total flow to the storage tank (T_A) is 216 m^3/h. Assuming the ore pulp density equal to 2267 kg/m^3 and that the proposed control strategy took 5 h more to overflow, under these operating conditions, it can be concluded that it is possible to avoid the waste of approximately 2448.36 tons of ore pulp that would be sent to the tailings dam.

5.2. Case 2: Disturbances

The proposed control strategy is now evaluated under process disturbances. Disturbances in the flow rate and density of the fed material are illustrated in Figure 11. At time $t = 0$ h, an increase of 10% in the thickener feed rate (q_A) is introduced into the system. At $t = 1$ h, the feed flow returns to the initial value, and a 10% decrease occurs at time $t = 2$ h. At $t = 3$ h, the feed flow returns to the initial value and remains there until the end of the simulation. At $t = 4$ h, the feed density (ρ_A) increases by 4% before returning to the initial value at $t = 5$ h.

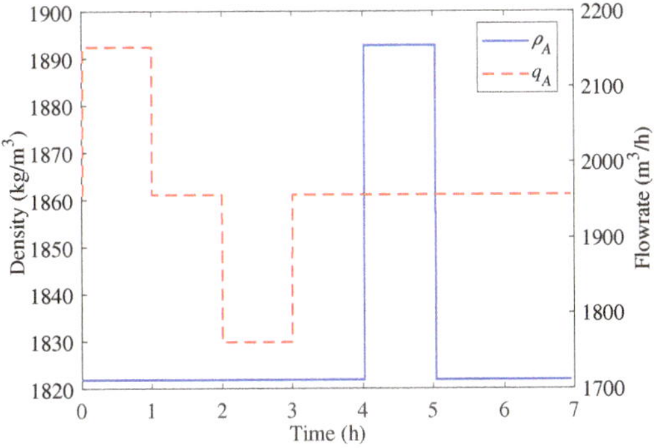

Figure 11. Thickener feeding disturbance.

Figure 12 shows the main process variables under the proposed control strategy and current control. The MVs are illustrated in Figure 13. The variable u_s is used to highlight the signal actually selected and applied to the actuators. Again, for the sake of simplicity, the signals of the loops for tanks TP_2 ($u_{8,9,10}$) and TP_3 ($u_{11,12,13}$) are omitted for the same reasons already discussed.

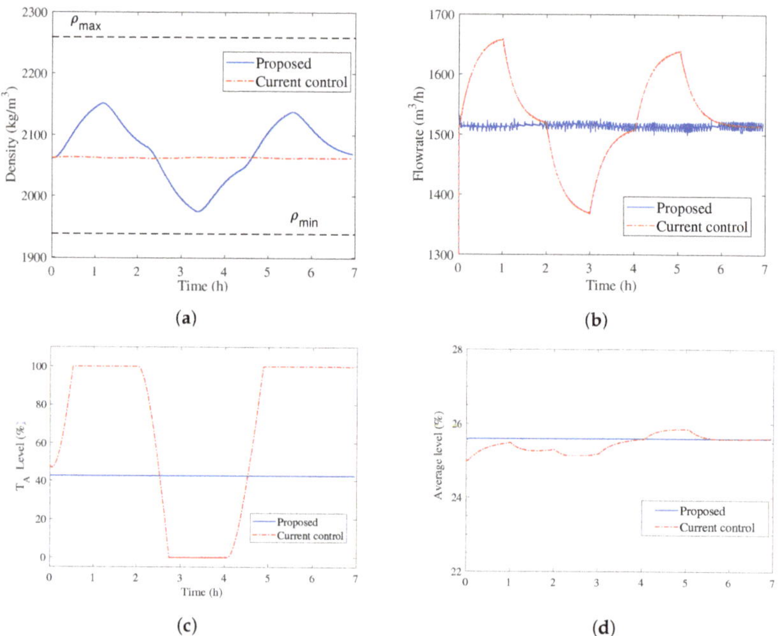

(c) (d)

Figure 12. Process variables—thickener feeding disturbance, (**a**) density of the underflow, (**b**) flow rate of the underflow, (**c**) level—storage tank, (**d**) average level—buffer tanks.

Variations in the flow rate and density of the thickener cause disturbances in the density of the underflow. Because the density of the underflow is a constraint variable, for the proposed control strategy, the controllers of loops DIC01 and DIC02 are not used at any time, as the maximum and minimum density restrictions have not been reached (see Figure 12a). The maximum and minimum density limits are defined by specialists in the plant operation.

The underflow density presents an oscillatory characteristic due to the integrated control strategy. Even so, according to Figure 12c, storage tank T_A does not overflow due to the disturbance. However, for the current and nonintegrated control, the storage tank level overflows according to the variation in the feed flow rate. According to Figure 12b, the underflow flow rate does not change under the proposed control strategy for any disturbances. For the current and nonintegrated control, it is observed that the flow rate varies by approximately 150 m^3/h, which causes variations in the average level of the buffer tanks, as shown in Figure 12d.

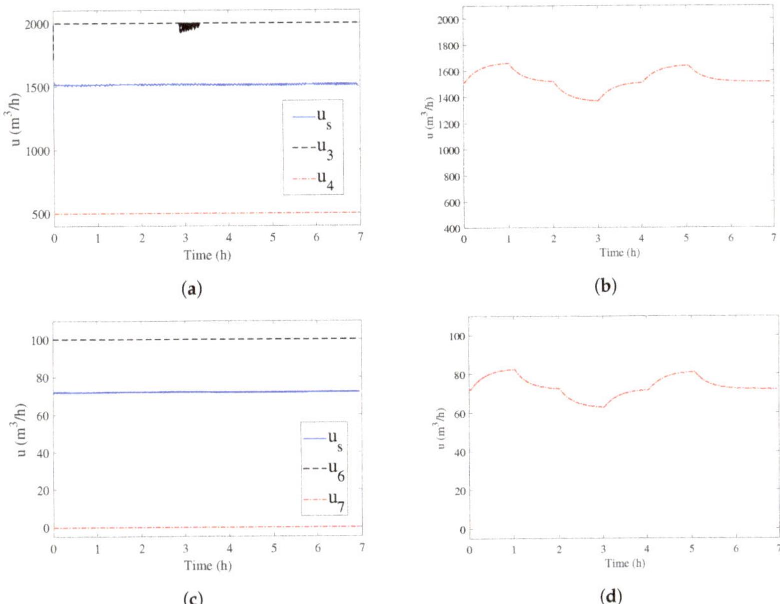

Figure 13. Manipulated variables—thickener feeding disturbance, (**a**) setpoint of loop FIC02—proposed control, (**b**) setpoint of loop FIC02—current control, (**c**) BP-005 frequency (TP_1)—proposed control, (**d**) BP-005 frequency (TP_1)—current control.

The MVs for the process under disturbance are shown in Figure 13. Note that for the proposed control strategy, there are no significant variations in the frequency of pumps BP-001 and BP-005. On the other hand, despite the variations in these variables, the current and nonintegrated strategies are not able to avoid storage tank overflow.

6. Conclusions

In this paper, a control strategy for an iron ore concentrate dewatering process was developed and evaluated by employing a dynamic model. The proposed strategy, based on ARC, could extend the operation period of the dewatering plant under disc filter failure. In the two-filter failure scenario, the time to storage tank overflow increases by approximately 5 h. This result indicates that the proposed control strategy has the potential to avoid material waste of approximately 2448.36 tons. In addition, the control strategy was also able to compensate for disturbances in the thickener feed rate and thickener feed density. Comparing the proposed strategy with the current and nonintegrated control strategies, it was observed that the proposed approach was able to control the average level of the buffer tanks by performing smoother manipulations in the underflow flow rate of the thickener. Smooth changes in the underflow flow rate allow for a more stable thickener operation, reducing the chance of low levels of the thicker layer and fewer variations in flocculant dosage. In addition, abrupt variations in density can impair the performance of filters that directly correlate to the density of the feed pulp. As demonstrated in the results, the proposed control strategy was able to reduce the amount of pulp that goes to the tailings dam. However, this control only postpones the overflow of the tanks, not completely eliminating material losses. This will only be achieved with physical changes to the process that require high investment. As future work, we intend to apply advanced control strategies based on artificial intelligence in order to anticipate the effect of filter failure. In addition, we will consider the effect of flocculant dosage on the thicker layer level.

Author Contributions: Conceptualization, Ê.L.J. and T.A.M.E.; methodology, Ê.L.J., M.T.d.S. and T.A.M.E.; software, Ê.L.J.; validation, Ê.L.J. and M.T.d.S.; formal analysis, M.T.d.S. and Ê.L.J.; investigation, E.L.J; resources, M.T.d.S.; data curation, M.T.d.S.; writing—review and editing, M.T.d.S., T.A.M.E.; visualization, M.T.d.S.; supervision, T.A.M.E.; project administration, T.A.M.E.; funding acquisition, T.A.M.E. All authors have read and agreed to the published version of the manuscript.

Funding: This research was funded by the Brazilian agencies CAPES (Finance Code 001) and CNPq (grants 444425/2018-7 and 306394/2021-9).

Institutional Review Board Statement: Not applicable.

Informed Consent Statement: Not applicable.

Data Availability Statement: Not applicable.

Acknowledgments: The authors are grateful for the support provided by the Vale S.A and the Instituto Tecnológico Vale.

Conflicts of Interest: The authors declare no conflict of interest.

References

1. Chaedir, B.A.; Kurnia, J.C.; Sasmito, A.P.; Mujumdar, A.S. Advances in dewatering and drying in mineral processing. *Dry. Technol.* **2021**, *39*, 1667–1684. [CrossRef]
2. Gupta, A.; Yan, D.S. *Mineral Processing Design and Operations: An Introduction*; Elsevier: Amsterdam, The Netherlands, 2016.
3. Tripathy, S.K.; Murthy, Y.R.; Farrokhpay, S.; Filippov, L.O. Design and analysis of dewatering circuits for a chromite processing plant tailing slurry. *Miner. Process. Extr. Metall. Rev.* **2021**, *42*, 102–114. [CrossRef]
4. Michaux, B.; Hannula, J.; Rudolph, M.; Reuter, M.; van den Boogaart, K.; Möckel, R.; Kobylin, P.; Hultgren, M.; Peltomäki, M.; Roine, A.; et al. Water-saving strategies in the mining industry–the potential of mineral processing simulators as a tool for their implementation. *J. Environ. Manag.* **2019**, *234*, 546–553. [CrossRef] [PubMed]
5. Diaz, P.; Salas, J.C.; Cipriano, A.; Núñez, F. Random forest model predictive control for paste thickening. *Miner. Eng.* **2021**, *163*, 106760. [CrossRef]
6. Bergh, L.; Ojeda, P.; Torres, L. Expert Control Tuning of an Industrial Thickener. *IFAC-PapersOnLine* **2015**, *48*, 86–91. [CrossRef]
7. Zhang, J.; Yin, X.; Liu, J. Economic MPC of deep cone thickeners in coal beneficiation. *Can. J. Chem. Eng.* **2016**, *94*, 498–505. [CrossRef]
8. Jia, R.; Zhang, B.; He, D.; Mao, Z.; Chu, F. Data-driven-based self-healing control of abnormal feeding conditions in thickening–dewatering process. *Miner. Eng.* **2020**, *146*, 106141. [CrossRef]
9. Magalhães, S.; Euzébio, T. Supervisory fuzzy controller for thickener underflow solids concentration on a simulated platform. In Proceedings of the 6th International Congress on Automation in Mining. GECAMIN, Santiago, Chile, 9–11 May 2018.
10. Sbarbaro, D.; Ortega, R. Averaging level control: An approach based on mass balance. *J. Process. Control.* **2007**, *17*, 621–629. [CrossRef]
11. Reyes-Lúa, A.; Backi, C.J.; Skogestad, S. Improved PI control for a surge tank satisfying level constraints. *IFAC-PapersOnLine* **2018**, *51*, 835–840. [CrossRef]
12. Chai, T.; Li, H.; Wang, H. An intelligent switching control for the intervals of concentration and flow-rate of underflow slurry in a mixed separation thickener. *IFAC Proc. Vol.* **2014**, *47*, 338–345. [CrossRef]
13. Zhao, D.; Chai, T.; Wang, H.; Fu, J. Hybrid intelligent control for regrinding process in hematite beneficiation. *Control. Eng. Pract.* **2014**, *22*, 217–230. [CrossRef]
14. Lopes, E.; Ferreira, A.E.; Moreira, V.S.; Euzebio, T.A.M. Fuzzy Fault Tolerant Controller Applied in an Iron Ore Concentrate Dewatering Plant. In Proceedings of the 6th Congress on Automation in Mining; Congress on Automation in Mining, GECAMIN, Santiago, Chile, 9–11 May 2018.
15. Olivier, L.E.; Craig, I.K. A survey on the degree of automation in the mineral processing industry. In Proceedings of the IEEE AFRICON, Cape Town, South Africa, 18–20 September 2017; pp. 404–409.
16. Spitzer, D.W. *Advanced Regulatory Control: Applications and Techniques*; Momentum Press: New York, NY, USA , 2009.
17. Muller, C.J.; Craig, I.K. Energy reduction for a dual circuit cooling water system using advanced regulatory control. *Appl. Energy* **2016**, *171*, 287–295. [CrossRef]
18. Wiid, A.; le Roux, J.; Craig, I. Pressure buffering control to reduce pollution and improve flow stability in industrial gas headers. *Control. Eng. Pract.* **2021**, *115*, 104904. [CrossRef]
19. Rajoria, A. Advanced process control for mineral processing operations. In *Innovative Exploration Methods for Minerals, Oil, Gas, and Groundwater for Sustainable Development*; Elsevier: Amsterdam, The Netherlands, 2022; pp. 357–369.
20. Sbárbaro, D.; Del Villar, R. *Advanced Control and Supervision of Mineral Processing Plants*; Springer: Berlin/Heidelberg, Germany, 2010.
21. Seborg, D.E.; Mellichamp, D.A.; Edgar, T.F.; Doyle III, F.J. *Process Dynamics and Control*; John Wiley & Sons: Hoboken, NJ, USA, 2010.

22. Euzébio, T.A.M.; Silva, M.T.D.; Yamashita, A.S. Decentralized PID Controller Tuning Based on Nonlinear Optimization to Minimize the Disturbance Effects in Coupled Loops. *IEEE Access* **2021**, *9*, 156857–156867. doi: 10.1109/ACCESS.2021.3127795. [CrossRef]
23. Skogestad, S. Simple analytic rules for model reduction and PID controller tuning. *J. Process. Control.* **2003**, *13*, 291–309. [CrossRef]

 sustainability

Article

Roof Fractures of Near-Vertical and Extremely Thick Coal Seams in Horizontally Grouped Top-Coal Drawing Method Based on the Theory of a Thin Plate

Guojun Zhang [1,2,3,*], Quansheng Li [1], Zhuhe Xu [1] and Yong Zhang [2,3,*]

1 State Key Laboratory of Water Resource Protection and Utilization in Coal Mining, National Institute of Clean and Low Carbon Energy, Beijing 102209, China
2 School of Energy and Mining Engineering, China University of Mining and Technology, Beijing 100083, China
3 Beijing Key Laboratory for Precise Mining of Intergrown Energy and Resources, China University of Mining and Technology, Beijing 100083, China
* Correspondence: 20039430@ceic.com (G.Z.); johnzy68@hotmail.com (Y.Z.)

Abstract: During the mining process of the near-vertical seam, there will be movement and collapse of the "roof side" rock layer and the "overlying coal seam," as well as the emergence of the "floor side" rock layer roof which is more complicated than the inclined and gently inclined coal seams, which causes problems with slippage or overturning damage. With the increase of the inclination of the coal seam, the impact of the destruction of the immediate roof on the stope and roadway gradually becomes prominent, while the impact of the destruction of the basic roof on the stope and the roadway gradually weakens. The destruction of the immediate roof of the near-vertical coal seam will cause a large area of coal and rock mass to suddenly rush to the working face and the two lanes, resulting in rapid deformation of the roadway, overturning of equipment, overturning of personnel, and even severe rock pressure disaster accidents, all of which pose a serious threat to coal mine safety and production. It is necessary to carry out research on the mechanical response mechanism of the immediate roof of near-upright coal seams, to analyse the weighting process of steeply inclined thick coal seam sub-level mining. A four fixed support plate model and top three clamped edges simply supported plate model for roof stress distribution are established before the first weighting of the roof during the upper and lower level mining process. The bottom three clamped edges simply supported plate model and two adjacent edges clamped on the edge of a simply supported plate model are established for roof stress distribution before periodic weighting of the roof during the upper and lower level mining process. The Galerkin method is used to make an approximate solution of deflection equation under the effect of sheet normal stress, and then roof failure criterion is established based on the maximum tensile stress strength criterion and generalized Hooke law. This paper utilizes FLAC^{3D} finite element numerical simulation software, considering the characteristics of steeply inclined thick coal seam sub-level mining. It undertakes orthogonal numerical simulation experiment in three levels with different depths, coal seam angles, lateral pressure coefficient, and orientation of maximum horizontal principal stress, and translates roof stress of corresponding 9 simulation experiment into steeply inclined roof normal stress. We conclude that the distribution law of normal stress along dip and dip direction of a roof under the circumstance of different advancing distances and different sub-levels. The caving pace of first weight and periodical weight were counted under the effect of the roof uniform normal stress. It can better predict the weighting situation of the working face and ensure the safe, efficient, and sustainable mining of coal mines.

Keywords: near-vertical coal seams; horizontally grouped top-coal drawing method; fracture modes; coordinate conversion; the maximum tensile stress; the first fracture span; periodic fractures span

Citation: Zhang, G.; Li, Q.; Xu, Z.; Zhang, Y. Roof Fractures of Near-Vertical and Extremely Thick Coal Seams in Horizontally Grouped Top-Coal Drawing Method Based on the Theory of a Thin Plate. *Sustainability* **2022**, *14*, 10285. https://doi.org/10.3390/su141610285

Academic Editor: Baoqing Li

Received: 3 June 2022
Accepted: 16 August 2022
Published: 18 August 2022

1. Introduction

With the gradual depletion of coal resources in mines in Eastern China and the contradiction between the coal resource exploitation and environment in the mines in Central China, the focus of coal resource exploitation was changed from Eastern China to Western China. Western China has a larger proportion of near-vertical (the seam inclination is greater than 45° in China) and extremely thick coal seams (the thickness of coal seams is greater than 8 m in China) [1,2], and the proportion continues to increase year by year. The coal seams with an inclination angle of 85° to 90° are near-vertical coal seams, which are widely found in Xinjiang, Anhui, Gansu, Inner Mongolia, Asturias in Spain, and the Lorraine coal areas in France [3]. The amplitude is bent, twisted, upright, or inverted, forming a nearly upright coal seam occurrence. The near-vertical coal seams are formed after many tectonic movements and evolutions, and their geological conditions and stress distributions are complex. The mining of a near-vertical coal seams forms a special coal rock structure, which is significantly different from other inclined coal seams.

The research on the mechanisms of deformation and fracture of a roof is divided into two approaches. In one approach, the goaf roof is simplified as an elastic beam, and the theory of the elastic beam is used to study the deformation and fracture mechanisms [4]. Qian [5] established voussoir beam theory based on the overlying strata and discussed the sliding-rotation stability condition of the voussoir beam structure. Song [6] analysed the basic top movement and breaking law of the horizontal sub-section top coal caving working face by using the rock slab theory and concluded that the pressure on the bottom side of the working face is lower than the pressure on the roof side and has a hysteresis phenomenon. Shi [7] analysed the difference of the coal caving process between the steeply inclined coal seam and the gently inclined coal seam and carried out an experimental study on the release performance of the broken top coal in the horizontal segmented top coal mining of the steeply inclined and extra-thick coal seam, appropriately increasing the segment height. Wang [8] studied the horizontal section height of steeply inclined fully mechanized caving face and listed the problems and solutions after reasonably raising the horizontal height. Cheng [9] used numerical calculation and similar simulation methods to compare the rock movement laws in the process of top coal caving mining with inclined layers and horizontal sections and believed that inclined layered mining could reduce the collapse of surrounding rock and the intensity of surface movement. Lai [10] obtained the range of pre-blasting and support by monitoring the loosening range of the top coal and comparing the stress-strain law of the top coal in the advanced pre-blasting. Dai [11] studied the rock movement mechanism of steeply inclined horizontal staged mining and established a prediction model of surface movement. It is believed that with the mining face being arranged layer by layer from top to bottom, the surface mobile basin has the value of expansion and subsidence to the roof side, as well as the characteristics of continuous accumulation. Cui [12] analysed the continuous disturbance effect of the sub-section caused by the horizontal section mining, and it is believed that increasing the thickness of the sub-section is conducive to weakening the impact on the lower sub-section. Shabanimashcool [13] presented two analytical approaches for studying voussoir beams by considering the horizontal loading condition of the beams. He [14] studied the elastic foundation coefficient, the span, and the stiffness of the main roof effect on the first fracture of the main roof with elastic foundation boundary. Guo [15] carried out effective support resistance and roof support technology of a fully mechanized mining face with hard roof conditions and thick coal seams. Yang [16] carried out evolution characteristics of overburden caving and void during multi-horizontal sectional mining in steeply inclined coal seams. He [17] carried out mechanism and prevention of rock burst in steeply inclined and extremely thick coal seams for fully mechanized top-coal caving mining. Su [18] conducted an experimental study on the rockburst and plate-pressing process of granite using a true triaxial test system and obtained the acoustic emission (AE) precursor characteristics of the instability of coarse-grained hard rock. Dong [19,20] confirmed that the reduction of isotropic components and the increase of double even numbers can serve as precursors for rock fracturing development and proposed that the anisotropic characteristics of wave velocity changes and AE event rates are useful supplements for identifying rock fracturing. Chen [21] carried out collapse behaviour and control of hard roofs in steeply inclined coal seams. Kong [22] carried out a stability analysis of a coal face based on coal face-support-roof system in a steeply inclined coal seam. He [23] carried out a true triaxial test system to conduct experiments on the rockburst and plate-pressing process of granite and obtained the mechanisms and precursors of slip and fracture of coal-rock parting-coal structure.

Although many experts and scholars have done a great deal of research on the mining technology and equipment of steeply inclined and extra-thick coal seams, and have achieved a series of results, there are few studies on the roof problems faced in the process of mining of steeply inclined and extra-thick coal seams. Mining steeply inclined and extra-thick coal seams safely and efficiently have become urgent problems to be solved in the development of the Chinese coal industry. Many scholars generally believe that steeply inclined and extra-thick coal seams have the following characteristics in the process of mining. Firstly, during the mining process of the steeply inclined coal seam working face, with the increase of the coal seam inclination angle, the sliding force of the surrounding rock along the tangential direction (coal seam inclination direction) increases, and the normal vertical stress decreases. Secondly, when the working face inclination angle exceeds the natural repose angle of a caving rock accumulation, the caving broken rock will slide down and roll along the floor of the working face, thus forming a migration law different from the near-horizontal and gently inclined coal seam. Thirdly, along the working face inclination, there is a large difference in the filling degree (the lower part is filled and compacted, the middle part is filled, the upper part is not filled), and the gangue migration law. Fourthly, along the inclined direction, the bearing pressure distribution shows asymmetrical features with less stress because the pressure along the middle and upper part of the working face is large, while the lower part of the working face is not filled. After the steeply inclined coal seam working face is advanced for a certain distance, the floor rock strata within a certain range will intensify and move to the mined space, resulting in an increase in the trend of deformation and damage of the floor rock mass. If it is not restrained when the floor is damaged, slips may occur and this slip damage area develops upward along the inclined direction. This causes large-scale instability of the floor rock mass, resulting in a higher likelihood of the destruction of the support system composed of the "roof-support-floor" of the working face.

At present, the research on the equipment, technology, and surrounding rock control of the horizontally grouped top-coal drawing method (HGTC) of steeply inclined and extra-thick coal seams in China is not mature enough. The steeply inclined and extra-thick coal seam has the characteristics of an area with a large amount of coal mined. The stress concentrates around the stope and the higher degree and the mine pressure appears severe. With the continuous increase of coal mining depth, the problems of mine pressure and floor are more prominent, mainly in the following aspects: the side pressure of the roadway is large and the maintenance of the mining roadway is difficult. In particular, the roof and floor are unstable in large areas, and a large area of coal and rock mass can suddenly rush to the working face and two roadways, which will cause rapid deformation of the roadway, overturning of equipment, overturning of personnel, and even serious ground pressure disaster accidents. This is can make coal mines unsafe because high-efficiency production brings serious threats, therefore, it is very necessary to study the causes, forms and depths of roof damage in steeply inclined coal seams.

2. Diagrams of the Logical Relation

In this work, the roof fracture regularity of the near-vertical and extremely thick coal seams with HGTC was studied using the theory of an elastic thin plate. Four mechanical models of the immediate roof were established based on the different mining conditions before the roof fractures with the immediate roof assumed to be in a state of elasticity.

The orthogonal experiment designs were used to analyse the normal stress distribution of the immediate roof. The immediate roof normal stress distributions of the FLAC^{3D} are substituted in the four thin plate models. By introducing the fracture criteria, the four models can be applied to simulate the fracture process of the immediate roof with HGTC. The diagram of the logical relation of the article analysis process is shown in Figure 1.

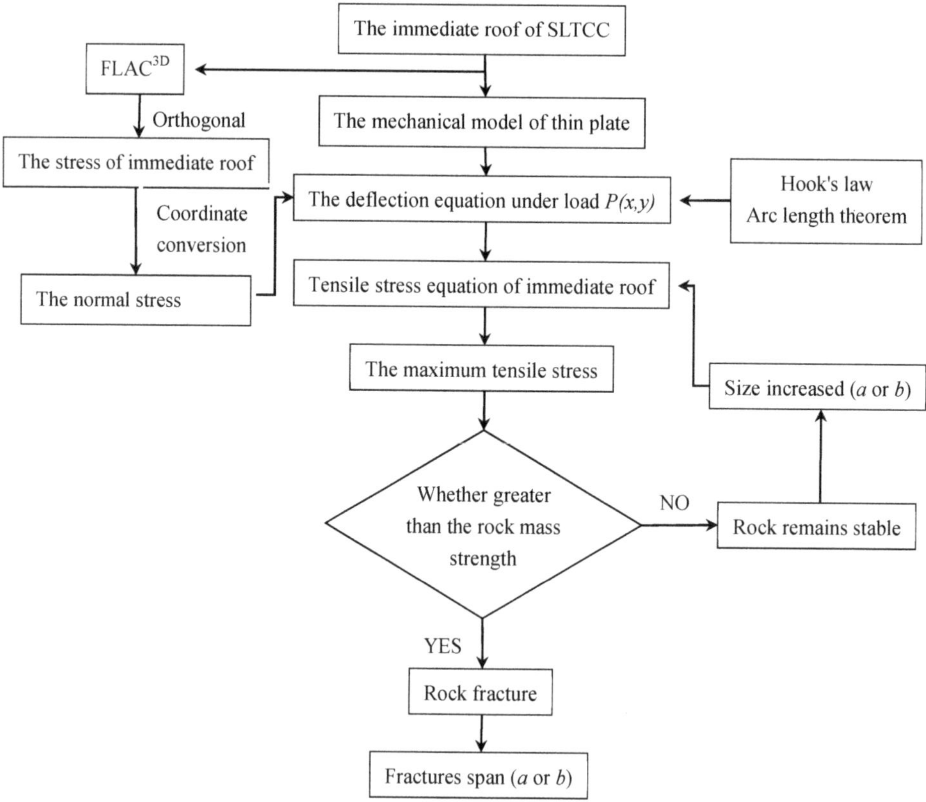

Figure 1. The diagrams of the logical relation.

3. Engineering Background

This work takes the Adaohai Mine as the research engineering background. Adaohai Mine is in the southern margin of the Daqingshan Coalfield of the Yinshan Mountains, 15 km away from Salazi Station and 7 km away from Beijing-Tibet highway in Inner Mongolia in China. The topography of the Adaohai Mine is very complicated, with steep cliffs, high mountains, deep valleys, and V-shaped gully development. The bedrock is exposed, the vegetation is very rare, and the coverage rate is low. The coal seams in the mining area are nearly vertical, and the ground collapse area formed in the goaf is small, only a narrow strip. However, with the gradual deepening of the mining depth, the range of ground collapse expands. More than 70% of the mining area is dominated by conglomerate and clastic rocks, with frequent phase changes, alternating fluvial facies and peat swamp facies, large changes in coal seam thickness, complex structure, and difficult to compare coal layers. The coal measure strata in the mining area include two coal seam groups, CU2 and CU4. The average dip angle of coal seams is 75°, and the maximum is 86°. The average thickness of the coal seam is 26 m. The height of subsection is 16 m, shown in Figure 2.

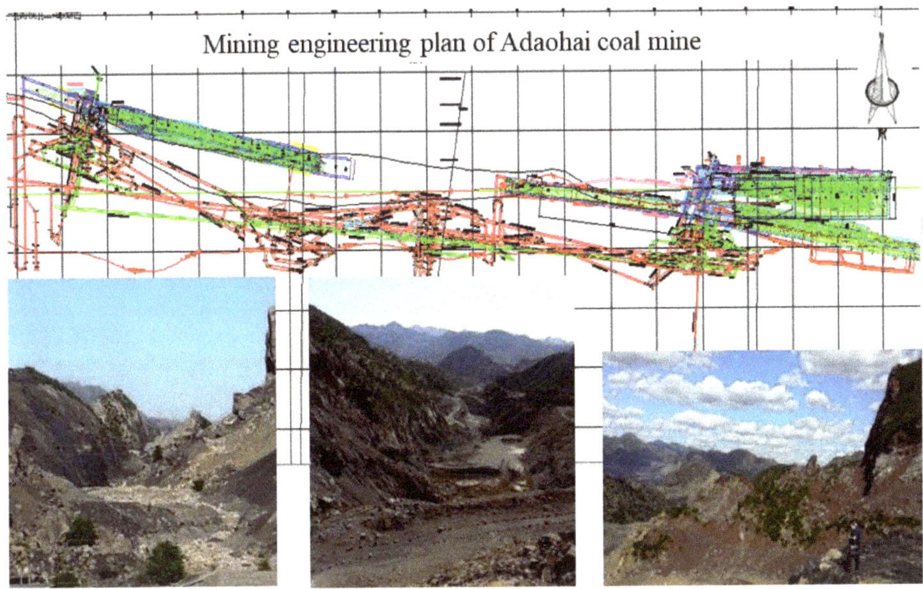

Figure 2. Mining pattern and surface subsidence of Adaohai coal mine in China.

The coal mining method is horizontally grouped top-coal drawing method mining, which developed from the flat slicing mining method. The difference between the two is that the height of each group for the former is several times that of each slice for the latter. The coal cutting method (mechanical cutting or blasting), similar to the method used in flat slicing, is accepted at the bottom of each group of coal, while the top coal caves by the aid of gravity. For a compact coal seam, which is hard to cave, a vibratory blast is used before drawing the top coal. This mining method possesses apparent advantages, like high monthly output for each working face and low divagate of roadways and so on, as shown in Figure 3.

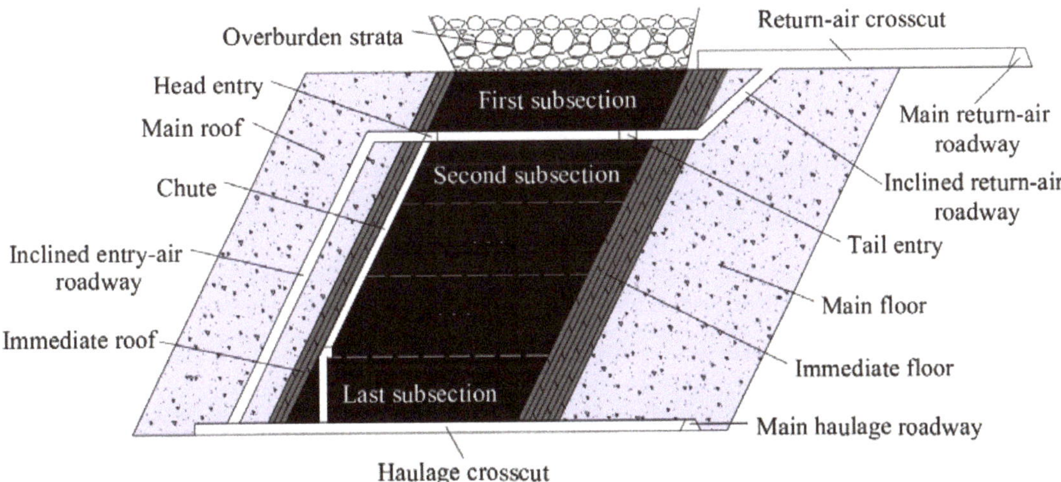

Figure 3. The diagrammatic sketch of HGTC.

The lithology and thickness of the roof and the bottom are shown in Table 1.

Table 1. The conditions of the roof and the bottom.

Roof and Bottom	Lithology	Thickness
Main roof	Conglomerate	25 m
Immediate roof	Sandstone	2.0 m
Floor	Kaolin	4.4 m
Basic bottom	Pebbly sandstone, Conglomerate	28 m

The ratio of the immediate roof thickness to the horizontal subsection height is 0.125. This ratio, less than 0.2 and more than 0.01, satisfies the assumptions of the thin plate theory [24]. Therefore, the thin plate mechanics models are introduced to analyse the stress distribution of the roof in various HGTC mining conditions.

4. Mechanical Model for the Roof of the HGTC Mining Face

4.1. Pressure Characteristics of the Near-Vertical Coal Seam Mining Process

The top of the inclined thick seam working face is a composite roof which is composed of rock stratums and coal seams. Because the strength of the coal seam is lower than the strength of the two sides of the rock, the destruction, caving, and release of the roof coal seam is preceded by roof rock, a large space is formed above the goaf. If the ratio of the goaf filling is low, the rock activity above the working face is more intense. Because the coal seam dip angle is large, the forces in the normal direction of the rock stratum (the interaction forces between rock stratums) have great variation after coal seam mining. The normal stress of the immediate roof adjacent to the exploitation space has an obvious change. The stress in inclination direction and trend direction of rock stratum varies small, and the rock layer above the top coal can form a relatively stable "articulated rock plate" and "pressure arch" structure. The space of the goaf will gradually enlarge with the enlargement of mining space, the "articulated rock plate" and "pressure arch" structure will gradually change from a steady state to an unstable state and then will be destroyed. The main fractures that form are bending break, rotation wreck, overall instability slipping, and horizontal movement and so on. The change of stress state in the immediate roof will directly influence the failure characteristics of the immediate roof.

The destruction of the rock layer is a combination of various forms of destruction in the working face mining of the inclined thick seam. The rock layers near the mining space (immediate roof and immediate floor) are more complicated, and the rock layers far away from the mining space (main roof and main floor) are relatively simple. The reason is that the stress field around the rock has become increasingly small with an increasing distance from the mining space. Therefore, the damage of the immediate roof will directly influence strata behaviour of the stope and laneway. Meanwhile, the working face pressure of the inclined thick seam is more complex and more intense than other conditions of occurrence.

4.2. The Thin Plate Mechanics Model in Different Mining Conditions

When the first subsection was excavated, the immediate roof experienced the first fracture and periodic fractures in sequence [25]. The four-edges clamped plate model (model A) was applied to calculate the stress distribution before the first fracture phase, as shown in Figure 4a. The three-edges clamped and forth-simple support plate model (model B) was applied to calculate the stress distribution during the periodic phase, as shown in Figure 4b.

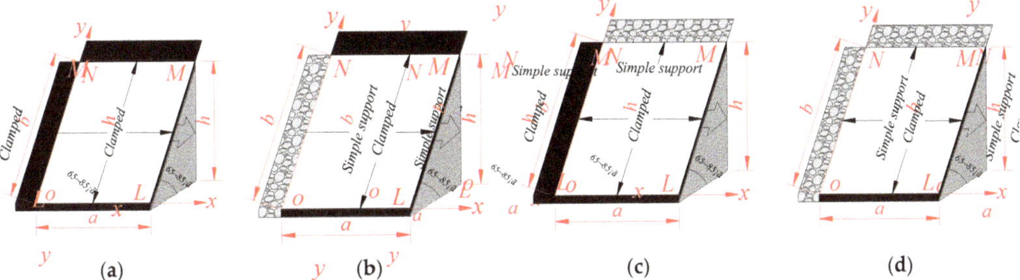

Figure 4. The mechanical model of the roof under different mining conditions (h is the subsection height, a is mining length, and b is the width of immediate roof). (**a**) model A; (**b**) model B; (**c**) model C; (**d**) model D.

When the first subsection was completed, mining began in the other subsections. The upper edge of roof, close to the goaf of upper subsection, was simplified as simple support. The immediate roof also experienced the first fracture and periodic fractures in turn. The first model is the bottom-three-clamped edges simply supported plate model (model C), and it was applied to calculate the stress distribution before the first fracture phase, as shown in Figure 4c. The other model has two adjacent edges clamped on the edge of a simply supported plate model (model D), and it was applied to calculate the stress distribution during the periodic phase, as shown in Figure 4d.

Model A and model C boundary conditions correspond to the first subsection and other subsection immediate roof first fracture event, respectively, and the first fracture pace of the immediate roof is equal to the length of "a". Model B and model D boundary conditions correspond to the first subsection and other subsection of the immediate roof, respectively, between immediate roof periodic fractures. The periodic fracture pace of the immediate roof is equal to the length of "a".

4.3. Disturbance Equation of the Mechanics Models

4.3.1. Galerkin Method

The bending problem of elastic thin plates can be solved directly and indirectly only in a few cases. The Galerkin method is another energy method based on an extreme value principle alongside a trigonometric series method [26].

The strain energy density of three-dimensional linear elastomer is [27]:

$$W = \frac{1}{2}(\sigma_{11}\varepsilon_{11} + \sigma_{22}\varepsilon_{22} + \sigma_{33}\varepsilon_{33} + \sigma_{12}\gamma_{12} + \sigma_{23}\gamma_{23} + \sigma_{31}\gamma_{31}) \tag{1}$$

where W is the strain energy density of linear elastomer, J; σ_{11}, σ_{22}, σ_{33}, σ_{12}, σ_{22}, σ_{31} is stress, Pa; ε_{11}, ε_{22}, ε_{33} is volume strain; γ_{12}, γ_{23}, γ_{31} is torsional strain.

For the small deflection bending of thin plates, according to the basic assumptions of thin plate theory ε_{33}, ε_{32}, ε_{31}, so its strain energy can be expressed as [27]

$$U = \iiint\limits_{v} W\mathrm{d}V = \frac{1}{2}\iiint\limits_{v} (\sigma_{11}\varepsilon_{11} + \sigma_{22}\varepsilon_{22})\mathrm{d}x_1\mathrm{d}x_2\mathrm{d}x_3 \tag{2}$$

where U is the strain energy of linear elastomer, J.

The relationship between the stress and strain of the thin plate and the deflection of the plate can be obtained from the elasticity and Hooke's Law [27]:

$$\left.\begin{array}{l}\varepsilon_{11} = -x_3\frac{\partial^2 w}{\partial x_1^2}, \sigma_{11} = -\frac{E}{1-v^2}x_3\left(\frac{\partial^2 w}{\partial x_1^2} + v\frac{\partial^2 w}{\partial x_2^2}\right) \\ \varepsilon_{22} = -x_3\frac{\partial^2 w}{\partial x_2^2}, \sigma_{22} = -\frac{E}{1-v^2}x_3\left(\frac{\partial^2 w}{\partial x_2^2} + v\frac{\partial^2 w}{\partial x_1^2}\right) \\ \gamma_{12} = -x_3\frac{\partial^2 w}{\partial x_1 \partial x_2}, \sigma_{22} = -\frac{E}{1+v}x_3\frac{\partial^2 w}{\partial x_1 \partial x_2}\end{array}\right\} \tag{3}$$

where, w is the deflection of elastic thin plate; v is Poisson's ratio; E is the elastic modulus, N/m^2.

Substituting Equation (3) into Equation (2), the strain energy expressed by deflection w after finishing is:

$$U = \frac{E}{2(1-v^2)}\iiint_v x_3^2\left(\frac{\partial^2 w}{\partial x_1^2} + \frac{\partial^2 w}{\partial x_2^2}\right)^2 - 2(1-v)\left[\frac{\partial^2 w}{\partial x_1^2}\frac{\partial^2 w}{\partial x_2^2} - \left(\frac{\partial^2 w}{\partial x_1 \partial x_2}\right)^2\right] dx_1 dx_2 dx_3 \tag{4}$$

Integrate Equation (4) along the direction of plate thickness h to obtain

$$U = \frac{D}{2}\iint_F x_3^2\left(\frac{\partial^2 w}{\partial x_1^2} + \frac{\partial^2 w}{\partial x_2^2}\right)^2 dx_1 dx_2 - D(1-v)\iint_F \left[\frac{\partial^2 w}{\partial x_1^2}\frac{\partial^2 w}{\partial x_2^2} - \left(\frac{\partial^2 w}{\partial x_1 \partial x_2}\right)^2\right] dx_1 dx_2 \tag{5}$$

where, F is the length and width range of the elastic thin plate; D is the bending stiffness of the elastic thin plate, $D = Eh^3/12(1 - v^2)$, N/m; h is elastic thin plates thickness, m.

For a rectangular plate with a fixed boundary, when $w = 0$ at the surrounding boundary, the second term integral on the right of Equation (5) is 0, so the strain energy of the rectangular plate with a surrounding boundary deflection of $w = 0$ is

$$U = \frac{D}{2}\iint_F x_3^2\left(\frac{\partial^2 w}{\partial x_1^2} + \frac{\partial^2 w}{\partial x_2^2}\right)^2 dx_1 dx_2 \tag{6}$$

If the plate is only subjected to the normal load $P(x,y)$, the external force potential energy is

$$V = -\iint_F P(x,y) w dx_1 dx_2 \tag{7}$$

where, V is the potential energy under the action of external force, J; $P(x,y)$ is the load distribution function of the elastic plate, Pa.

The combination of Equations (6) and (7) can obtain that the elastic potential energy of the whole plate is

$$\Pi = U + V = \frac{D}{2}\iint_F x_3^2\left(\frac{\partial^2 w}{\partial x_1^2} + \frac{\partial^2 w}{\partial x_2^2}\right)^2 dx_1 dx_2 - \iint_F p(x,y) w dx_1 dx_2 \tag{8}$$

where Π Is the elastic potential energy of elastic thin plate, J.

Potential energy of Equation (8) Π the extreme value w is the solution of the thin plate bending problem. When the boundary condition of the thin plate is complicated, it is difficult to directly solve the partial differential equation of the plate. The Galerkin method uses the energy method to solve the partial differential equation of the thin plate. In principle, the Galerkin method is the equivalent of applying the method of variation of parameters to a function space by converting the equation to a weak formulation. Typically, one then applies some constraints on the function space to characterize the space with a finite set of basic functions [26]. The Galerkin method provides a powerful numerical solution for differential equations. To calculate the bending thin plate with the Galerkin method, the deflection of thin plate can be written as

$$\omega = \sum_{m=1}^{\infty} \sum_{n=1}^{\infty} A_{mn} \varphi_{ij} \tag{9}$$

where φ_{ij} is the displacement function that meets all the displacement boundary conditions and stress boundary conditions. A_{mn} indicates undetermined constants that satisfy the following:

$$\iint_F \left[D\nabla^4 \omega - p(x_1, x_2) \right] \varphi_{ij} dx_1 dx_2 = 0 \tag{10}$$

The deflection equation of the four mechanical models is expressed in the space Cartesian coordinate system (SCCS), in which the *X*-axis is along the strike direction of the immediate roof, the *Y*-axis is along the inclined direction of the immediate roof, and the *Z*-axis is along the normal direction of the immediate roof.

4.3.2. Disturbance Equation of Model A

Model A is shown in Figure 4a using the SCCS. The *OL*, *LM*, *ON*, and *MN* are clamped by substance coal and rock. The boundary conditions of model A can be expressed as

$$\begin{cases} x = 0, x = a, \omega = 0, \frac{\partial \omega}{\partial x} = 0 \\ y = 0, y = b, \omega = 0, \frac{\partial \omega}{\partial y} = 0 \end{cases} \tag{11}$$

where ω is the deflection of thin plate, and a and b can be seen as constant.

According to the theory of elasticity [27,28] and the theory of plates and shells [29,30], the deflection of the thin plate of model A should satisfy

$$\frac{\partial^4 \omega}{\partial x^4} + 2\frac{\partial^4 \omega}{\partial x^2 \partial x^2} + \frac{\partial^4 \omega}{\partial y^4} = \frac{p(x, y)}{D} \tag{12}$$

where $p(x, y)$ is the normal stress loading on the thin plate, N.

According to the Galerkin method and the boundary conditions of the four-edges clamped plate, the deflection equation can be given by

$$\omega_1 = \sum_{m=1}^{\infty} \sum_{n=1}^{\infty} A_1 \left(1 - \cos\frac{2m\pi x}{a}\right) \left(1 - \cos\frac{2m\pi y}{b}\right), \ (m, n \in \mathrm{N}^+) \tag{13}$$

where A_1 is the undetermined constant of the model A, and m and n are positive integers.

Deriving Equation (12), substituting it into Equation (13), and then substituting the result into Equation (10), the undetermined constants of the model A can be obtained as

$$A_1 = \frac{a^3 b^3}{4\pi^4 D} \cdot \frac{\int_0^a \int_0^b p(x, y) \left(1 - \cos\frac{2\pi x}{a}\right) \left(1 - \cos\frac{2\pi y}{b}\right) dx dy}{3a^4 + 2a^2 b^2 + 3b^4} \tag{14}$$

4.3.3. Disturbance Equation of Model B

Model B is shown in Figure 4b using the SCCS. The *OL*, *LM*, and *MN* are clamped by substance coal and rock, and the *ON* is simple support. The boundary conditions of model B can be expressed as

$$\begin{cases} x = 0, \omega = 0, \frac{\partial^2 \omega}{\partial x^2} = 0 \\ x = a, \omega = 0, \frac{\partial \omega}{\partial x} = 0 \\ y = 0, y = b, \omega = 0, \frac{\partial \omega}{\partial y} = 0 \end{cases} \tag{15}$$

According to the Galerkin method and the boundary conditions of model B, the deflection equation can be given by

$$\omega_2 = \sum_{m=1}^{\infty} \sum_{n=1}^{\infty} A_2 \left(1 - \cos \frac{2n\pi y}{b}\right) \sin^3 \frac{m\pi x}{a} \tag{16}$$

where A_2 is the undetermined constant of model B.

Deriving Equation (12) and substituting it into Equation (16), and then substituting the result into Equation (10), the undetermined constants of model B can be given by

$$A_2 = \frac{32a^3b^3}{\pi^4 D m^3 n^3} \cdot \frac{\int_0^a \int_0^b p(x,y)\left(1 - \cos \frac{2n\pi y}{b}\right) \sin^3 \frac{m\pi x}{a} dxdy}{135(a/m)^4 + 72(a/m)^2(b/n)^2 + 80(b/n)^4} \tag{17}$$

4.3.4. Disturbance Equation of Model C

Model C is shown in Figure 4c using the SCCS. The *ON*, *LM*, and *MN* are clamped by substance coal and rock, and the *OL* is simple support. The boundary conditions of model A can be expressed as

$$\begin{cases} y = 0, \omega = 0, \frac{\partial^2 \omega}{\partial y^2} = 0 \\ x = 0, x = a, \omega = 0, \frac{\partial \omega}{\partial x} = 0 \\ y = b, \omega = 0, \frac{\partial \omega}{\partial y} = 0 \end{cases} \tag{18}$$

According to the Galerkin method and the boundary conditions of model C, the deflection equation can be given by

$$\omega_3 = \sum_{m=1}^{\infty} \sum_{n=1}^{\infty} A_3 \left(1 - \cos \frac{2m\pi x}{a}\right) \sin^3 \frac{n\pi y}{b}, \ (\mathrm{m},\ \mathrm{n} \in N^+) \tag{19}$$

where A_3 is the undetermined constant of model C.

Deriving Equation (12) and substituting it into Equation (19), and then substituting the result into Equation (10), the undetermined constants of model C can be obtained as

$$A_3 = \frac{32a^3b^3}{\pi^4 D m^3 n^3} \cdot \frac{\int_0^a \int_0^b p(x,y)(1 - \cos \frac{2m\pi x}{a}) \sin^3 \frac{n\pi y}{b} dxdy}{135(a/m)^4 + 72(a/m)^2(b/n)^2 + 80(b/n)^4} \tag{20}$$

4.3.5. Disturbance Equation of Model D

Model D is shown in Figure 4d using the SCCS. The *LM* and *MN* are clamped by substance coal and rock, and the *OL* and *ON* are simple support. The boundary conditions of model A can be expressed as

$$\begin{cases} x = 0, \omega = 0, \frac{\partial^2 \omega}{\partial x^2} = 0; y = 0, \omega = 0, \frac{\partial^2 \omega}{\partial y^2} = 0 \\ x = a, \omega = 0, \frac{\partial \omega}{\partial x} = 0; y = b, \omega = 0, \frac{\partial \omega}{\partial y} = 0 \end{cases} \tag{21}$$

According to the Galerkin method and the boundary conditions of model D, the deflection equation can be given by

$$\omega_4 = \sum_{m=1}^{\infty} \sum_{n=1}^{\infty} A_4 \sin^3 \frac{m\pi x}{a} \sin^3 \frac{n\pi y}{b}, (\mathrm{m},\ \mathrm{n} \in N^+) \tag{22}$$

where A_4 is the undetermined constants of model D.

Deriving Equation (12) and substituting it into Equation (22), and then substituting the result into Equation (10), the undetermined constants of model D can be obtained as

$$A_4 = \frac{256a^3b^3}{9\pi^4 D m^3 n^3} \cdot \frac{\int_0^a \int_0^b p(x,y) \sin^3 \frac{m\pi x}{a} \sin^3 \frac{n\pi y}{b} dxdy}{25(a/m)^4 + 18(a/m)^2(b/n)^2 + 25(b/n)^4} \tag{23}$$

4.4. The Roof Fracture Criterion

When the elastic plate is deflected in the normal direction, the plate will be bent. In this case, the surface of the plate can be simplified into a space curved surface. Because the edge of the sheet is fixed, the arc length in each direction of the plate is greater than the length of the two fixed edges. Because the length of the plate was increased, the stress state of the plate will be mainly tensile stress. Generally, a rock has a high compressive strength and a low tensile strength, and rock shear strength marks the middle of compressive strength and tensile strength. Therefore, the theory of maximal tension stress was used to determine the roof strata fracture.

$$\sigma_u = \sigma_1 \tag{24}$$

where σ_u is the ultimate tensile strength of rock, Pa; σ_1 is the maximum tension stress of the rock, Pa.

In this part, only the small deformations that occur in engineering structures will be considered. The small displacements of particles of a thin plate will usually be resolved into components dx and dy parallel to the coordinate axes x and y. These components are assumed to be very small quantities varying continuously over the volume of the body. Consider a small element dx and dy of a thin elastic plate (Figure 5). If the thin plate undergoes a deformation and dx and dy are the components of the displacement of the point 0, the displacement in the x-direction of adjacent point A on x-axis is ds_x. In the same manner, it can be shown that the displacement in the x-direction of adjacent point B on x-axis is ds_y. According to the principle of calculating the arc length, the ds_x and ds_y can be expressed as:

$$\begin{cases} ds_x = \sqrt{1+w'^2}dx = \sqrt{1+\left(\frac{\partial w}{\partial x}\right)^2}dx \\ ds_y = \sqrt{1+w'^2}dy = \sqrt{1+\left(\frac{\partial w}{\partial y}\right)^2}dy \end{cases} \tag{25}$$

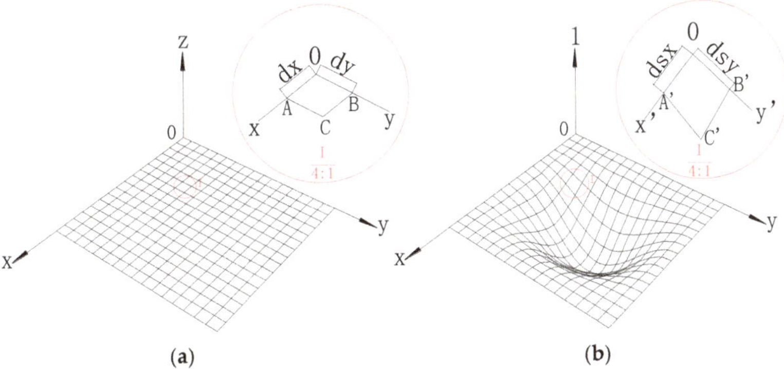

Figure 5. The deformation of the elastic thin plate, (**a**) without deformation; (**b**) small deformation.

The authors shall use the letter ε for unit elongation. To indicate the directions of strain, the same subscripts to these letters will be used as those for the stress components. The unit elongation of the x-direction and the y-direction can be expressed as

$$\begin{cases} \varepsilon_x = \frac{ds_x - dx}{dx} = \sqrt{1 + \left(\frac{\partial w}{\partial x}\right)^2} - 1 \\ \varepsilon_y = \frac{ds_y - dy}{dy} = \sqrt{1 + \left(\frac{\partial w}{\partial y}\right)^2} - 1 \end{cases} \tag{26}$$

Based on the generalized Hooke's law and the theory of elastic thin plate, the normal strain of the thin plate can be ignored, i.e., the values of ε_z, τ_{xy} and τ_{yz} are zero; thus, the stresses of the x-direction and the y-direction can be expressed as

$$\begin{cases} \sigma_x = \frac{E}{1-v^2}\left(\varepsilon_x + v\varepsilon_y\right) \\ \sigma_y = \frac{E}{1-v^2}\left(\varepsilon_y + v\varepsilon_x\right) \\ \tau_{xy} = \frac{E}{2(1+v)}\gamma_{xy} \end{cases} \tag{27}$$

Substituting Equation (26) into Equation (27), the relationship between the deflection of the thin plate and the stress at the directions of the x-axis and the y-axis can be expressed as

$$\begin{cases} \sigma_x = \frac{E}{1-v^2}\left[\left(\sqrt{1 + \left(\frac{\partial w}{\partial x}\right)^2} - 1\right) + v\left(\sqrt{1 + \left(\frac{\partial w}{\partial y}\right)^2} - 1\right)\right] \\ \sigma_y = \frac{E}{1-v^2}\left[\left(\sqrt{1 + \left(\frac{\partial w}{\partial y}\right)^2} - 1\right) + v\left(\sqrt{1 + \left(\frac{\partial w}{\partial x}\right)^2} - 1\right)\right] \\ \tau_{xy} = \frac{E}{2(1+v)}\gamma_{xy} \end{cases} \tag{28}$$

Based on the theory of maximal tension stress, substituting the deflection equation into Equation (28), and solving for the values of σ_x and σ_y, the immediate roof fracture criterion can be expressed as the following:

If the max $[\sigma_x, \sigma_y] < \sigma_u$, then the immediate roof is intact.

If the max $[\sigma_x, \sigma_y] = \sigma_u$, then the immediate roof is the critical level of the fracture and is intact.

If the max $[\sigma_x, \sigma_y] > \sigma_u$, then the immediate roof is fractured. The value of letter 'a' or letter 'b' is the fracture span of immediate roof. The high of sub-level is a constant, therefore the value of letter 'a' can be seen as the first or periodic fracture span of immediate roof.

5. The Stress Situation of the Roof through Numerical Simulation Analysis

5.1. Numerical Simulation Model

The boundary conditions and stress conditions were both required when the four mechanics model defection equation was calculated. However, because the stress characteristic of immediate roof was difficult to obtain from field measurement tests, FLAC^{3D} was used to calculate the normal stress in the immediate roof. To simulate the normal stress distribution of the four proposed models, the hybrid boundary conditions are applied at the boundary of the model. To study the normal stress distribution of the immediate roof in HGTC, the models are created to simulate the excavation process of HGTC based on the geological condition of the Adaohai Coal Mine.

The immediate roof consists of two or more strata with different lithological characters. The stratums that have the similar mechanical properties are considered as one layer in the simulation model. The Mohr–Coulomb criterion was used to determine the failure of materials in the numerical simulation. Based on the lithological characteristics of Adaohai Coal Mine, the model was simplified to 13 coal or rock layers, as listed in Table 2.

Table 2. Primary mechanical parameters of the coal or rock stratum.

NO.	Lithology	Thick-ness/m	Density Kg/m^3	Bulk Modu-lus/GPa	Shear Modulus/GPa	Friction Angle/ Degree (°)	Cohesion/ MPa	Tensile Strength/GPa
1	Loose layer	20	2100	7.0	3.5	25°	5.5	1.6
2	Sandy mudstone	Variable	2600	8.1	6.0	36°	18.8	3.5
3	Mudstone	48	2470	2.6	2.0	38°	4.5	1.0
4	Sandy mudstone	24	2450	8.1	6.0	36°	18.8	3.5
5	Medium sandstone	16	2430	10.9	6.9	31°	39.5	5.1
6	Sandstone	8	2600	4.9	3.7	30°	27.2	6.1
7	Mudstone	2	2430	2.6	2.0	38°	4.5	1.0
8	Coal seam	28	1330	1.2	0.8	28°	4.2	0.9
9	Mudstone	4	2400	2.6	2.0	38°	4.5	1.0
10	Sandstone	12	2450	4.9	3.7	30°	27.2	6.1
11	Medium sandstone	16	2650	10.9	6.9	31°	39.5	5.1
12	Sandy mudstone	24	2500	8.1	6.0	36°	18.8	3.5
13	Coarse sandstone	48	2500	12.5	9.4	35°	35.6	3.5

The normal stress distribution of the immediate roof is affected by many factors. Many experiments are required to analyse these many factors. To simplify the simulation workload, the normal stress distribution of the immediate roof was analysed by the method of orthogonal experiment design [31–33]. Three levels with different depths, coal seam angles, lateral pressure coefficients, and maximum principal stress directions were considered in the orthogonal experiment design to better illustrate the stress distribution of immediate roof in HGTC, as listed in Table 3.

Table 3. The simulations of nine representative combinations, based on the orthogonal array L9 (34).

NO	Seam Depth	Coal Seams Angle	Lateral Pressure Coefficient	Maximum Principal Stress Direction	Length	Width	Height	Blocks Number	Grid Points Number
1	100 m	65°	1	0°	277 m	200 m	120 m	553,750	586,921
2	100 m	75°	1.25	45°	257 m	200 m	120 m	516,250	550,685
3	100 m	85°	1.5	90°	239 m	200 m	120 m	515,000	549,155
4	300 m	65°	1.5	45°	277 m	200 m	120 m	553,750	586,921
5	300 m	75°	1	90°	257 m	200 m	120 m	516,250	550,685
6	300 m	85°	1.25	0°	239 m	200 m	120 m	515,000	549,155
7	500 m	65°	1.25	90°	277 m	200 m	120 m	553,750	586,921
8	500 m	75°	1.5	0°	257 m	200 m	120 m	516,250	550,685
9	500 m	85°	1	45°	239 m	200 m	120 m	515,000	549,155

The variety of mining depths, lateral pressure coefficients and the maximum horizontal principal stress were simulated by applying the different boundary conditions to the simulation model.

A positive correlation was found between the vertical stress and buried depth [34], and different pressures were applied on top of model to simulate various mining depths. The pressure was calculated using equation $\sigma_H = \gamma H$, where σ_H is the pressure loading on the top boundary of the model, H is the buried depth, and γ is the bulk density of stratum (average value is 2.5 × 104 N/m^3).

Based on the Mohr–Coulumb criterion and the mechanics of materials [35], the maximum principal stress and its orientation can be calculated as

$$\begin{cases} \sigma_1 = \frac{1}{2}(\sigma_x + \sigma_y) + \frac{1}{2}\sqrt{(\sigma_x - \sigma_y)^2 + 4\tau_{xy}{}^2} \\ \sigma_2 = \frac{1}{2}(\sigma_x + \sigma_y) - \frac{1}{2}\sqrt{(\sigma_x - \sigma_y)^2 + 4\tau_{xy}{}^2} \\ \alpha_0 = \frac{1}{2}\arctan\left(\frac{-2\tau_{xy}}{\sigma_x - \sigma_y}\right) \end{cases} \tag{29}$$

where σ_1 and σ_2 are the maximum principal stress and minimum principal stress in the plane, Pa; σ_x and σ_y are the stress loading along the x-axis and y-axis, Pa; τ_{xy} is the shear stress in the plane, Pa; α_0 is the orientation of the maximum horizontal principal stress, degree.

The different maximum principal stress and its orientation angle were simulated using the different boundary conditions. The relationship between the maximum principal stress and the boundary stresses of the simulation model can be expressed as

$$\begin{cases} \sigma_x = \sigma_1 \cos^2 \alpha_0 + \sigma_2 \sin^2 \alpha_0 \\ \sigma_y = \sigma_1 \sin^2 \alpha_0 + \sigma_2 \cos^2 \alpha_0 \end{cases} \tag{30}$$

where σ_x and σ_y are the stress loading on the left and right boundary of model, Pa.

To determine the relationship between the "σ_1" and "σ_2", the statistics of six mines, a total of 12 positions, and the in situ stress parameters were counted. According to the field in situ stress test, the ratio of "σ_1" to "σ_2" was 1.74 to 1.96, with an average of 1.89, as listed in Table 4; thus, the ratio of "σ_1" and "σ_2" in the orthogonal experiment design is 1.89.

Table 4. The in situ stress of the field measurement case in some mines in China.

In-Situ Stress Test Position	Depth/m	σ_1/MPa	σ_2/MPa	σ_H/MPa	σ_1/σ_2
The rock cross-cut at mining level + 1126 at Adaohai Mine	367.5	18.03	10.9	9.25	1.95
The winch chamber at mining level + 1228 at Adaohai Mine	206.9	12.26	6.84	6.27	1.96
The No. 2-1022 tail entry at Ganhe Mine	461	16.18	8.38	11	1.93
The head entry of scope 2 at Ganhe Mine	529	14.78	7.92	12.77	1.87
The No. 2151 head entry at Tuanbai Mine	33	9.27	5.32	7.8	1.74
The No. 310 tail entry at Tuanbai Mine	405	12.37	6.7	9.67	1.80
The No. 10-1021 head entry at Huipodi Mine	367.9	9.32	4.99	8.77	1.87
The haulage roadway of east scope at Huipodi Mine	355.1	10.32	5.57	8.33	1.85
The central of 1051 lane yard at Pangpangta Mine	490	9.63	5.1	12.26	1.89
The No. 1092 tail entry at Pangpangta Mine	592	11.68	6.16	14.8	1.90
The main haulage roadway at Changping Mine	348.1	10.81	5.6	8.7	1.93
The first contact alley at Changping Mine	343.9	9.64	4.99	8.6	1.93

In this paper, the stress transferred method was used to simulate the real in situ stress conditions. The boundary conditions for different models are shown in Figure 6.

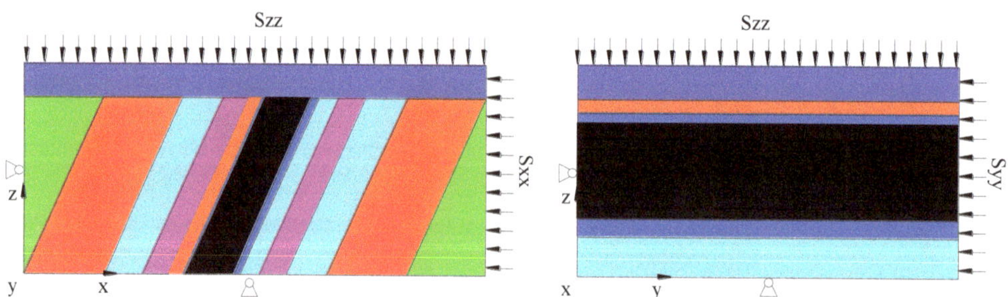

Figure 6. The schematic diagram of the boundary condition.

The detailed boundary conditions of the numerical simulation model are shown in Table 5.

Table 5. The boundary conditions of the numerical model.

NO.	The Bottom of Models	The Upper of Models (Mpa)	The Negative of x-Axis	The Direction of x-Axis (Mpa)	The Negative of y-Axis (Mpa)	The Direction of y-Axis (Mpa)
1	Fixed	0.25	Fixed	0.32	Fixed	0.18
2	Fixed	0.25	Fixed	0.31	Fixed	0.31
3	Fixed	0.25	Fixed	0.27	Fixed	0.48
4	Fixed	5.25	Fixed	7.88	Fixed	7.88
5	Fixed	5.25	Fixed	6.78	Fixed	3.72
6	Fixed	5.25	Fixed	8.47	Fixed	4.65
7	Fixed	10.25	Fixed	9.09	Fixed	16.54
8	Fixed	10.25	Fixed	19.85	Fixed	10.90
9	Fixed	10.25	Fixed	10.25	Fixed	10.25

The distance between the excavation zone and the boundary is 60 m to eliminate the influence of boundaries. The working face was excavated for 4 m each time, for a total of 80 m. The first subsection and second subsection were excavated in turn.

5.2. Coordinate Conversion

The stress data calculated by FLAC^{3D} could not be directly used in the proposed thin plate mechanics models because these two methods are based on different coordinate systems. Data conversion was required between these two coordinate systems. The stress information includes the vertical stress (σ_z), the horizontal stress (σ_x, σ_y), and the shear stress (τ_{xy}, τ_{xz}, τ_{yz}). According to the stress status at a point in different Cartesian coordinate system conversion relationships [36], the authors converted the stress data of FLAC^{3D} to data in the new Cartesian coordinate system, for which the x-axis is the working face mining distance, the y-axis is the inclined direction of immediate roof, and the z-axis is the normal direction of immediate roof. The new Cartesian coordinate system was obtained by rotating the numerical simulation Cartesian coordinate system by θ degrees along the y-axis, as shown in Figure 7.

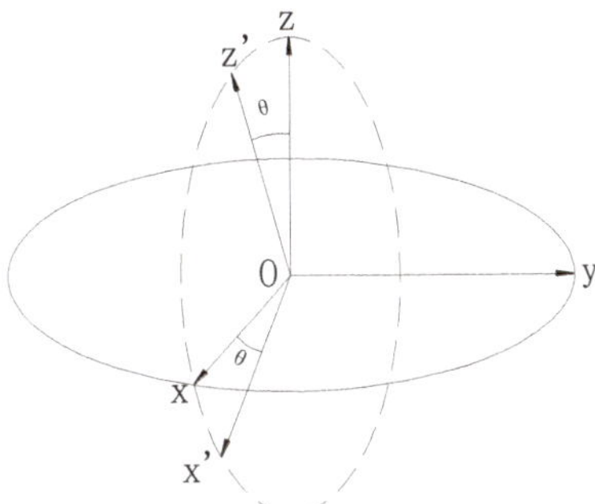

Figure 7. The diagram of the SCCS rotation.

The relationship between the different Cartesian coordinate systems can be expressed as

$$\begin{cases} \sigma_{x'} = \sigma_x \cos^2\theta + \sigma_z \sin^2\theta - \tau_{xz}\sin 2\theta \\ \sigma_{z'} = \sigma_x \sin^2\theta + \sigma_z \cos^2\theta + \tau_{xz}\sin 2\theta \\ \tau_{x'y'} = -\tau_{yz}\sin\theta + \tau_{xy}\cos\theta \\ \tau_{y'z'} = \tau_{yz}\cos\theta + \tau_{xy}\sin\theta \\ \tau_{x'z'} = (\sigma_x - \sigma_z)\sin\theta\cos\theta + \tau_{xz}\cos 2\theta \end{cases} \quad (31)$$

where $\sigma_{x'}$, $\sigma_{z'}$, $\tau_{x'y'}$, $\tau_{y'z'}$, and $\tau_{x'z'}$ are stress and strain in the new Cartesian coordinate system.

5.3. The Normal Stress Distribution of the Immediate Roof

Normal stress distribution, in which the z-axis is the normal stress direction, is given by Equation (31). The normal stress of the immediate roof along the direction of dip with different excavation conditions is shown in Figure 8.

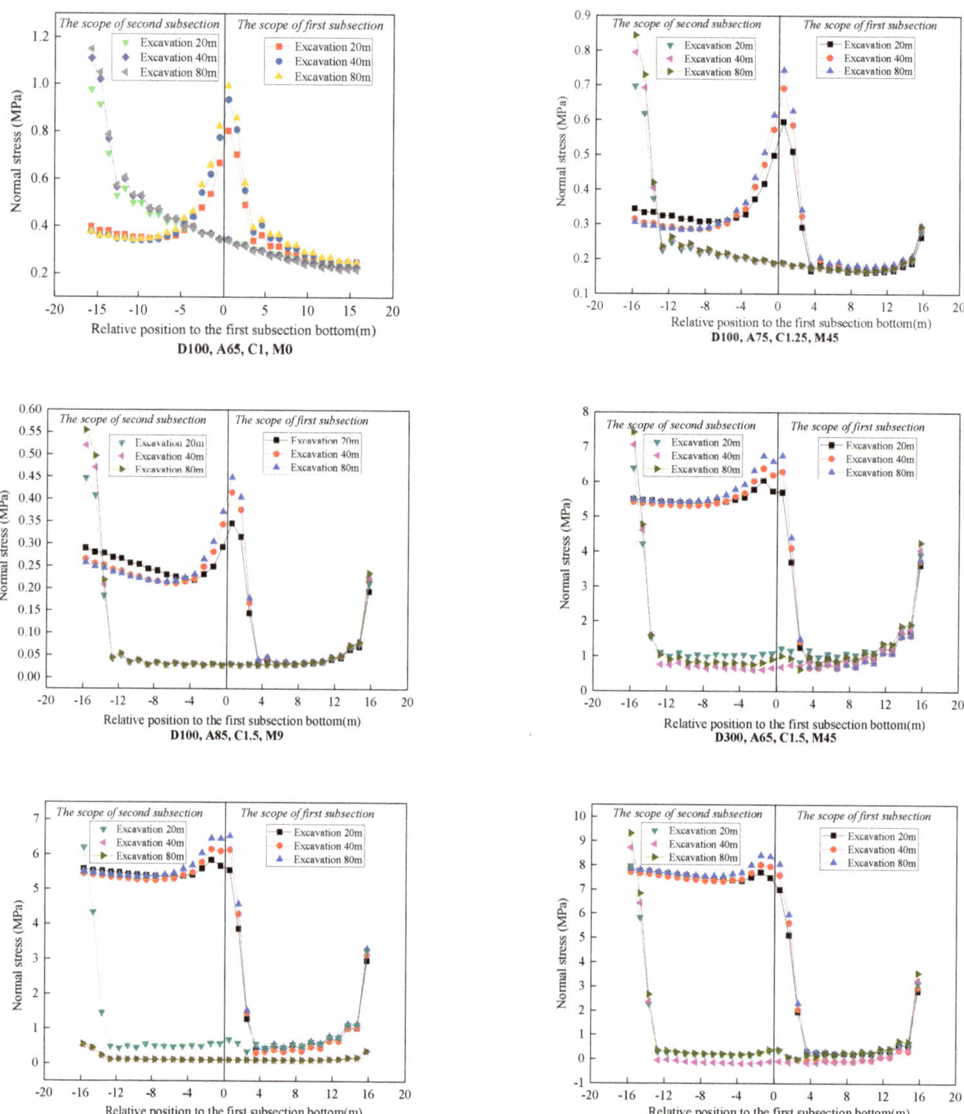

Figure 8. *Cont.*

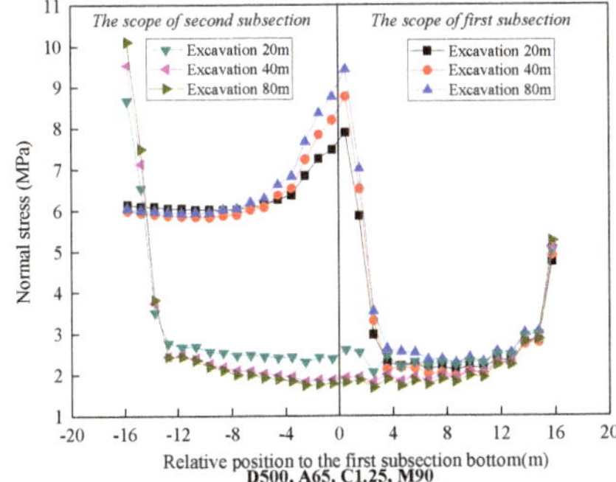

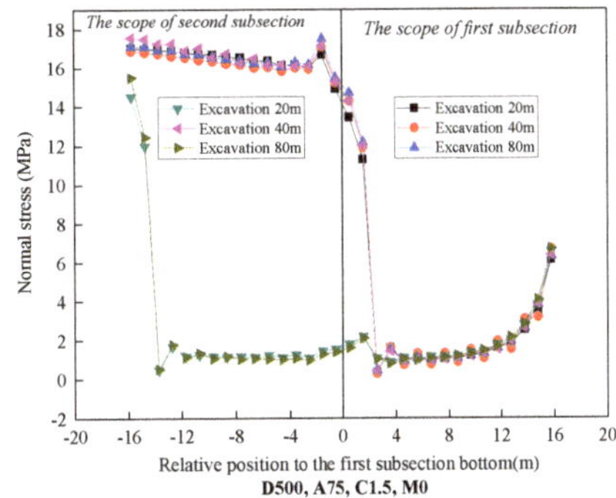

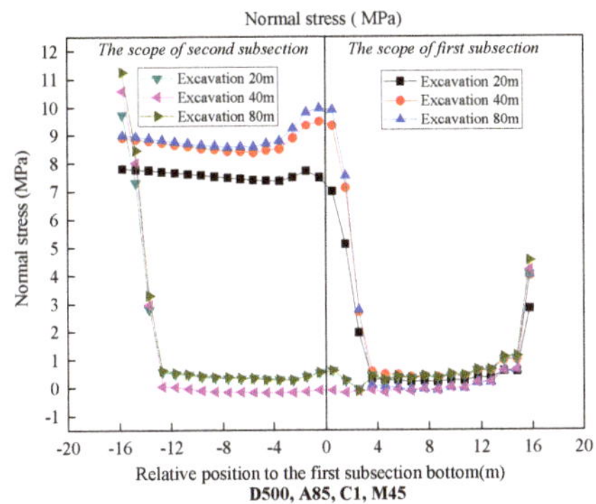

Figure 8. The normal stress distribution of the immediate roof in the direction of dip (D is the seam depth, A is the coal seam angle, C is the lateral pressure coefficient, and M is the maximum principal stress direction).

Figure 8 shows the normal stress distribution of the immediate roof when the working face was excavated 20 m, 40 m, and 80 m in the first subsection and the second subsection. The following normal stress regularity of the immediate roof can be drawn: the normal stress distribution of the first and second subsection immediate roof can be approximately considered as a uniform distribution; the normal stress is positively associated with the seam depth, the lateral pressure coefficient, and the maximum principal stress direction (the angle is between 0 and 90), whereas it is negatively associated with the coal seam angle. The influences in descending order of the four orthogonal factors on the simulation results in a linear distribution scope of immediate roof is as follows: the coal seam angle, the seam depth, the maximum principal stress direction, and the lateral pressure coefficient. The influence in descending order of the four orthogonal factors on the simulation results of the uniform distribution scope of immediate roof is as follows: the maximum principal stress direction, the coal seam angle, the seam depth, and the lateral pressure coefficient.

Considering the effect of four factors including the depths, coal seam angles, lateral pressure coefficients, and maximum principal stress directions on the normal stress distribution of immediate roof, although the above four factors have a significant difference in the normal stress, the distribution of normal stress is similar and can be regarded as uniform load.

5.4. The Fractures Span of the Immediate Roof in HGTC

Through the above study, the deflection equation of four thin models, the normal stress distribution of the immediate roof, and the failure criterion were obtained. Based on these results, the fractures span of the immediate roof in HGTC can be obtained.

In this part, the fractures span of model A is analysed and demonstrated. According to the results of the numerical simulation, the normal stress distribution can be simplified as uniform loading: $p(x,y) = p$. In addition, the deflection equation of model A under the uniform load can be expressed as

$$\omega_1 = A_1\left(1-\cos\frac{2\pi x}{a}\right)\left(1-\cos\frac{2\pi y}{b}\right) = \frac{a^4b^4p}{4\pi^4 D(3a^4+2a^2b^2+3b^4)}\left(1-\cos\frac{2\pi x}{a}\right)\left(1-\cos\frac{2\pi y}{b}\right) \tag{32}$$

Substituting Equation (32) into Equation (28), the tensile stress along the x-axis (the strike of immediate roof) and the y-axis (the dip of immediate roof) can be expressed as

$$\begin{cases} \sigma_{tx1} = \frac{E}{1-v^2}\left[\left(\sqrt{1+\frac{16A_1{}^2\pi^2}{a^2}\sin^4\frac{\pi y}{b}\sin^2\frac{2\pi x}{a}}-1\right)+v\left(\sqrt{1+\frac{16A_1{}^2\pi^2}{b^2}\sin^4\frac{\pi x}{a}\sin^2\frac{2\pi y}{b}}-1\right)\right] \\ \sigma_{ty1} = \frac{E}{1-v^2}\left[\left(\sqrt{1+\frac{16A_1{}^2\pi^2}{b^2}\sin^4\frac{\pi x}{a}\sin^2\frac{2\pi y}{b}}-1\right)+v\left(\sqrt{1+\frac{16A_1{}^2\pi^2}{a^2}\sin^4\frac{\pi y}{b}\sin^2\frac{2\pi x}{a}}-1\right)\right] \end{cases} \tag{33}$$

where σ_{tx1} and σ_{ty1} are the tensile stress of the x-axis and y-axis, Pa.

The maximum value of Equation (33) was calculated, where the corresponding point (x, y) denotes the position of the maximum tensile stress in the immediate roof before first fracture.

When the value of "x" was "$0.3524a$" or "$0.6476a$" and the value of "y" was "$\frac{b}{\pi}\arctan\frac{2va}{b}$" or "$b-\frac{b}{\pi}\arctan\frac{2va}{b}$", the maximum tensile stress ($\sigma_{tx1\max}$) in the direction of the x-axis can be obtained:

$$\sigma_{tx1\max} = \frac{E}{1-v}\left(\sqrt{1+\frac{256A^2\pi^2v^4a^2}{(4v^2a^2+b^2)^2}}-1\right) \tag{34}$$

where the value of "x" was "$\frac{a}{\pi}\arctan\frac{2vb}{a}$" or "$a-\frac{a}{\pi}\arctan\frac{2vb}{a}$", m; the value of "$y$" was "$0.3524b$" or "$0.6476b$ ",m; the maximum tensile stress ($\sigma_{ty\max}$) in the direction of the y-axis can be obtained by

$$\sigma_{ty1\max} = \frac{E}{1-v}\left(\sqrt{1+\frac{256A^2\pi^2v^4b^2}{(4v^2b^2+a^2)^2}}-1\right) \tag{35}$$

The symbol b represents the inclined extent of the first subsection and is a constant. When $\sigma_{tx\max}$ or $\sigma_{ty\max}$ reaches the ultimate tensile strength of immediate roof rock, the immediate roof is fractured. The symbol a is the first fracture span of the immediate roof in the first subsection.

Similarly, the periodic fractures span of the first subsection and the first and periodic fractures span of the second subsection can be obtained.

For Model B, when the value of "x" was "$\frac{a}{\pi}\arctan\frac{3b}{2av}$", m; the value of "$y$" was "$0.25b$" or "$0.75b$", m; the maximum tensile stress ($\sigma_{tx2\max}$) in the direction of x-axis can be obtained by

$$\sigma_{tx2\max} = \frac{E}{(1-v^2)}\left[\left(\sqrt{1+\frac{2916A_2{}^2b^4a^2v^2}{(9b^2+4a^2v^2)^3}}-1\right)+v\left(\sqrt{1+\frac{2916A_2{}^2b^6}{(9b^2+4a^2v^2)^3}}-1\right)\right] \tag{36}$$

In addition, when the value of "x" was "$\frac{a}{\pi}\arctan\frac{3bv}{2a}$, m; the value of "$y$" was "0.25$b$" or "0.75$b$", m; the maximum tensile stress ($\sigma_{ty2\max}$) in the direction of the y-axis can be obtained by

$$\sigma_{ty2\max} = \frac{E}{1-v^2}\left[\left(\sqrt{1+\frac{2916A_2{}^2b^6v^6}{(9b^2v^2+4a^2)^3}}-1\right)+v\left(\sqrt{1+\frac{2916A_2{}^2b^4a^2v^2}{(9b^2v^2+4a^2)^3}}-1\right)\right] \quad (37)$$

For Model C, when the value of "x" was "0.25a" or "0.75a, m; the value of "y" was "$\frac{b}{\pi}\arctan\frac{3av}{2b}$", m; the maximum tensile stress ($\sigma_{tx3\max}$) in the direction of the x-axis can be obtained by

$$\sigma_{tx3\max} = \frac{E}{1-v^2}\left[\left(\sqrt{1+\frac{2916A_2{}^2a^6v^6}{(9a^2v^2+4b^2)^3}}-1\right)+v\left(\sqrt{1+\frac{2916A_2{}^2a^4b^2v^2}{(9a^2v^2+4b^2)^3}}-1\right)\right] \quad (38)$$

In addition, when the value of "x" was "0.25a" or "0.75a", m; the value of "y" was "$\frac{b}{\pi}\arctan\frac{3a}{2bv}$", m; the maximum tensile stress ($\sigma_{ty3\max}$) in the direction of the y-axis can be obtained by

$$\sigma_{ty3\max} = \frac{E}{(1-v^2)}\left[\left(\sqrt{1+\frac{2916A_2{}^2a^4b^2v^2}{(9a^2+4b^2v^2)^3}}-1\right)+v\left(\sqrt{1+\frac{2916A_2{}^2a^6}{(9a^2+4b^2v^2)^3}}-1\right)\right] \quad (39)$$

For Model D, when the value of "x" was "0.9a", m; the value of "y" was "0.9b", m; the maximum tensile stress ($\sigma_{tx4\max}$ and $\sigma_{ty4\max}$) in the directions of the x-axis and the y-axis can be obtained by

$$\sigma_{tx4\max} = \frac{E}{1-\nu^2}\left\{\left[\sqrt{1+\frac{9A_4{}^2m^2\pi^2(9-3\sqrt{5})}{4a^2}}-1\right]+\nu\left(\sqrt{1+\frac{9A_4{}^2n^2\pi^2(9-3\sqrt{5})}{4b^2}}-1\right)\right\} \quad (40)$$

$$\sigma_{ty4\max} = \frac{E}{1-\nu^2}\left\{\left(\sqrt{1+\frac{9A_4{}^2n^2\pi^2(9-3\sqrt{5})}{4b^2}}-1\right)+\nu\left[\sqrt{1+\frac{9A_4{}^2m^2\pi^2(9-3\sqrt{5})}{4a^2}}-1\right]\right\} \quad (41)$$

In the actual engineering application process, the Poisson's ratio "v" of the immediate roof, the rock layer inclination "θ", the segmental height "$b\sin\theta$", the elastic modulus "E," and the normal stress of the immediate roof can be regarded as constants, so the difference between the advancing distance and the working face can be obtained. The maximum tensile stress "$\sigma_{tx\max}$" and "$\sigma_{ty\max}$" pull in different directions on the immediate roof. At the same time, the tensile strength "σ_u" and strength reduction coefficient "β" of the immediate roof can also be obtained through laboratory and field sonic tests. The maximum tensile stress of the immediate roof obtained by theoretical calculation and the maximum tensile stress of the rock obtained by the test are compared and analysed. According to the rock formation failure criterion established above, the stability of the immediate roof is judged. When the internal tensile stress of the immediate roof reaches the tensile strength of the rock mass, the tensile failure of the immediate roof will occur at the maximum tensile stress, and the internal cracks of the immediate roof will expand. The direction is perpendicular to the direction of tensile stress, and the corresponding advancing distance of the working face is the first or periodic failure span of the immediate roof rock stratum.

6. Discussion

From the perspective of rock depositional age, the rock layer above the coal seam and formed after the formation of the coal seam is generally called the roof. From the perspective of engineering practice, the rock layer above the coal seam is generally called the roof. The roof of the near-upright extra-thick coal seam is composed of three types of coal-rock masses: the "roof side" rock layer, the "overlying coal seam," and the "floor side" rock layer. The roof is no longer a single rock layer roof in the traditional sense. This is the essential difference between the near-vertical and extra-thick coal seam roof and the traditional gentle and inclined coal seam roof [37]. In the process of mining near-vertical and extra-thick coal seams, not only will the "roof side" rock layers and "overlying coal seams" move and collapse, but also the "bottom side" rock layers that are more complicated than inclined and gently inclined coal seams. There is a problem of slippage or overturning failure of the roof. With the increase of the inclination of the coal seam, the influence of the damage of the direct roof on the stope and the roadway is gradually highlighted, and the influence of the damage of the basic roof on the stope and the roadway is gradually weakened. The damage of the direct roof of the near-upright coal seam will cause a large area of coal and rock mass to suddenly rush to the working face and two roadways, resulting in rapid deformation of the roadway, overturning of equipment, overturning of personnel, and even serious ground pressure disaster accidents, which will threaten the safety of coal mines. Efficient production poses a serious threat. Therefore, it is of great significance to study the breaking position of the first (periodic) breaking of the direct roof and the evolution process of breaking during the mining of near-vertical coal seams.

In this work, the four-edge-clamped plate model and the top-three-clamped-edges simply supported plate model are introduced to calculate the deflection of the immediate roof before the immediate roof first fracture of the first subsection or the other subsections. The bottom-three-clamped-edges simply supported plate model and the two-adjacent-edges clamped on the edge of a simply supported plate model are introduced to calculate the deflection of the immediate roof before the immediate roof periodic fracture of the first subsection or the other subsections. The immediate roof is assumed to be an elastic thin plate before the first and periodic fracture, and the Galerkin method is used to calculate the solution of deflection equation under the effect of normal stress. By introducing the new fracture criteria, the maximum tensile stress strength criterion and generalized Hooke's law, the fracture processes of the immediate roof of these four models are analysed. To verify the normal stress distribution of the four thin plate models, the FLAC^{3D} numerical simulation software was used to simulate the direct top stress distribution characteristics of the horizontal segmented top coal caving in the near-upright extra-thick coal seam. According to the simulation results, the normal stress distribution of the immediate roof is uniform. Thus, the deflection equations can be calculated under the uniform normal stress. The most dangerous position of the immediate roof and the first or periodic fracture span can be obtained by using the thin plate model and the new fracture criteria. It can better predict the weighting situation of the working face, and ensure the safe, efficient, and sustainable mining of coal mines.

This paper has academic and practical significance for further research on the roof stability of the near-upright extra thick coal seam. However, some deficiencies need to be pointed out. Due to the limitations of experimental conditions, in the process of solving the thin plate mechanical model, only the influence of the stress on the normal line of the thin plate on the deflection of the plate is considered. The influence of the tangential stress on the deflection of the thin plate is ignored. The result has certain limitations. The number of coal mining faces that are near-vertical and have an extra-thick coal seam using the horizontally grouped top-coal drawing method is relatively small in China. This is not conducive to the development of on-site measurement work, and the theoretical model lacks the verification of on-site engineering practice, which needs more targeted research in the future.

7. Conclusions

The objective of this work was to investigate the roof fracture regularity of near-vertical and extremely thick coal seams with HGTC. The Galerkin method, the theory of maximal tension stress, and orthogonal experiment design were used to calculate the roof before the first and periodic fractures span. The following main conclusion can be drawn:

(1) Four mechanical models and the deflection equations for roof fractures in near-vertical and extremely thick coal seams have been established, which cover the possibility of stress on the roof when the roof is gradually damaged by a horizontally grouped top-coal drawing method in near-vertical and extremely thick coal seams.
(2) The influence of the seam depth, the coal seam angle, the lateral pressure coefficient, and the maximum principal stress direction on the normal stress distribution of the immediate roof with HGTC were analysed by orthogonal experiment design and FLAC^{3D}, and the normal stress could be simplified as uniform loading. The load could be set as a uniform load according to the numerical simulation results, when calculating the maximum tensile stress of the roof under the four states.
(3) A calculation method for roof stress distribution based on Hooke's law and arc length theorem is proposed. Taking the maximum tensile stress strength criterion as the near-vertical immediate roof fracture criterion, the first fracture and periodic fractures span could be obtained, which can better predict the weighting situation of the working face, and ensure the safe, efficient, and sustainable mining of coal mines.

Author Contributions: Conceptualization, G.Z. and Z.X.; methodology, G.Z.; investigation, G.Z. and Q.L.; writing-original draft, G.Z., Q.L. and Z.X.; formal analysis, G.Z., Z.X. and Y.Z.; writing—review & editing, G.Z. and Y.Z. All authors have read and agreed to the published version of the manuscript.

Funding: This research was funded by the state Key Research and Development Program of China (2018YFC0604501), the Science and Technology Innovation Project of China Energy Investment Corporation (SHGF-16-24), and the National Natural Science Foundation of China (51904303, 52004291).

Institutional Review Board Statement: Not applicable.

Informed Consent Statement: Not applicable.

Data Availability Statement: The data used to support the findings of this study are available from the corresponding author upon request.

Conflicts of Interest: The authors declare no conflict of interest.

References

1. Tu, H.; Tu, S.; Yuan, Y.; Wang, F.; Bai, Q. Present situation of fully mechanized mining technology for steeply inclined coal seams in China. *Arab. J. Geosci.* **2015**, *8*, 4485–4494. [CrossRef]
2. Yang, S.; Zhao, B.; Li, L. Coal wall failure mechanism of longwall working face with false dip in steep coal seam. *J. China Coal Soc.* **2019**, *44*, 367–376.
3. Qin, D.; Wang, X.; Zhang, D.; Guan, W.; Zhang, L.; Xu, M. Occurrence characteristic and mining technology of ultra-thick coal seam in Xinjiang, China. *Sustainability* **2019**, *11*, 6470. [CrossRef]
4. Pan, Y.; Wang, Z.; Liu, A. Analytic solutions of deflection, bending moment and energy change of tight roof of advanced working surface during initial fracturing. *Chin. J. Rock Mech. Eng.* **2012**, *31*, 32–41.
5. Qian, M.; Miao, X.; He, F. Analysis of key block in the structure of voussoir beam in longwall mining. *J. China Coal Soc.* **1994**, *19*, 557–563.
6. Song, Y. Discussion on the law of old top pressure in steeply inclined horizontal section top coal mining. *Coal Sci. Technol.* **1997**, *25*, 35–38.
7. Shi, P.; Qi, T.; Zhang, J.; Wang, S.; Tian, Y. Scientific analysis on sublevel caving in thinner of the steep close and thick seam. *Chin. J. Coal* **2004**, *29*, 385–387.
8. Wang, N. Discussion on Reasonably Raising the Horizontal Section Height of Steeply Inclined Fully-Mechanized Caving Face. *West. Prospect. Eng.* **2007**, *10*, 149–153.
9. Cheng, W.; Xie, J.; Wang, G.; Xie, P. Comparison of strata movement between section and slicing of top coal mining in steep and thick seams. *Chin. J. Coal* **2009**, *34*, 478–481.
10. Lai, X.; Qi, T.; Jiang, D.; Cui, F.; Ma, J.; Shan, P. Comprehensive determination of dimension of segment pre-blasting of sub-level top coal caving in steep seams. *Chin. J. Coal* **2011**, *36*, 718–721.
11. Dai, H.; Guo, J.; Yi, S.; Wang, G.; Liu, A.; Kong, B.; Zou, B. The mechanism of strata and surface movements induced by extra-thick steeply inclined coal seam applied horizontal slice mining. *Chin. J. Coal* **2013**, *39*, 1109–1115.
12. Cui, F.; Lai, X.; Cao, J.; Shan, P. Mining disturbance of horizontal section full-mechanized caving face in steeply inclined coal seam. *Chin. J. Min. Saf. Eng.* **2015**, *32*, 610–616.
13. Shabanimashcool, M.; Li, C.C. Analytical approaches for studying the stability of laminated roof strata. *Int. J. Rock Mech. Min. Sci.* **2015**, *79*, 99–108. [CrossRef]
14. He, F.; Chen, D.; Xie, S. The kDL effect on the first fracture of main roof with elastic foundation boundary. *Chin. J. Rock Mech. Eng.* **2017**, *36*, 1384–1399.

15. Guo, W.B.; Wang, H.S.; Dong, G.W.; Li, L.; Huang, Y.G. A case study of effective support working resistance and roof support technology in thick seam fully-mechanized face mining with hard roof conditions. *Sustainability* **2017**, *9*, 935. [CrossRef]
16. Yang, W.H.; Lai, X.P.; Cao, J.T.; Xu, H.C.; Fang, X.W. Study on evolution characteristics of overburden caving and void during multi-horizontal sectional mining in steeply inclined coal seams. *Therm. Sci.* **2020**, *24*, 3915–3921. [CrossRef]
17. He, S.; Song, D.; Li, Z.; He, X.; Chen, J.; Zhong, T.; Lou, Q. Mechanism and Prevention of Rockburst in Steeply Inclined and Extremely Thick Coal Seams for Fully Mechanized Top-Coal Caving Mining and Under Gob Filling Conditions. *Energies* **2020**, *13*, 1362. [CrossRef]
18. Su, G.S.; Gan, W.; Zhai, S.B.; Zhao, G.F. Acoustic emission precursors of static and dynamic instability for coarse-grained hard rock. *J. Cent. South Univ.* **2020**, *27*, 2883–2898. [CrossRef]
19. Dong, L.; Zhang, Y.; Ma, J. Micro-crack mechanism in the fracture evolution of saturated granite and enlightenment to the precursors of instability. *Sensors* **2020**, *20*, 4595. [CrossRef]
20. Dong, L.; Chen, Y.; Sun, D.; Zhang, Y. Implications for rock instability precursors and principal stress direction from rock acoustic experiments. *Int. J. Min. Sci. Technol.* **2021**, *31*, 789–798. [CrossRef]
21. Chen, D.; Sun, C.; Wang, L. Collapse behavior and control of hard roofs in steeply inclined coal seams. *Bull. Eng. Geol. Environ.* **2021**, *80*, 1489–1505. [CrossRef]
22. Kong, D.; Xiong, Y.; Cheng, Z.; Wang, N.; Wu, G.; Liu, Y. Stability analysis of coal face based on coal face-support-roof system in steeply inclined coal seam. *Geomech. Eng.* **2021**, *25*, 233–243.
23. He, Z.L.; Lu, C.P.; Zhang, X.F.; Guo, Y.; Wang, C.; Zhang, H.; Wang, B.Q. Research on Mechanisms and Precursors of Slip and Fracture of Coal–Rock Parting–Coal Structure. *Rock Mech. Rock Eng.* **2022**, *55*, 1343–1359. [CrossRef]
24. Kirchhoff, G. Uber das Gleichewicht and die Bewegung einer elastischen Scheibe. *J. Reine Angew. Math.* **1850**, *40*, 51–88.
25. Xie, G.X.; Chang, J.C.; Yang, K. Investigations into stress shell characteristics of surrounding rock in fullymechanized top-coal caving face. *Int. J. Rock Mech. Min. Sci.* **2009**, *46*, 172–181. [CrossRef]
26. Shamshiri, R.; Ismail, W.W. Implementation of Galerkin's method and modal analysis for unforced vibration response of a tractor suspension model. *Res. J. Appl. Sci. Eng. Technol.* **2014**, *7*, 49–55. [CrossRef]
27. Yang, G. *An Introduction to Elastic-Plastic Mechanics*; Tsinghua University Press: Beijing, China, 2004.
28. Timoshenko, S.; Goodier, J.N. *Theory of Elasticity*; McGraw-Hill Book Company, Inc.: New York, NY, USA, 1951.
29. Timoshenko, S.; Woinowsky-Krieger, S. *Theory of Plates and Shells*; McGraw-Hill Book Company, Inc.: New York, NY, USA, 1959; pp. 79–104.
30. Huang, K.; Xia, Z. *Theory of Plates and Shells*; Tsinghua University Press: Beijing, China, 1987; pp. 1–5.
31. Kharaghani, H. Arrays for orthogonal designs. *J. Comb. Des.* **2000**, *8*, 166–173. [CrossRef]
32. Wu, A.; Huang, M.; Han, B. Orthogonal design and numerical simulation of room and pillar configurations in fractured stopes. *J. Cent. South Univ.* **2014**, *21*, 3338–3344. [CrossRef]
33. Vieira, H., Jr.; Sanchez, S.; Kienitz, K.H.; Belderrain, M.C.N. Generating and improving orthogonal designs by using mixed integer programming. *Eur. J. Oper. Res.* **2011**, *215*, 629–638. [CrossRef]
34. Brady, B.H.G.; Brown, E.T. *Rock Mechanics: For Underground Mining*, 3rd ed.; Springer Science & Business Media: Beilin, Germany, 2006; pp. 142–161.
35. Sun, X.; Fang, X.; Guan, L. *Mechanics of Materials*, 5th ed.; Higher Education Press: Beijing, China, 2009; pp. 216–217.
36. Zhong, W. *New Solution System for Elasticity*; Dalian University of Technology Press: Dalian, China, 1995; pp. 7–9.
37. Zhang, G.; Zhang, Y. Immediate roof first fracture characteristics of suberect and extremely thick coal. *J. China Coal Soc.* **2018**, *43*, 1220–1229.

MDPI

Article

Evaluation of Compressive Geophysical Prospecting Method for the Identification of the Abandoned Goaf at the Tengzhou Section of China's Mu Shi Expressway

Shukun Zhang *, Peng Jiang, Lu Lu, Shuai Wang and Haohao Wang

School of City and Architecture Engineering, Zaozhuang University, Zaozhuang 277160, China
* Correspondence: zhangshukun@lntu.edu.cn or 4254423@163.com

Abstract: Subsidence deformation of abandoned goafs can induce cracking, distortion and even collapse of surface buildings (structures), and thus, subsidence deformation poses a great threat. Accurate detection of the abandoned goaf location and overburden morphology is an important prerequisite for stability evaluation and scientific management of surface buildings (structures), and effective detection methods are bottlenecks for accurate detection. Taking the abandoned goaf in the Tengzhou section of China's Mu Shi expressway as an engineering example, step-by-step detection, traditional detection and combination methods are used to determine the location of the underlying abandoned goaf and overburden morphology. First, we conduct disaster investigation on the expressway and surface within the affected area of the abandoned goaf and initially determine the detection area. Then, according to the principle that the detection range can be examined step-by-step from large to small, the high-density resistivity method is used for detection, and the high-resolution seismic method is further selected to analyze the target area. Then, based on the results of the resistivity method, the position of the abandoned goaf is evaluated with the high-resolution seismic method, and the distributions of the overburden subsidence, the water-filled fractured zone and the caving zone (the three belts) are determined. Finally, boreholes are drilled deep into the bottom of the abandoned goaf at specific locations and the distributions of the abandoned goaf and three belts are verified and corrected with drilling data, acoustic detection and borehole TV imaging technology, thereby providing accurate data on abandoned goafs for highway stability evaluation.

Keywords: abandoned goaf; expressway; progressive detection; step-by-step detection; exploration and combination; comprehensive detection

Citation: Zhang, S.; Jiang, P.; Lu, L.; Wang, S.; Wang, H. Evaluation of Compressive Geophysical Prospecting Method for the Identification of the Abandoned Goaf at the Tengzhou Section of China's Mu Shi Expressway. *Sustainability* **2022**, *14*, 13785. https://doi.org/10.3390/su142113785

Academic Editors: Longjun Dong, Yanlin Zhao and Wenxue Chen

Received: 30 August 2022
Accepted: 18 October 2022
Published: 24 October 2022

1. Introduction

After the mining of underground mineral resources, a large number of goafs will be left behind. These abandoned goafs can become deformed and degraded under the influence of internal and external factors, which will cause cracks, distortion and even collapse of surface structures [1–3]. With the development of China's economy, there are more roads, railways and large buildings, and the available land resources tend to be scarce. It is difficult to avoid construction work at sites with an underlying abandoned goaf [4]. To realize the construction of buildings (structures) at the site of an abandoned goaf, it is necessary to accurately evaluate and dispose of the abandoned goaf. There are already many successful cases [5–7]. However, the implementation of these successful cases must be based on accurately grasping the location of the abandoned goaf and overburden morphology [8–10]. Most abandoned goafs are often characterized by an advanced age and a lack of mining data. However, abandoned goafs are still subject to illegal prospecting and random mining [11,12]. It is difficult to meet the stability evaluation requirements of the building (structure) foundation by acquiring detailed information about the abandoned gob area only based on mining information collected by several parties [13,14]. The use of reasonable and efficient detection methods to achieve accurate detection of the abandoned goaf location and overburden morphology is particularly important.

At present, the detection methods for abandoned goafs mainly include electrical detection, electromagnetic induction detection, seismic detection and radioactive detection. The electrical detection and electromagnetic induction methods acquire electric distribution characteristics of the underground rock and soil by an electric field or magnetic field to determine the problem area and have the characteristics of low cost and high efficiency [3,12]. Seismic exploration mainly analyses the elastic wave transfer laws of the rock and soil to examine the characteristics of the underground rock and soil and has a high detection precision. Seismic exploration is commonly used in the determination of the goaf, especially the determination of the three belts [15–17]. Radioactive detection technology has high sensitivity and high precision and is more frequently applied to radioactive mining areas. Geophysical exploration methods have certain flaws. For example, the electrical detection and electromagnetic induction methods are more susceptible to external electromagnetic fields, and seismic exploration is less accurate at large depths [18]. Therefore, to improve the detection accuracy of abandoned goaf information, a variety of methods are often used for comprehensive detection to eliminate multiple solutions and overcome the limitations of various geophysical methods as much as possible, while large-area deep drilling is adopted in the verification stage. The use of multiple geophysical exploration methods can achieve the purpose of complementary geophysical techniques, but the approach also greatly increases the cost of geophysical exploration [19]. Even in this case, it is difficult to provide true verification data, and only densely spaced drilling verification with depth provides true verification data. When intense verification drilling is conducted, the process can also destroy the stable structure of the abandoned goaf, which can easily cause the abandoned goaf to become reactivated and induce secondary disasters. It is necessary to use a combination of fewer but reasonable geophysical techniques according to the situation of the abandoned goafs and supplementary effective verification methods to comprehensively determine the exact location and state of abandoned goafs.

At present, the detection of abandoned goafs mainly relies on combining as many geophysical methods as possible to improve the accuracy of the detection results, and more often, in-depth detection and superficial verification methods are implemented [20,21]. The objective of this paper is to take the Tengzhou Section of China's Mushiexpressway as an engineering example without detailed knowledge of the mining materials, and the step-by-step detection method is used to determine the location and filling status of the abandoned goaf that affects the expressway area. At the same time, the locations of the three belts of the overburden in the abandoned goaf are determined, which provides a scientific basis for stability evaluation of abandoned goafs and post-highway construction control.

2. Methodology

The highway with underlying abandoned goaf in the study area is located in the southeast direction of Tengzhou City, China. The terrain in this area is generally gentle, and there are local steep ridges. An expressway is a linear structure, and detection of the underlying abandoned goaf is mainly conducted along the direction of the expressway and the two sides of the adjacent roadbed [13]. Step-by-step detection is used to gradually reduce the detection target area, and a combination of several detection methods is used to determine the position of the abandoned goaf and the overburden state.

Preliminary detection areas are selected through investigation and analysis. In the absence of accurate mining data, relevant information is collected from abandoned mining areas by relevant departments, such as the Mining Bureau. At the same time, disaster investigation is carried out around the affected expressway. Specifically, disaster investigation includes examining the topography and landform characteristics, determining the distribution characteristics of surface cracks, and studying the spatial position relationship of the highway's subsidence-induced distortion. Using the high-density resistivity method to achieve a high detection efficiency along the line length, a long measurement line is laid in the preliminary selected area for basic coverage. The long-line measurement results are interpreted, and the detection area is further reduced according to the high-resistance characteristics of the abandoned goaf; short-line detection is then carried out. The high-density resistivity method is used for the final interpretation results, thereby providing detection targets for high-resolution seismic methods. The use of high-resolution seismic methods that are based on the arrival time and axis distribution characteristics of different geotechnical bodies can effectively determine the position and shape of the abandoned goaf in the detection target area and the distribution range of the three belts. Verification holes are drilled into the goaf at the maximum subsidence in the target area, and the position and shape of the goaf are determined using the high-resolution seismic method and the distribution range of the three belts are verified and corrected through drilling core data, sound wave detection data and borehole TV imaging data; the location and status of the abandoned goaf are thus accurately determined. The process is shown in Figure 1.

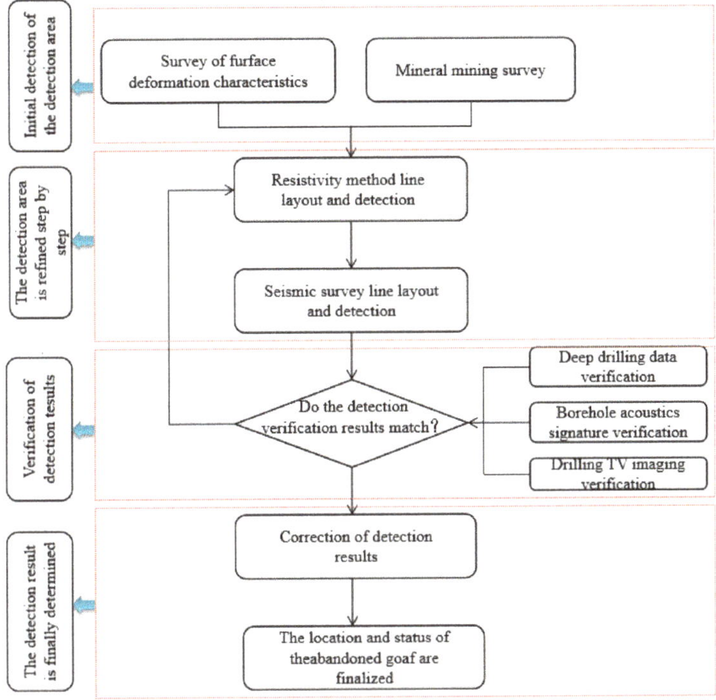

Figure 1. Abandoned goaf detection verification process.

The distribution range of the three belts is further corrected, and the acoustic detection of the borehole is carried out. The data processing of the borehole wave velocity test is mainly to calculate the wave velocity of rock mass. The borehole wave velocity test adopts a double receiving probe, and the rock wave velocity between the two receiving probes is [22]:

$$v = \frac{\Delta l}{t_1 - t_2} \tag{1}$$

In the above equation, Δl refers to the spacing between the two receiving probes, t_1 refers to the arrival time of the longitudinal wave at probe 1 and t_2 refers to the arrival time of the longitudinal wave at probe 2.

3. Case Study

3.1. Regional Survey Analysis

(1) Investigation and analysis of surface disaster characteristics

The expressway with the underlying abandoned goaf is located in the southeast of Tengzhou, China. The terrain in this area is generally flat with scarps. Under the abandoned goaf, the mileage range of the expressway distortion is K3+480~K4+280, the guardrail is very distorted, and the subgrade and pavement sinking deformation are very large. Since 1990, the maximum vertical displacement of the subgrade midline has been lower than the original design elevation of 1.5 m. As indicated by surface disaster survey data, the main types of disasters on both sides of the highway above the abandoned goaf are surface subsidence deformation, collapse pits and ground cracks. The location of the route space and the disaster characteristics are shown in Figure 2.

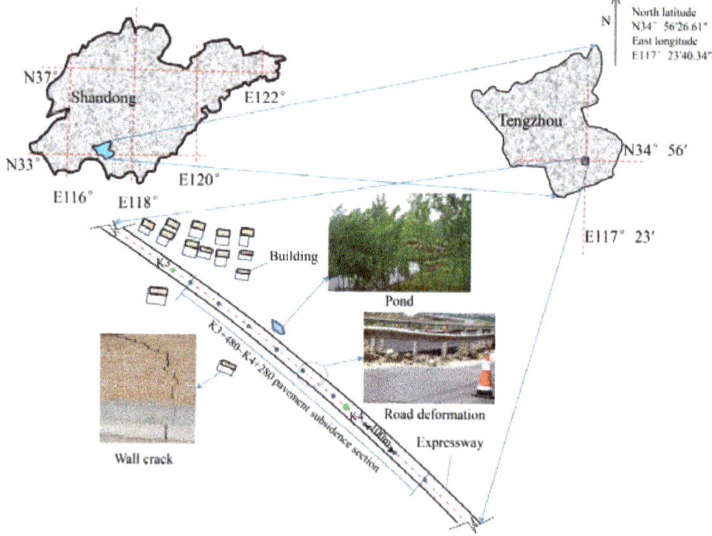

Figure 2. Highway and surface disaster location relationship.

Surface subsidence deformation and collapse pit investigation: In Figure 2, on the north side of the expressway, an elliptical pond has been formed due to surface subsidence. The short axis of the pond is 10.5 m long, and the long axis is 38.9 m long. The direction of the long axis is basically parallel to the direction of the expressway. The position of the pond is close to the maximum deformation position of highway subsidence, and the edge of the pond exhibits different degrees of tensile cracks. On the two sides of the highway K3+480~K4+280, there are a number of small collapse pits, and the edge of the pit near K3+760 also has circular cracks with different widths. At present, most of the subsidence areas are still likely to be active, thereby forming new concave areas.

Residential deformation survey: In Figure 2, a private residence is located at K3+780 on the south side of the expressway, and the cross-range of the highway subgrade is approximately 30 m. A residential housing survey revealed that the wall has a crack of approximately 2 cm. Upon inquiry to the residents of the household, it was learned that the wall had gradually cracked and the crack had continuously expanded since 1997; in addition, crack growth was the fastest after 2000. The residents were evacuated for safety reasons. The cracks continued to expand from 2000 to 2006 and have basically stabilized since 2006. Other walls have cracks of varying degrees.

Highway crack investigation: Since the highway pavement was repaired several times at irregular intervals, no distinct cracks were observed on the expressway surface, and only twist deformation was observed. There are clear cracks on the side of the highway subgrade. The cracks are narrow in the upper part and wider in the lower part, and some intersect the circular cracks that have developed on the ground. The most cracks are observed near K3+754.

(2) Investigation and analysis of regional coal mining

Coal mining-related materials were collected at the Municipal Mining Bureau and the Municipal Highway Bureau. The coal mine is located in Tengzhou city, Shandong Province. The coal mine is equipped with a pair of main and auxiliary inclined shafts near the outcrop of the No. 3 coal seam. The location corresponds to K3+760 on the north side of the expressway, and the cross-range of the subgrade is approximately 50 m. The No. 3 coal seam is inclined to the east and southeast, and the mining direction is basically the same as that of the expressway. The mining thickness of the coal seam is 1.3~13.0 m, and the thickness of most mining areas is between approximately 2 and 8 m. The length of the coal seam projection is approximately 1000 m, which tends to the southeast, and the inclination angle is 22°, while the local inclination angle is 23°~26°. The No. 3 coal seam is mainly mined via longwall mining, and the roof is mined via direct mining. In 1984, mining in the area was stopped. From 1985 to the end of 2005, the coal mine still used residual and wear-through mining and direct roof mining of the No. 3 coal seam. The mining location and the cross-range of the expressway are 240~360 m. The mining range is relatively far from the expressway. The mining depth is 130~176 m, the average mining thickness is 2~8 m and the dip angle is 23°~26°.

The coal mines in the area are old and operated by private companies and corporations. The mined areas have limited information, and there are no mining drawings. It is impossible to determine the detailed mining conditions. The roof of the stratum and coal seam in this area is broken, the strength is low, and the burial depth of the coal seam roadway is shallow. Temporarily supported roadways and goafs collapse due to the roof rock mass, which will inevitably lead to the development of geological disasters on the surface [3], such as the collapse of the two sides of the highway, deformation of the expressway and destruction of the surrounding houses.

According to the shallow drilling data collected in the previous period, quaternary sediments are the main cover within 10 m of the stratum in the affected area of highway subsidence; in the depth range of 10~30 m, the sediments are mainly composed of fully weathered to strongly weathered sandstone and shale interbeds. Among these strata, the developed coal seam is visible at a depth of 25 m at K3+684, and the weak and slightly weathered rock mass occurs below 30 m; there are no detailed drilling data on the deeper layers.

3.2. Detection and Analysis via the High-Density Resistivity Method

Detection with the DUK-2A high-density resistivity method. The method has the characteristics of a simple measurement and layout, automatic data collection and result graph generation, high efficiency, etc. [12]. A total of 60 electrodes were used for data acquisition, and a Winner device was used. The maximum isolation factor was 16 and the distance between the electrodes was 10 m. The specific layout scheme is shown in Figure 3. The measuring electrode distance is 5 m, and the specific arrangement is shown in Figure 3. Under normal circumstances, the goaf abnormal resistance is divided into two cases. If the goaf is not filled with any collapsed rock or groundwater, the goaf will appear as an anomalous high-resistance area; if part of the goaf is filled with collapsed rock or groundwater, with some of voids remaining, the goaf will exhibit high-resistance anomalies in certain parts. When the goaf is completely filled with collapsed rock or groundwater, the goaf will contain conductive ions such as Ca^{2+} and Mg^{2+}, making the filled area appear as a low-resistance anomaly. Considering the distribution characteristics of surface disasters and to acquire mining data on the abandoned goaf in the early stage, survey lines 1 and 2 are laid out. Survey line 1 is arranged at the foot of the roadbed slope on the north side of the expressway (section K3+480~K3+960), and survey line 2 is arranged at the foot of the slope on the south side of the highway subgrade (section K3+470~K3+950). Based on the interpreted results of lines 1 and 2, survey line 3 is arranged 30 m away on the north side of survey line 1 and parallel with survey line 1 (section K3+750~K3+905).

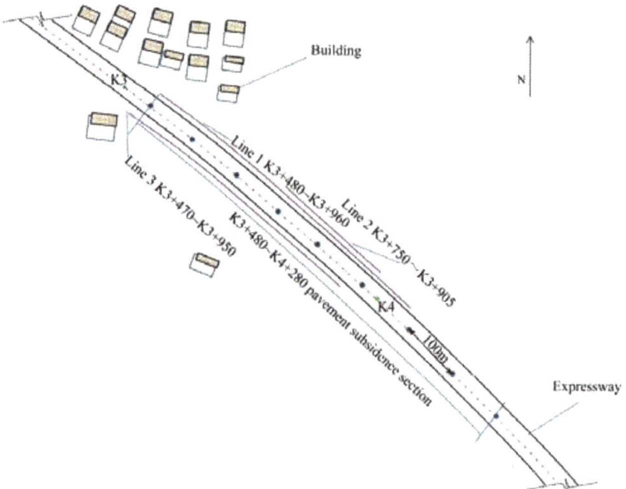

Figure 3. Electrical method line layout diagram.

In survey line 1, for the K3+480~K3+960 segment, reciprocating measurements are taken. As electrical profiling reflects that the test depth is shallow, the results of the measured profile extracted using the two-section double-pole device are shown in Figure 4. From top to bottom, the actual acquisition apparent resistivity profile, the calculated apparent resistivity profile, the interpreted profile and the follow-up line measurement profile are similar. The interpreted profile shows that the apparent resistivity of the rock and soil within a depth of 15 m is less than 10 Ωm. This layer is mainly a quaternary cover layer, which is low in resistivity due to atmospheric precipitation. The burial depth is between 15 and 50 m, and the apparent resistivity is between 10 and 600 Ωm. This layer is mainly composed of sandstone and shale with different degrees of weathering. The K3+640~K3+760 and K3+820~K3+900 segments exhibit an anomalous high-resistivity area at a depth of 50 m, and the apparent resistivity is between 700 and 7000 Ωm. According to the preliminary drilling data, the layers are mostly siltstone and sandstone. Generally, the resistivity of sedimentary rocks is within 1000 Ωm, and it is preliminarily concluded that there are well-developed fractures in the rock mass below 50 m. When the detection depth is increased to 88 m, the high-resistance anomalous area does not form a closed area, and the distribution characteristics of the exposed goaf are not distinct. It is inferred that the abandoned goafin in this section is likely to be filled by collapsed rock.

In survey line 2, for the reciprocating measurement in the K3+630~K3+790 section, the measurement direction is from large to small pile numbers, and the detection results are shown in Figures 5 and 6. From the detection results, it can be seen that the high-resistance characteristics of the goafs in the K3+650~K3+780 and K3+795~K3+820 sections of line 2 are distinct. Since the electrical profile reflects a shallow depth, the data of the measured cross-section of the double-pole device are extracted to clarify the electrical anomalous reaction in the goaf. The area below a depth of 60 m is also characterized by high-resistance goaf characteristics, and the exploration depth is increased to 80 m. There is no distinct anomalous area in the goaf distribution, indicating that the deep goaf is filled with collapsed rock.

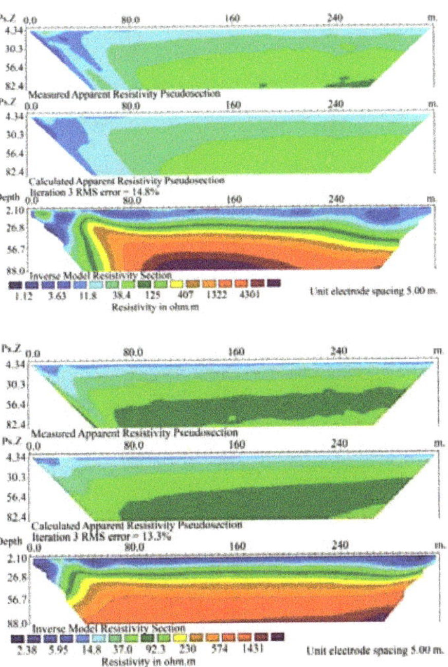

Figure 4. Resistivity profiles of the K3+480~K3+800 and K3+640~K3+960 section diodes in line 1.

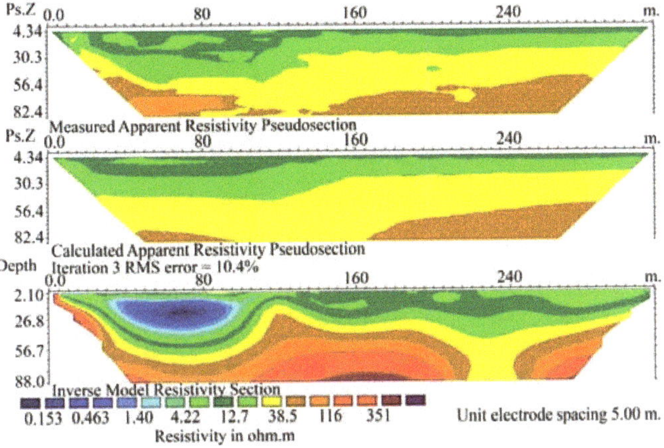

Figure 5. Line 2 K3+790-K3+470 section detection results.

Survey line 3 is arranged in section K3+750~K3+905, the most severely distorted section of the expressway, based on detection results of survey lines 1 and 2, and measurements are obtained from large to small pile numbers. As shown in Figure 7, the shallow depth is 40 m, and the internal resistivity is less than 50 Ωm. The survey line is located 30 m on the south side of the roadbed. Considering that there is a subsidence pond at this location, the water system is connected due to the development of bed cracks, and the apparent resistivity is significantly reduced. In the K3+865~K3+835 and K3+820~K3+795 sections, at a burial depth from 56~76 m, there is a semi-closed anomalous high-resistivity area, which is inferred to be the impact range of the goaf; the boundary top interface is indicated with dashed circles in the figure.

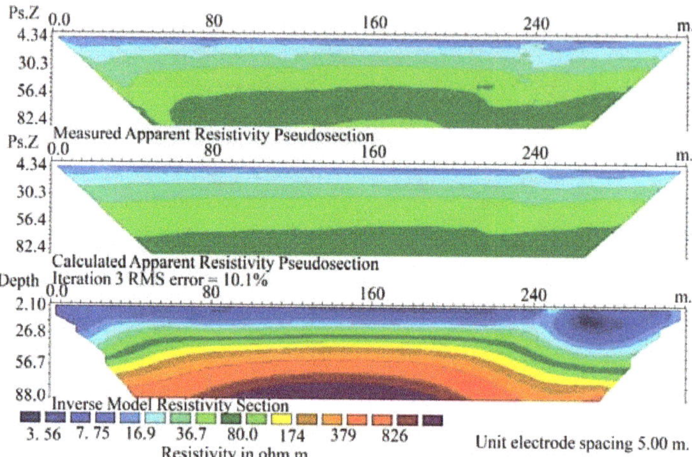

Figure 6. Line 3 K3+950~K3+630 section detection results.

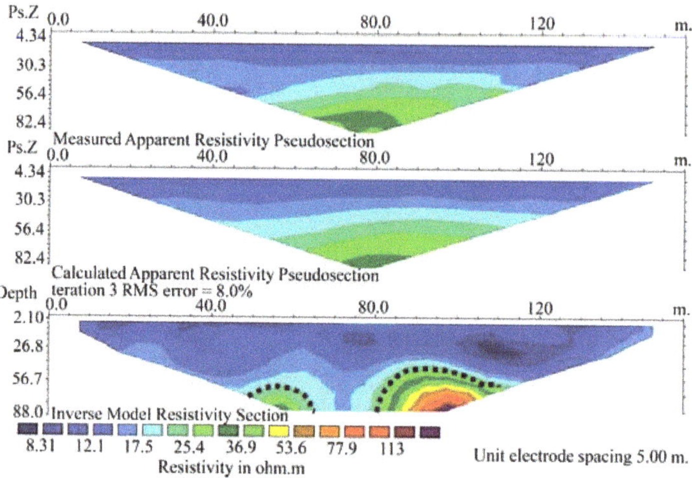

Figure 7. Line 3 K3+905~K3+750 section detection results.

The approximate position of the goaf under the expressway is measured by the high-density resistivity method using line 1 and sections K3+640~K3+760 and K3+820~K3+900, line 2 and sections K3+650~K3+780 and K3+795~K3+820, and line 3 and section K3+865~K3+835. According to the thickness, depth, dip angle of the goaf and location of the abandoned goaf, the extent of the influence of goaf subsidence on the surface can be inferred, and the general range of deformation of the expressway is K3+480~K4+280. It is preliminarily determined that most of the abandoned goaf is filled with collapsed rock. The burial depth is within 0~50 m, which is the range of the subsidence and water-filled fractured zones. The burial depth of 50~88 m is the impact range of the goaf. Due to the small difference in resistivity of the shallow geotechnical bodies, it is difficult to distinguish the areas of the subsidence and water-filled fractured zones. Considering the detection error and objective detection condition constraints, the high-resolution seismic method is used to comprehensively determine the target area of K3+470~K4+080 and further examine the position of the goaf and the distribution of the three belts.

3.3. High-Resolution Seismic Method Target Detection

The curved subsidence zone and the water-filled fractured zone of the abandoned goaf are difficult to clearly distinguish on the resistivity profile. Based on the high-density resistivity detection results, high-resolution seismic reflection wave characteristic analysis is carried out to further determine the three-belt distribution characteristics of the goaf [23]. Seismic lines are arranged in the determined detection target area K3+470~K4+080. The main acquisition parameters of the engineering seismograph are: the time window is 182 ms, the sampling points are 1820 and the sampling interval is 0.1 ms, with an offset of 8 m and a dot pitch of 1 m. The hammer is excited. The test results are shown in Figures 8 and 9. Survey line 4 is arranged at the shoulder position on the north side (section K3+480~K4+080), and survey line 5 is arranged at the shoulder position on the south side (section K3+470~K4+070). Since interpretation analysis of lines 4 and 5 is conducted the same way, only line 4 is used as an example for the analysis.

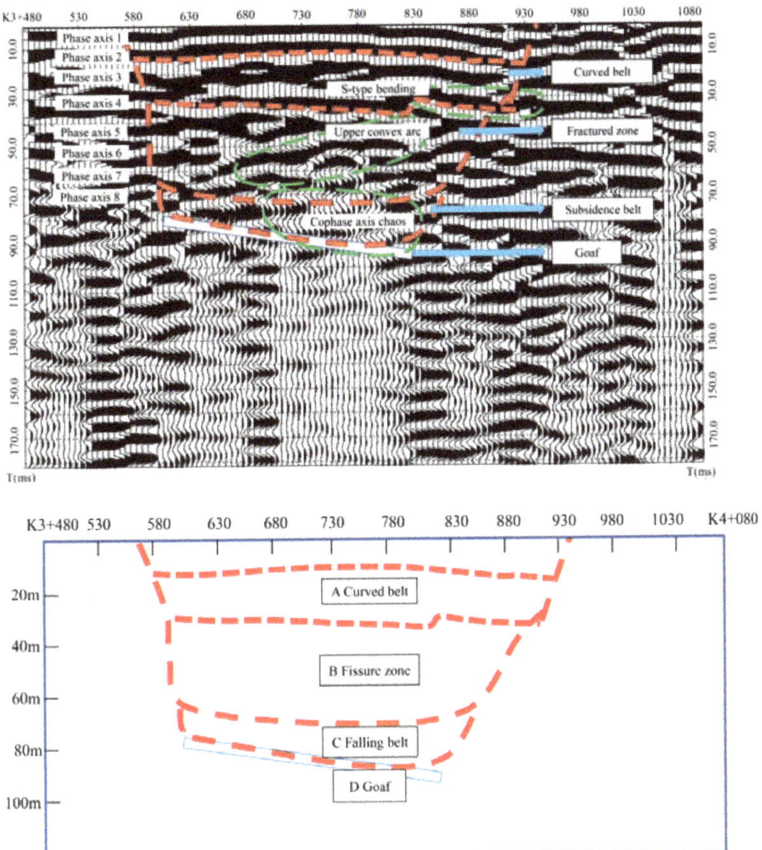

Figure 8. Line 4 detection results.

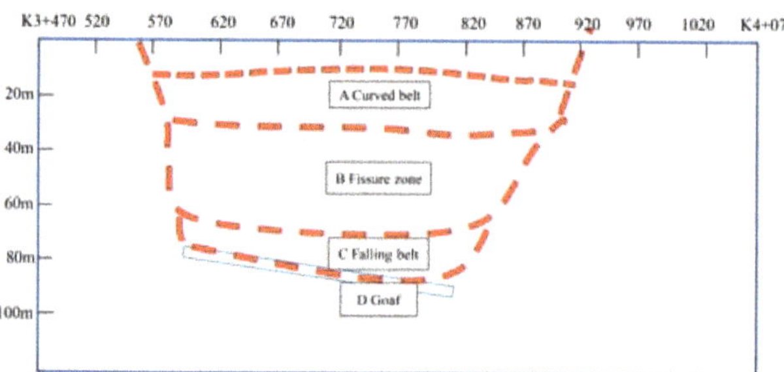

Figure 9. Line 5 detection results.

Interpretation of the detection results: The reflected waves in the goaf have the following general behavior of the seismic image: for the same curve, fractures or discontinuities occur along the same phase axis, and the waveform of the reflection profile appears to be oscillating or periodically increasing. From the test results in Figure 8, the in-phase axis group is relatively continuous, the in-phase axis is clearly identifiable, and there are many features. The arrival times of the first, second and third in-phase axes are approximately 5 ms, 15 ms and 25 ms, respectively. The in-phase axis is continuous and has irregular bending. This phenomenon indicates that there are many sets of sedimentary layers in the quaternary overburden and weathered rock strata, and the reflection characteristics of the bedrock surface are very notable; the regional lithology is consistent, and the in-phase axis is continuous. It is inferred that this part is the curved subsidence zone affected by the goaf, and the seismic test results are consistent with the resistivity test results.

The wave arrival time of the fourth in-phase axis is approximately 35 ms, the phase is relatively continuous, and there is an S-bend between K3+830 and K+930. The arrival times of the 5th and 6th in-phase axes are 45 ms and 55 ms, respectively, and the in-phase axis is relatively continuous overall, but it is more disordered; an upper convex arc is present between K3+730 and K3+830. Goaf collapse has caused many cracks in the rock formation, and the fissure zone can attenuate the energy of the emitted wave, resulting in the emission wave's in-phase axis disorder. It is inferred that this part is the water-filled fractured zone affected by the goaf.

The wave arrival times of the 7th and 8th in-phase axes are 65 ms and 75 ms, respectively, and an upper convex arc is also present between K3+630 and K3+730; in addition, the in-phase axis is very disordered. The depth of exploration continues to increase, the in-phase axis of the same phase is more disordered and difficult to identify, and there is in-phase axis loss between K3+680 and K3+800. This finding indicates that the site is affected by mining activities, resulting in a large number of gravel collapses and cavities. The broken rock mass and the unfilled cavity form a strong reflection interface and a region in which the reflected wave energy is rapidly depleted. The multiple interface reflections and dissipation of reflected wave energy lead to chaotic observations and loss of in-phase axis. It is inferred that this part is the goaf zone. This finding is consistent with the actual mining data.

According to the characteristics of the seismic reflection waves and the results of the above electrical method, combined with the general position correspondence of the three belts, a comprehensive analysis is carried out to determine the location of the abandoned goaf and the specific limits of the three belts. The range of the abandoned goafs that have a significant impact on the expressway is as follows: along the expressway, with regards to the K3+580~K3+830 section on the north side and the K3+580~K3+900 section on the south side, the lateral distance of the north-south sides along the expressway axis is within 30 m. Among these sections, the burial depth is approximately 35 m, which is a subsidence and deformation zone. The burial depth is between 35 and 75 m, which is the water-filled fractured zone. The subsidence zone is between 75 and 90 m, and the unfilled cavity of the goaf is approximately 2~8 m.

3.4. Probe Verification

To further verify the accuracy of detection, the deformation of the expressway and the detection results are comprehensively considered, and the K3+580~K3+830 section is selected, which covers the anomalous high-resistivity area; the resistivity method and the seismic method are used to examine the missing in-phase axis area via deep drilling. As shown in Figure 10, the No. 1, No.2 and No.3 holes are characteristic boreholes with a drilling distance of 40 m, and the boreholes are drilled deep into the goaf. To avoid verification results that are not representative and based on the characteristic drilling data of the three boreholes, the No.2 borehole sound wave detection and TV imaging data are used for verification.

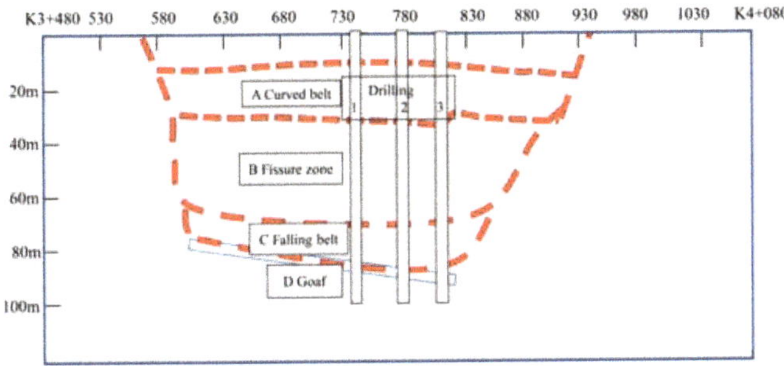

Figure 10. Drilling position map.

(1) Drilling and sound wave detection verification

Drilling data are shown in Table 1. The relative drilling position to seismic line 4 is shown in Figure 10. According to the drilling data, the depth of the borehole is basically the same, indicating that the cavity formed by caving is not excessively inclined. The hole height detected by the drill pipe is different, indicating that the goaf is in a filled or semi-filled state with no clear regularity. According to the drilling data and related experiments, the 30 m depth is mainly composed of quaternary sediments and strongly weathered mudstones and sandstones. Its saturated uniaxial compressive strength is less than 0.26 MPa and the softening coefficient is less than 0.17, indicating a significant water-weakening effect and a certain degree of expansion and relative water repellency; the depth between 30 and 80 m mainly consists of weak weathered siltstone and sandstone, and below 78 m, the drilling phenomenon appears. The thickness of the drill pipe is approximately 3~5 m, and the goaf is filled with groundwater due to the existence of a large number of connected fissures. The semi-filled goaf is currently in equilibrium, and the drilling quality is basically consistent with the position of the three belts determined from the geophysical results. According to the drilling data, it is observed that the range of the fractured and caving zones is slightly larger compared to the geophysical exploration results, which may be caused by different degrees of disturbance to the rock sample in drilling sampling.

Table 1. Drilling data.

Point Number	Hole Depth/m	Drop Depth/m	Drill Rod Detection Cavity Height/m
1: K3+732	78.0	74.2	3.8
2: K3+800	84.0	80.0	3.2
3: K3+767	85.0	80.5	5.1

The distribution range of three belts is further corrected, and the acoustic detection of boreholes is carried out. The data processing of the borehole wave velocity test is based on Equation (1), discussed in the Methodology section. The No.2 borehole is selected as the test hole to perform acoustic wave detection and then the acoustic wave test data are processed to plot the acoustic wave test curve, as shown in Figure 11. The rock mass velocity at a depth of 50 m in the borehole is basically between 1800 and 3500 ms^{-1}, which is a normal value range. This section is mainly composed of quaternary sedimentary cover and strongly weathered mudstone and sandstone. The abovementioned detection results show that this area mainly belongs to the curved subsidence zone in the three belts of the goaf. The rock mass velocity below the 50 m depth drops sharply to between 1200 and 3000 ms^{-1}, which is mainly composed of broken, weak weathered siltstone and sandstone. However, the wave velocity of generally weakly weathered siltstone should be above 4000 ms^{-1}, indicating that a large number of fissures and unfilled voids are present in the rock mass of this section, which is consistent with the characteristics of the caving zone. The characteristics of the No. 2 hole wave velocity along with the depth are basically consistent with the distribution characteristics of the overlying strata in the goaf detected by the high-resolution seismic method.

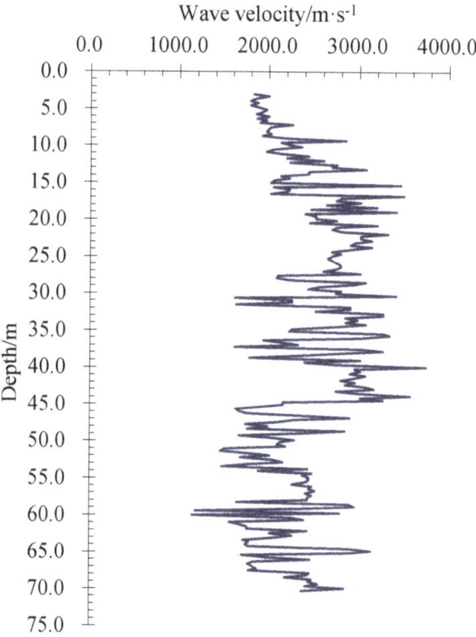

Figure 11. Acoustic wave detection curve [24].

(2) Borehole TV imaging technology verification

The borehole TV imaging method is more intuitive [25]. The No. 2 drilling feature position image is analyzed and the feature boundary image is captured, as shown in Figure 12. Figure 12a shows that when the probe enters the hole at 30.5 m, the rock wall of the hole is notably rough and uneven, and the weathering is severe. The probe penetrates to 46.5 m, as shown in Figure 12b, and although the wall of the hole is smooth, cracks in the rock are clearly visible. The probe gradually descends, and the wall of the cracked hole is observed; the borehole wall is uneven. Figure 12c shows a separation void formed by rock mass collapse at 68 m, indicating that groundwater has reached the goaf and that the groundwater is visible. The pores of the fracture zone are developed, and the pores are filled with groundwater. As shown in Figure 12d, the drilling probe penetrated into the groundwater-filled area. The water is turbid due to disturbance, and it is difficult to see the surrounding environment despite the intense lighting. After approximately 1 h at a stationary position, the probe continues to descend, and it is observed that there is no round hole shape, which is accompanied by a clearly visible boundary line. It is determined that the boundary line should be the boundary between the roof and the two gangues of the goaf, as shown in Figure 12e. Based on the other observed probe angles, it is found that there are many roadway collapse zones filled with gravel and goaf filled with groundwater, and the field of view is not very clear, as shown in Figure 12f. At a depth of 85.0 m, the descent cannot be continued because the bottom has been reached. The drilling details in the goaf are clearly recorded by borehole TV imaging technology. The verification results show that the range of rock formation changes is consistent with the geophysical results; although goaf collapse is severe, and the goaf is in a semi-filled state and filled with groundwater, it still has distinct roadway features.

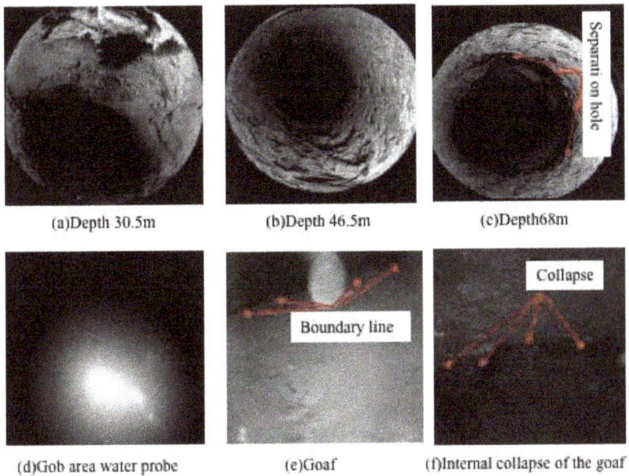

Figure 12. No. 2 borehole TV images [24].

3.5. Verification and Correction Analysis

It has been verified by the deep drilling, sound wave detection and TV imaging data of the feature location that the goaf is full of groundwater, and most of the voids are filled by the collapsed rock mass; in addition, locally there are still clear roadway features. The abandoned goafs can become deformed and degraded under the influence of internal and external factors, which will cause cracks, distortion and even collapse of surface structures. Especially, the water and rock mechanical properties will influence the stability of abandoned goafs. The abandoned goaf has remained stable for a long time, but disturbance will likely lead to secondary activation of the abandoned goaf, especially due to digging, which suddenly releases the groundwater in the goaf of surface expressways, causing catastrophic hazards [26–28]. The proven results of the resistivity and seismic methods are more accurate. Since the abandoned goaf has an irregular shape, and the error distribution is used to further correct the heights of the fracture zone, the caving zone and the goaf. The average size determined via verification technology is considered the average value, and then the average value of the geophysical results is calculated. Finally, the range of the goaf is distributed within 30 m on the north side of the expressway axis, namely, the K3+580~K3+830 section, and the south side is mainly the K3+580~K3+900 section. The curved subsidence zone has a depth of approximately 35 m, the water-filled fractured zone has a burial depth of 35~75 m, the subsidence zone is between 75 and 90 m, and the average height of the abandoned goaf is approximately 4.5 m.

4. Conclusions

Taking the abandoned goaf in the Tengzhou section of China's Mu Shi expressway as an engineering example, step-by-step detection, traditional detection and combination methods are used to determine the location of the underlying abandoned goaf and overburden morphology. The following conclusions were reached:

(1) In the absence of detailed mining data on abandoned goafs, step-by-step detection, traditional detection and combination probing methods should be applied. The target area is gradually reduced and determined, which improves the location of the abandoned goaf and the efficiency of overburden morphology detection; mitigating the lack of effective joint detection with necessary verification techniques is scientific and reasonable for determining the location and status of abandoned goafs.

(2) Traditional re-detection and superficial verification methods are improved. It is proposed to verify and correct the joint geophysical findings by using sound wave detection, TV imaging technology and deep drilling data. The approach also intuitively shows the structural characteristics of the three belts in the goaf and the filling characteristics of the goaf, thereby effectively avoiding multiple detection solutions of the abandoned goaf.

(3) Because the abandoned goaf has an irregular shape, the error distribution is used to correct the heights of the fracture zone, the caving zone and the goaf, which provides a basis for the subsequent evaluation of the abandoned goaf stability and the expressway subgrade.

(4) The detection results show that the K3+580~K3+900 section is within 30 m on both sides of the central axis of the expressway, and the goaf is widely distributed in the depth range of 75~90 m. The depths of the three belts are determined: the curved subsidence zone has a depth of approximately 35 m, the water-filled fractured zone has a burial depth of 35~75 m, the caving zone is between 75 and 90 m, and the average height of the abandoned goaf is approximately 4.5 m. Taking the abandoned goaf in the Tengzhou section of China's Mushi expressway as an engineering example, the location and filling state of the abandoned goaf affecting the expressway area are determined, and the location of the overburden in the abandoned goaf is determined, which provides a scientific basis for stability evaluation of abandoned goafs and post-highway construction control.

Author Contributions: Conceptualization, S.Z.; methodology, P.J.; software, S.W.; validation, H.W.; formal analysis, S.W.; investigation, H.W. data curation, L.L.; writing—original draft preparation, L.L.; writing—review and editing, S.Z.; funding acquisition, S.Z. All authors have read and agreed to the published version of the manuscript.

Funding: This research received no external funding.

Informed Consent Statement: Not applicable.

Data Availability Statement: Not report any data.

Acknowledgments: Thanks to the China Zaozhuang City Highway Bureau for providing engineering geological conditions and seismic image data.

Conflicts of Interest: The authors declare no conflict of interest.

References

1. Wang, S.; Su, Y.; Wang, C. Analysis of stress and strain of surrounding rock in goaf based on GIS. *Arab. J. Geosci.* **2021**, *14*, 1843. [CrossRef]
2. Tan, Y.L.; Zhao, T.B.; Xiao, Y.X. Researches on floor stratum fracturing induced by antiprocedure mining underneath close-distance goaf. *J. Min. Sci.* **2010**, *46*, 3. [CrossRef]
3. Ma, H.; Wang, J.; Wang, Y. Study on mechanics and domino effect of large-scale goaf cave-in. *Saf. Sci.* **2012**, *50*, 689–694. [CrossRef]
4. Xu, P.; Mao, X.; Zhang, M.; Zhou, Y.; Yu, B. Safety analysis of building foundations over old goaf under additional stress from building load and seismic actions. *Int. J. Min. Sci. Technol.* **2014**, *24*, 713–718. [CrossRef]

5. Díaz-Fernández, M.E.; Álvarez-Fernández, M.I.; Álvarez-Vigil, A.E. Computation of influence functions for automatic mining subsidence prediction. *Comput. Geosci.* **2010**, *14*, 83–103. [CrossRef]
6. Du, M.; Wang, X.; Zhu, J. Analysis on Stability of Edge Roadbed in Large Mined-Out Regions Where the High-Speed Railway Pass. *Geotech. Geol. Eng.* **2021**, *39*, 1849–1860. [CrossRef]
7. Wang, L.; Li, N.; Zhang, X.-N.; Wei, T.; Chen, Y.-F.; Zha, J.-F. Full parameters inversion model for mining subsidence prediction using simulated annealing based on single line of sight D-InSAR. *Environ. Earth Sci.* **2018**, *77*, 161. [CrossRef]
8. Li, L.; Wu, K.; Zhou, D.-W. Evaluation theory and application of foundation stability of new buildings over an old goaf using longwall mining technology. *Environ. Earth Sci.* **2016**, *75*, 763. [CrossRef]
9. Wang, C.; Lu, Y.; Qin, C.; Li, Y.; Sun, Q.; Wang, D. Ground disturbance of different building locations in old goaf area: A case study in China. *Geotech. Geol. Eng.* **2019**, *37*, 4311–4325. [CrossRef]
10. Mccay, A.T.; Valyrakis, M.; Younger, P.L. A meta-analysis of coal mining induced subsidence data and implications for their use in the carbon industry. *Int. J. Coal Geol.* **2018**, *192*, 91–101. [CrossRef]
11. Krishnamurthy, N.; Rao, V.A.; Kumar, D.; Singh, K.K.K.; Ahmed, S. Electrical resistivity imaging technique to delineate coal seam barrier thickness and demarcate water filled voids. *J. Geol. Soc. India* **2009**, *73*, 639–650. [CrossRef]
12. Das, P.; Pal, S.; Mohanty, P.R.; Priyam, P.; Bharti, A.K.; Kumar, R. Abandoned mine galleries detection using electrical resistivity tomography method over Jharia coal field, India. *J. Geol. Soc. India* **2017**, *90*, 169–174. [CrossRef]
13. Helm, P.; Davie, C.; Glendinning, S. Numerical modelling of shallow abandoned mine working subsidence affecting transport infrastructure. *Eng. Geol.* **2013**, *154*, 6–19. [CrossRef]
14. Du, Z.; Ge, L.; Ng, A.H.-M.; Li, X. Investigation on mining subsidence over Appin–West Cliff Colliery using time-series SAR interferometry. *Int. J. Remote Sens.* **2018**, *39*, 1528–1547. [CrossRef]
15. Nie, L.; Zhang, M.; Jian, H. Analysis of surface subsidence mechanism and regularity under the influence of seism and fault. *Nat. Hazards* **2013**, *66*, 773–780. [CrossRef]
16. Yan, W.; Chen, J.; Tan, Y.; Zhang, W.; Cai, L. Theoretical Analysis of Mining Induced Overburden Subsidence Boundary with the Horizontal Coal Seam Mining. *Adv. Civ. Eng.* **2021**, *2021*, 6657738. [CrossRef]
17. Zhou, B.; Hatherly, P.; Sun, W. Enhancing the detection of small coal structures by seismic diffraction imaging. *Int. J. Coal Geol.* **2017**, *178*, 1–12. [CrossRef]
18. Gan, F.; Han, K.; Lan, F.; Chen, Y.; Zhang, W. Multi-geophysical approaches to detect karst channels underground—A case study in Mengzi of Yunnan Province, China. *J. Appl. Geophys.* **2017**, *136*, 91–98. [CrossRef]
19. García-Yeguas, A.; Ledo, J.; Piña-Varas, P.; Prudencio, J.; Queralt, P.; Marcuello, A.; Ibañez, J.M.; Benjumea, B.; Sánchez-Alzola, A.; Pérez, N. A 3D joint interpretation of magnetotelluric and seismic tomographic models: The case of the volcanic island of Tenerife. *Comput. Geosci.* **2017**, *109*, 95–105. [CrossRef]
20. Yu, G.; Li, H.; Zhang, D. Simulation Study on Electromagnetic Response Characteristics of Unfavorable Geological Body Under Complex Conditions. *Geotech. Geol. Eng.* **2021**, *39*, 3371–3382. [CrossRef]
21. Nogueira, P.V.; Rocha, M.P.; Borges, W.R.; Silva, A.M.; Assis, L.M.D. Study of iron deposit using seismic refraction and resistivity in Carajas Mineral Province, Brazil. *J. Appl. Geophys.* **2016**, *133*, 116–122. [CrossRef]
22. Lei, T.; Sinha, B.K.; Sanders, M. Estimation of horizontal stress magnitudes and stress coefficients of velocities using borehole sonic data. *Geophysics* **2012**, *77*, WA181–WA196. [CrossRef]
23. Arosio, D.; Longoni, L.; Papini, M.; Zanzi, L. Seismic characterization of an abandoned mine site. *Acta Geophys.* **2013**, *61*, 611–623. [CrossRef]
24. Zhang, S.; Zhang, X.; Li, Y. Old goaf detection and verification techniques under highway. *J. Cent. South Univ. (Sci. Technol.)* **2015**, *46*, 3361–3367.
25. Chatterjee, R.; Gupta, S.D.; Mandal, P.P. Fracture and stress orientation from borehole image logs: A case study from Cambay basin, India. *J. Geol. Soc. India* **2017**, *89*, 573–580. [CrossRef]
26. Liu, J.; Zhao, Y.; Tan, T.; Zhang, L.; Zhu, S.; Xu, F. Evolution and modeling of mine water inflow and hazards characteristics in southern coalfields of China: A case of Meitanba mine. *Int. J. Min. Sci. Technol.* **2022**, *32*, 513–524. [CrossRef]
27. Liu, Q.; Zhao, Y.; Tang, L.; Liao, J.; Wang, X.; Tan, T.; Chang, L.; Luo, S.; Wang, M. Mechanical characteristics of single cracked limestone in compression-shear fracture under hydro-mechanical coupling. *Theor. Appl. Fract. Mech.* **2022**, *119*, 103371. [CrossRef]
28. Zhao, Y.; Zhang, L.; Wang, W.; Wan, W.; Ma, W. Separation of elastoviscoplastic strains of rock and a nonlinear creep model. *Int. J. Geomech.* **2017**, *18*, 04017129. [CrossRef]

MDPI

Article

Coal Wall Spalling Mechanism and Grouting Reinforcement Technology of Large Mining Height Working Face

Hongtao Liu *, Yang Chen, Zijun Han, Qinyu Liu, Zilong Luo, Wencong Cheng, Hongkai Zhang, Shizhu Qiu and Haozhu Wang

School of Energy and Mining Engineering, China University of Mining and Technology (Beijing), Beijing 100083, China
* Correspondence: 108925@cumtb.edu.cn; Tel.: +86-(150)-0105-5568

Abstract: To control the problem of coal wall spalling in large mining height working faces subject to mining, considering the Duanwang Mine 150505 fully mechanized working face, the mechanism of coal wall spalling in working faces was investigated by theoretical analysis, numerical simulation and field experiment. Based on analysis of coal wall spalling in the working face, a new grouting material was developed. The stress and plastic zone changes affecting the coal wall, before and after grouting in the working face, were analyzed using numerical simulation and surrounding rock grouting reinforcement technology was proposed for application around the new grouting material. The results showed that: (1) serious spalling of the 150505 working face was caused by the large mining height, fault influence and low roof strength, and (2) the new nano-composite low temperature polymer materials used have characteristics of rapid reaction, low polymerization temperature, adjustable setting time, high strength and environmental protection. Based on analysis of the working face coal wall spalling problem, grouting reinforcement technology based on new materials was proposed. Industrial tests were carried out on the working face. Field monitoring showed that the stability of the working face coal wall was significantly enhanced and that rib spalling was significantly improved after comprehensive anti-rib-spalling grouting measures were adopted. These results provide a basis for rib spalling control of working faces under similar conditions.

Keywords: roof subsidence; grouting reinforcement; roadway support; destruction of surrounding rock; dynamic pressure roadway

Citation: Liu, H.; Chen, Y.; Han, Z.; Liu, Q.; Luo, Z.; Cheng, W.; Zhang, H.; Qiu, S.; Wang, H. Coal Wall Spalling Mechanism and Grouting Reinforcement Technology of Large Mining Height Working Face. *Sensors* **2022**, *22*, 8675. https://doi.org/10.3390/s22228675

Academic Editor: Fabio Bovenga

Received: 11 September 2022
Accepted: 7 November 2022
Published: 10 November 2022

Publisher's Note: MDPI stays neutral with regard to jurisdictional claims in published maps and institutional affiliations.

1. Introduction

Comprehensive mechanized mining is often used in coal mining where the mining height of the fully mechanized mining face is significant, the strata pressure is relatively violent, and coal wall spalling is prone to occur [1,2]. Soft coal quality increases the probability of rib spalling. Following rib spalling, end-face roof caving can occur, which prejudices coal mine safety and affects the normal advancement of the working face, seriously restricting the efficient production of the mine [3,4].

The problem of coal wall spalling is related to the surrounding rock strength. Many researchers have investigated the strength of the rock mass through rock mechanics experiments, the results being of great importance for evaluation of the stability of the surrounding rock [5–7]. There are generally two major ways to prevent or reduce coal wall spalling [8–11]. First, the stress state of the coal wall of the working face can be improved, so that the unidirectional force of the coal body is avoided as far as possible to prevent stress concentration in the coal wall of the working face. This is mainly achieved by accelerating the advancing speed of the working face, timely beating of the guard panel, moving the frame with pressure, and increasing the initial support force of the support. Since the initial support force of the support cannot be infinitely increased, and the force provided by the guard panel is also limited, this first approach to preventing coal wall spalling is ineffective.

A second approach is to alter the physical and mechanical properties of the coal, increasing the cohesion and strength of coal and ensuring the integrity of the coal. This approach is mainly effected through cement grouting and chemical grouting [12–14]. Therefore, it is important to identify low-temperature grouting materials which have good permeability, strong adhesion, low reaction heat, are of low cost, and can be widely applied in coal mines. Such materials are of great value for the prevention of coal wall spalling, improving coal production and increasing the economic benefits of coal-mining enterprises [15,16].

With respect to grouting support technology, based on problems associated with the large deformation of surrounding rock, serious fragmentation of the coal side and failure of the support body in a kilometer deep well, Kang Hongpu [17] proposed the use of high pressure anchor grouting-spraying synergistic control technology for a soft coal body and achieved good results in the field. Seeking to address the problem of large deformation of a deep coal seam, Xie Shengrong [18] proposed the use of external anchor-internal unloading collaborative control technology, which significantly improved the stress environment of the surrounding rock.

Using the 150505 working face of the Duanwang Mine as the engineering context, the coal wall spalling mechanism of a working face of significant mining height was studied through field investigation and numerical simulation. Based on analysis of the coal wall spalling problem, a new type of low-temperature polymer grouting reinforcement material was developed and use of surrounding rock grouting reinforcement technology suitable for the 150505 working face was proposed. The results of application of the proposed approach indicated that progress on the working face was enhanced, leading to economic benefits for the mine.

2. Engineering Background and Failure Characteristics of the Working Face

2.1. Engineering Overview

The coal seam of the 150505 fully mechanized mining face in the Wang Mine of the Shouyang Section of Shanxi Province is $15^{\#}$ coal. The strike length of the working face is 288 m, the inclination length is 159 m and the mining area is 49,109.9 m^2. The 15# coal type is a lean semi-bright-type coal, forming blocks and occurring where the coal seam is stable. The minimum thickness of the coal seam is 3.8 m, the maximum thickness is 4.25 m and the average thickness is 4.1 m. The minimum dip angle of the working face coal seam is 0°, the maximum dip angle is 15° and the average dip angle is 6°, so the occurrence of the working face coal seam is relatively stable. A plane layout of the working face and a histogram of the roof and floor of the rock are shown in Figure 1.

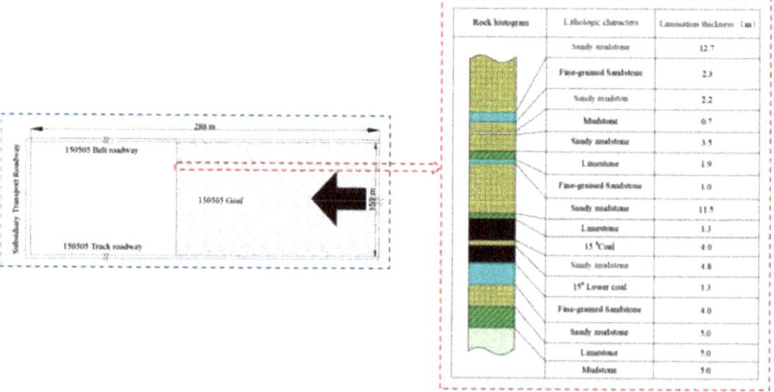

Figure 1. 150505 working face layout relationship and comprehensive histogram.

2.2. Overview of Coal Wall Spalling at the Working Face

The rib spalling problem has been a common problem in working faces with significant mining height, often leading to the support toppling and skewing, the scraper conveyor becoming stuck and the working face toppling, affecting the advancing speed of the working face. In severe cases, it may even cause significant casualties.

Mining the working face of a 15# coal seam is associated with coal seam occurrence stability and the thickness change is not substantial. The working face is a working face with large mining height. Since the roof of the coal seam is mainly sandy mudstone and the strength of mudstone is low, the support does not connect to the roof when the mining height is high and roof management is very difficult. When the support force is too large, the support is prone to mudstone collapse, spalling, roof fall and the occurrence of other accidents. Rib spalling of a coal wall in the process of working face advancement is shown in Figure 2.

(a)

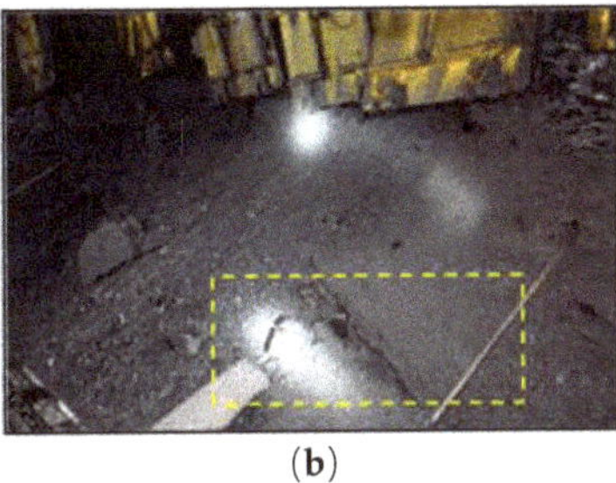

(b)

Figure 2. Failure diagram of working face. (**a**) Block collapse; (**b**) Support falling.

3. New Polyurethane Composite Materials and Complete Technology for Coal and Rock Reinforcement

3.1. Development of New Grouting Materials

With respect to the causes of underground surrounding rock failure, the failure modes of the surrounding rock can be divided into two categories: failure induced by a structural plane or geological weak plane and failure induced by original rock stress. Soft rock failure and strength weakening are the fundamental causes of surrounding rock deformation. The surrounding rock shows different deformation characteristics at different stages. Therefore, strengthening the surrounding rock to improve the mechanical properties of the surrounding rock has become a key area of research. In light of the causes of weakening and failure of surrounding rock strength, grouting reinforcement technology applied to surrounding rock is strengthened in stages, and the corresponding point, line and surface reinforcements are carried out at different stages. The advantage of this approach is that it can improve the mechanical properties of the weak surface of the surrounding rock combined structure, enabling it to form a bearing structure and to ensure the self-stability of the surrounding rock. Moreover, the grout consolidation body can seal cracks and prevent the internal rock mass from soaking up water and gas, which is of great importance for maintaining the mechanical properties of the surrounding rock and ensuring its lasting stability [19].

Many types of chemical grouting materials have been developed with different grouting materials having different defects. Some materials have the disadvantages of polluting the environment, having a long cementation time, brittleness, irritating smell, flammability and of being harmful to the human body. As an ideal chemical grouting material, it should have the following characteristics [20]: good injectability, good durability, low reaction temperature, short curing time, good tensile and compressive strength, good permeability resistance, non-toxicity, odorlessness, low price and convenience of transportation. Up to now, most slurries do not fully meet the above requirements. In addition, in current applications of grouting materials, coal seam fire and smoke accidents have occurred in some mining areas during grouting. This is due to the high temperature produced by the rapid oxidation of the surrounding coal and rock during the chemical reaction of the grouting materials. Therefore, the grouting material must reduce the maximum reaction temperature of the coal mine reinforcement material grouting process.

Based on the characteristics of the above grouting materials, the new polyurethane composite material developed in this paper represents a new, low-temperature and safe liquid two-component environmentally friendly polymer material. It is produced by the mixing reaction of A and B two-component materials at low temperature with a volume ratio of 1:1. The main constituent of the A component is polyether polyol, while the main constituent of the B component is polymethylene polybasic polyisocyanate. The composition parameters are shown in Table 1.

Table 1. Material composition table.

Component A: Raw Material Name	Weight/%	Component B: Raw Material Name	Weight/%
Polyol	0–98	Isocyanate	65–100
Foam over stabilizer	0–2.0	Polyol	0–10
Catalyst	0.4–0.8	Plasticizer	0–35

The new material is produced as follows: First, polyether polyols and some isocyanate groups are reacted to generate prepolymers with high molecular weight with isocyanate groups at the end. Then, the prepolymers with isocyanate groups at the end are reacted with polyether polyol to form polycarbamate. The material has the following characteristics: when gelled, it is less affected by underground groundwater, has low reaction heat release, high adhesion, rapid strength increase, and has strong permeability resistance, abrasion resistance, stamping resistance and aging resistance. The grouting material is mainly suitable for the reinforcement of loose broken coal rock mass, roadway, small coal pillars and roadway in disrepair in the underground mining face. Advance grouting reinforcement to prevent top coal caving during excavation of reserved top coal is required.

The two-component materials are poured using a high-pressure grouting pump; the components A and B react quickly to form low exothermic and totally non-flammable elastomers, which are injected into the coal seam or rock strata by high pressure and can be extended to voids and cracks along the cracks of the soft coal and rock strata. The material can solidify in a short time and produce high strength cementitious materials to achieve the purpose of reinforcement; at the same time, the formed gel is subjected to secondary expansion and the loose coal seam in the non-grouting area is extruded so that the soft foundation coal seam forms a hard matrix to ensure the effective and safe mining of the shearer heading.

The structure of this material as shown under different high-resolution electron microscopes is shown in Figure 3.

To determine the physical and mechanical properties of the grouting materials, physical and mechanical tests were carried out in the laboratory. The standard test conditions in the laboratory were temperature 23 $\pm$ 2 °C and relative humidity 50 $\pm$ 5%. Before the tests, the samples were kept under standard test conditions for 24 h. The laboratory test diagram is shown in Figure 4 and the measured physical and mechanical parameters of the grouting materials are shown in Table 2.

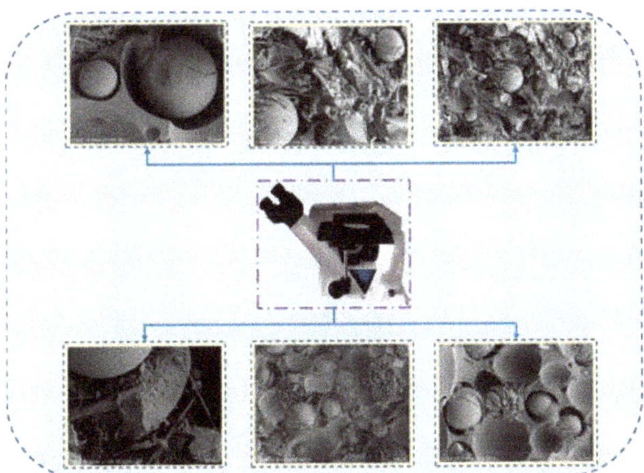

Figure 3. Images of new polyurethane composite materials under electron microscope.

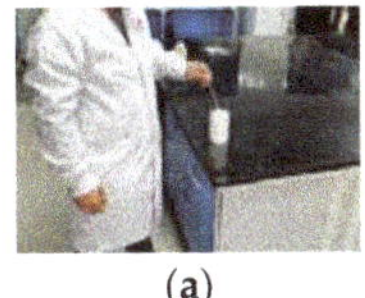
(a)

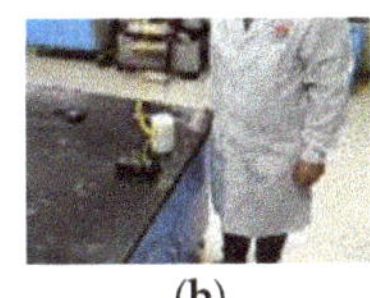
(b)

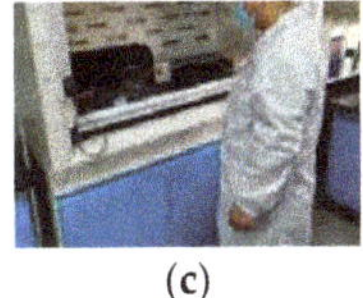
(c)

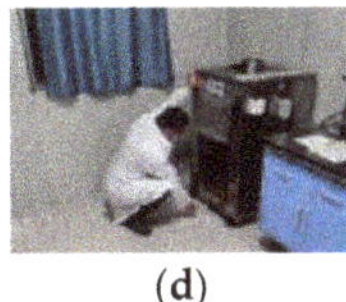
(d)

Figure 4. Laboratory test diagram. (**a**) Curing time test; (**b**) reaction temperature; (**c**) flashpoint test; (**d**) flame-retardant performance test.

Table 2. Technical index of grouting material.

Project		Standard Requirements	Test Results	
			Component A	Component B
Appearance		Homogeneity, No caking	Homogeneity; No caking	
Flashpoint /°C		≥100	225	Not detected
Curing time (23 ± 2) °C/s		——	40~100	
Expansion multiple /times		≥1.0	2.0~3.0	
Ant aging performance (80 °C, 168 h)		No surface change, No mass loss	No surface change; No mass loss	
Maximum reaction temperature /°C		≤140	80~90	
Bond strength /MPa		≥3.5	4.2	
Flame-retardant and anti-static properties		Meet standards MT113	Non-combustible; anti-static performance meets the requirements of standard MT113	
Harmful	Benzene /g/kg	≤5.0	Not detected	
Material	Toluene + Xylene /g/kg	≤150	Not detected	
Limit	Total volatile organic compounds /g/L	≤700	44	
Quality guarantee period (At room temperature/month)		——	≥6	≥6

3.2. Grouting Equipment and Grouting Technology

The supporting parts of the grouting equipment are shown in Figure 5. It is mainly composed of a grouting flower tube, hole packer, grouting tube, special injection gun, high pressure hose, grouting pump, A component material and B component material.

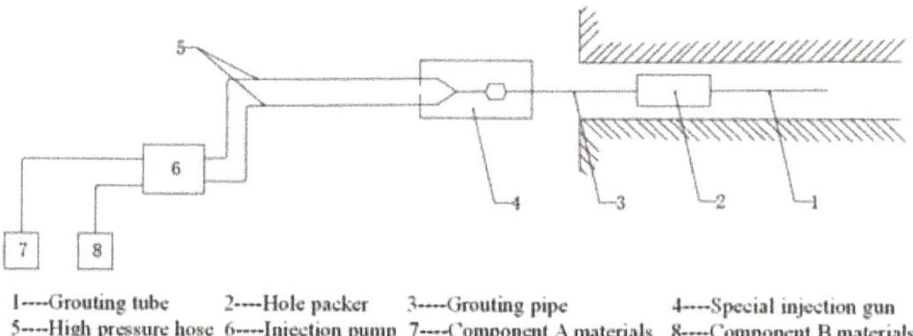

Figure 5. Part drawings of grouting equipment.

The grouting construction technology mainly includes: (1) the design of the hole position; (2) drilling; (3) installing the injection tube; (4) connecting the equipment air supply pipeline and grouting pipeline; (5) starting grouting; (6) stopping grouting; (7) replacement grouting; and (8) cleaning equipment and accessories after grouting. The detailed grouting process diagram is shown in Figure 6.

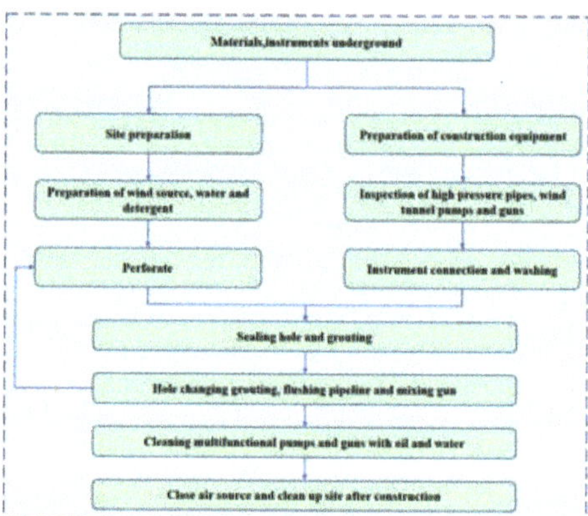

Figure 6. Grouting process flowchart.

4. Numerical Simulation Verification of Grouting Effect

FLAC^{3D} 5.0 is a finite difference software package for studying geotechnical problems. In this paper, through numerical simulation, the stress and plastic zone changes of the coal wall before and after grouting were observed and compared, the peak position of the advanced abutment pressure of the coal body before and after grouting were determined and the failure range of the coal body in front of the coal wall were investigated.

4.1. Model Design Size

The model inclination length (X-direction) is 240 m, the strike length (Y-direction) is 200 m, and the height (Z-direction) is 67 m. According to the geological histogram of #15 coal in the Duanwang Mine, the model establishes 16 strata including the coal seam, a total of 3,245,547 nodes, and 3,168,000 grids with a minimum grid size of 1 m $\times$ 1 m $\times$ 1 m. The Mohr–Coulomb constitutive model is used in the model. Vertical stress is applied on the model boundary, which is calculated according to a depth of 230 m at the upper boundary of the model. The vertical stress is 6.5 MPa and the lateral pressure coefficient is

1.2. In addition to the upper boundary, the normal displacement constraint is applied on five boundaries. The numerical simulation model is shown in Figure 7. The mechanical parameter table of the surrounding rock mass is shown in Table 3.

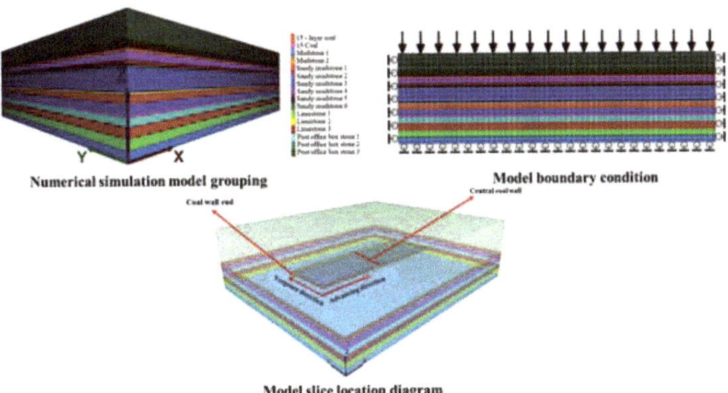

Figure 7. Numerical simulation model diagram.

Table 3. Rock mechanics parameters.

Serial Number	Lithologic Characters	Lamination Thickness (m)	Total Thickness (m)	Bulk Modulus (GPa)	Shear Modulus (GPa)	Cohesion (MPa)	Angle of Internal Friction (°)	Tensile Strength (MPa)
16	sandy mudstone	12.70	67.20	7.01	4.50	1.93	29	2.50
15	fine-grained sandstone	2.30	54.50	8.95	5.89	2.12	29	3.00
14	sandy mudstone	2.20	52.20	7.01	4.50	1.93	29	2.50
13	mudstone	0.70	50.00	4.01	3.50	1.50	28	2.00
12	sandy mudstone	3.50	48.30	7.01	4.50	1.93	29	2.50
11	limestone	1.90	44.80	11.95	8.87	2.51	32	4.0
10	fine-grained sandstone	1.00	42.90	8.95	5.89	2.12	29	3.00
9	sandy mudstone	11.50	41.90	7.01	4.50	1.93	29	2.50
8	limestone	1.30	20.40	11.95	8.87	2.51	32	4.00
7	#15 coal	4.00	29.10	3.98	2.50	1.03	22	1.50
6	sandy mudstone	4.80	25.10	7.01	4.50	1.93	29	2.50
5	#15 lower coal	1.30	20.30	3.98	2.50	1.03	22	1.50
4	fine-grained sandstone	4.00	19.00	8.95	5.89	2.12	29	3.00
3	sandy mudstone	5.00	15.00	7.01	4.50	1.93	29	2.50
2	limestone	5.00	10.00	11.95	8.87	2.51	32	4.00
1	mudstone	5.00	5.00	4.01	3.50	1.50	28	2.00

The specific simulation steps are as follows: First, the vertical force is applied under the original rock stratum to simulate the pressure of the upper rock stratum. The upper boundary is a free constraint and the other boundary simulates the bottom layer with fixed constraints; the original stress is set horizontally. After the model is stable, the excavation simulation of the #15 coal is carried out. From x = 40 m and y = 75 m, the coordinate point, the working face with 160 m inclined length is excavated along the x-axis and the working face advances 50 m to the y-axis. Then, the model is calculated and, to be stable, the end and middle of the working face are sliced and the plastic zone and its stress nephogram before and after grouting are observed.

4.2. *Comparative Analysis before and after Grouting*

4.2.1. Comparison of Advanced Abutment Pressure before and after Grouting

The distribution of the advanced abutment pressure before grouting is shown in Figure 8. It can be seen from Figure 8 that the advanced abutment pressure of the working face can be roughly divided into four regions:

1. The stress reduction area is 0–3 m in front of the working face, which is the most significant area of coal fragmentation in front of the working face. The cracks in the coal body are crisscross and there is a risk that the whole coal body will fall off.
2. Severely affected area, 3–10 m ahead of the working face. The area is the face of the advanced abutment pressure severely affected area. In this area, the stress value increases sharply to the peak stress, and then falls sharply. The stress on the coal within the scope of the region is the strongest with high stress causing the coal to be broken.
3. The slowly decreasing area is 10–30 m in front of the working face. This area is the front part of the working face advanced abutment pressure influence area. The deformation increases slowly and the stress value decreases slowly, which has a continuous effect on the coal body in the area and the coal fracture is small in area.
4. In the original rock stress area, beyond 30 m in front of the work, the stress value of coal in this area gradually decreases to the original rock stress and the integrity of coal in this area is good.

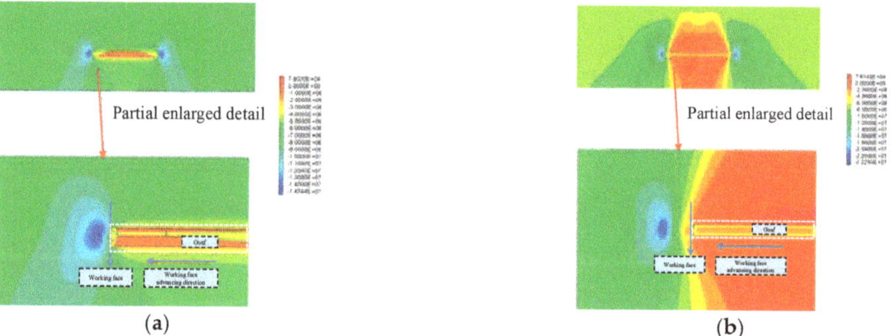

Figure 8. Local stress nephogram before grouting. (**a**) End section; (**b**) Middle section.

The advanced abutment pressure of the working face after grouting is shown in Figure 8. It can be seen from Figures 9 and 10 that the advanced abutment pressure of the coal wall is consistent with the distribution before grouting, which first increases and then decreases. The coal and rock mass is relatively broken within 0–3 m and the abutment pressure is small. The stress concentration area is formed within 3–10 m. The maximum supporting pressure is 24.5 MPa, which is 0.5 MPa higher than that before grouting. The maximum peak point after grouting is about 3 m ahead of the working face and 2 m ahead of the peak point before grouting. From this point of view, grouting has obviously improved the strength and bearing capacity of the coal wall in the working face, strengthened the mechanical properties of the broken coal affected by mining, and improved the coal strength greatly. After grouting reinforcement, the cracks in the coal body are filled and the self-supporting and self-stability ability is strengthened. The failure conditions of the coal body are transformed from the original weak strength condition of the crack to the strength condition close to the coal body.

Figure 9. Local stress nephogram after grouting. (**a**) End section; (**b**) Middle section.

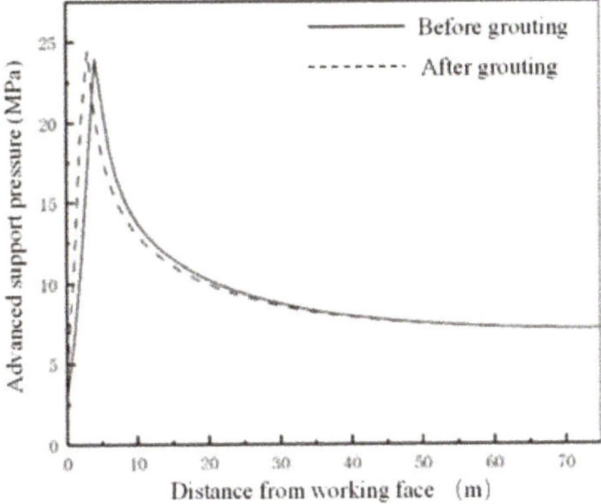

Figure 10. Comparison of advanced abutment pressure before and after grouting.

4.2.2. Plastic Zone Analysis of Working Face before and after Grouting

Figure 11 is the plastic zone map of the working face before grouting. It can be seen from Figure 11 that the strength of the coal and rock mass is small; the failure range of the plastic zone in the coal wall of the working face is 2–5 m. The failure mode includes the ongoing tensile and shear failure that has occurred. The failure degree of the plastic zone in the roof rock is relatively complex and the failure range is highly covered to the upper rock. Near the left and right sides of the working face, the width of the plastic zone of the coal wall is small, being only 2 m. The working face tends to the middle position and the width of the plastic zone of the coal wall reaches 5 m. The width of the plastic zone of the coal wall is related to the advanced abutment pressure. The boundary of the plastic zone of the coal wall is the position where the advanced abutment pressure reaches a maximum value.

Figure 12 shows the distribution of the plastic zone at the end and middle of the coal wall of the working face after grouting. It can be seen from Figure 12 that, after grouting, due to the increase in coal strength at the coal wall of the working face, the failure width of the plastic zone of the coal wall is significantly reduced compared with that before grouting. The failure width of the plastic zone at the end is reduced from 2 m to 1 m and the failure width of the central plastic zone is reduced from 5 m to 3 m. The plastic failure region still includes ongoing tensile shear failure that has occurred. After the coal wall grouting, due to the reinforcement of the coal body, the stress concentration at the crack tip is weakened, the stress state of the rock mass near the crack is improved, and the strength and integrity of the coal body are improved, so that the overall bearing capacity of the coal wall of the working face is improved and the rib spalling of the working face is improved.

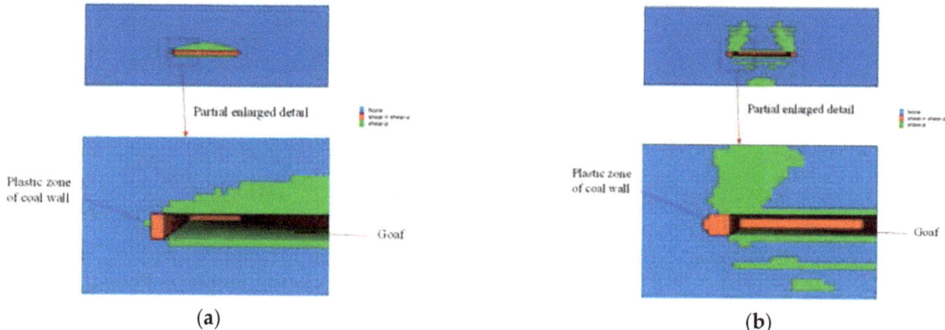

(a) (b)

Figure 11. Plastic zone of coal wall before grouting. (**a**) Plastic zone of end coal wall; (**b**) Plastic zone of central coal wall.

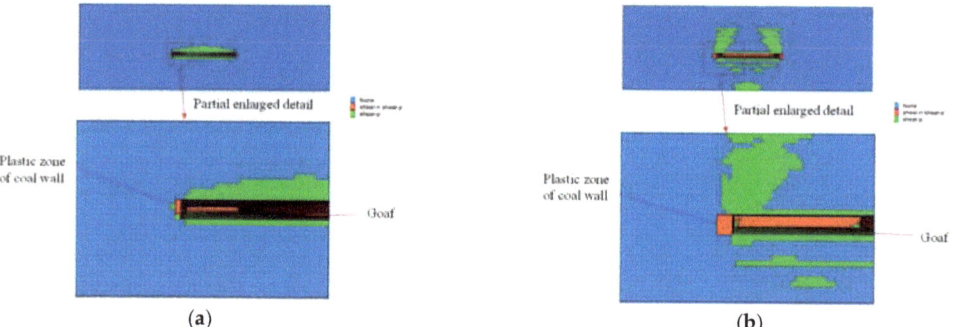

(a) (b)

Figure 12. Plastic zone of coal wall after grouting. (**a**) Plastic zone of end coal wall; (**b**) Plastic zone of central coal wall.

5. Grouting Scheme and Field Engineering Practice

5.1. Coal Wall Spalling Mechanism Analysis of Working Face

Based on investigation of the geological condition and numerical simulation of the 150505 fully mechanized mining face in the Duanwang Mine, the main causes of coal wall spalling in working face can be summarized as follows:

(1) Large Mining Height of Working Face

Since the working face is a fully mechanized working face with one-time mining, the working face has high mining height and high mining intensity, which makes the advanced abutment pressure of the working face larger and causes the coal body in the stress-increasing area to be broken. When the working face is pushed to the broken coal body position, the coal wall is easily broken under the extrusion of the support panel, resulting in a large area of rib caving. The probability and severity of coal wall spalling increase with mining height.

(2) Working Face Affected by Geological Structure

The geological exploration results of the geological structure in front of the working face show that there were nine faults in the working face, which influenced the continuous mining of the working face. When the working face passes the fault, the roof dislocation is greater and the support cannot provide sufficient support resistance to prevent roof subsidence. If the fault-affected area is large, the roadway roof cannot be effectively supported for a long time, which will generate significant security risks and affect the stability of the working face [21,22].

(3) Influence of Working Face Roof Lithology

The coal seam roof of the working face is composed of 1.31 m direct roof strata (limestone) and 11.47 m old roof strata (sandy mudstone). The strength of the mudstone is low, easily weathered in water and prone to collapse. Because the strength of the mudstone is low and it is easily broken, the support cannot effectively connect to the roof, which leads to coal wall spalling and roof caving.

(4) Influence of Roof Pressure on Coal Wall Spalling

Under the action of self-weight and roof pressure, the coal wall will show two failure modes of tensile fracture failure and shear failure. In the brittle and hard coal body, transverse tensile stress that cannot be alleviated or released by the deformation of the coal body is generated in the coal wall. When the tensile stress reaches the failure strength of the coal body, coal wall failure spalling occurs. The transverse deformation in the soft coal seam will relieve or release the tensile stress caused by compression; finally, shear sliding failure occurs because the shear stress in the coal wall is greater than the shear strength. In the large mining height working face, the mine pressure behavior is more intense. When the roof is near the initial pressure and periodic pressure, the abutment pressure of the coal wall of the working face reaches a peak value; the spalling is more likely to occur under the action of high roof pressure. After the rapid advance, the roof pressure is alleviated and the depth and length of the coal wall spalling will be reduced accordingly.

5.2. Grouting Scheme

In the working face, hole grouting is needed in the treatment stage. The working face is 4 m high, 2.5 m below the roof, 6 m deep, and 45° upward. In 150505, the return airway from the machine tail is 6 m, 2.5 m below the roof hole grouting, with a hole depth of 6 m, and an elevation angle 45° upward for the hole. The hole packer is placed 1–1.5 m away from the orifice. The pore distance preliminary design is 4 m with single hole grouting quantity control in 20 groups. A schematic diagram of the site grouting is shown in Figure 13. The diameter of the grouting borehole is 32 mm. To prevent the slurry from leaking in the roadway wall caused by excessive grouting pressure, the grouting pressure is 2–5 MPa.

(**a**)

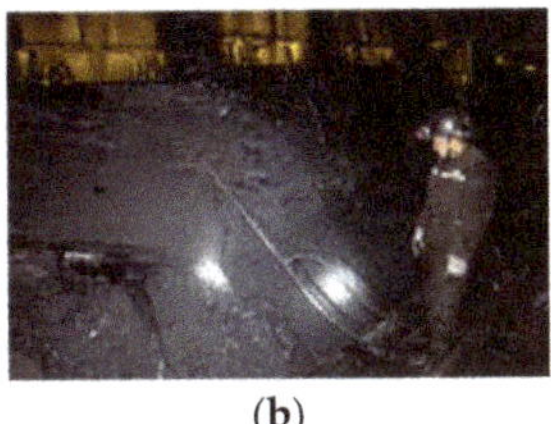
(**b**)

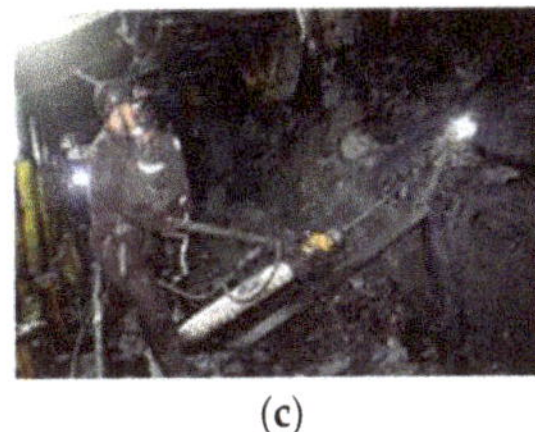
(**c**)

Figure 13. On-site grouting schematic diagram. (**a**) Grouting pump and connecting injection gun; (**b**) Connection between grouting pipe and grouting gun; (**c**) Perforate.

5.3. Field Effect Analysis

By comparing the rib spalling of the working face before and after grouting, it was found that the number of rib spalling events in the coal wall after grouting was significantly reduced and that the average size of the rib spalling coal body was also significantly reduced, indicating that coal injection helped to prevent rib spalling of the coal wall. When the working face was pushed to the grouting section, the coal wall of the working face was observed and it was found that the coal wall formed a bonding layer; the grouting solution appeared around the obvious cracks, indicating that the slurry had created a bonding effect on the coal body. In addition, with advancement of the working face, the number and depth of coal wall spalling events in the working face were gradually reduced. During the final mining of the working face, the coal wall was basically flat and there was no coal wall spalling with large spalling depth. The grouting reinforcement effect was obvious and the efficient production of the working face was achieved while ensuring safety.

In addition, when the roof is broken, more than 10 people need to be assigned and production is stopped to maintain the roof. After application of this material, the construction is carried out during the maintenance shift, which does not affect the coal cutting of the production shift. Only three to four people are needed to organize the construction and labor productivity is increased by 150%. After grouting reinforcement, there is no longer a requirement to stop production to maintain the roof so that the coal mining work on the working face can occur continuously and as normal. On a monthly basis, over 1500 tons of raw coal mining were saved and raw coal production increased by 15%. The loss of coal due to the fault fracture zone was reduced and the recovery rate increased by 8.9%.

6. Conclusions

The main conclusions relating to coal wall spalling of the 150505 fully mechanized working face in Duanwang Mine are as follows:

1. The working face has a large mining height, there are many faults and folds in the mining area and the roof of the coal seam is weak.
2. A new nano-composite low temperature reinforcement material was developed, which has the characteristics of rapid reaction, low polymerization temperature, adjustable setting time, high strength and environmental protection.
3. Through field observation and numerical simulation, the strength difference of the coal wall in the 150505 fully mechanized working face of Duanwang Mine, before and after grouting, was analyzed. It was found that the stability of the coal wall in the working face was significantly enhanced and rib spalling was significantly improved after comprehensive rib spalling prevention grouting measures were adopted.
4. Based on investigation of the 150505 fully mechanized working face in the Duanwang Mine, a comprehensive prevention and control scheme of rib grouting reinforcement was proposed and industrial tests were carried out. The grouting reinforcement effect obtained was clear.

Author Contributions: Conceptualization, H.L., Y.C., Z.H. and Q.L.; methodology, Z.H. and W.C.; software, H.Z.; validation, Z.L. and S.Q.; formal analysis, Z.L. and Q.L.; investigation, Z.H. and H.W.; data curation, Z.H., W.C. and S.Q.; writing—original draft preparation, H.L. and Z.H.; writing—review and editing, Z.L. and H.Z.; supervision, H.L.; funding acquisition, H.L. All authors have read and agreed to the published version of the manuscript.

Funding: This research was funded by the National Natural Science Foundation of China grant number 52004289 And the Fundamental Research Funds for the Central Universities was funded by 2022XJNY01.

Institutional Review Board Statement: Not applicable.

Informed Consent Statement: Not applicable.

Data Availability Statement: Not applicable.

Acknowledgments: The authors are thankful to the anonymous reviewers for their kind suggestions.

Conflicts of Interest: The authors declare no conflict of interest.

References

1. Yu, X.Y.; Wang, Z.S.; Yang, Y.; Mao, X. Discrete element numerical simulation of overburden movement law in fully mechanized top-coal caving mining with large mining depth. *J. Min. Strat. Control Eng.* **2021**, *3*, 01353.
2. Wang, H.B. Study on Mining Pressure Behavior and Coal Wall Reinforcement Technology of Soft Coal Island Face with Large Dip Angle. Ph.D. Thesis, China University of Mining and Technology (Beijing), Beijing, China, 2014.
3. Liu, Q. Study on Coal Wall Spalling Control Based on Working Condition of Fully Mechanized Mining Support with Large Mining Height. Ph.D. Thesis, Taiyuan University of Technology, Taiyuan, China, 2019.
4. Zhang, J.X.; Ju, Y.; Zhang, Q.; Ju, F.; Xiao, X.; Zhang, W.; Zhou, N.; Li, M. System and method of low damage mining of mine ecological environment. *J. Min. Strat. Control Eng.* **2019**, *1*, 013515.
5. Zhao, Y.L.; Zhang, L.Y.; Liao, J.; Wang, W.J.; Liu, Q.; Tang, L. Experimental study of fracture toughness and subcritical crack growth of three rocks under different environments. *Int. J. Geomech.* **2020**, *20*, 04020128. [CrossRef]

6. Liu, J.H.; Zhao, Y.L.; Tan, T.; Zhang, L.Y.; Zhu, S.T.; Xu, F.Y. Evolution and modeling of mine water inflow and hazard characteristics in southern coalfields of China: A case of Meitanba mine. *Int. J. Min. Sci. Technol.* **2022**, *32*, 513–524. [CrossRef]
7. Zhao, Y.L.; Liu, Q.; Zhang, C.S.; Liao, J.; Lin, H.; Wang, Y.X. Coupled seepage-damage effect in fractured rock masses: Model development and a case study. *Int. J. Rock Mech. Min. Sci.* **2021**, *144*, 104822. [CrossRef]
8. Zhu, S.T.; Feng, Y.; Jiang, F.X.; Liu, J. Mechanism and risk assessment of overall-instability-induced rockbursts in deep island longwall panels. *Int. J. Rock Mech. Min. Sci.* **2018**, *106*, 342–349. [CrossRef]
9. Zhu, S.T.; Feng, Y.; Jiang, F.X. Determination of abutment pressure in coal mines with extremely thick alluvium stratum: A typical kind of rockburst mines in China. *Rock Mech. Rock Eng.* **2016**, *49*, 1943–1952. [CrossRef]
10. Wang, B.; Zhu, S.T.; Jiang, F.X.; Liu, J.; Shang, X.; Zhang, X. Investigating the Width of Isolated Coal Pillars in Deep Hard-Strata Mines for Prevention of Mine Seismicity and Rockburst. *Energies* **2020**, *13*, 4293. [CrossRef]
11. Wang, J.C. Coal scientific mining based on mining strata control. *J. Min. Strat. Control Eng.* **2019**, *1*, 013505.
12. Xu, L.F.; Yao, W.; Huang, T.L.; Deng, G.L.; Lin, S.C.; Yang, X. Research on optimization of filling material proportion in Yangjiagou Bauxite Mine. *J. Min. Strat. Control Eng.* **2022**, *4*, 92–99.
13. Xie, L.; Zhang, D.B.; Liang, S. Optimisation and simulation of the effect of grouted cable bolts as advanced support in longwall entries. *J. Min. Strat. Control Eng.* **2022**, *4*, 50–60.
14. Qie, L.; Shi, Y.N.; Liu, J.G. Experimental study on grouting diffusion of gangue solid filling bulk materials. *J. Min. Strat. Control Eng.* **2021**, *3*, 124–133.
15. Cai, M.F. Key Theory and Technology of Surrounding Rock Stability and Strata Control in Deep Mining. *J. Min. Strat. Control Eng.* **2020**, *2*, 033037.
16. Zhu, T.Y. Study on Strata Behavior and Control of Fully Mechanized Mining Face with Deep Ultra-Large Mining Height in Dahaize Coal Mine. Master's Thesis, Inner Mongolia University of Science and Technology, Baotou, China, 2019.
17. Kang, H.P.; Jiang, P.F.; Yang, J.W.; Wang, Z.G.; Yang, J.H.; Liu, Q.B.; Wu, Y.Z.; Li, W.Z.; Gao, F.Q.; Jiang, Z.Y.; et al. Roadway soft coal control technology by means of grouting bolts with high pressure-shotcreting in synergy in more than 1000 m deep coal mines. *J. China Coal Soc.* **2021**, *46*, 747–762.
18. Xie, S.R.; Wang, E.; Chen, D.D.; Jiang, Z.S.; Li, H.; Liu, R.T. Collaborative control technology of external anchor-internal unloading of surrounding rock in deep large-section coal roadway under strong mining influence. *J. China Coal Soc.* **2022**, *47*, 1946–1957.
19. Zhou, X.Q.; Fang, Y.Z. *Manager of Coal Mine Safety Production (Retraining Textbook)*; Coal Industry Press: Beijing, China, 2005.
20. Ma, Z.; Pang, H.; Yang, Y.L.; Xu, Y.L. Review progress of chemical grouting materials. *Guangzhou Chem.* **2014**, *1*, 9–13.
21. Wang, Z.H.; Yang, J.H.; Meng, H. Mechanism and controlling technology of rib spalling in mining face with large cutting height passing through fault. *J. China Coal Soc.* **2015**, *40*, 42–49.
22. Jiang, J.Q.; Wu, Q.L.; Qu, H. Evolutionary characteristics of mining stress near the hard-thick overburden normal faults. *J. Min. Saf. Eng.* **2014**, *31*, 881–887.

Article

Design Method and Application of Stope Structure Parameters in Deep Metal Mines Based on an Improved Stability Graph

Xingdong Zhao and Xin Zhou *

Laboratory for Safe Mining in Deep Metal Mine, Northeastern University, Shenyang 110819, China
* Correspondence: 2010397@stu.neu.edu.cn

Abstract: Deep mining has become an inevitable trend of mining development. Previously conducted studies have established that reasonable stope structure parameters are the premise to ensure the safe and efficient production of deep mines. In order to ensure the safety of deep mining, in this paper, we systematically review the existing stope structure parameter design methods, and then put forward a deep stope structure design method based on the stability of mining rock mass. Based on rock mass quality classification, this method uses a critical span graph and an improved stability graph, and fully considers the influence of joint occurrence and mining stress on the stability of surrounding rock, to design the stope structural parameters. Taking into consideration the deterioration of the quality of deep rock mass, we collect mining data at home and abroad, improve the stability graph, and make it suitable for the design of stope structural parameters with different mining methods. The design process of stope structural parameters is expounded through field engineering cases, and it has specific guiding significance for the design of stope structural parameters in deep metal mines.

Keywords: safety mining; deep mining; stope structural parameters; improved stability graph; engineering application

Citation: Zhao, X.; Zhou, X. Design Method and Application of Stope Structure Parameters in Deep Metal Mines Based on an Improved Stability Graph. *Minerals* **2023**, *13*, 2. https://doi.org/10.3390/min13010002

Academic Editor: Yosoon Choi

Received: 8 November 2022
Revised: 30 November 2022
Accepted: 17 December 2022
Published: 20 December 2022

1. Introduction

With the gradual depletion of shallow metal mineral resources, the development of deep mineral resources has become an inevitable trend. After entering deep mining, the in situ stress increases and the mining technical conditions and environment deteriorate seriously, which presents significant challenges to the safety production of deep metal mines [1–3]. Reasonable stope structure parameters are an effective means to ensure safe and efficient production in deep metal mines.

At present, the design methods of stope structural parameters include the engineering analogy method, theoretical analysis method, numerical analysis method, and comprehensive analysis method; the latter two methods are widely used. Qiu et al. analyzed the stope stability of the Bainiuchang Mine based on the pillar area bearing theory, and they optimized the stope structure parameters using ANSYS numerical simulation [4]. Based on precision finite element modeling and simulation, Zhang et al. determined a reasonable width range and interval value of the strip for strip mining with subsequent filling [5]. Khayrutdinov et al. used the FLAC 3D modeling software and changed the filling strength to control the stress–strain behavior of rock mass, to reduce the impact of underground mining on the surface, and to improve mining safety [6,7]. Li et al. proposed a dynamic cross layout model based on IDZs, which could dynamically adjust the sublevel height and drift spacing according to the ore-rock bulk flow parameters, economic indicators, ore body occurrence conditions, drilling machine, etc. [8]. Taking the Chengchao Iron Mine as the engineering background, Tan et al. used the method of combining a theoretical calculation, a numerical simulation, and a physical similarity experiment to sublevel height, production drift spacing, and drawing space [9].

The design of stope structural parameters is a complex nonlinear problem that usually involves multiple decision variables such as rock mechanics, mining ground pressure, loss, and the dilution index. It is difficult to express the relationships among the parameter variables by using exact mathematical and mechanical expressions, and correlations among the variable parameters are weak. With continuously increasing mining depth and the continuous expansion of mining scale, the geological environment and stress conditions of underground mining sites become more complex [10], and as a result the existing methods for designing stope structural parameters are no longer applicable due to their limitations, as summarized in Table 1.

Table 1. Design methods for designing stope structural parameters.

Method	Definition	Limitations
Engineering analogy method	According to the existing mine data, it is applied to similar engineering objects, and then the corresponding stope structure parameters of the research object are obtained.	With continually increasing underground mining conditions, it becomes more difficult to find underground mining structures under the same mining conditions.
Analytical method	Taking some mines as the research object, and aiming at the specific mining conditions and environment of mines, the formula calculation method of structural parameter design of underground stope is constructed by using mechanical theory.	The structure of deep rock mass is complex, uncertain, and fuzzy. Using the ideal mathematical model may have good results in a specific mine, and can not reasonably and comprehensively describe all situations.
Numerical analysis method	Using numerical simulation software to establish the geometric model of the stope, and its excavation process is simulated to comprehensively compare and analyze the stress, displacement, and plastic zone distribution of the stope after excavation, and therefore, to obtain better parameters of the stope.	Optimization is carried out on the given stope structural parameters, and there is no method to determine the initial structural parameters. At the same time, the optimization analysis process of the modeling cycle is very cumbersome and heavy workload.
Comprehensive analysis method	By studying the relevant factors to determine the stope structural parameters, the corresponding evaluation system model or mathematical formula is established by using some emerging Sciences (neural network, genetic algorithm, and artificial intelligence) to optimize the stope structural parameters.	The calculation results are based on the size of sample data and the means and methods used in modeling, and the results of different methods are quite different.

Deep stope is a special mining technical condition with strong mining disturbance under high stress. When designing stope structural parameters, we should consider the occurrence of ore body, geological structure, and rock mass quality as well as fully analyze the impact of mining stress induced by deep mining on surrounding rock stability. In particular, it is necessary to break through the stope structure design based on the "experience method" and the "engineering analogy method" and change the design method for stope structure parameters based on deep mining ground pressure response. The design method for stope structural parameters that is proposed in this paper is based on rock mass quality classification, and the structural parameters of deep stope are designed by using an improved critical span graph and an improved stability graph. This method fully considers the key influencing factors such as the occurrence of mining ore body, joint development degree, and mining stress, to ensure the stability of ore and rock in the mining process and to achieve the purpose of safe and efficient production.

2. Materials and Methods

2.1. Design Principle of Stope Structural Parameters

From the actual production conditions of mines at home and abroad, on the one hand, the stope structure parameters play a decisive role in the stability of the stope, and on the other hand, they also affect the economic benefits of mining [11,12], as shown in Figure 1.

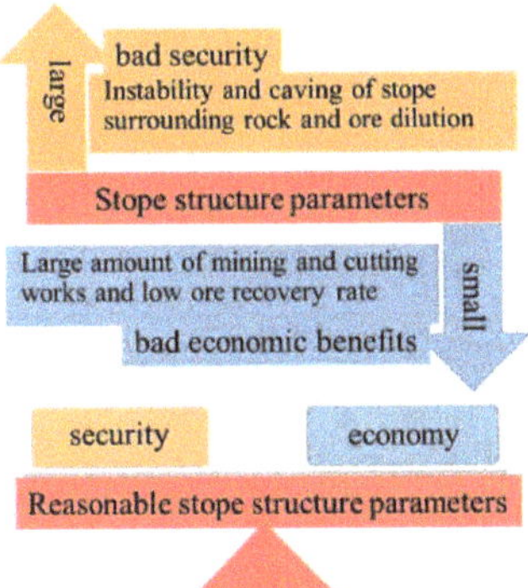

Figure 1. Influence of stope structural parameters on safety and the economy.

When the structural parameters of the stope are too large, it leads to instability and caving of ore and rock, increases the loss and dilution, and makes it impossible to operate normally and safely. When the stope structural parameters are too small, it leads to large mining and cutting quantities and low ore recovery, which reduces the economic benefits. Therefore, optimization of stope structure parameters should be considered from two aspects, i.e., safety production and economic benefit. The relationship between these two aspects should be balanced in the design, and the stope structure parameters should be increased as much as possible on the premise of ensuring safety in order to improve economic benefit.

2.2. Design Process of Stope Structural Parameters

The design of structural parameters of deep stope is related to the occurrence of a mining ore body, the degree of joint development, in situ stress, and other natural factors, as well as the selected mining method, mining sequence, and ground pressure control method. The underground stope can be roughly seen as a cuboid, which is composed of three elements: length, width, and height; its model is shown in Figure 2.

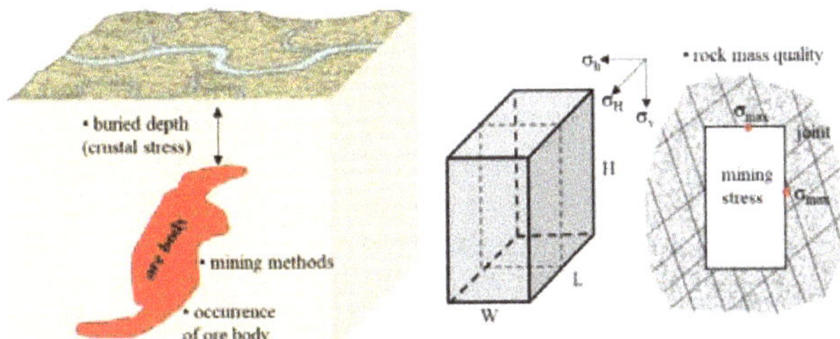

Figure 2. Mechanical model of underground stope.

The specific process of stope structure parameter design is shown in Figure 3. The general idea is based on the technical and economic conditions of deposit mining, and a critical span graph and an improved Mathews stability graph are used as tools to incorporate into the stope structure parameter design the factors affecting stope stability such as rock mass quality, joint occurrence, and mining stress.

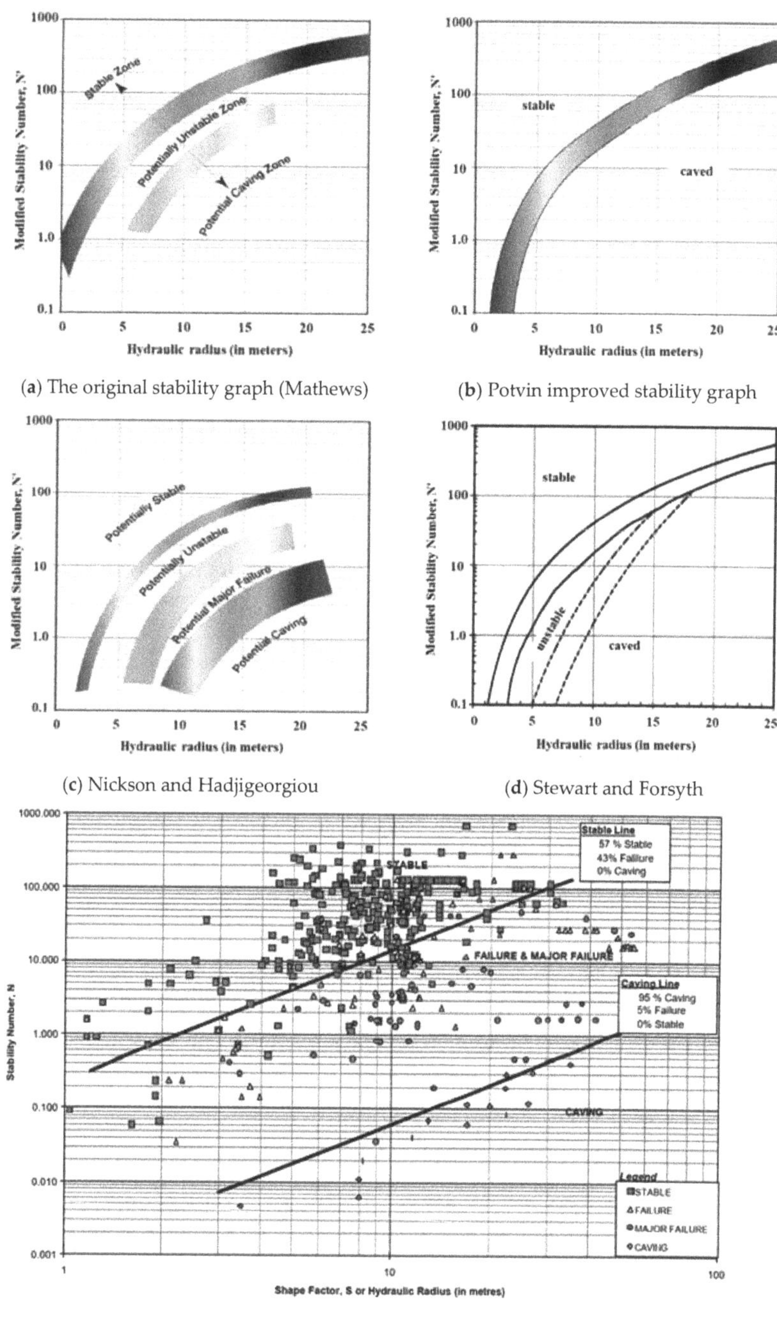

(a) The original stability graph (Mathews)

(b) Potvin improved stability graph

(c) Nickson and Hadjigeorgiou

(d) Stewart and Forsyth

(e) Mawdesley

Figure 5. Stability graph development process. (**a**) The original stability graph (Mathews); (**b**) Potvin improved stability graph; (**c**) Nickson and Hadjigeorgiou improved stability graph; (**d**) Stewart and Forsyth improved stability graph; (**e**) Mawdesley improved stability graph.

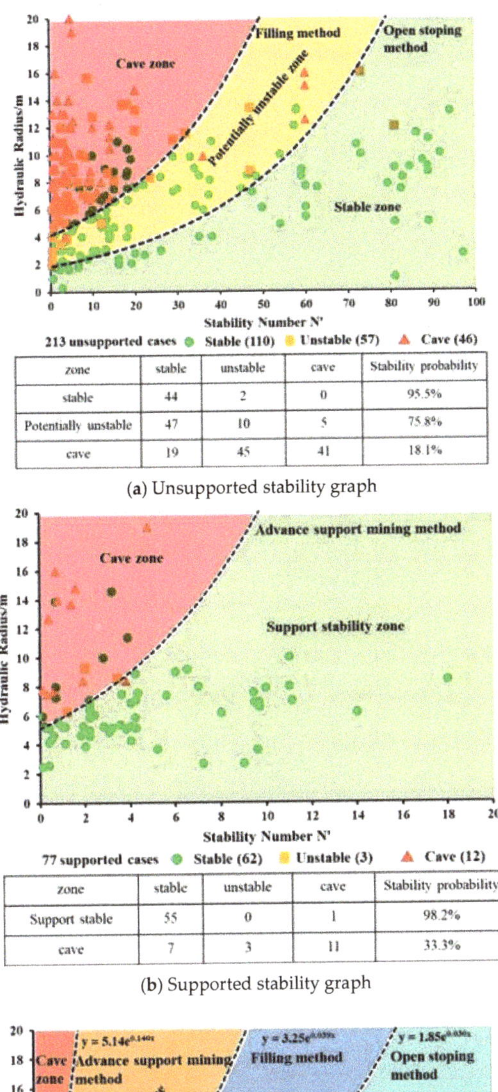

zone	stable	unstable	cave	Stability probability
stable	44	2	0	95.5%
Potentially unstable	47	10	5	75.8%
cave	19	45	41	18.1%

(**a**) Unsupported stability graph

zone	stable	unstable	cave	Stability probability
Support stable	55	0	1	98.2%
cave	7	3	11	33.3%

(**b**) Supported stability graph

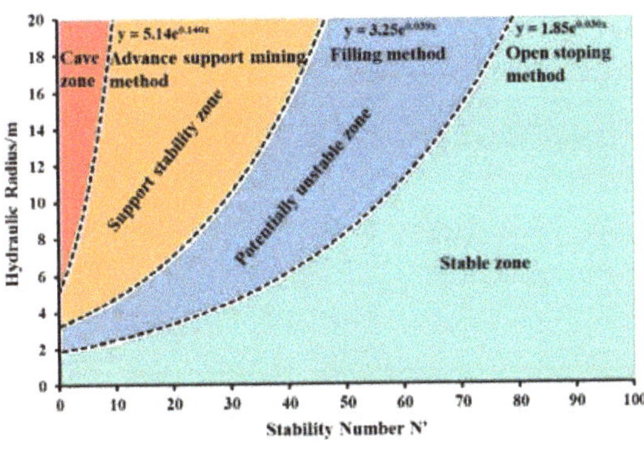

(**c**) Improved stability graph

Figure 6. Improved stability graph. (**a**) Unsupported stability graph; (**b**) Supported stability graph; (**c**) Improved stability graph.

The data in Figure 6a,b is integrated to form a complete stability graph, as shown in Figure 6c. The boundary between the stable zone and the potentially unstable zone can be used as the maximum allowable hydraulic radius of the open stoping method. The boundary between the potentially unstable zone and the support stability zone can be used as the maximum allowable hydraulic radius for filling mining. The boundary between the support stability zone and the cave zone can be used as the maximum allowable hydraulic radius of the advance support mining method.

3. Case Study

3.1. Project Profile

The Sanshandao gold mine is located in Laizhou City, Shandong Province. The ore body mainly occurs in the Sanshandao fault zone. The lithology is mainly pyrite sericitized cataclastic rock and sericitized granite, and the rock mass stability is relatively poor [25]. The three-dimensional model of the ore body is shown in Figure 7. The average dip angle of the ore body is 50° (Figure 7a), and the design stope is located in the middle section from −915 m to approximately −960 m, with an average thickness of about 45 m (Figure 7b). Therefore, it belongs to an inclined thick large ore body.

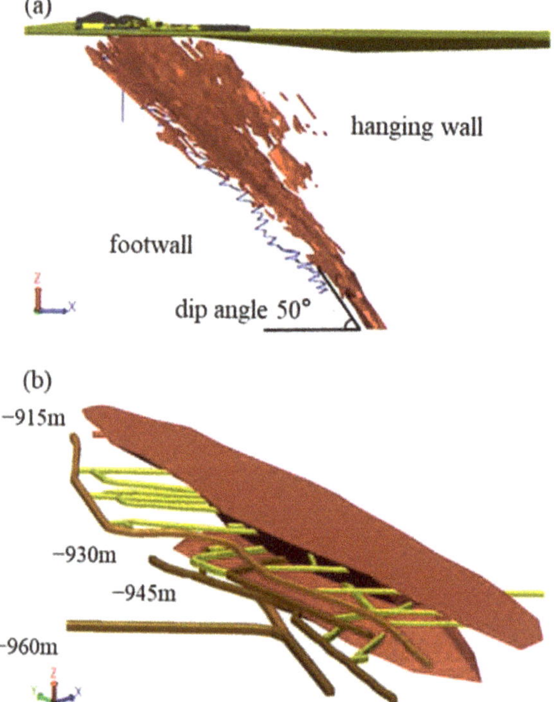

Figure 7. Three-dimensional model of ore body. (**a**) Overall three-dimensional model of ore body; (**b**) Three dimensional model of −915m ~ −960m middle section ore body.

In order to ensure the safety and efficiency of mining, it is proposed to adopt the sublevel open stope and subsequent filling mining method. The three views of the mining method are shown in Figure 8.

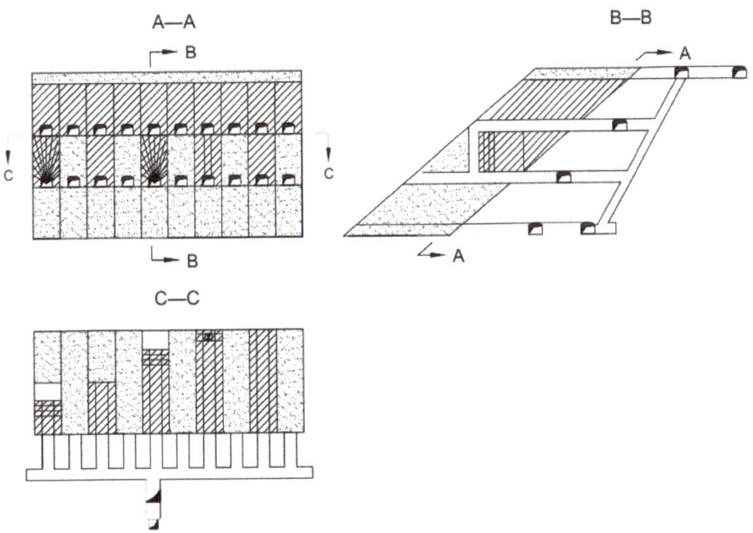

Figure 8. Sublevel open stope and backfilling.

The current stage height of the mine is 45 m. According to the existing development project, three sections are set for mining, and the stope height is determined to be 15 m. Then, the limit span (width) of stope stability is determined with the help of the RMR stability graph; on this basis, the stope length is designed by using the improved stability graph.

3.2. Rock Mass Quality Estimation

Barton rock mass quality classification (Q) and rock mass geomechanics classification (i.e., RMR) are common methods for quantitative evaluation of rock mass quality grade [14,26]. Before classification, the rock quality index (RQD), joint conditions, uniaxial compressive strength of intact rock, in situ stress conditions, and other evaluation indexes are collected.

The rock quality index (RQD) determines the range of core belonging to the ore body according to the spatial position relationship between exploration drilling and ore body as shown in Figure 9, and calculates the ratio of the total length of the part exceeding 10 cm in the core to the total length of the core. Finally, the RQD value is 23.92.

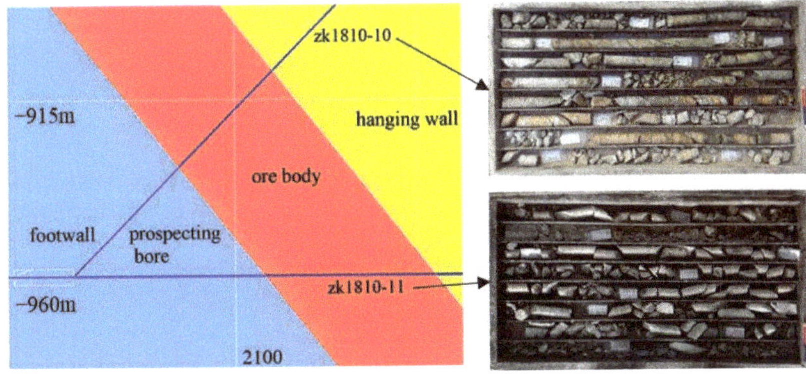

Figure 9. RQD calculation from exploration drill core.

The survey line method is used to collect the occurrence information of joints in the exploration roadway in the middle section of −960 m. The occurrence of joints is grouped by Dips software and the occurrence of dominant joints is determined, as shown in Figure 10. The joints are well developed and penetrating, with an opening of less than 1 mm. The roughness of the joint surface is general, partially filled with mud, slightly weathered, and wet.

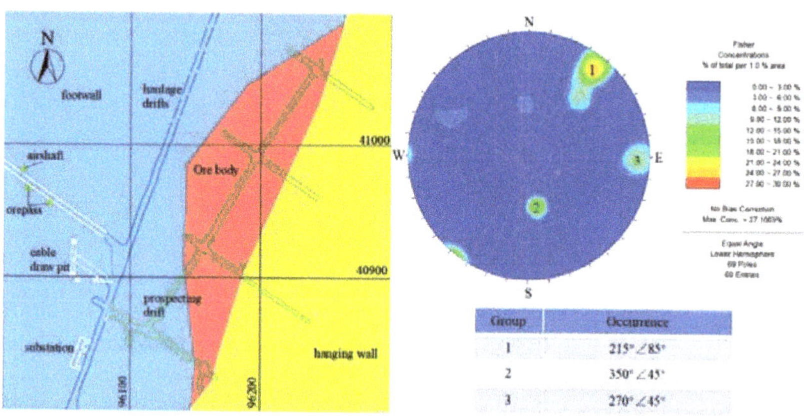

Figure 10. Survey joint occurrence in prospecting drift and determination of dominant joints with Dips software.

Rock samples were selected for the rock mechanics experiment, and the uniaxial compressive strength of intact rock was measured to be 86.74 MPa. The Q and RMR rock mass quality evaluation methods were used to evaluate the ore and rock quality, as shown in Tables 2 and 3.

Table 2. Quality classification of Barton rock mass.

Evaluating Indicator	RQD	J_n	J_r	J_a	J_w	SRF	Grade	Rating	Description
Score	23.92	6	1.5	3	1	2	1.00	IV	poor

Table 3. Rock mass geomechanics classification (i.e., RMR).

Evaluating Indicator	Uniaxial Compressive Strength	RQD	Joint Spacing	Joint Surface Condition	Groundwater	Joint Occurrence	Grade	Rating	Description
Score	7	3	15	15	10	−10	40	IV	poor

3.3. Stope Structure Parameter Design

The RMR critical span graph is used to determine the maximum allowable span (width) of the stope. When the RMR score is 40, the maximum unsupported span is 8 m, and therefore, the width of the stope is 8 m.

According to the above-determined stope width of 8 m and stope height of 15 m, and combined with the indoor rock mechanics experiment and field joint investigation results, the stability coefficient N′ of stope roof and two walls is calculated.

The stability number N′ is calculated as follows [27]:

$$N' = Q' \times A \times B \times C \tag{1}$$

where Q′ is the index of rock mass quality, which is the modified Q system classification method; A incorporates the effects of mining stress; B is the joint occurrence adjustment coefficient; C is the gravity adjustment coefficient.

Q′ can be represented by the following equation:

$$Q' = \frac{RQD}{J_n} \times \frac{J_r}{J_a} \quad (2)$$

where RQD is the rock quality designation, J_n is the joint set number, J_r is the joint roughness number, and J_a is the joint alteration number.

3.3.1. A Is the Influence Coefficient of Mining Stress

The value of A is determined by two parameters: mining stress and uniaxial compressive strength of intact rock. The calculation formula are as follows:

$$\sigma_c/\sigma_1 < 2;\ A = 0 \quad (3)$$

$$2 \leq \sigma_c/\sigma_1 \leq 10;\ A = 0.1125(\sigma_c/\sigma_1) - 0.125 \quad (4)$$

$$\sigma_c/\sigma_1 > 10;\ A = 1.0 \quad (5)$$

where σ_c is the uniaxial compressive strength of intact rock and σ_1 is the mining-induced stress.

The mining-induced stress (σ_1) required for calculating A value is calculated using the RS2 numerical simulation software, and the elastic Mohr Coulomb constitutive model is adopted. The displacement boundary condition is adopted, the horizontal principal stress is 40.7 MPa, and the vertical stress is 25.1 MPa. The calculation results are shown in Figure 11, and the calculation results of A value are shown in Table 3.

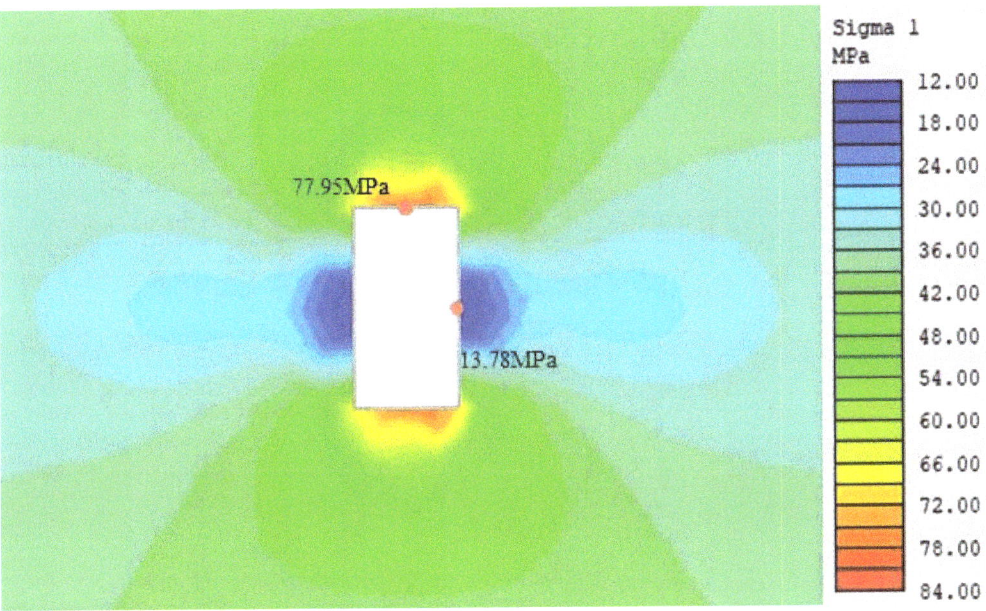

Figure 11. Mining-induced stress solution results.

3.3.2. B Is the Joint Occurrence Adjustment Coefficient

B is the joint orientation factor. According to the statistical results of joint occurrence, the main dip angle is 45°. Combined with Figure 12, the B value is 0.5.

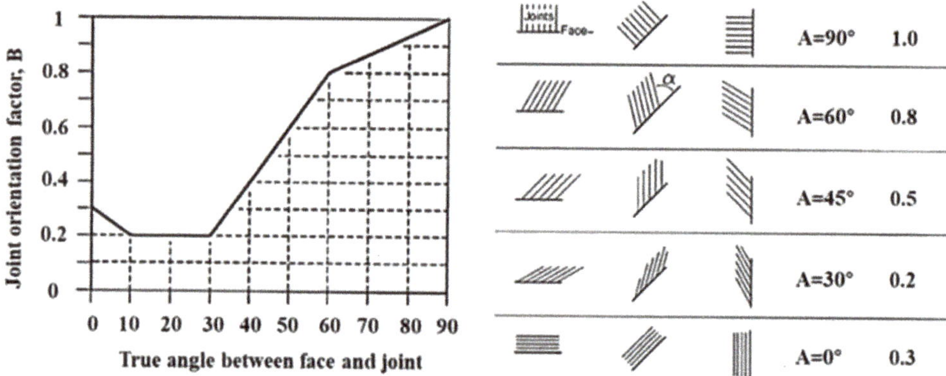

Figure 12. Joint occurrence adjustment coefficient.

3.3.3. C Is the Gravity Adjustment Coefficient

The failure modes of stope roof and two sides under the action of gravity are mainly considered, such as stope roof caving, slope slicing, slope sinking, and rock mass structural plane sliding. Potvin (1988) believed that gravity-induced failure and spalling failure depended on the dip angle of the mining surface, and the two-slope sinking and sliding failure mainly depended on the dip angle of the stable joint controlling the ore. The value of C can be estimated by the following empirical formula:

$$C = 8-6\cos\alpha \tag{6}$$

For stope roof, α is 0° and C value is 2; for stope wall, α is 90° and C value is 8.

According to the rock mechanics experiment, the joint occurrence investigation results, and empirical formula, the calculation results of stability coefficient N′ is shown in Table 4.

Table 4. Calculation statistics of stability coefficient N′.

Position	Q′	A	B	C	N′
Roof	1.99	0.1	0.5	2	0.19
Wall	1.99	0.58	0.5	8	4.61

The improved stability graph (Figure 6) is used to determine the hydraulic radius of the roof and two walls, and then the stope length is calculated, according to Formulas (7)–(9), and the results are listed in Table 5:

$$L_{roof} = \frac{2HR_{roof}W}{W-2HR_{roof}} \tag{7}$$

$$L_{wall} = \frac{2HR_{wall}H}{H-2HR_{wall}} \tag{8}$$

Stope length L

$$L = \min\left(L_{roof}, L_{wall}\right) \tag{9}$$

where L_{roof} is the maximum allowable length of stope roof; HR_{roof} is the maximum allowable hydraulic radius of stope roof; L_{wall} is the maximum allowable length of stope wall; HR_{wall} is the maximum allowable hydraulic radius of stope wall; W is stope width; H is stope height; L is stope length.

Table 5. Calculation statistics of stope length.

Stope Height/m	Stope Width/m	Position	Hydraulic Radius/m	Stope Length/m
15	8	roof	3.27	36.09
		wall	3.89	16.16

According to the calculation results in Table 5, when the stope width is 8 m and the stope height is 15 m, the stope length is restricted by the stability of the two walls, and the maximum length is 16.16 m. The average thickness of the ore body is 45 m, which can be mined in three times, and the length of each time is 15 m. Finally, the stope structure parameter is determined as 8 m × 15 m × 15 m.

4. Conclusions

The design of the structural parameters of deep stope is an optimization problem. On the premise of ensuring production safety, larger structural parameters of the stope should be selected to maximize the economic benefits of mines.

Based on the stability graph, in this paper, we outline the design process of stope structural parameters, and improve the stability graph. Although the application scope of the improved stability graph has been reduced, its data concentration is improved, the zoning is more accurate, and it is more suitable for the deterioration of the quality of deep rock mass. In addition, each zone corresponds to the corresponding mining method, which is more targeted in the design of stope structure parameters.

At the same time, taking the stope structure parameter design of −960 m middle section of the Sanshandao Gold Mine as an example, in this paper, we expound the design process of stope structure parameters in detail, which has specific guiding significance for the design of stope structure parameters.

There are many factors affecting stope stability. In this paper, for the design process of stope structure parameters, we considered the key elements such as rock mass quality, joint occurrence, and mining stress; however, we did not discuss the influence of blasting vibration on stope stability. At the same time, the determination of stope self-stability time will be the next research direction.

Author Contributions: Writing—original draft preparation and editing, investigation, software, X.Z. (Xin Zhou); funding acquisition, writing—review, X.Z. (Xingdong Zhao); translation, X.Z. (Xingdong Zhao) and X.Z. (Xin Zhou). All authors have read and agreed to the published version of the manuscript.

Funding: Project U1806208 supported by the National Natural Science Foundation-Shandong Jointed Fundation of China; projects 52130403 supported by the Key Program of National Natural Science Foundation of China; and project N2001033 supported by the Fundamental Scientific Research Business Expenses of Central Universities.

Data Availability Statement: The data presented in this study are available on request from the corresponding author.

Conflicts of Interest: The authors declare no conflict of interest.

References

1. Xie, H.-P.; Gao, F.; Ju, Y. Research and development of rock mechanics in deep ground engineering. *Chin. J. Rock Mech. Eng.* **2015**, *34*, 2161–2178. [CrossRef]
2. Li, X.; Zhou, J.; Wang, S.; Liu, B. Review and exploration on the exploitation of deep solid resources. *Chin. J. Nonferrous Met.* **2017**, *27*, 1236–1262. [CrossRef]
3. Zhao, X.; Zhou, X.; Zhao, Y.; Yu, W. Research status and progress of prevention and control of mining disasters in deep metal mines. *J. Cent. South Univ.* **2021**, *52*, 2522–2538. [CrossRef]
4. Qiu, H.-Y.; Huang, M.-Q.; Weng, Y.-J. Stability Evaluation and Structural Parameters Optimization of Stope Based on Area Bearing Theory. *Minerals* **2022**, *12*, 808. [CrossRef]
5. Zhang, L.; Hu, J.-H.; Wang, X.-L.; Zhao, L. Optimization of stope structural parameters based on mathews stability graph probability model. *Adv. Civ. Eng.* **2018**, *2018*, 1754328. [CrossRef]

6. Khayrutdinov, A.; Kowalik, T. Stress-strain behavior control in rock mass using different-strength backfill. *Min. Inf. Anal. Bull.* **2020**, *10*, 42–55. [CrossRef]
7. Rybak, J.; Khayrutdinov, M.M.; Kuziev, D.A.; Kongar-Syuryun, C.B.; Babyr, N. Prediction of the geomechanical state of the rock mass when mining salt deposits with stowing. *J. Min. Inst.* **2022**, *253*, 61–70. [CrossRef]
8. Li, G.; Ren, F.; Ding, H.; Liu, H.; Sun, M.; Li, G. A Dynamic Intersecting Arrangement Model Based on Isolated Draw Zones for Stope Structure Optimization during Sublevel Caving Mining. *Math. Probl. Eng.* **2021**, *2021*, 6669558. [CrossRef]
9. Tan, Y.; Guo, M.; Hao, Y.; Zhang, C.; Song, W. Structural Parameter Optimization for Large Spacing Sublevel Caving in Chengchao Iron Mine. *Metals* **2021**, *11*, 1619. [CrossRef]
10. Cai, M.-F.; Xue, D.-L.; Ren, F.-H. Current status and development strategy of metal mines. *Chin. J. Eng.* **2019**, *41*, 417–426. [CrossRef]
11. Zhao, X.; Niu, J.a. Method of Predicting Ore Dilution Based on a Neural Network and Its Application. *Sustainability* **2020**, *12*, 1550. [CrossRef]
12. XU, S.; AN, L.; LI, Y.; WU, J. SOM-based optimization of stope structural parameters of deep & large-sized orebody. *J. Min. Saf. Eng.* **2015**, *32*, 883–888. [CrossRef]
13. Lowson, A.; Bieniawski, Z. Critical assessment of RMR based tunnel design practices: A practical engineer's approach. In Proceedings of the SME, Rapid Excavation and Tunnelling Conference, Washington, DC, USA, 24 June 2013; pp. 23–26.
14. Bieniawski, Z. *Rock Testing and Site Characterization*; Elsevier: Amsterdam, The Netherlands, 1993; pp. 553–573.
15. Pakalnis, R. *Design Methods 2015: Proceedings of the International Seminar on Design Methods in Underground Mining*; Australian Centre for Geomechanics: Perth, Australia, 2015; pp. 37–56.
16. Mawdesley, C.; Trueman, R.; Whiten, W.J. Extending the Mathews stability graph for open-stope design. *Min. Technol.* **2001**, *110*, 27–39. [CrossRef]
17. Mortazavi, A.; Osserbay, B. The Consolidated Mathews Stability Graph for Open Stope Design. *Geotech. Geol. Eng.* **2021**, *40*, 2409–2424. [CrossRef]
18. Delentas, A.; Benardos, A.; Nomikos, P. Analyzing Stability Conditions and Ore Dilution in Open Stope Mining. *Minerals* **2021**, *11*, 1404. [CrossRef]
19. Mathews, K.E.; Hoek, E.; Wylle, D.C.; Stewart, S. *Prediction of Stable Excavation Spans for Mining at Depths below 1000 Meters in Hard Rock*; Golder Associates: Vancouver, BC, Canada, 1981.
20. Potvin, Y.; Hudyma, M.; Miller, H. *The Stability Graph Method for Open Stope Design*; University of British Columbia: Vancouver, BC, Canada, 1988.
21. Stewart, S.V.; Forsyth, W. The Mathew's method for open stope design. *CIM Bull.* **1995**, *88*, 45–53.
22. Sheshpari, M. Failures in backfilled stopes and barricades in underground mines. *Electron. J. Geotech. Eng.* **2015**, *1*, 191–212.
23. Suorineni, F.T. The stability graph after three decades in use: Experiences and the way forward. *Int. J. Min. Reclam. Environ.* **2010**, *24*, 307–339. [CrossRef]
24. Mawdesley, C.A. Using logistic regression to investigate and improve an empirical design method. *Int. J. Rock Mech. Min. Sci.* **2004**, *41*, 756–761. [CrossRef]
25. Zhao, X.-D.; Zhou, X.; Wei, H. Structural Parameter Design of Sublevel Open Stope and Backfilling in Sanshandao Gold Mine. *Met. Mine* **2022**, *51*, 101. [CrossRef]
26. Barton, N.; Lien, R.; Lunde, J.J.R.M. Engineering classification of rock masses for the design of tunnel support. *Rock Mech.* **1974**, *6*, 189–236. [CrossRef]
27. Mitri, H.S.; Hughes, R.; Zhang, Y. New rock stress factor for the stability graph method. *Int. J. Rock Mech. Min. Sci.* **2011**, *48*, 141–145. [CrossRef]

Review

Block Caving Mining Method: Transformation and Its Potency in Indonesia

Sari Melati [1,2,3], Ridho Kresna Wattimena [1,3], David Prambudi Sahara [4,5,*], Syafrizal [3,6], Ganda Marihot Simangunsong [1,3], Wahyu Hidayat [5,7], Erwin Riyanto [5,8] and Raden Roro Shinta Felisia [9]

1 Geomechanics Research Group, Faculty of Mining and Petroleum Engineering, Institut Teknologi Bandung, Bandung 40132, Indonesia
2 Mining Engineering Study Program, Faculty of Engineering, Universitas Lambung Mangkurat, Banjarbaru 70714, Indonesia
3 Mining Engineering Study Program, Faculty of Mining and Petroleum Engineering, Institut Teknologi Bandung, Bandung 40132, Indonesia
4 Global Geophysics Research Group, Faculty of Mining and Petroleum Engineering, Institut Teknologi Bandung, Bandung 40132, Indonesia
5 Geophysical Engineering Study Program, Faculty of Mining and Petroleum Engineering, Institut Teknologi Bandung, Bandung 40132, Indonesia
6 Earth Resources Exploration Research Group, Faculty of Mining and Petroleum Engineering, Institut Teknologi Bandung, Bandung 40132, Indonesia
7 Geophysical Engineering Study Program, Faculty of Technology Mineral, UPN Veteran, Yogyakarta 55283, Indonesia
8 Geoengineering Division, PT Freeport Indonesia, Mimika 99968, Indonesia
9 English Literature Study Program, Faculty of Letters and Cultures, Universitas Gunadarma, Jakarta 16452, Indonesia
* Correspondence: david.sahara@gf.itb.ac.id

Abstract: The block caving mining method has become increasingly popular in the last two decades. Meanwhile, Indonesia has several potential ore bodies which have not yet determined suitable mining methods. The references to block caving mining projects worldwide and the potency of metal deposits in Indonesia were reviewed to determine the requirements of ore bodies suitable for mining using the transformed block caving method. This method can be applied on a blocky ore body with a thickness of 200–800 m, various rock mass strengths until 300 MPa, from low to high (from 0.3% Cu until more than 1.0% Cu), but of uniform grade and at a depth from 500 to 2200 m. The technical specifications for running block caving mines have been synthesized, including preparation methods, undercutting strategy, mine design, mining equipment and monitoring. Considering the requirements and the successful practice of the block caving project in the Grasberg Caving Complex as a role model, the Indonesian government should concentrate on the detailed exploration of porphyry deposits and feasibility studies on applying the method to the prospective ore bodies, i.e., Onto, Tambulilato, Tumpangpitu and Randu Kuning. In addition, the exploration method, cost, operation, environment, mining policy and social geology are important aspects worth noting.

Keywords: caving; Indonesia; mining; porphyry; underground

Citation: Melati, S.; Wattimena, R.K.; Sahara, D.P.; Syafrizal; Simangunsong, G.M.; Hidayat, W.; Riyanto, E.; Felisia, R.R.S. Block Caving Mining Method: Transformation and Its Potency in Indonesia. *Energies* **2023**, *16*, 9. https://doi.org/10.3390/en16010009

Academic Editors: Longjun Dong, Yanlin Zhao and Wenxue Chen

Received: 11 October 2022
Revised: 13 November 2022
Accepted: 26 November 2022
Published: 20 December 2022

1. Introduction

Mining production has increased significantly in the last two decades, especially driven by technological mastery in mineral exploration, mining and processing. Mining escalation started in 2002 after digitalization and communication rapidly spread worldwide. Excluding the slight decline in mining production in 2009 and 2016 due to the global crisis, the general trend in mining production is increasing with an average annual increase of 20.7% [1]. This was mainly because of the high demand for raw materials for electronic devices, communications and interconnection networks. Therefore, investment in the search for new resources and reserves, as well as replenishment in funding for upgrading productivity, is becoming more necessary than ever. This momentum contributes to advancements in the mining sector in general, including improving the reliability and efficiency of mining methods.

The exploration of safe and clean mining on Earth and asteroids has become an important issue lately, especially in preventing safety accidents and environmental damage. Micro-seismic monitoring, along with stress and blast vibration monitoring, is a reliable technology both in open pit and underground mining for early detection of safety hazards and building an effective emergency rescue system [2]. One of the important achievements in micro-seismic monitoring is that the dynamic fracture formation process around the longwall mine can be explained by clustering methodology. The accurate cluster location can be determined by dividing a continuous group of mining seismic events and relating them to some parts of the rock mass [3]. These technological developments in mining practice are expected to increase the opportunities for mining sustainability. The combined solution of the Triple Helix Model (THM) between government, industry and university, as well as the Open innovation (OI) concept and Environmental, Social, and Governance (ESG), should enable sustainable development in a specific country and globally [4].

Most of the highly productive mines in the world are excavated on the surface. Surface mining is more profitable as it has exceptional advantages in flexibility and mobilizing. Therefore, realization design for producing as much ore as possible is convenient. Unfortunately, surface mining has an economic limit when the remaining ore reserve becomes deeper. In that case, underground mining methods are the only option. In addition, underground mining is assumed to leave fewer environmental impacts than surface mining [5]. Among underground mining methods, block caving is the most cost-effective as it can produce 10,000–100,000 tons per day with a relative operating cost of USD 1 to 2.5 per ton [6,7]. In the last twenty years, existing block caving mines have varying ore body thickness ranges from 200 to 800 m. A study proposes that the cut and fills stopping method was the optimal underground mining method for deep mining (>800 m below the ground surface), compared with block caving and four other methods [8]. However, it is only relevant for the lead-zinc-silver Trepca mineralization deposit investigated in the study, which has an irregular shape and ore thickness of 30–100 m. Other studies on underground coal mining noted that non-pillar mining or Longwall Mining, a caving method for coal deposits, had the lowest environmental burden and was determined to be the optimal mining method [9]. Thus, the caving method can be favored as an underground mining method due to its high productivity and low environmental impact. The deposit size must be huge enough to justify the investment costs at the beginning of production.

In the caving mining method, mining is achieved by breaking most or all of the ore body. Caving mines are classified as Longwall, Sublevel Caving and Block Caving by requirement factors, including ore strength, rock strength, deposit shape, deposit dip, deposit size, ore grade, ore uniformity and depth. Longwall is recommended for any ore strength, weak/moderate rock strength, tabular deposit shape, low/flat deposit dip, thin/wide deposit size, moderate ore grade, uniform grade and moderate/deep depth [10]. Due to coal seams having a relatively flat dip, the longwall method is usually applied in coal mining. Longwall mining in coal seams is an underground mining technique where a tabular block longwall panel of coal with a typical length of 1.5–3.0 km, a typical width of 200–300 m and a typical height of 3.0–4.5 m is extracted. Two pairs of roadways are first driven outside the panel within the seam for access. Machines used for operations are drum shearer machine as coal cutter, belt conveyor for hauling and hydraulic-powered roof supports providing temporary support during coal cutting [11]. Longwall mining is applied intensively in Australia, China and America and also in Ukraine, India, Turkey, Bangladesh and Poland, [11–28]. In Indonesia, the longwall mining method is applied in Kutai Kertanegara, East Kalimantan, at the mining concession of PT Gerbang Daya Mandiri (GDM). GDM recoverable sub-bituminous coal reserves are approximately 29.2 million tons and one million tons of annual production have been planned [29–33].

The second caving method, Sublevel Caving, is an underground mining method proposed for moderate and strong strength of ore, weak rock strength, tabular/massive deposit shape, steep deposit dip, large thick deposit size, moderate ore grade, moderate uniformity grade and moderate depth [10]. Sublevel caving is a mass mining method in which the ore is drilled and blasted while the waste rock caves and fills the space created by the extraction of ore. The ore body is divided into vertical intervals called sublevel intervals. The ore within each sublevel interval is drilled in a fan-shaped design at a constant horizontal distance along the production drift. Load Haul Dump machines load muck pile from the draw point [34]. Dilution becomes the issue in this method due to the strength of the ore body and rock mass. Sublevel caving is applied in iron mines in Ukraine, iron mines in Sweden, iron oxide mines in Norway, coal mines in Spain, coal mines in India and gold mines in Australia [34–42].

The last caving method, block caving, is recommended for moderate and weak ore and rock strength, weak rock strength, tabular/thick deposit shape, steep deposit dip, very thick deposit size, moderate ore grade, moderate uniformity grade and moderate depth [10]. The first documented underground mine which applied the block caving method was the Pewabic iron mine in 1895 [43]. Although the block caving method has been known for over a decade, its massive deployment has only occurred in the last twenty years. During that time, the number of cave mining projects increased almost four times, from 17 to over 50 [44,45]. Not only in number, but caving mines were also getting larger and deeper. In the beginning, footprint areas and block heights were less than 100,000 m^2 and 200 m, respectively, with the maximal overburden depth of 600 to 700 m. Recent cave mines have footprint areas greater than 400,000 m^2 and block heights exceeding 400 m, with an overburden depth of up to 1200 m [46,47]. The thickness of the ore body mined using the block caving method varies between 200 and 800 m. As an indication, the ore body widths of Northparkes, Ridgeway, Stornoway, Palabora, Grasberg, Oyu Tulgui and El-Teniente are 200 m, 200 m, 225 m, 250 m, 400 m, 500 m and 500–800 m, respectively [47–57].

The most significant evolution in the block caving method is related to the strength of the caved rock mass. Regarding ore and rock strength criteria, block caving was designed for weak or moderate rocks [10]. The International Society of Rock Mechanics (ISRM) Commission on the Classification of Rocks and Rock Masses in 1981 classified rock mass with uniaxial compressive strength of 6–60 MPa as suitable for block caving. Thanks to massive underground mining technology innovation, block cave is now implemented in competent rock; for instance, rock mass with the highest UCS (138 MPa) in Cadia East, 144 MPa in Northparkes, 157 MPa in Deep Mill Level Zone, 170 MPa in El Teniente and 300 MPa in Palabora [58–60]. This condition is classified as a high-strength rock. Some blocks mentioned above in caving fields took advantage of rock mass preconditioning to accelerate production, as well as to increase the safety of the workers. Rock mass preconditioning aims to generate new cracks or to elongate the extent of in-situ cracks using hydraulic fracturing and/or destress blasting on the competent rock. Some block caving in competent rock mass performed the rock preconditioning during the preparation and the development zone before the regular block caving processes [58–60].

All these improvements have enabled the block caving method to majorly contribute to the supply of ores from underground mines. Five of the ten underground ore mines with the highest production levels in 2020–2021 use the megatons' block caving method [61,62]. These are copper, gold and silver mines at Grasberg Operations (PT Freeport, Indonesia), Cadia Valley (Newcrest Mining, Australia), Padcal (Philex Mining, Philippines) and New Afton (New Gold, BC, Canada). In addition, diamond mines are located at Udachy (Alrosa, Russia), where the total ore processed per year is between 3.39–51.53 million tons.

In Indonesia, PT Freeport Indonesia introduced the block caving method. There are four of five ore bodies in the PT Freeport Indonesia underground complex, which are mined by the block caving method, i.e., Deep Ore Zone (DOZ), Deep Mill Level Zone (DMLZ), Grasberg Block Cave (GBC) and Kucing Liar (KL). This method is designed to produce 20,000 to 130,000 tons of ore per day. Grasberg caving complex is categorized as super caves with Chuquicamata and New Mining Levels at the El Teniente project in Chile, Oyu Tolgoi projects in Mongolia and the Resolution Copper project in Arizona [63].

Located in an active tectonic region, Indonesia is famous for hosting large rock mass intrusion bodies suited to the block caving method. However, only a few have been operated because the application takes time for technology readiness, feasibility studies, and development. As a first step, the potential deposits must be explored in detail. Since the block caving mining practice plays an essential role in maintaining or even upgrading the total national capacity of rock mass mining in Indonesia, the development and potential growth of the block caving method application in Indonesia have been reviewed. The discussions on the transformation and potency of block caving in Indonesia are based on four aspects, i.e., technological advancement in block caving, geologic and tectonics analysis in Indonesia, existing operations, and non-technical aspects. It is expected that the opportunities to spread the block caving method in mineral exploitation, especially in Indonesia, can fulfill the growth in energy demand faced today and in the following decades. This study is important in dissemination for practitioners, engineers and academics regarding the block caving method, which is the most efficient method of modern and future mining for huge ore bodies. Studies in Indonesia are necessary to provide recommendations for the potential application of this mining method in utilizing the country's natural resources (especially metal deposits).

2. Block Caving Method

2.1. Initial Block Caving for Weak to Moderate Rock

Block caving becomes attractive when deposits near the surface appropriate for the open pit mining method are harder to find. Furthermore, open pit mining has an economical depth limit of the ore body that can be extracted. Because of this limitation, ore bodies with low-grade levels at a greater depth cannot be mined using an open pit. On the contrary, block caving has relatively high productivity and inexpensive production costs, but with a higher risk in terms of technology and safety.

In the underground mining classification system, block caving is a method for mining weak to moderate rock and ore (Table 1). Block caving can extract rock on a large scale following the geometry of the cave propagation. As the cave propagates, ore mass at the top of the cave and the edge of the abutment will be heavily fragmented and eventually free fall to be excavated at the production level. As block caving can excavate large rock mass, it can be applied to almost all rock grades, low to high. The method is applied to mineral deposits such as iron ore, copper, molybdenum mineralization and diamond-bearing kimberlite pipes [64]. Recently, it has been used: on porphyry deposits, kimberlite deposits, skarn and porphyry-related deposits, asbestos mines, iron ore deposits, sedimentary exhalative (sedex) deposits, volcanogenic massive sulfide (VMS), stratiform deposits and others [65].

Figure 1 shows the general description of the block caving method. The accumulation of stresses at the top of the cave due to gravity and internal stress is enough to frack and break rock mass naturally if some conditions are fulfilled, e.g., regarding the radius of the cave and rock strength. The first stage is the production or extraction level development below the ore body. The undercut level was commonly excavated about 10–20 m above the production level, depending on the thickness of the overburdened rock and in-situ stress conditions. Pillars between the undercut drifts are then drilled and blasted to build slots below the orebody. This activity is known as undercutting. Draw bells are constructed between the production and undercut levels to accumulate the broken ore that had fallen from the top of the cave. Sometimes, secondary fragmentation using a jackhammer or secondary blasting is required if the block size of broken ore is larger than the requirement of the processing stages. Well fragmented ore is transferred to draw points at the production level through the ore pass.

Table 1. Underground Mining Classification Characteristic (Hartman, *Introductory Mining Engineering*, 1987).

Underground Method	Unsupported				Supported			Caving	
Factors	Shrinkage stoping	Sublevel stoping	Stope and pillar	Room and pillar	Cut and fill stoping	Square set stoping	Longwall	Sublevel caving	Block caving
Ore strength	Strong	Moderate/ strong	Moderate/ strong	Moderate/ strong	Moderate/ strong	Weak	Any	Moderate/ strong	Weak/ moderate
Rock strength	Strong	Fairly strong	Moderate/ strong	Moderate/ strong	Weak	Weak	Weak/ moderate	Weak	Weak/ moderate
Deposit shape	Tabular/ lenticular	Tabular/ lenticular	Tabular/ lenticular	Tabular	Tabular/ lenticular	Any	Tabular	Tabular/ massive	Tabular/ thick
Deposit dip	Fairly steep	Fairly steep	Low/ moderate	Low/ flat	Fairly steep	Any	Low/ flat	Fairly steep	Fairly steep
Deposit size	Thin/ moderate	Thick/ moderate	Any	Large/ thin	Thin/ moderate	Usually, small	Thin/ wide	Large thick	Very thick
Ore grade	Fairly high	Moderate	Low/ moderate	Moderate	Fairly high	High	Moderate	Moderate	Low
Ore uniformity	Uniform	Uniform	Variable	Uniform	Variable	Variable	Uniform	Moderate	Uniform
Depth	Shallow/ moderate	Moderate	Shallow/ moderate	Shallow/ moderate	Moderate/ deep	Deep	Moderate/ deep	Moderate	Moderate

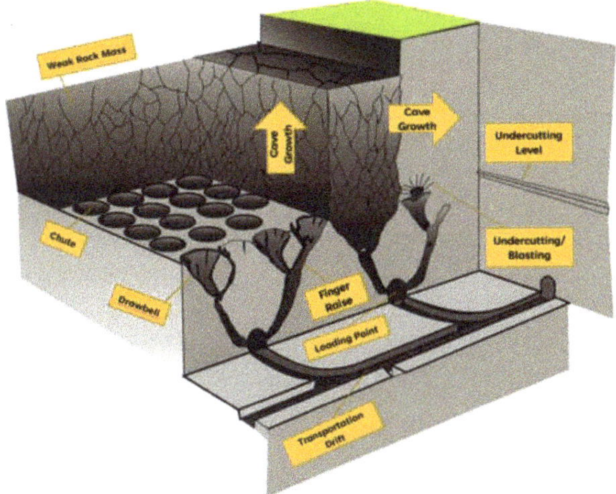

Figure 1. Layout of initial block caving. Undercutting blasting is generated to facilitate cave growth through weak rock. Draw bells and Finger raises are used to transport the falling ore to the loading point. The ore is transported to the transportation drift.

The drawing of blasted ore induces the flow of caving material and removes the support of the cave back. The cave back fails, and the muck pile fills the space formed by undercutting activity. The cave will continue to propagate upward if the ratio between the area and the circumference of the cave has satisfied the in-situ rock strength. In this case, blasted ore falls to draw points at the production level. This continuous caving process is the expected flow production in the block caving method. The last mining cycle is loading and hauling from the production level to the processing facility on the surface. The footprints of cave mining are usually built in several thousands of square meters. Typically, the development takes about ten years from the first access drifting until the first production when ore mass reaches the cave propagation and collapses.

In the early underground mining references, e.g., Pewabic iron mine, the block caving mining method was suited to ore bodies with the characteristics below [10,66].

1. A weak ore body can easily be fractured or fail and be separated around the block.

2. A weak wall rock breaks into bigger boulders than an ore fragment, where the pressure helps to break the ore body below.
3. Homogeneous deposit shape is required, as it is impossible to conduct selective mining. Should eye catching characteristics cause physical differentiation between ore body and capping, dilution at the draw point can be avoided. Ore body should be difficult to react with air. Therefore, this method is not appropriate for sulfidation deposits.
4. Dip of the deposit is not a problem. However, a dip > 65° is favorable if it is a vein.
5. Deposit thickness > 3 m with height > 35 m.
6. The grade of ore should not be high.
7. The depth is moderate.

The grade of ore should not be high (point 6) as the uniformity of grades in the ore body is more important than a certain percentage of grades, because in this method almost all parts of deposits can be recovered. Considerations of mine feasibility depend more on the volume of ore, along with the other contained minerals, that will determine the life of the mine. The high-grade Cu deposits containing low-grade other minerals is equivalent with the low-grade Cu deposits containing high-grade other minerals. As quantitative values for comparison, Ridgeway Deeps has 0.38% Cu and 1.80 g/t Au [49–51], while El-Teniente ore body has 0.62–0.98% Cu, 0.019% Mo, 0.005 g/t Au and 0.5 g/t Ag [55,66]. Both are mined economically using the block caving method.

Later, with the more quantitative approach, Miller-Tait, 1995 [67] stated that the block caving method is most appropriately applied to mine deposits with these characteristics:

1. A massive ore body with a thickness of more than 100 m, a dip of more than 55° and depth of more than 100 m.
2. Grade distribution is relatively uniform.
3. Very low-quality ore body (Rock Mass Rating, RMR = 0–20), wall rock is from very weak to moderate (RMR = 0–60).
4. The ore body and wall rock's uniaxial compressive strength (σ_c) are very weak. Compared with major principal stress (σ_1), the ratio σ_c/σ_1 is lower than 5.

2.2. Transformation to Competent Rock

To accelerate the technology and safety of block caving, caving practicians routinely discuss the current block caving technology, paradigm-shifting and recent developments in block caving at the International Conference on Block and Sublevel Caving. These events were held five times in Cape Town (2007), Perth (2010), Santiago (2014), Vancouver (2018) and Adelaide (2022). Together with other cave mining methods, panel caving and sublevel caving, block caving transforms itself in order to improve its viability, safety, cost, production and profitability. Research and practice were performed to develop solutions that reduce lead times and capital investment for the near future.

Industry research and innovation mostly contribute to upgrading the implementation of recent technology in block caving practice and benchmarking. The experiences of mining consultants and contractors in servicing different sites enrich a mature and flexible operation. University or research institutes support a profound understanding of rock behavior and its response during mining, develop modified formulation or classification systems and deploy new approaches to solve the latest problems. Collaborative research between all interested parties is a valuable process.

The newest update in block caving transformation announced in 2022 is the modified block caving method called Raise Caving. The fundamental difference with the conventional block caving method is that this system uses a hoisting system for mucking the ore through the raise, which is developed vertically at the center of the ore body from the top to the bottom. Boring, charging of explosives and supporting the raise are conducted on the platform placed in the hoisting system. This platform is removed from the raise during blasting activity and placed at the station's level to avoid damage [68–72]. The method consists of two phases, the de-stressing phase, and the production phase. The de-stressing stages consist of the slot rise, the start slots and the slots that are developed from top to bottom of the ore body. In the production phase, de-stressed rock masses are blasted and fall to the draw points. The caving rises from bottom to top. Ladinig described raising caving as a hybrid method starting from a pillar-supported method in the early phase, converting to an artificially supported shrinkage-stopping method during the production phase and ending up in a caving method as stopes are drawn empty. Raise caving was successfully tested for a pre-feasibility study on Kiruna Mine. It reduces approximately 50% of the cost of infrastructure development compared with the conventional caving method [71].

In the last two decades, block caving underwent a series of transformations in terms of mining equipment and technologies. It impacted the escalation of production rates and scale. At the start of the block cave mine in 1898, the production rate was about 2–8 kilo tons per day as it was only implemented on a weak rock at shallow depths. There was no significant improvement for 80 years, during which production rates increased steadily to 10–20 kilo tons per day. The mechanized system began to be utilized in the 1970s. All active cave mines nowadays use Load Haul Dump (LHD) for material handling that could achieve 20–40 kilotons per day. LHD is like conventional loaders but was developed for the toughest rock mining applications, taking overall production economy, safety and reliability into consideration. In addition, induced micro-seismicity due to mineral extraction conducted in deep underground mining is continuously monitored to minimize the seismic hazard. Thanks to the improvement in block caving technology in the last decade, the production capacity increased to 80–160 kilotons per day.

Technical improvements in mining extraction and safety have allowed block caving to be applied on moderate to strong rock [73–78]. In some block caving, hydraulic fracturing, and or destress blasting were implemented before the undercutting process started [58,60]. Hydraulic fracturing aims to decrease rock mass strength by stimulating new fractures and extending existing fractures. These preparation stages are called preconditioning. Figure 2 shows the scheme of block caving on a large scale, where preconditioning rock mass is conducted on the next excavating process target.

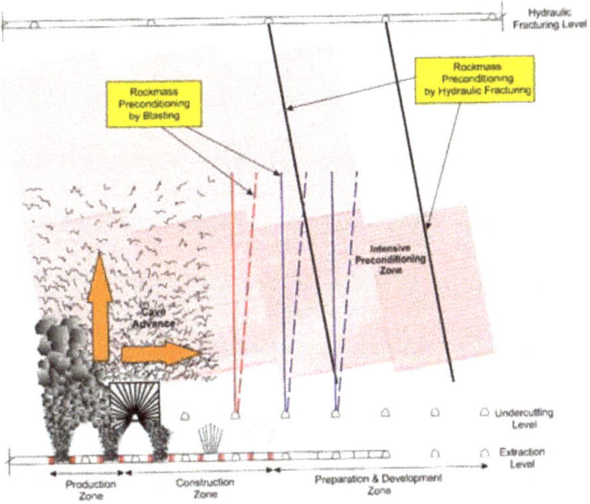

Figure 2. Layout of modern block caving with hydrofracturing and destress blasting for preconditioning the competent rock mass. In this example, rock mass preconditioning by blasting is carried out through a well drilled from the undercutting level (blue and red lines) and hydraulic fracturing through a well drilled from hydraulic fracturing above the cave. The falling material from the propagated cave is mucked from the extraction level.

The layout of intensive preconditioning, where hydraulic fracturing and de-stress blasting were combined, is also shown in Figure 2. These techniques enabled competent rock to be mined using block caving because of rock fragmentation before production. The hydraulic radius of the block caving is also increased due to the higher rock strength of the competent rock. For instance, in a weak rock required hydraulic radius is only a maximum of 50 m. However, now block caving mines in competent rock reach continuous caving at minimum hydraulic radius of 60–70 m and an undercutting area of around 280 m × 280 m.

In mining production technology, autonomous LHD is an excellent technology that has been relied on for the prodigious rate of major block caves productivity. This improves mining safety and, therefore, enables block caving in low-grade and larger deposits. A tele-remote system was experimented with in the 1980s with the main goal of removing the operator from the mucking process at the draw points. The recent version of autonomous LHD enables one operator to control 2–3 LHDs from a central control station. Henderson mine, Colorado, Magma mine, Arizona, and the Andes, Chuquicamata, Chile, Grasberg, Indonesia and Oyu Tulgui, Mongolia are cave projects that have been utilizing this autonomous system [79].

In terms of modeling, numerical modeling offers a fast and representative rock mechanics assessment when several designs need to be analyzed. Numerical modeling was used for the first time in 1973 to study cave propagation behavior at the El Teniente Mine in Chile [80]. In the beginning, numerical modeling of block caving is limited to two-dimensional, elastic behavior and a continuum model. Along with developing tools and applications, it upgrades to the three-dimensional, elastoplastic behavior, and discontinuous or hybrid model. Most of the numerical modeling is exploited to simulate the stress–strain condition, rock mass behavior around the cave, cave propagation mechanism, and subsidence behavior, or compared to the other results of empirical, physical, or analytical methods. Discontinuous and hybrid models are developed specially to accommodate the modeling of heterogeneous or defected rock and its flow in the caving process [81,82]. Cellular automata, an artificial intelligence with iterative, stochastic, and discrete mathematical models coupled with its predecessor numerical tools, is the recommended approach for block caving modeling. Caving mechanics models were then utilized for production scheduling [83–88].

Mining companies in the USA and Australia, Freeport McMoran, Newcrest Mining and Rio Tinto, are the leading companies in block caving mining practice. Their success in operating several sites becomes a reference for other mining companies in practicing this method. They noted that the geotechnical setting is always a determinant factor of mine planning and design. Freeport McMoran successfully redesigned the mine and modified ground support in long- and short-term planning at Grasberg Block Caving when faced with heterogeneous rock masses [56]. Newcrest Mining considered fragmentation, hydraulic fracturing, mine design, cave management and the stress abutment zone as geotechnical considerations for mine planning and design in the Cadia Valley Operation [49]. Rio Tinto, with its recent experiences in Grasberg, Palabora, Northparkes, Argyle and present projects in Oyo Tolgui and Resolution Mine, but cautioned that detailed measurement and monitoring are the most important steps in the design, construction and operation of block caving mines [57].

Productivity, grade and mining value are sensitive aspects in operating block caving projects [89]. As the most promising advanced underground mining method after open pit closure, it is necessary to consider whether the exposure on the surface and the risk of slope failure of the open pit affect the grade of the underground ore body [90]. Regarding this issue, the concept of "Cave to Mill" was introduced to provide consistent tonnage and grade feed which consists of (i) better characterization of the material reporting to draw points, (ii) measurement of the variation in metal content that is delivered from draw points and (iii) a bulk sorting system which offers flexibility and control [91]. Sensor-based sorting applications and automation of an online grade analyzer are helpful tools for intensifying the capacity of material handling and the efficiency of the quality control process [89–91].

Understanding the actual condition of rock mass is vital in mine stability analysis. In terms of actualization aspects, monitoring is the best method for quantifying rock mass and is used as a reference for validating and verifying predictive models. Micro-seismic monitoring is also the most effective method for block caving compared to other monitoring methods (displacement monitoring with dilatometer, multiple borehole extensometer, borehole camera, scanner, radar, or aerial photographs). This condition is due to its capacity, which covers a large area and records the spatio-temporal alteration. In addition to being sensitive to movement, the data obtained is very comprehensive and flexible for various advanced analysis purposes. Micro-seismic monitoring methods have been applied in several caving mines, using both passive and active seismic sources. Based on our previous study, the placement of the seismometer network significantly affects the resolution and level of data certainty [92]. The collaboration of analytical methods and iterative solutions has helped to localize the source of seismic events accurately. The accuracy of the location of the source of this event is essential for modeling geological structures in rock mass [93].

3. Requirements and Technical Specifications for Block Cave Mines Globally

Currently, there are over 50 cave mining projects in various stages of study and development in the world (Figure 3). They have spread mostly in North America, Australia, Africa, and South America. Some valuable projects are also located in Asia, i.e., in Indonesia, the Philippines and Mongolia. The several specifications of the existing block caving mines in the world are reported in Appendix A, Table A1. The block caving method has transformed from the original limited version to the recent version that is more productive, more efficient, with higher technology, and safer. The block caving method in the future may become competitive in terms of its development process if it can keep following the current rapidly evolving trend, be more flexible in applying varied depths and grades of the ore body and maintain minimum risks associated with mining hazards. Based on the best practice and available references for block caving [94], e.g., Palabora and Cullinan in South Africa; Grasberg Caving Complex in Indonesia; Northparkes Mine, Argyle, Ridgeway Deeps, Cadia East and Carrapateena in Australia; Andina and El Teniente in Chile; Renard and Red Chris in Canada; Resolution Mine in Arizona; Salvador in Central America; Jwaneng in Botswana; Padcal in the Philippines; and Oyu Tolgoi in Mongolia, block cave mines' requirements and technical specifications could be summarized as follows.

Mine planning becomes crucial to achieve the best tactical short-term and long-term strategies, since it ensures the life of the robust block caving on strong rock masses. In the development stage of block caving, especially if using autonomous hauling, the haul distance from the production level at the bottom of the ore body to the processing plant on the surface must be designed as efficiently as possible to ensure the smooth flow of transportation equipment. In addition, structural design, functional design and maintenance management system design should deal with realistic and appropriate design limits [95,96].

In the undercutting stage, the optimization of the undercut is the main parameter in design. It imposes stress perturbation, hydraulic radius, cave propagation, caving direction and the stability of pillars at the extraction level [97–99]. The main target of undercutting competent rock mass is to achieve a large hydraulic radius while keeping the level stable. To do so, the height of the undercut cannot be less than 10 m and requires wider pillars and powerful rock support on the extraction level. This mining process makes the block caving work like an underground rock factory [99].

Block caving has several risks related to rock mass instability. Rock burst and strain burst are the main unexpected risks in cave establishment, as the constructed void is big enough to disturb the equilibrium condition of the rock mass. Common geotechnical hazards include subsidence, mud rush, rock fall, caving stall, caving hazards, collapse, flying rock and other uncontrolled material movements. Air blasts must be encountered where the cave intersects any excavation.

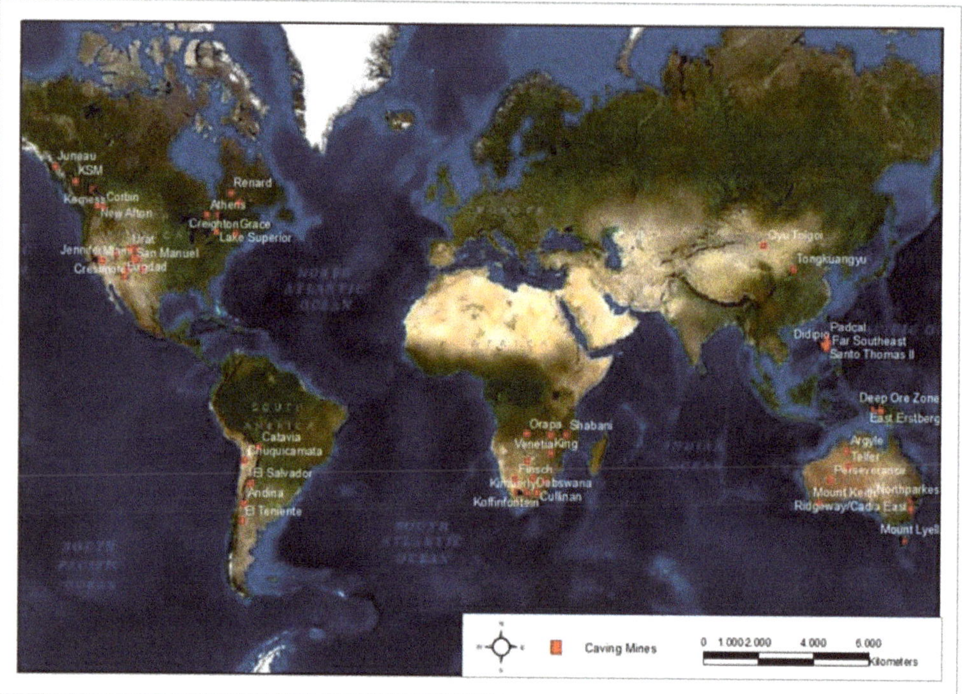

Figure 3. Distribution of caving mining globally compiled from previous studies, some of them listed in Table A1 (Appendix A).

It is also worth noting that the block caving applied on strong-hard-competent rock masses may face the problem of rock fragmentation. If the rock is deformed in coarse fragmentations, it might cause production loss and impede the continuity of ore flow at the draw point. Optimum fragmentation is a must to avoid material being stuck at draw points. One needs to understand the geological settings and structures of the in-situ rock: rock's strength, mineral composition, joint conditions, and natural fractures, to plan the mining strategies: pre-conditioning, undercutting, cave initiation, cave propagation, drawing and ore handling process, since these conditions achieve successful fragmentation and draws controlling. In some block caving, fragmentation problems could be avoided by a combination of comprehensive modeling of geological settings, statistical analysis and forecasting algorithms, digital imaging, laboratory testing, or numerical simulation for fragmentation modeling or dilution prediction [100–104].

Based on the last transformation and the practice of existing and developing block cave mining projects, the following sum up the new/updated typical orebody parameters requirement for block caving:

1. Ore body dimension

Suitable ore body types for this method are porphyry or pipe, large with thick, wide, tabular, or blocky dimensions. There are several applications on other types of deposits, i.e., sedex, VMS, stratiform, and others. However, the most common is porphyry.

2. Ore body and rock mass quality

As preconditioning had been a casual practice in many caving projects, there are no limited requirements for ore strength of the ore body. The upper limit of the uniaxial rock strength of competent rock mass to be mined using block caving is 300 MPa.

3. Grade

Block caving is allowable for mining from low to high-grade ore bodies. The minimum cut-off grade varies depending on the specific site's engineering setting and other factors. However, the grade distribution of ore bodies should be relatively uniform. The exploited deposits have grade from 0.3% to >1.0% Cu.

4. Depth

On the latest block caving mining projects, orebody lies at depths from 500 up to 2200 m below the surface.

Technical specifications for running block caving mines:

1. Preparation method

Technical specifications of hydraulic fracturing and/or de-stress blasting are defined by ore body and rock mass quality. Undercutting blasting is continued until the cave propagation is achieved by a certain hydraulic radius depending on the in-situ rock strength.

2. Undercutting strategy

The advanced undercutting trend is the most optimum strategy by compromising pre-undercutting and post-undercutting. This strategy allows only a limited excavation of the draw bell in the draw horizon before the undercutting process. The remaining development in extraction level is continued in the de-stress condition area.

3. Mine Design

Both El Teniente and C-Cut are still used. There are no significant changes from the original version of the block caving mining method in undercut, production and draw bell configuration. Production rates accelerate to more than 80,000–130,000 tons per day.

4. Mining equipment

Mucking equipment from loading in the draw point to haulage level consists of Load Haul Dumps. They have been operated autonomously in some super cave projects.

5. Monitoring system

Micro-seismic monitoring, with its advancement in tomography technology, is a helpful and reliable tool for measuring damage and modelling stress distribution on the rock mass induced by block caving mines. It should provide a concise and precise description of the experimental results, their interpretation and the experimental conclusions that can be drawn.

4. Existing Caving Mine in Indonesia

4.1. Grasberg Caving Complex

Esrtberg prospective area was discovered by mining engineers, Jean Jacques Dozy and A. Colijn in 1936. However, this area was not explored in detail until Freeport discovered the report in 1960. An expedition led by Forbes Wilson and Del Flint rediscovered the Erstberg mineral deposit. By signing the Contract of Work (CoW) with the government of Indonesia in 1967, Freeport became the mining contractor for the Erstberg deposit. In 1973, Ertsberg was declared operational after completing the exploration and feasibility study. Grasberg deposits were discovered in 1988, three kilometers from the Erstberg mine [105]. Since the opening of Erstberg and Grasberg, several block caving mines have been operated by PT Freeport Indonesia (PTFI) to mine the ore body.

Orebodies on the CoW area of PTFI are extracted using surface and underground mining systems (Figure 4) [106]. Ertsberg East Skarn System (EESS) deposits were mined by block caving. Gunung Bijih Timur (GBT) block caving mine started production in 1980 through to 1993, from an elevation of 3474 m to 3626 m, recovering 60 million tons of ore. Intermediate Ore Zone (IOZ) or Erstberg Stockwork Zone block caving was in operation from 1994 until 2003, with total production of over 50 million ore. IOZ mine connected vertically with the GBT mine. The footprint of the IOZ orebody was 330 m long by 220 m wide, with ore column heights of 150–220 m. DOZ mine block cave mine is located about 320 m below IOZ mine and about 1200 m below the surface.

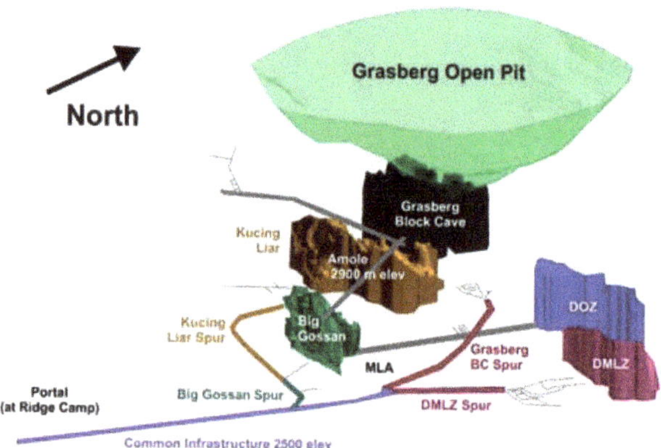

Figure 4. Ore bodies in Grasberg Block Cave Mine. There are five existing blocks in the Grasberg Block Cave Mine, i.e., Grasberg Block Cave, Kucing Liar, Big Gossan, DOZ, DMLZ. Each block is accessed through a spur towards the Portal at Ridge Camp.

DOZ block was at the elevation of 3126 m to 3476 m. The undercut level is 3146 m or 20 m above the production level. DOZ block cave was designed to be more than 1000 m long and 500 m wide. The mine had 857 draw points, designed in a column footprint of 15 × 18 m. There were three main specific geological units that were encountered in the DOZ mine: diorite and magnetite skarn, forsterite skarn and dolomite-marble and highly altered localized ore (HALO). They had a specific gravity of about 2.22–2.9, Rock Quality Designation (RQD) of 22–93%, Unconfined compressive strength (UCS) from 21–130 MPa, and Rock Mass Rating (RMR) 25–90, classifieds from poor to fair up to very good. The mining operation was completed after being targeted to produce 25 kilotons of ore per day [107].

Deep Mill Level Zone (DMLZ) is the deepest block caving below DOZ. DMLZ, located at elevation 3.125–2.590 msl, has a reserve of about 472 million tons, 0.85% copper and 0.72 g/ton gold. DMLZ targeted a diorite intrusion body bounded by dolomite. The diorite is similar that in Ertsberg, which was also altered by potassic phyllite and endo-skarn garnet intersected by quartz vein and anhydride [108–111]. Diorite in Extraction Level and Undercut Level has UCS 156.5 MPa and middle RQD 70–100% [59]. This strong, intact rock and competent rock mass characteristic make cave mining practice more challenging. DMLZ will have 700 active draw points in total production to complete the mine operation in 2041 [56].

Grasberg Block Cave (GBC) is the underground mine using the block caving method that continued production of Grasberg orebody after the Grasberg Open Pit Mine was completed in 2018 [109,110]. GBC began to develop in 2004 by drifting the first access. Caving was started by undercutting in September 2018 and draw-belling in December 2018. Grasberg orebody is a porphyry copper-gold deposit formed by a multiphase dioritic intrusion replenished in the center of a volcanic breccia complex. The mineralization is about 1600 m in length vertically and from about 200 m to over one kilometer in width. The extraction level using El Teniente-style layout with production panel drifts space of 30 m are at 2830 m elevation. The draw point spacing of 20 m was designed and resulted in a total of 2400 draw points. The undercut level was developed 20 m above the extraction level. The overall 700,000 m^2 large and about one kilometer in diameter footprint of the GBC orebody is sectioned into eight blocks. Operating multiple blocks simultaneously to meet the production target is exceptionally challenging due to its special condition, where a large block cave sits below the largest open pit in an area of high rainfall. The mine closure of the cave was targeted in 2041 with a production of 130,000–160,000 tons per day. Character of rock mass in this cave mine is more complicatedly heterogenous than DOZ, classified as fair to very good ground and poor to fair ground regarding geotechnical domain. The first domain consists of 7 lithologies and has RQD 72–88% with UCS 80–140 MPa. The second comprises five (5) lithologies and has RQD 11–90% with UCS 5–80 MPa [56,105–113]. Kucing Liar block deposit, the next operated large block cave mine, was developed in 2022 and targeted to begin production in 2027.

No detailed technical notes related to blasting practice specifications at Grasberg Caving Complex were documented. Based on the underground ring blasting design (long hole blasting techniques applied in block caving), the most common diameter ranges from 64 to 115 mm. For draw belling, a fan of 102 mm diameter consists of 9 holes with the length of blast hole drilled in the roof range from 8 to 22.9 m. By burdening 2.5–2.6 m, it consumes about 58 to 188 kg emulsion (1.1 gr/cc) per hole: the smaller hole diameter, 89 mm, is used on a narrow inclined undercut ring to avoid dilution. The burden is designed to be closer, only ranging from 1.8 to 2.1 m, with 5 holes in a fan. It needs from 8.7 to 69 kg emulsion (1.0 gr/cc) per hole [114]. As the draw belling and undercutting blasting output, the broken ore was mucked from the draw point using Load Haul Dump (LHDs) operated remotely, then delivered via loading chute to a rail haulage system. Wire mesh and shotcrete were applied to support the draw point and undercut level. Blasting conducts for two rings (fan series) every day and results in about 80–100 kilotons of Cu-Au fragmented ore.

4.2. Semi Caving in Pongkor

The second case of a block cave mine in Indonesia presented in this paper is a small block caving modified from a typical underground mining method. Initially, all primary veins were mined using cut and fill as a common mining method. However, some ore mass collapses were found in sill drift during development. Therefore, in 2004, the semi-caving system was successfully applied in Pongkor underground gold mine. Given the classification of weak ore observed, caving mining was selected to be operated on those ore masses. Based on the in-situ rock strength, a small hydraulic radius of approximately 3.3 m was required for cave propagation. During application for four months with the undercut, the dimension caused by the continuous caving of the ore body was approximately 20 × 10 m and ore production increased to 95.7%. This case could be referred to by miners regarding how to modify the underground mining method as semi-caving if the ore body or rock mass is extremely weak for maintaining excavation [115].

5. Mineralization Type and Potential Cave Mines in Indonesia

In mine planning and design, ore body parameters are fixed technical and economic considerations. Mining method classified by ore body condition is the main aspect of planning and design. The mineralization style includes all the ore body's basic parameters. The geological setting and geotechnical characteristics of the ore body and the surrounding rocks are fundamental to mine design. This condition will influence tunnel support requirements and productivity. The depth below the surface, regional stresses, and geothermal gradients can all have significant impacts on aspects of mine design, performance, and cost. The commodity may also influence how the ore is mined, treated, or transported [65].

This means that the style of mineralization is a fundamental control in the life of a mine. This condition is in accordance with the laws and regulations in Indonesia. As written in the Indonesian Mining Law, 2020, mining is a part of or all of the activities stages to manage and undertake minerals or coal, which includes general study, exploration, feasibility study, construction, mining, processing and/or refinement, development and/or utilization, transportation and selling, and also post-mining activity.

This law had been detailed by the Ministerial Decree of Energy and Mineral Resources in 2018. It is explained within it that mine planning is arranged at the stage of the mine feasibility study. Mining method and system as an aspect of mine planning consists of mining system (surface or underground), mining method, mine schedule, production rates, and life of the mine. The mining system must be suited to special and geotechnical conditions, ore bodies, mine environmental considerations, and mining technology. Based on this guidance, the inventory of ore deposits and their style of mineralization is the first step in assessing the appropriateness of block caving as a mining method in a specific site.

As the path of the world's ring of fire, the Indonesian archipelago was formed due to the interaction and collision of the gigantic crustal blocks of the Eurasian, Indian, Australian, and Pacific plates. The process was induced by ultrabasic rocks containing rich mineral resources distributed extensively in eastern Indonesia, while in western Indonesia most of the orebodies explored are associated with the active volcano-plutonic arc or the stable mass of the Sunda Shelf [116–119].

Indonesia's geological condition is promising since the magmatic arc is strongly associated with copper and gold mineralization. Gold mineralization in Indonesia was formed in the andesitic arc. The andesitic arc occurs in the Cretaceous range to the Pliocene (3–20 million years), especially in the Cenezoic age. At the time, Indonesia's plates started to experience subduction and actively generated a certain zonation of magmatic arcs. The identified gold deposits in Indonesia are copper-gold porphyry, skarn, high and low epithermal sulfidation system, sediment-hosted gold, gold-silver-baril and base metals deposits and kelian type, a transition from porphyry to an epithermal system [116–119]. Based on tectonic activity along the magmatic arc, the eastern part of Indonesia is dominated by porphyry and skarn formations, as well as a small percentage of hydrothermal sulfide deposits and hosted sediment. In Western regions of Indonesia, mineralization tends to be epithermal deposits. Low sulfidation was generated in relatively shallow areas of Sunda Land [119,120]. Regarding the relationship with magmatic arcs, deposits in Indonesia are related to andesitic magmatic arcs that formed rapidly during magmatic activity. This shows that this mineralization is related to the subduction of the ocean floor. Epithermal deposits were formed along the continental arc, which was the island arc joining the Sunda Shelf during the period of mineralization due to crust thickening and intensive elongation. Porphyry gold occurred both in the environment of the island arc and continental arc [119–121].

Porphyry and epithermal deposits form in the upper crust. They are related to sulfur and water-rich intermediate to silicic magmatic sources of hydrothermal fluids that move upward and produce extensive hydrolytic and alkali wall-rock alteration, quartz veins and sulfides. Figure 5 shows that the porphyry body is formed above magma chambers where fluids hydro fracture rock at 700–350 °C and pressures range from supra-lithostatic to supra-hydrostatic. The formation depth ranges from 2 to 10 km [119,120].

Block caving method had been implemented on these ore body genesis: kimberlite pipes, skarn deposits, porphyry deposits, asbestos deposits in peridotite rock, iron ore deposits and stratified sedimentary horizons. Thirty-one of 80 mines that used the block caving method were porphyry deposits, while 6 were skarn and porphyry-related deposits, including Grasberg Caving Complex in Indonesia [65]. It is proven that this type of porphyry deposit is very suitable for using this underground mining method, considering the relatively large shape of the ore body. Moreover, the block caving mines in porphyry deposits are practically a transition of the open pit above it, in terms of extending their mine life. Observing the existing copper and gold mines in Indonesia, the block caving mining method was only implemented economically in the porphyry type. In contrast, the other quartz veins were mined using the stopping method. Thus, it is highly recommended to explore porphyry deposits and declare them as prospective deposits for the application of the block caving method.

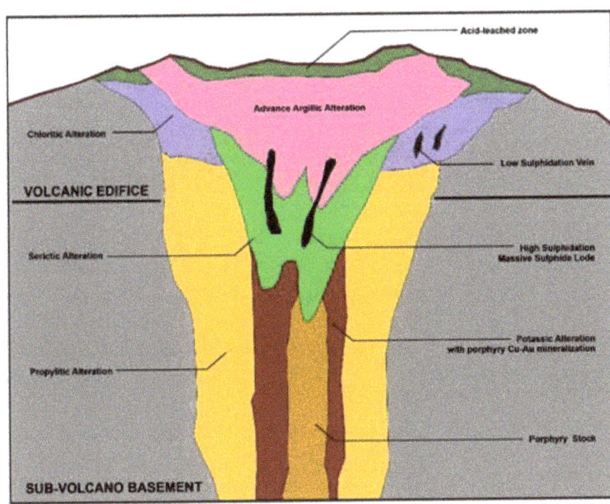

Figure 5. The schematic of porphyry ore body. Different colors indicate the different mineralization and alteration of the intrusion body.

Figure 6 shows the location of porphyry deposits in Indonesia. They are distributed mostly in eastern Indonesia. Tapada, Bulagidun, Tombulilato and Motomboto porphyry deposits were generated on Sulawesi East Mindanao Arc. Grasberg porphyry deposits and skarn system related were generated on Medial Irian Jaya Arc. The genesis of the Kaputusan porphyry deposit is associated with Halmahera Arc. Onto, Elang, Batu Hijau, Randu Kuning and Tumpang Pitu porphyry deposits are located in southern Indonesia, which is associated with the Sunda-banda Arc.

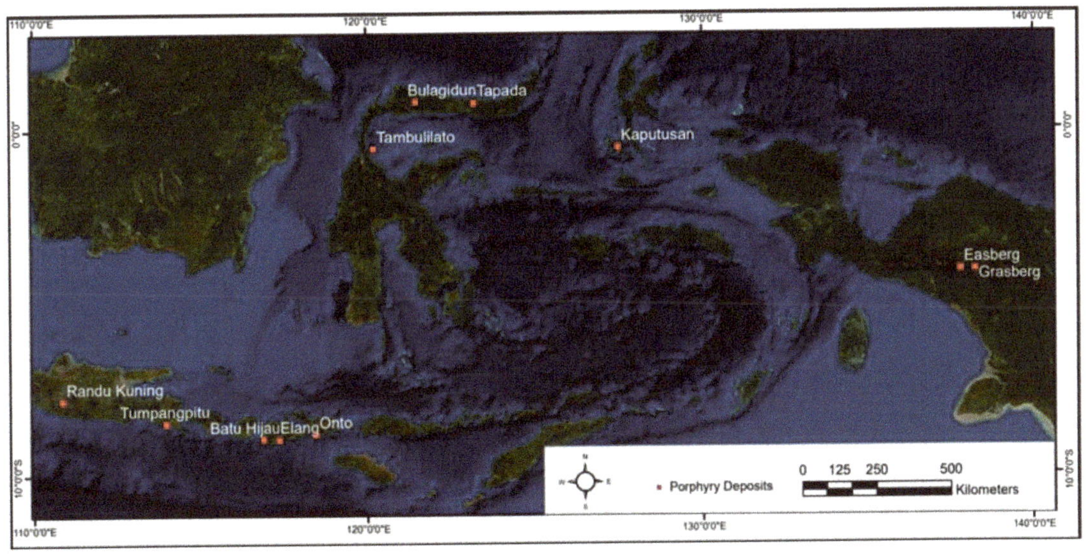

Figure 6. Porphyry deposits in Indonesia, compiled from previous studies in Table A2 (Appendix A).

Table A2 (Appendix A) shows that the style of mineralization of economic deposits in Indonesia is dominated by porphyry and quartz vein. Ore mineralization deposits in Indonesia are generally classified as primary and secondary deposits. A primary deposit is formed from the magmatic process. Porphyry, stockwork, alteration zone, or skarn system are related to hydrothermal phases. In epithermal phases, gold enrichment fulfills the empty structure or voids in the rock mass, where Au/Ag/Cu/Mo quartz veins are concentrated. Secondary deposit such as alluvial placer transported and sedimented on river meander is extracted by local miners using artisanal or traditional mining methods.

A large dimension is a fixed requirement for implementing the block caving method. The large ore bodies explored in Indonesia with potential to be mined using the block caving method are the Onto deposit in Eastern Sumbawa, Cabang Kiri deposit in Gorontalo, Tumpangpitu-Tujuh Bukit Porphyry in East Java and Randu Kuning deposit. Most global block caving projects have been mining the Cu Porphyry orebodies. Indonesia needs to be concerned about performing a feasibility study on this ore body type.

Copper and gold mineralization and their associated minerals, such as silver and molybdenum, is related to magmatic arc. Generally, Indonesia's copper and gold mineralization results in various deposits of porphyry, high-sulfidation epithermal deposits, low-epithermal deposits, Au-Ag-Cu±base metal mineralization, skarn and sediment-hosted. Based on the tectonic events that occurred along magmatic arcs, eastern Indonesia is dominated by porphyry and skarn, some high sulfidation hydrothermal deposits and sediment-hosted. Western Indonesia has mineralization that consists of low-sulfidation epithermal deposits in the shallow Sunda arc [116–121].

Porphyry gold deposits can be formed on both island arc and continental arc [65,114–119]. Magmatic arcs in Indonesia where porphyry deposits are suspected to be found are Sunda-Banda Arc (Au-Cu porphyry), Aceh Arc (Cu-Mo porphyry), Central Kalimantan Arc (transition from epithermal to porphyry), Sulawesi-East Mindanao Arc (Au-Cu porphyry) and Central Irian Jaya Arc (porphyry and skarn orebody). Halmahera arc has not yet been explored. However, the mineralization type is hypothesized in the form of Cu-Au porphyry.

- Onto deposit

Onto deposit is a large Cu-Au deposit discovered in 2013 on eastern Sumbawa Island. Cu occurred as covellite and pyrite-covellite veinlets in a tabular block. The block dimension is at least 1.5 × 1 km and the vertical height is ≥1 km. In 2013, a diamond drill program tested an extensive advanced argillic alteration litho-cap within the Hu'u project on eastern Sumbawa Island, Indonesia. A very large and blind copper-gold deposit (Onto) was discovered, in which copper occurs largely as disseminated covellite with pyrite and as pyrite-covellite veinlets in a tabular block measuring at least 1.5 × 1 km, with a vertical thickness of ≥1 km. The porphyry intrusions were emplaced at shallow depth (≤1.3 km), with A-B–type quartz veinlet stockworks developed over a vertical interval of 300 to 400 m between ~100 and 500 m below sea level (BSL), 600 to 1000 m below the present surface, which is at 400 to 600 m above sea level. Although the greatest amount of copper occurs as para-genetically late covellite deposit during the formation of the advanced argillic alteration, approximately 60% of the resource at 0.3% Cu cut-off still occurs within the porphyry stocks, indicating that porphyry stocks are a fundamental control on mineralization [121].

- Porphyry deposits on Tambulilato

Porphyry Cu-Au mineralization on Tambulilato, North Sulawesi, is present at Cabang Kiri, Sungai Mak, Kayubulan Ridge and Cabang Kanan. Hypogene Cu-Au mineralization is typically associated with magnetite-bearing K-silicate assemblages, partially obliterated by sericite, illite and chlorite. Copper as chalcopyrite and bornite and Au show a positive correlation. Multiphase intrusions and alteration-mineralization events are commonplace. Biotite-bearing K-silicate alteration at Cabang Kiri and Kayubulan Ridge yields K-Ar ages of 2.93 ± 0.06 and 2.36 ± 0.05 Ma, respectively. Cabang Kiri possesses a gold-rich zone (>1.5 ppm Au) open at depth and the bulk of the mineralization at Sungai Mak is contained in a supergene chalcocite blanket [122].

- Tumpangpitu, Tujuh Bukit porphyry deposit

The inferred hypogene sulfide resource estimate for the Tumpangpitu Cu-Au-Mo porphyry deposit is 1.9 billion tons at 0.45% Cu, 0.45 g/t Au and 90 ppm Mo, equating to 28.1 Moz Au, 19 Blbs Cu and 400 Mlbs Mo at a 0.2% Cu cut off. The total measured, indicated, and inferred resource for the high-sulfidation Au-Ag oxide is 71.4 million tons at 0.80 g/t Au and 26.3 g/t Ag with 1.9 Moz Au and 60.3 Moz Ag at 0.3 g/t Au cut-off grade. The advanced exploration resulted in the Tujuh Bukit porphyry deposit having a total of 1.9 billion tons of inferred global resource of ore at an average grade of 0.45% copper and 0.45 g per ton of gold, containing 19 billion pounds of copper and 28 million ounces of gold. In early 2018, work began on the underground development of the exploration decline. It progressed encouragingly throughout the year toward developing a long-life and low-cost block cave porphyry copper and gold mine [123].

- Randu Kuning deposit

The Randu Kuning prospect is a part of the East Java Southern Mountain Zone, occupied mostly by plutonic and volcanic igneous rocks, volcaniclastic, siliciclastic and carbonate rocks. Magmatism-volcanism products were indicated by the abundance of igneous and volcaniclastic rocks of Mandalika and Semilir Formation and many dioritic intrusive rocks of the Late Eocene-Early Miocene magmatism. Porphyry Cu-Au and intermediate sulfidation epithermal Au-base metals mineralization at Randu Kuning have strong genetic correlation with the magmatism and volcanism processes. The mineralized dioritic intrusive rocks in the area are distributed at the center of the depression of an ancient volcanic crater. Many intermediate sulfidation epithermal prospect areas surround the Randu Kuning porphyry Cu-Au. Most mineralization, including porphyry and epithermal environments, is associated with quartz-sulfide veins. However, not all porphyry vein types contribute to copper and gold mineralization. The early quartz-magnetite veins (particularly A and M vein types) generally do not contain Cu-Au or are barren. In contrast, the later sulfide-bearing veins, such as quartz-sulfide (AB type) veins, chalcopyrite-pyrite (C type) veins and quartz-sulfides-carbonate (D type) veins are mineralized. Mineralization contains copper and gold deposits in the range of about 0.66–5.7 g/t Au and 0.04–1.24% Cu. On the epithermal level, mineralization is mostly related to pyrite+sphalerite+chalcopyrite+quartz+carbonate veins and hydrothermal breccias. The epithermal veins and breccia lead to the occurrence of silver, zinc and lead mineralization. It commonly contains around 0.4–1.53 g/t Au, 0.8–8.5 g/t Ag, 0.17–0.39% Cu, 0.003–0.37% Zn, 00089–0.14% Pb [124].

The block caving method is a promising underground mining method for mineral production in Indonesia, specifically for Cu-Au-Ag-Mo porphyry-type deposits. It is needed a comprehensive and high certainty of orebody knowledge. For thinner deposit shapes like quartz veins, semi-caving is reliable as an alternative method to modify the main method–usually stopping group type (cut and fill, shrinkage, sublevel stopping)–while the ore body has a weak or poor quality.

6. Opportunities and Challenges of Block Caving Method in Indonesia

The number of prospective ore bodies to be mined using block caving methods mentioned above will increase the exploration program. Presently, the new targets of exploration in Indonesia are determined by scholars and investors who spend their resources to study prospective areas on Sunda Arc, such as Tangse, Gunung Subang, Ojolali and Miwah [125–128].

6.1. Exploration

The exploration method should be by type of deposits to ensure the success of exploration activities. Valuable deposits in Indonesia are mostly porphyry and epithermal. The most useful geophysical techniques in exploration for these are high-resolution magnetics and electrical surveys. Airborne magnetic and electromagnetics surveys are fast and cost-effective, particularly in areas of rugged topography. Regional magnetics, gravity, remote sensed data, and topographic data can also identify major structures, intrusive complexes and alterations. Radiometric surveys can be useful in mapping geology and alteration. Table A3 shows the recommended exploration methods classified by the type of deposits summarized [120]. Geophysical exploration methods benefit the prospecting and general exploration stage or scope study. After prospecting has been plotted, another detailed method should be conducted, such as detailed topography, core drilling and geophysical logging. Laboratory testing is conducted, including rock mass characterization and ore body quality. This aims to arrive at the best understanding of ore body conditions. Therefore, feasibility studies can be held using data with high confidence.

6.2. Cost

Investment capital cost to develop block caving method is expensive. For representation, the pre-feasibility cost of Tujuh Bukit Porphyry was about $58 million [123,129]. The development cost of GBC and DMLZ plus the infrastructure was $7.8 billion, taking place from 2008 to 2021 [130]. Ridgeway Deeps was developed at the cost of A$525 million [131]. Capital investment for Carrapateena block cave development of about A$1.2 billion to A$1.3 billion was weighted toward 2025 to 2027. It has been analyzed that incentive pricing of USD 3 to 3.5 per lb ore is required to upscale the currently proposed caving projects into the production stage [132]. Of course, this consideration is a special challenge for the investor. The joint venture between the government mining company and national or global world-class mining company investment hopefully will be helpful in accelerating the operation of the block caving method in Indonesia. Investment capital cost on block caving projects usually involves the construction of a block cave, underground crusher, automated remote loaders, modifications to the processing plant and deployment of the monitoring system. Raise caving, introduced recently, should be considered for its feasibility in reducing the cost of infrastructure development.

6.3. Operational

Block caving is a mining method with large productivity/mass mining. Remote-control systems for ore drawing and mucking are unavoidable to improve safety and pursue production targets. The deployment and mastery of this remote-control technology system should be supported by providing training/internships for prospective operators working in block caving mines. Mastery of the technology of controlling remote mining will be beneficial for the availability of Indonesian human resources competent in operating block caving mining systems. The possibility of adopting a block caving or even the raise caving method should be considered for optimum mining practice.

6.4. Environment

Block caving does not cause uncontrollable environmental impacts, such as changes in landscape on the surface, especially in block caving mines in strong deep rocks. In some instances where the deposited deposits result from an open pit mine, it is necessary to conduct a comprehensive study to establish the optimal crown pillar geometry [133,134]. In relatively shallow deposits, block caving can trigger subsidence at ground level. The maximum area, angle and subsidence depth should be predicted at the feasibility study stage and monitored during mining to ensure no facilities on the surface are affected [135,136]. Other risks to worry about during the operational phase of mining that may impact workers are rock stability, mud rush and air blast. With a good understanding of the geological structure, hydrogeology, and fluid mechanics within rock masses, the best mining designs and practices can be applied to control unexpected impacts. At the stage of detailed exploration, the three aspects must be appropriately characterized and the influence of mining on all three must be considered [137,138]. In addition to the development of science and technology, mastery of human resources to understand the condition of the rock mass is absolutely necessary in the context of implementing an environmentally friendly block caving mine in the future.

6.5. Policy

Indonesia has been experiencing a tortuous journey in mineral resource management. At the beginning of the independence era (1945–1965), President Soekarno spread a strong spirit of nationalism and anti-imperialism. In this era, the Indonesian government nationalized mining companies that the Dutch regime controlled then. Investment and aid from western countries were rejected [139]. In the New Order regime (1965–1997), foreign investments were facilitated by Mining Law 1967 with the mining work contract (Kontrak Karya) scheme. The reformation era had been inspired by the decentralization that had impacted the rapid growth of mining permits owned by local businesses. The government's lack of readiness for the monitoring system and the experience of mine owners in good mining practices left some "troublesome homework" until the present day. The biggest problem is that of the voids of former mines and their unfinished land arrangement, as their function is written in the mine closure documents. Finally, after almost half a century, Indonesia's government issued a new Mining Law in 2009 (Mineral and Coal Mining Law No.4 2009). It reorganized the superintendence of the mining industry, specifically, responsibility for carrying out product down streaming and prioritizing the fulfillment of domestic demand. After the legalization of the newest Mining Law (Mining Law No.3 2020–The Amendment of Mineral and Coal Mining Law No.4 2009) and Omnibus Law 2021, the Indonesian government remained open to global investors with commensurate authorization and benefits. This condition is more suitable for the investment climate and promises a brighter future for the mining industry in Indonesia. Therefore, there is a legal opportunity to apply the block caving method in Indonesia by experienced foreign companies.

6.6. Social Geology

Mining industries have specific characteristics: high return, high risk, high technology and high investment. Mining is an industry operating over long time periods, especially in the block caving method, where the development and production time is relatively long due to the large deposit volume. This requires engagement with community growth on the local scale and country development on a national scale. As echoed in the last decade, social geology has become a required aspect that must be fulfilled [139–141]. Besides being technically feasible and economically profitable, mining projects must also be socially acceptable. Every new investment that comes in must be able to make a real contribution to society. The issue has become a crucial and sensitive issue, especially due to the impact of the social gap issue since Grasberg and Erstberg reserves have been exploited until the present. The appropriateness of profit sharing between the mining company and Indonesia has been criticized. This special requirement is known as a "social license" to operate.

The new mode of block caving mining methods uses autonomously operated and machine-intensive means for producing a large volume of ore. It is possible to assume that this reduced need for human labor means that this mining project provides only a low positive impact on people's employment. In addition, Indonesian miners are not familiar with this method, so their participation will be restricted. The contribution of the block caving mining project must be realized in another form, for instance, technological transfer by training Indonesian workers to become advanced in the application of these methods. Indonesia should mirror settled mining countries such as Australia and USA, who have made block caving an underground mining method supplying a large total volume of their national ore production.

7. Conclusions and Recommendations

The new requirements of ore bodies, suitably mined using the transformed block caving method, have been successively identified. This method can be applied to a blocky ore body with a thickness of 200–800 m, various rock mass strengths until 300 MPa, from low to high (from 0.3% Cu until more than 1.0% Cu), but uniform in grade and at a depth from 500 to 2200 m. The technical specifications for running block caving mines, including preparation methods, undercutting strategy, mine design, mining equipment and monitoring systems, have been synthesized. The potential ore bodies mined using the transformed block caving method have been recommended. The block caving method is promising for underground mining metal deposits in Indonesia, especially Cu-Au porphyry deposits. Although it is favorable for rock mass of various strengths and ore bodies at any grade level, a large dimension of the ore body is the fixed requirement. This condition is due to the need for fast and continuous production to comply with the specific pay-back period and settle the investment costs. Therefore, porphyry is the appropriate ore body type to be mined using this method. Considering the requirements and the successful practice of the block caving project in Grasberg Caving Complex as a role model, the Indonesian government has been suggested as the main responsible body, assisted by national or global investors. The suggestion is to concentrate on the detailed exploration of porphyry deposits and feasibility studies applying the method to the prospective ore bodies, i.e., Onto, Tambulilato, Tumpangpitu and Randu Kuning. In addition, exploration method, cost, operation, environment, mining policy and social geology are important aspects worth noting. The infrastructure development stage of block caving needs 3–5 years until continuous production. The existing block caving projects have a life for production ranging from 12 to 35 years. Thus, a long-term concession is proposed, of about 15 until 40 years, covering development and operational production stages.

Author Contributions: Conceptualization, S.M., R.K.W. and D.P.S.; methodology, S.M., R.K.W. and D.P.S. and S.; software, S.M., W.H. and E.R.; validation, R.K.W., D.P.S., S., G.M.S. and E.R.; formal analysis, R.K.W., D.P.S., S.M., G.M.S. and S.; investigation, S.M., R.K.W., D.P.S., W.H.; resources, R.K.W., D.P.S., S.M. and G.M.S.; data curation, S.M., E.R., D.P.S. and G.M.S.; writing—original draft preparation, S.M., R.K.W., D.P.S.; writing—review and editing, S.M., R.K.W., D.P.S., S., R.R.S.F.; visualization, S.M., W.H.; supervision, R.K.W. and D.P.S.; project administration, S.M.; funding acquisition, R.K.W., D.P.S., S.M. All authors have read and agreed to the published version of the manuscript.

Funding: This study was supported by the Indonesian Education Scholarship from Puslapdik and LPDP, the Ministry of Education and Culture, awarded to Sari Melati.

Data Availability Statement: The main text includes all the data supporting this article.

Acknowledgments: We appreciate all miners, scholars, and practitioners in the block caving method for their devotion to developing this mass underground mining method. We would also like to express our gratitude to the researchers in our groups: Rock Mechanics, Global Geophysics and Earth Resources Exploration. Their invaluable knowledge and competence assisted us in dealing with this new issue. We thank the staff and technicians at Geomechanics and Mining Equipment Laboratory, Vulcanology and Geothermal Laboratory, Geophysical Instrumentation and Electronics Laboratory, Mineralogy, Microscopy and Geochemistry Laboratory, for their assistance in providing us with the resources needed to run the research. This study is supported by research funding from ITB entitled *Quantification of competent rock mass on block caving mining using micro-seismic monitoring and numerical modeling*, awarded to Ridho K. Wattimena.

Conflicts of Interest: The authors declare that the research runs without any commercial or financial relationships that could be construed as a potential conflict of interest.

Appendix A

Table A1. Block caving projects in the world.

Mines	Ore Bodies and Types of Mineralization	Rock Mass	Depth	Production Rates and Reserves	Footprint	Mine Design	Equipment	Time	Ref.
Argyle, Australia	Diamond pipe. Volcanic vent intrusion of magmatic lamproite and lamproitic tuff.	Granite, dolerite, basalt and metamorphosed quartzite and mudstone. UCS 35–104 MPa, RMR 45–59	$\sigma_1 = 2\sigma_v$, $\sigma_2 = 1.5\sigma_v$, $\sigma_3 = 0.027z$	18,000 tpd (Lift 1). It has produces 800M carats	75,000 m^2	an advanced undercut technique using a W-incline undercut design	Real-time LHD dispatch	2008 (under-cutting), 2015 (development complete), 2020 (final production)	[142,143]
Cadia East, Australia	Monzonite porphyry, Au-Cu porphyry deposits	Andesit, monzonite, quartz. UCS 132–140 MPa, E 65–67 GPa, FF < 15.	63:42:36 @1200; 72:48:41 @1400;	26 Mta	Width orebody 700 m	El Teniente, Drawbell Spacing: 32 × 20 m	Load-Haul-Dump (LHD) operation	2000 (production)	[47,77,99]
Carrapateena, South Australia	copper-gold deposit	brecciated granite complex	500 m	10,000–120,000 t of copper and 110,000–120,000 oz	70,000 m^2	El Teniente, draw point spacing of 32 m × 22	LHD., Jaw Gyratory crusher, crushed-ore-bin, conveyor system	2020 (Prefeasibility study), 2026 (Production), 2045 (final year)	[132,144–146]
Cullinan, South Africa	Kimberlite pipe	UCS Kimberlite: 80–130 (Grey), 73–193 (Hypabyssal) Mpa; UCS Country rock: 140–220 (Norite), 60–240 (Metasediments), Hydraulic radius: 30; Mining Rock Mass Rating: 30–50 (grey), 25–35 (contacts, internal dykes and shear zones), 40–60 (Hypabissal)	630–732 mbs	3.9 Mt/a; Reserves: 38.6 Mt, grade 38.8, 14.97 Mt	32 ha	Centenary-Cut; Undercut tunnels 4 m wide and 4 m height, 16 m spacing; Extraction level 4.2 m wide by 4.2 m high, spacing 16–18 m; Tunnel spacing in the production level of 32 m; Drawpoint spacing 18 m	Tamrock Toro–LHD	1980–2037 (operation)	[147]

Table A1. *Cont.*

Mines	Ore Bodies and Types of Mineralization	Rock Mass	Depth	Production Rates and Reserves	Footprint	Mine Design	Equipment	Time	Ref.
El Teniente, Chile	copper-molybdenum deposit	Andesite, dacite, diorite, braden pipe; UCS 120, 110, 140, 90 MPa; RMR 53–59, 59–66, 64–66	2200 m. $\sigma_1 = 0.0328z + 16$, $\sigma_2 = 0.0283z + 5$, $\sigma_1 = 0.0265z$	Productions 140,000 tons per day. Measured resources 1128 million tons 0.985%Cu	500–800 m	El Teniente, hydraulic radius 26 m	Load-Haul-Dump (LHD)	1997 (pre-undercutting), 2032 (planned final production)	[66,148]
Grasberg, Indonesia	Cu-Au Porphyry, Skarn	Fair to the very good ground: 80–140 MPa; Poor to fair ground: 5–80 MPa	1200 m	60,000–100,000 tpd; 160,000 (planned for 2026)	Area: 700,000 m^2	El Teniente, Drawbell Spacing: 20 × 30 m	Load-Haul-Dump (LHD) operation, rail haulage system	2004 (construction), 2018 (production)	[59,109,110,112,149,150]
Jwaneng, Botswana	diamond-bearing kimberlite complex	Sand, calcrete, laminated shale, carbonaceous shale, quarzitic shale, chert pebble conglomerate-bevets, carbonaceous shale and dolomite. UCS 25 MPa (weak kimberlite), >250 MPa (very competent dolomite)	~1000 m; $\sigma_1 = 0.9–1.1\sigma_v$, $\sigma_2 = 0.5\sigma_v$, $\sigma_3 = 0.027z$	No data found	No data found	No data found	No data found	2032 (construction)	[151,152]
Northparkes Mine, Australia	Trachyandesites (Volcanics) and finger-like monzonite porphyry (MP) intrusions, potassic alteration and occurs predominately in stockwork quartz veins.	Gypsum and quartz. MRMR's in Lift 1 ranged from 33 to 54.	>800 m	16,000 tpd (E26 Lift 1, Lift 2, Lift 2 N); 18,000 tpd (E48 Lift 1). Reserves 27 million tons of Ore.	Width vein 200 m, height 800 m. 196 meters long by 180 meters wide.	Northparkes layout style, Hydraulic radius 20–25	Load Haul Dump	2002 (production)	[48]

Table A1. *Cont.*

Mines	Ore Bodies and Types of Mineralization	Rock Mass	Depth	Production Rates and Reserves	Footprint	Mine Design	Equipment	Time	Ref.
Oyu Tolgoi, Mongolia	copper-gold-molybdenum mineralization	volcanic and quartz monzo-diorite (QMD); Dacite tuff, breccia (IGN), basalt flows and minor volcaniclastic strata (Va). Dikes: rhyolitic, hornblende biotite andesite, dacite and basalt. MRMR < 20	1385 m	95,000 tpd (Hugo North Lift 1)	Hugo deposit height 900 m, length 1.8 km, width 500 m.	El Teniente draw point layout on 31 × 18 m spacing	underground trucking system, gyratory crushers, conveyor system, concentrator	2015 (construction), 2020 (production)	[57,153]
Padcal, Philippines	Cu-Au Porphyry	0.18% Cu, 0.27 g/t Au; 56 Mlbs Cu, 166.700 oz Au	No data found	Production 70,000 m^2	No data found	No data found	No data found	2020 (exploration)	[154]
Palabora, South Africa	Magmatic-hydrothermal deposit.	Carbonatite, 139 MPa (intact), 111 MPa (rock mass)	1200–1800 m.	Production 30,000–82,000 tpd; Reserve 960 Mt	250 × 650 m	off-set herringbone style, 20 cross-cut	LHD, crusher	2000–2014	[50,51,104]
Ridgeway Deeps, Australia	Au-Cu porphyry	Cadia Valley Monzonite (93–155 MPa), Forrest Reef Volcaniclastics (87–150 MPa) and Weemalla Sediments (88–144 MPa). Average density 2.85 t/m^3	1100 m. $\sigma_1 = 65$ MPa, $\sigma_2 = 47$ MPa, $\sigma_3 = 32$ MPa	101 mt at 1.8g/t Au and, 0.38% Cu for 2.6 Moz Au and 380 kt Cu	500 × 200 m^2	Offset Herringbone layout and consists of 15 extraction drives, 250 drawpoints	Load haul Dump	2005–2017 (production)	[50]
Shabanie, Zimbabwe	Asbestos	Dunite Sill intruding Precambrian Gneisses	No data found	No data found	No data found	No data found	No data found	1970 (production)	[146]

Table A1. *Cont.*

Mines	Ore Bodies and Types of Mineralization	Rock Mass	Depth	Production Rates and Reserves	Footprint	Mine Design	Equipment	Time	Ref.
Stornoway Diamonds' Renard Mine, Quebec, Canada	Kimberly pipe	Pyroclastic, granitoid and gneissic host rock, UCS 4.5–26 MPa	600 m. $\sigma_1 = 0.9$–$1.1\sigma_v$, $\sigma_2 = 0.5\sigma_v$, $\sigma_3 = \rho.g.h$	3000–5000 tpd	225 m	Herringbone. Drawpoints are 5.3 m wide, distance between center 15 m	Load haul Dump	2018 (production)	[55]
Lvivvuhillia SE Mine, and Ukraine	Coal, carbonous formation	Sandstone, Argillite, Aleurite	Sandy shale 23.2–31.1 MPa	100 ktons per months	its average mining thickness is 1.24 m.	10.3–10.6 m^2 for boundary entry	Coal shearers, Scraper, Oil-pumping station	2020 (production)	[25,26]
the 10th Anniversary of Kazakhstan's Independence Mine, and Kazakh-stan	Chromite deposits	Peridotite and Serpentinite, UCS 17.1–64.5 MPa	Depth 900 m, $\sigma_1 = \sigma_3 =$ and $\sigma_z = 24.8$ MPa	No data found	180 m	Undercut-caving system, Drawpoint spacing 12–24 m	No data found	Development (2021)	[155]

Table A2. Ore deposit and type of mineralization in Indonesia.

Location	Ore Genesis	Type	Size, Dip	Grade, Volume	Rock Mass	Mining Method	Status	References
Awak Mas, Latimojong, South Sulawesi	Hydrothermal	albite-ankerite-pyrite alteration halo	up to ~75 m width	Indicate and inferred resource of 38.4 Mt at 1.41 gr/t Au~1.74 Moz Au	Phyllite and schist.	-	Exploration	[156]
Batu Hijau	Epithermal	Porphyry	a zone 300 m × 900 m containing > 0.3 wt % Cu	> 0.1 wt % Cu, > 0.1 wt % Cu, Mo (> 30 ppm)	Diorite, metavolcanic rock	Open Pit, Block Caving (in panning)	Production Planning	[157,158]
Beruang Kanan, Kalimantan Tengah	Epithermal	quartz vein, porphyry	vein direction is N 312° E/43°	Not explore yet	Dasite, diorite, silica sand	-	2017 (Exploration)	[159]
Bombana, Southeast Sulawesi	Secondary (placer) in Langkolawa in Wumbubangka derived from orogenic gold	Gold-bearing quartz vein	2 cm–2 m	grades <0.005 g/t to 134 g/t	mica schist, phyllite, metasandstone and marble)	Placer mining -artisanal and small-scale gold mining	2011 (study)	[160,161]
Bulagidun	Hydrothermal	a copper, gold and tourmaline bearing porphyry and breccia system	up to 500 m lateral distance, veins up to 2 m true width,	more than 14.4 Mt at 0.68 ppm Au and 0.61 wt.% Cu	early diorite to quartz diorite to late tonalite and post-mineral andesitic dykes.	-	Geological Study	[162]
Cibaliung, Banten	Epithermal	Quartz vein	Dyke 1 to 120 m wide, 20 to >300 m long.	1.3 Mt 10.42 g/t Au, 60.7 g/t Ag 3 g/t cut-off; 435,000 ounces of Au and 2.54 Mounces Ag	UCS 16.85 MPa, Tensile strength 0.69 MPa	Cut and fill	2001 (Exploration); 2010 (production	[163,164]
Cikidang (Cikotok)	Low sulfidation epithermal adularia	Quartz vein	Thickness 0.7–3 m, dip 60–86°	74.9 g/t Au, 1.2–225 g/t Ag	Lapilli tuff, breccia andesite, claystone, limestone, Sandstone	Underhand stall-stopping method	1998 (production),	[165]
Elang	Epithermal	Porphyry	undescribed	300 t Au, >5 Mt Cu	Volcanoclastic and esitic	-	Exploration	[123,129]

Table A2. *Cont.*

Location	Ore Genesis	Type	Size, Dip	Grade, Volume	Rock Mass	Mining Method	Status	References
Ertsberg	Contact metasomatism	Skarn system	length > 1.1 km, 4–60 m thick, depth >700 m	2.69 percent Cu, 1.02 g/t Au and 16 g/t.	dolomitic sediments	Block Caving	Production	[105]
Gosowong, Halmahera	Epithermal	Quartz vein, porphyry	Thickness 30–40 m, dip 35–70°	0.99 million metric tons (Mt) at 27 g/t Au and 38 g/t Ag	Volcaniclastic and pyroclastic	Open pit	1996 (exploration)	[118]
Grasberg	Contact metasomatism	Porphyry	1.2 km (pit)	over 32 Mt of Cu and 3 kt of Au	Diorite, limestone	Open Pit, Block Caving	Production	[56,109,110, 150]
Gunung Subang, West Java	Epithermal	Gold-bearing minerals	0.01–0.2 m; 40–81°	Au 0.22–14.49 ppm, Ag 17–21.40 ppm, Cu 8.25–34515 ppm, Pb 107.69–2226 ppm, 35.36–7335 ppm	Andesite, tuff, breccia	-	2018 (Prospection)	[126]
Kelian, East Kalimantan	Epithermal	Au-Ag mineralization	0.25–5 m	240 t Au	Rhyolite	Open pit	2003 (Mine closure)	[166]
Kencana	Epithermal	Au deposit	Thickness 12 m, 45°	39 g/ton gold	RMR 25–55 and esite lavas,	Underhand cut and fill	Production	[167]
Malala, Northwest Sulawesi	Hydrothermal	fluorine-poor (quartz monzonite or differentiated monzogranite) class of molybdenum deposits	50 m	estimated resource of 100 Mt at 0.14% MoS.	granites and granodiorites	-	1993 (Geological study)	[168]
Miwah, Aceh	Hydrothermal	high-sulfidation Au–Ag deposit	>60°	inferred total resource of 3.13 million oz (Moz) of Au at a cut-off grade of 0.2 g/t Au	silicified rocks, breccia	-	2019 (Prospection)	[128]

Table A2. *Cont.*

Location	Ore Genesis	Type	Size, Dip	Grade, Volume	Rock Mass	Mining Method	Status	References
Ojolali, Lampung	Epithermal	Tambang Vein:Ag-Au intermediate sulfidation deposit; Bukit Jambi Vein: low sulfidation Au-Ag deposit	<50 m, ~50°	inferred resource 167 g/t Ag and 0.7 g/t Au, forms a total of 40 Moz Ag and 170,000 oz Au	Basalt and esite	-	2014 (Geological study)	[127]
Pani JV Project, Hulawa, Gorontalo	Hydrothermal	Open vein and breccia	No description	Resources 72.7 mt, 0.98 g/t, 2.3 mlb Au	UCS 21.42 MPa, UTS 2.06 MPa	Breccia, granodiorite and dasite	Conceptual study (2020)	[123,129]
Pongkor	Hydrothermal alteration	Vein	Thickness 2–24 m	2.1 million metric tons at 13.63 ppm gold and 163.24 ppm silver (proven ore reserve)	volcanic breccia, lapilli tuff and esite lava and siltstone	Cut and Fill Stopping, Semi Caving	Production	[115–117]
Tambulilato: Cabang Kiri, Sungai Mak, Kayubulan and Cabang Kanan	Hydrothermal	Poprhyry (Cabang Kiri, Sungai Mak, Kayu Bulan, Cabang Kanan), high-sulfidation epithermal Au-Ag (Motomboto); low-sulfidation epithermal Au-Ag (Kaidundu)	Various wide of veins and porphyry	392.3 million tons, 0.49%Cu, 0.43 g/t/Au, 1.65 g/t Ag.	Dacite, vulcanic, diorite	Stopping underground mining	Production (until 2052)	[122]
Tangse, North Sumatra	Hydrothermal	Cu-Mo porphyry deposit	Not explored yet	Not explore yet	Diorite	-	2018 (Prospecting)	[125]
Toguraci, Halmahera	Epithermal	Quartz vein, porphyry	-	0.41 Mt, 27 g/t Au	Andesitic lava, UCS 80 MPa	Under Hand Cut and Fill (UHCF) and Open Stope (Sub Level–Blind Stope)	1996 (exploration)	[169]

Table A2. *Cont.*

Location	Ore Genesis	Type	Size, Dip	Grade, Volume	Rock Mass	Mining Method	Status	References
Tujuh Bukit	Hydrothermal	Porphyry	described	Inferred resources 1.9 bt, 0.45% Cu, 0.45 g/t, 8.7 mt Cu, 28 mlb Au	sedimentary and andesitic volcanic rocks	Open Pit	Production (2021)	[123,129,170]
Tumpangpitu, East Java	Epithermal	Porphyry	Mineralization > 800 m	1.9 Gt, 0.45% Cu, 0.45 g/t Au	Diorite and esite, breccia	-	Exploration	[123,129,170]
Underground Tujuh Bukit	Hydrothermal	Porphyry (high level porphyry copper-gold-molybdenum deposit (sulfide)		Inferred resources 1.9 bt, 0.45% Cu, 0.45 g/t, 8.7 mt Cu, 28 mlb Au	sedimentary and andesitic volcanic rocks	Underground mining (undetermined)	Pre-feasibility study (2021)	[123]
Wetar, Pulau Wetar, Southwestern moluccas	Volcanic- hosted massive sulfide (VMS)	Primarily pyrite	~150,100.70 m and ~1,209,030 m,	20 mt, 38% S, 33% Fe, host Cu, Au, Ag, Zn	Basaltic and andesite	Open pit	Production (2010)	[171]

Table A3. Recommended exploration methods for the ore body type.

Type of Deposit	Characteristic	Recommended Exploration Methods & Tools	Rationalization
Porphyry Cu-Au deposits	Commonly associated with magnetite that can produce strong discrete magnetic anomalies. Strong charge abilities due to sulfides are typically associated with porphyry systems.	High-resolution magnetic survey	Porphyry is usually within a zone of magnetite-destructive alteration. Magnetic surveys are also valuable for defining regional structure and geology in the porphyry environment.
		Gravity, radio metrics, remote sensing and topography	Mineralization and clay-pyrite alteration can produce strong anomalies and late-stage and post-mineral intrusions can be mapped as low chargeability within the system. These systems may be more conductive than the host rock because of clay-pyrite alteration and sulfide veining and airborne electromagnetic can be helpful in locating and defining their extent.
High sulfidation epithermal system	Gold is commonly associated with massive silica alteration.	Resistivity and airborne electromagnetic survey	This alteration results in resistivities in the order of thousands of ohmmeters compared with background resistivities of tens of ohm-meters in argillic and propylitic alteration. Alteration in high sulfidation epithermal deposits is magnetite destructive over a large area, although it does not appear to have a large vertical extent as the subdued characterization of the underlying lithologies can be observed.
Low sulfidation epi-thermal system	Gold in this deposit is in thin quartz veins associated with major structures. Some deposits are associated with broad zones of magnetite destruction, which is apparent in the regional magnetics.	High-resolution magnetics, resistivity surveying	The alteration associated with the veins is magnetite destructive and high-resolution magnetics can be beneficial and cost-effective technique to map the structures and alteration. Generally, the high resistivity zones are due to silicification are coincident with the structure identified in the magnetics.

References

1. Reichl, C.; Schatz, M. *World Mining Data 2022. International Organizing Committee for the World Mining Congresses*; Federal Ministry of Agriculture, Regions and Tourism: Vienna, Austria, 2022.
2. Dong, L.; Sun, D.; Shu, W.; Li, X. Exploration: Safe and clean mining on Earth and asteroids. *J. Clean. Prod.* **2020**, *257*, 120899. [CrossRef]
3. Lurka, A. Spatio-temporal hierarchical cluster analysis of mining-induced seismicity in coal mines using Ward's minimum variance method. *J. Appl. Geophys.* **2021**, *184*, 104249.
4. Brodny, J.; Tutak, M. Challenges of the polish coal mining industry on its way to innovative and sustainable development. *J. Clean. Prod.* **2022**, *375*, 134061. [CrossRef]
5. Sahu, H.; Prakash, N.; Jayanthu, S. Underground Mining for Meeting Environmental Concerns–A Strategic Approach for Sustainable Mining in Future. *Procedia Earth Planet. Sci.* **2015**, *11*, 232–241. [CrossRef]
6. Harraz, H. Underground mining. In *Mining Methods*; Geology Department, Faculty of Science: Tanta, Egypt, 2011.
7. Orellana, L.; Castro, R.; Hekmat, A.; Arancabia, E. Productivity of a Continuous Mining System for Block Caving. *Rock Mech. Rock Eng.* **2016**, *50*, 657–663.
8. Ibishi, G.; Yavuz, M.; Genis, M. Underground mining method assessment using decision-making techniques in a fuzzy environment: Case study, Trepça mine, Kosovo. *Min. Miner. Depos.* **2020**, *14*, 134–140. [CrossRef]
9. Wu, H.; Yin, Z.; Zhang, Y.; Qi, C.; Liu, X.; Wang, J. Comparizon of underground coal mining methods based on life cycle assessment. *Front. Earth Sci.* **2022**, *10*, 933. [CrossRef]
10. Hartman, H. *Introductory Mining Engineering*; John Wiley & Sons: New York, NY, USA, 1987.
11. Adhikary, D.P.; Guo, H. Modelling of longwall mining-induced strata permeability change. *Rock Mech. Rock Eng.* **2015**, *48*, 345–359. [CrossRef]
12. Mitchell, G.W. *Longwall Mining*; Australasian Coal Mining Practice: Melbourne, Australia, 2009; pp. 340–373.
13. Palchik, V. Formation of fractured zones in overburden due to longwall mining. *Environ. Geol.* **2003**, *44*, 28–38. [CrossRef]
14. Lechner, A.M.; Baumgartl, T.; Matthew, P.; Glenn, V. The impact of underground longwall mining on prime agricultural land: A review and research agenda. *Land Degrad. Dev.* **2016**, *27*, 1650–1663. [CrossRef]
15. Bai, Q.; Tu, S. A general review on longwall mining-induced fractures in near-face regions. *Geofluids* **2019**, *2019*, 3089292. [CrossRef]
16. Mark, C. Coal bursts in the deep longwall mines of the United States. *Int. J. Coal Sci. Technol.* **2016**, *3*, 1–9. [CrossRef]
17. Shabanimashcool, M.; Li, C.C. Numerical modelling of longwall mining and stability analysis of the gates in a coal mine. *Int. J. Rock Mech. Min. Sci.* **2012**, *51*, 24–34. [CrossRef]
18. Islavath, S.R.; Deb, D.; Kumar, H. Numerical analysis of a longwall mining cycle and development of a composite longwall index. *Int. J. Rock Mech. Min. Sci.* **2016**, *89*, 43–54. [CrossRef]
19. Guo, H.; Adhikary, D.P.; Craig, M.S. Simulation of mine water inflow and gas emission during longwall mining. *Rock Mech. Rock Eng.* **2009**, *42*, 25–51. [CrossRef]
20. Gao, Y.; Liu, D.; Zhang, X.; He, M. Analysis and optimization of entry stability in underground longwall mining. *Sustainability* **2017**, *9*, 2079. [CrossRef]
21. Wang, G.; Pang, Y. Surrounding rock control theory and longwall mining technology innovation. *Int. J. Coal Sci. Technol.* **2017**, *4*, 301–309. [CrossRef]
22. Islam, M.R.; Hayashi, D.; Kamruzzaman, A.B.M. Finite element modeling of stress distributions and problems for multi-slice longwall mining in Bangladesh, with special reference to the Barapukuria coal mine. *Int. J. Coal Geol.* **2009**, *78*, 91–109. [CrossRef]
23. Hoseinie, S.H.; Ataei, M.; Khalokakaie, R.; Ghodrati, B.; Kumar, U. Reliability analysis of drum shearer machine at mechanized longwall mines. *J. Qual. Maint. Eng.* **2012**, *18*, 98–119. [CrossRef]
24. Najafi, A.B.; Saeedi, G.R.; Farsangi, M.A. Risk analysis and prediction of out-of-seam dilution in longwall mining. *Int. J. Rock Mech. Min. Sci.* **2014**, *70*, 115–122. [CrossRef]
25. Shavarskyi, I.; Falshtynskyi, V.; Dychkovskyi, R.; Akimov, O.; Sala, D.; Buketov, V. Management of the longwall face advance on the stress-strain state of rock mass. *Min. Miner. Depos.* **2022**, *16*, 78–85. [CrossRef]
26. Ukrinform. Ukrainians Miners Produces 2.6 Mtonnes of Coal in September. Available online: https://www.ukrinform.net/rubric-economy/3109854-ukrainian-miners-produced-26m-tonnes-of-coal-in-september.html (accessed on 21 October 2022).
27. Wojtecki, Ł.; Kurzeja, J.; Knopik, M. The influence of mining factors on seismic activity during longwall mining of a coal seam. *Int. J. Min. Sci. Technol.* **2021**, *31*, 429–437. [CrossRef]
28. He, M.; Gao, Y.; Yang, J.; Gong, W. An innovative approach for gob-side entry retaining in thick coal seam longwall mining. *Energies* **2017**, *10*, 1785. [CrossRef]
29. Pongpanya, P.; Sasaoka, T.; Shimada, H.; Hamanaka, A.; Wahyudi, S. Numerical study on effect of longwall mining on stability of main roadway under weak ground conditions in Indonesia. *J. Geol. Resour. Eng.* **2017**, *3*, 93–104.
30. Sasaoka, T.; Mao, P.; Shimada, H.; Hamanaka, A.; Oya, J. Numerical analysis of longwall gate-entry stability under weak geological condition: A case study of an Indonesian coal mine. *Energies* **2020**, *13*, 4710. [CrossRef]
31. Sasaoka, T.; Takamoto, H.; Shimada, H.; Oya, J.; Hamanaka, A.; Matsui, K. Surface subsidence due to underground mining operation under weak geological condition in Indonesia. *J. Rock Mech. Geotech. Eng.* **2015**, *7*, 337–344. [CrossRef]

32. Sasaoka, T.; Hamanaka, A.; Shimada, H.; Matsui, K.; Lin, N.Z.; Sulistianto, B. Punch multi-slice longwall mining system for thick coal seam under weak geological conditions. *J. Geol. Resour. Eng.* **2015**, *4*, 28–36.
33. Mao, P.; Sasaoka, T.; Shimada, H.; Hamanaka, A.; Wahyudi, S.; Oya, J.; Naung, N. Three-Dimensional Analysis of Gate-Entry Stability in Multiple Seams Longwall Coal Mine under Weak Rock Conditions. *Earth Sci. Res.* **2020**, *9*, P72. [CrossRef]
34. Skawina, B.; Greberg, J.; Salama, A.; Gustafson, A. The effects of orepass loss on loading, hauling, and dumping operations and production rates in a sublevel caving mine. *J. S. Afr. Inst. Min. Metall.* **2018**, *118*, 409–418. [CrossRef]
35. Kosenko, A.V. Improvement of sub-level caving mining methods during high-grade iron ore mining. Natsional'nyi Hirnychyi Universytet. *Nauk. Visnyk* **2021**, *1*, 19–25.
36. Yi, C.; Sjöberg, J.; Johansson, D. Numerical modelling for blast-induced fragmentation in sublevel caving mines. *Tunn. Undergr. Space Technol.* **2017**, *68*, 167–173. [CrossRef]
37. Blachowski, J.; Ellefmo, S. Numerical modelling of rock mass deformation in sublevel caving mining system. *Acta Geodyn. Geomater.* **2012**, *9*, 167.
38. Singh, R.; Singh, T.N. Investigation into the behaviour of a support system and roof strata during sublevel caving of a thick coal seam. *Geotech. Geol. Eng.* **1999**, *17*, 21–35. [CrossRef]
39. Shekhar, G.; Gustafson, A.; Hersinger, A.; Jonsson, K.; Schunnesson, H. Development of a model for economic control of loading in sublevel caving mines. *Min. Technol.* **2019**, *128*, 118–128. [CrossRef]
40. Navarro, J.; Schunnesson, H.; Ghosh, R.; Segarra, P.; Johansson, D.; Sanchidrian, J. Application of drill-monitoring for chargeability assessment in sublevel caving. *Int. J. Rock Mech. Min. Sci.* **2019**, *119*, 180–192. [CrossRef]
41. Abolfazlzadeh, Y.; Hudyma, M. Identifying and describing a seismogenic zone in a sublevel caving mine. *Rock Mech. Rock Eng.* **2016**, *49*, 3735–3751. [CrossRef]
42. Zhang, Z.X.; Wimmer, M. A case study of dividing a single blast into two parts in sublevel caving. *Int. J. Rock Mech. Min. Sci.* **2018**, *104*, 84–93. [CrossRef]
43. Ge, Q.; Fan, W.; Zhu, W.; Chen, X. Application and research of Block Caving in Pulang Copper Mine. In *IOP Conference Series: Earth and Environmental Science*; ESMA 2017; IOP Publishing: Beijing, China, 2018; pp. 1–8.
44. Eberhardt, E.; Woo, K.; Stead, D.; Elmo, D. Transitioning from open pit to underground mass mining: Meeting the rock engineering challenges of going deeper. In *ISRM Congress 2015 Proceedings—International Symposium on Rock Mechanics*; International Society for Rock Mechanics: Salzburg, Austria, 2015; ISBN 978-1-926872-25-4.
45. Potvin, Y.; Jakubec, J. *Preface. Caving 2018: Proceedings of the Fourth International symposium on Block and Sublevel Caving*; Australian Centre for Geomechanics: Vancouver, BC, Canada, 2018.
46. Woo, K.-S.; Eberhardt EElmo, D.; Stead, D. Empirical investigation and characterization of surface subsidence related to block cave mining. *Int. J. Rock Mech. Min. Sci.* **2013**, *61*, 31–42. [CrossRef]
47. Sainsbury, B.; Sainsbury, D.; Carroll, D. Back-analysis of PC1 cave propagation and subsidence behaviour at the Cadia East mine. In *Caving 2018: Proceedings of the Fourth International Symposium on Block and Sublevel Caving*; Australian Centre for Geomechanics: Vancouver, Australia, 2018; pp. 167–178.
48. Van As, A.; Jeffrey, R.G. Hydraulic Fracturing as a Cave Inducement Technique at Northparkes Mines. In Proceedings of the MassMin2000 Conference, Brisbane, QLD, Australia, 29 October–2 November 2000.
49. Cuello, D.; Newcombe, G. Key geotechnical knowledge and practical mine planning guidelines in deep, high stress, hard rock conditions for block and panel cave mining. In *Caving 2018: Proceeding of the Fourth International Symposium on Block and Sublevel Caving*; Australian Centre for Geomechanics: Vancouver, BC, Canada, 2018; pp. 17–35.
50. Lowther, R.J.; Capes, G.W.; Sharrock, G.B. A deformation monitoring plan for extraction level drives at Ridgeway Deeps block cave mine. In *Caving 2010*; Potvin, Y., Ed.; Australian Centre for Geomechanics: Perth, Australia, 2010.
51. Ferguson, G.; Cuello, D.; Moreno, P.; Potvin, Y.; Valdivia, E. Strategy for research and development in the cave mining industry. In *Caving 2018*; Potvin, Y., Jakubec, J., Eds.; Australian Centre for Geomechanics: Perth, Australia, 2018.
52. Westley-Hauta, R.L.; Meyer, S. Characterisation of seismic activity at a kimberlite block caving operation in a complex geological setting in Quebec, Canada. In *Caving 2022: Fifth International Conference on Block and Sublevel Caving, Adelaide, Australia, 30 August–1 September 2022*; Australian Centre for Geomechanics: Perth, Australia, 2022; pp. 1101–1120.
53. Groves, D.I.; Vielreicher, N.M. The Phalabowra (Palabora) carbonatite-hosted magnetite–copper sulfide deposit, South Africa: An end-member of the iron-oxide copper–gold–rare earth element deposit group? *Miner. Depos.* **2001**, *36*, 189–194. [CrossRef]
54. Moss, A.; Dianchenko, S. Townsend, Interaction between the block cave and the pit slopes at Palabora mine. *J. S. Afr. Inst. Min. Metall.* **2006**, *106*, 479–484.
55. Stegman, C.; van As, A.; Peebles, E. Past learning focus innovative solutions to future cave mining. In *Caving 2018: Proceeding of the Fourth International Symposium on Block and Sublevel Caving*; Australian Centre for Geomechanics: Vancouver, BC, Canada, 2018; pp. 37–41.
56. Campbell, R.; Banda, H.; Fajar, J.; Brannon, C. Optimising for success at the Grasberg Block Cave. In *Caving 2018*; Australian Centre for Geomechanics: Perth, Australia, 2018; Volume 3–15, pp. 91–106.
57. Moorcroft, T.H.; Simanjuntak, K.; Dorjsuren, O.; Sanaakhorol, M.; Enkhtaivan, E.; Watt, G.; Eickhoff, V.; Cerny, L.; Deasy, C.; Zimmermann, T. Oyu Tolgoi and Rio Tinto partnership with Palantir Technologies to provide effective geotechnical risk management. In *Caving 2022*; Potvin, Y., Ed.; Australian Centre for Geomechanics: Perth, Australia, 2022.

58. Catalan, A.; Onederra, I.; Chitombo, G. Evaluation of intensive preconditioning in block and panel caving-Part I, quantifying the effect on intact rock. *Min. Technol.* **2017**, *126*, 209–220. [CrossRef]
59. Widodo, S.; Anwar, H.; Syafitri, N.A. Comparative analysis of ANFO and emulsion application on overbreak and underbreak at blasting development activity in underground Deep Mill Level Zone (DMLZ) PT Freeport Indonesia. *IOP Conf. Ser. Earth Environ. Sci.* **2019**, *279*, 012001. [CrossRef]
60. Catalan, A.; Onederra, I.; Chitombo, G. Evaluation of intensive preconditioning in block and panel caving-Part II, quantifying the effect on seismicity and draw rates. *Min. Technol.* **2017**, *126*, 221–239. [CrossRef]
61. Els, F.; Jamasmie, C.; Stutt, A.; Leotaud, V.R.; Chen, J.; Venditti, B. Ranked: World's 10 Biggest Underground Mines by Tonnes of Ore Milled. 2021. Available online: https://www.mining.com/featured-article/ranked-worlds-10-biggest-underground-mines-by-tonnes-of-ore-milled/ (accessed on 3 September 2022).
62. Jacubec, J. Adressing the challenges and future of cave mining. SRK Consulting's International Newsletter. 2018. Available online: https://www.caveminingforum.com/addressing-challenges-future-cave-mining/ (accessed on 15 September 2022).
63. Fernberg, H. *Mining in steep orebody. Mining Methods in Underground Mining*; Atlas Copco: Nacka Municipality, Sweden, 2007; pp. 33–37.
64. Peele, R. *Mining Engineer's Handbook*, 3rd ed.; John Wiley & Sons: New York, NY, 1941.
65. Banks, C. *Relationships between Geology, Ore-Body Genesis, and Rock Mass Characteristics in Block Caving Mines*; The University of British Colombia: Vancouver, BC, Canada, 2009.
66. United States Geological Survey (USGS). El Teniente: Porphyry Copper (Cu-Mo) Deposit in Libertador General Bernardo O'Higgins, Chile. 2022. Available online: https://mrdata.usgs.gov/sir20105090z/show-sir20105090z.php?id=311 (accessed on 21 October 2022).
67. Miller-Tait LPakalnis, R.; Poulin, R. *UBC Mining Method Selection. Proc. Mine Planning and Equipment Selection*; Balkema: Rotterdam, The Netherlands, 1995.
68. Wagner, H.; Ladinig, T. Raise Caving Method for Mining an ore from an Ore Body, and a Mining Infrastructure, Monitoring System, Machinery, Control System and Data Medium Therefor. Classification: E21C41/22 92006.01—Methods of Underground Mining; Layouts Therefor for Ores, e.g. Mining Placers. World Intellectual Property Organization (WIPO). Applicant: Luossavaara Kiirunavaara AB. International Patent WO2021236000A1, 25 November 2021.
69. Ladinig, T. *Raise Caving-Cheaper and More Flexible Mining at Great Depths*; Mining Engineering and Mineral Economics, Montan Universitat: Leoben, Austria, 2021.
70. Ladinig, T.; Wagner, H. Raise Caving—A Hybrid Mining Method Addressing Current. *Berg Huettenmaenn Mon.* **2022**, *167*, 177–186. [CrossRef]
71. Ladinig, T.; Wimmer, M.; Wagner, H. Raise caving: A novel mining method for deep mass mining. In *Caving 2022: Fifth International Conference on Block and Sublevel Caving, Adelaide, Australia, 30 August–1 September 2022*; Australian Centre for Geomechanics, The University of Western Australia: Perth, Australia, 2022.
72. Karlsson, M.; Ladinig, T.; Grynienko, M. Test mining with raise caving mining method: One-time chance to prove the concept? In *Caving 2022: Fifth International Conference on Block and Sublevel Caving, Adelaide, Australia, 30 August–1 September 2022*; Australian Centre for Geomechanics, The University of Western Australia: Perth, Australia, 2022.
73. Chacon, E.B. Hydraulic fracturing used to precondition ore and reduce fragment size for block caving. In *MassMin*; Instituto de Ingenieros de Chile: Santiago, Chile, 2004.
74. Van As, A.; Jeffrey, R.; Chacon, E.; Barrera, V. Preconditioning by hydraulic fracturing for block caving in a moderately stressed naturally fractured orebody. In *MassMin*; Instituto de Ingenieros de Chile: Santiago, Chile, 2004.
75. Rogers, S.E. Understanding hydraulic fracture geometry and interactions in pre-conditioning through DFN and numerical modeling. In Proceedings of the 45th US Rock Mechanics/Geomechanics Symposium, San Francisco, CA, USA, 26–29 June 2011.
76. He, Q.S. Review of Hydraulic Fracturing for Preconditioning in Cave Mining. *Rock Mech. Rock Eng.* **2016**, *49*, 4893–4910. [CrossRef]
77. Catalan, A.; Dunstan, G.; Morgan, M.; Green, S.; Jorquera, M.; Thornhill, T.; Onederra, I.; Chitombo, G. How Can an Intensive Preconditioning Concept Be Implemented at Mass Mining Method? Application to Cadia East Panel Caving Project. In Proceedings of the 6th U.S. Rock Mechanics/Geomechanics Symposium, Chicago, IL, USA, 23–26 June 2012.
78. Flores, G.; Catalan, A. A transition from a large open pit into a novel "macroblock variant" block caving geometry at Chuquicamata mine, Codelco Chile. *J. Rock Mech. Geotech. Eng.* **2019**, *11*, 549–561. [CrossRef]
79. Fiscor, S. Problem solvers overcome adversity with advances in equipment and technology: Tracking 150 Years of Mining Engineering. *Eng. Min. J.* **2016**, *217*, 110.
80. Rafiee, R.; Ataei, M.; Khalookakaie, R.; Jalali, S.E.; Sereshki, F.; Norrozi, M. Numerical modeling of influence parameters in cavability of rock mass in block caving mines. *Int. J. Rock Mech. Min. Sci.* **2018**, *105*, 22–27. [CrossRef]
81. Bahrani, N.; Kaiser, K.; Corkum, A. Suggested methods for estimation of confined strength of heterogeneous (defected) rocks caving operation in a complex geological setting in Quebec, Canada. In *Caving 2022*; Potvin, Y., Ed.; Australian Centre for Geomechanics: Perth, Australia, 2018.
82. Sainsbury, D.; Carrol, D. Analysis of Caving Behavior Using a Synthetic Rock Mass—Ubiquitous Joint Rock Mass Modelling Technique. In *SHIRMS 2018*; Australian Centre for Geomechanics: Perth, Australia, 2018.

83. Cumming-Potvin, D.W. A re-evaluation of the conceptual model of caving mechanics. In *Caving 2018*; Australian Center for Geomechanics: Perth, Australia, 2018; pp. 179–190.
84. Cumming-Potvin, D.W.-C. Numerical simulation of a centrifuge model of caving. In *Caving 2018*; Australian Center for Geomechanics: Perth, Australia, 2018; pp. 191–206.
85. Fuenzalida, M. REBOP-FLAC3D hybrid approach to cave modelling. In *Caving 2018*; Australian Center for Geomechanics: Perth, Australia, 2018; p. 297.
86. Hebert, Y.; Sharrock, G. Three-dimensional simulation of cave initiation, propagation and surface subsidence using a coupled finite difference-cellular automata solusion. In *Caving*; Australian Center for Geomechanics: Perth, Australia, 2022; pp. 151–166.
87. Gomez, R.; Castro, R. Stress modelling using cellular automata for block caving applications. *Int. J. Rock Mech. Min. Sci.* **2022**, *154*, 105124. [CrossRef]
88. Arndt, A.B.; Bui, T.; Diering, T.; Austen, I.; Hocking, R. Integrated simulation and optimation tools for production scheduling using finite element analysis caving geomechanics simulation coupled with 3D cellular automata. In *Caving 2018: Proceedings of the Fourth International Symposium on Block and Sublevel Caving, Vancouver, BC, Canada, 15–17 October 2018*; Australian Center for Geomechanics: Perth, Australia, 2018; pp. 247–260.
89. Nadolski, S.; Klein, B.; Hart, C.J.R.; Moss, A.; Elmo, D. An approach to evaluating block and panel cave projects for sensor-based sorting applications. In *Caving 2018: Proceeding of the Fourth International Symposium on Block and Sublevel Caving, Vancouver, BC, Canada, 15–17 October 2018*; Australian Center for Geomechanics: Perth, Australia, 2018; pp. 133–140.
90. Henning, M.G. Grade control and segregation at new Gold's new Afton block cave operation, Kamloops, British Colombia. In *Caving 2018: Proceedings of the Fourth International Symposium on Block and Sublevel Caving, Vancouver, BC, Canada, 15–17 October 2018*; Australian Centre for Geomechanics: Perth, Australia, 2022; pp. 141–150.
91. Moss, A.; Klein, B.; Nadolski, S. Cave to mill: Improving valeu of caving operations. In *Caving 2018: Proceeding of the Fourth International Symposium on Block and Sublevel Caving*; Australian Centre for Geomechanics: Perth, Australia, 2018; pp. 119–132.
92. Hidayat, W.; Sahara, D.; Widiyantoro, S.; Suharsono, S.; Wattimena, R.K.; Melati, S.; Putra, I.R.A.; Prahastudhi, S.; Sitorus, E.; Riyanto, E. Testing the Utilization of a Seismic Network Outside the Main Mining Facility Area for Expanding the Microseismic Monitoring Coverage in a Deep Block Caving. *Appl. Sci.* **2022**, *12*, 7265. [CrossRef]
93. Dong, L.; Zou, W.; Li, X.; Shu, W.; Wang, Z. Collaborative localization method using analytical and iterative solutions for microseismic/acoustic emission sources in the rockmass structure for underground mining. *Eng. Fract. Mech.* **2019**, *210*, 95–112. [CrossRef]
94. Ross, I.T. Bencmarking and its application for caving projects. In *Caving 2018: Proceedings of the Foutrh International Symposium on Block and Sublevel Caving, Vancouver, BC, Canada, 15–17 October 2018*; Australian Centre for Geomechanics: Vancouver, BC, Canada, 2018; pp. 473–486.
95. Thompson, R.; Malekzehtab, H. Underground roadway design consideration for efficient autonomous hauling. In *Caving 2018: Proceeding of the Fourth International Symposium on Block and Sublevel Caving, Vancouver, BC, Canada, 15–17 October 2018*; Australian Centre for Geomechanics: Perth, Australia, 2018; pp. 337–350.
96. Paredes, P.; Leano, T.; Jauriat, L. Chuquicamata underground mine design: The simplification of the ore handling system of Lift 1. In *Caving 2018: Proceeding of the Fourth International Symposium on Block and Sublevel Caving, Vancouver, BC, Canada, 15–17 October 2018*; Australian Centre for Geomechanics: Perth, Australia, 2018; pp. 385–398.
97. Rogers, S.; Elmo, D.; Webb, G.; Catalan, A. Volumetric Fracture Intensity Measurement for Improved Rock Mass Characterisation and Fragmentation Assessment in Block Caving Operations. *Rock Mech. Rock Eng.* **2015**, *48*, 633–649. [CrossRef]
98. Liu, Y.; Nadolski, D.; Elmo, D.; Klein, B.; Scoble, M. Use of Digital Imaging Processing Techniques to Characterise Block Caving Secondary Fragmentation and Implications for a Proposed Cave-to-Mill Approach. In Proceedings of the 49th U.S. Rock Mechanics/Geomechanics Symposium, American Rock Mechanics Association (ARMA), San Francisco, CA, USA, 1–25 June 2015.
99. Castro, R.; Cuello, D. Hang-up analysis and modeling for Cadia East PC1-S1 and PC2-S1. In *Caving 2018: Proceeding of the Fourth International Symposium on Block and Sublevel Caving, Vancouver, BC, Canada, 15–17 October 2018*; Australian Centre for Geomechanics: Perth, Australia, 2018; pp. 233–246.
100. Dorador, L.; Eberhardt, E.; Elmo, D. Influence of rock mass veining and non-persistent joints on secondary fragmentation during block caving. In *Caving 2018: Proceedings of the Fourth International Symposium on Block and Sublevel Caving, Vancouver, BC, Canada, 15–17 October 2018*; Australian Centre for Geomechanics: Perth, Australia, 2018.
101. Parsons, J.; Hamilton, D.B.; Ludwicki, C. Non-vertical cave and dilution modeling at New Gold's New Afton Mine. In *Caving 2018: Proceeding of the Fourth International Symposium on Block and Sublevel Caving, Vancouver, BC, Canada, 15–17 October 2018*; Australian Centre for Geomechanics: Perth, Australia, 2018; pp. 323–336.
102. Sánchez, V.; Castro, R.L.; Palma, S. Gravity flow characterization of fine granular material for Block Caving. *Int. J. Rock Mech. Min. Sci.* **2019**, *114*, 24–32. [CrossRef]
103. Castro, R.; Arancibia, L.; Gomez, R. Quantifying fines migration in block caving through 3D experiments. *Int. J. Rock Mech. Min. Sci.* **2022**, *151*, 105033. [CrossRef]
104. Diering, T.; Ngidi, S.N.; Bezuidenhout, J.J.; Paetzold, H.D. Palabora Lift 1 block cave: Understanding the grade behaviour. In *Caving 2018: Proceeding of the Fourth International Symposium on Block and Sublevel Caving, Vancouver, BC, Canada, 15–17 October 2018*; Australian Centre for Geomechanics: Perth, Australia, 2018.

105. Rumbiak, U.; Laim, C.K.; Al Furqan, R.; Rosana, M.; Yuningsih, E.; Tsikoras, B.; Ifandi, E.; Malik Chen, H. Geology, alteration geochemistry, and exploration geochemical mapping of the Erstberg Cu-Au-Mo district in Papua, Indonesia. *J. Geochem. Explor.* **2022**, *232*, 106889. [CrossRef]
106. PT Freeport Indonesia. *Block Cave Underground Mine*; PT Freeport Indonesia: South Jakarta, Indonesia, 2020.
107. Sahupala, H.A.; Szwedzicki, T.; Prasetyo, R. Diameter of a draw zone-a case study from a block caving mine, Deep ore Zone, PT Freeport Indonesia. In *Caving 2018: Proceeding of the Fourth International Symposium on Block and Sublevel Caving, Vancouver, BC, Canada, 15–17 October 2018*; Australian Centre for Geomechanics: Perth, Australia, 2018; pp. 633–644.
108. Ruswanto, M.F.; Fatimah, M.R.; Yuningsih, E.T.; Purwariswanto, B.A. Minegrafi batuan Penyusun Tambang Deep Mill level Zone (DMLZ) PT Freeport Indonesia. *Bull. Sci. Contrib. Geol.* **2017**, *15*, 173–180.
109. Brannon, C.; Beard, D.; Pascoe, N.; Priatna, A. Development of and production update for the Grasberg Block Cave mine–PT Freeport Indonesia. In *MassMin 2020: Proceedings of the Eighth International Conference & Exhibition on Mass Mining-University of Chile, Virtual, 4–8 October 2020*; University of Chile: Santiago, Chile, 2020; pp. 747–760.
110. Beard, D.; Brannon, C. Grasberg Block Cave mine: Cave planning and undercut sequencing. In *Caving 2018: Proceeding of the Fourth International Symposium on Block and Sublevel Caving, Vancouver, BC, Canada, 15–17 October 2018*; Australian Center for Geomechanics: Perth, Australia, 2018; pp. 373–384.
111. De Beer, W.; Jalbout, A.; Riyanto, E.; Ginting ASullivan, M.; Colllins, D.S. The design, optimisation, and use of the seismic system at the deep and high-stress block cave Deep Mill Level Zone mine. *Undergr. Min. Technol.* **2017**, *2017*, 233–246.
112. Widijanto, E.; Wattimena, R.K.; Kramadibrata, S.; Rai, M.A. Geometry effects on Slope Stability at Grasberg Mine Through the Transsition from Open Pit to Underground Block Cave Mining. *Electron. J. Geotech. Eng.* **2017**, *22*, 1733–1745.
113. Meinert, L.D.; Hefton, K.K.; Mayes, D.; Tasiran, I. Geology, zonation, and fluid evolution of the Big Gossan Cu-Au skarn deposit, Ertsberg district, Irian Jaya. *Econ. Geol.* **1997**, *92*, 509–534. [CrossRef]
114. Onederra, I.; Chitombo, G. Design methodology for underground ring blasting. *Min. Technol.* **2007**, *116*, 180–195. [CrossRef]
115. Wattimena, R.K.; Sulistianto, B.; Risono; Matsui, K. A semi-caving method applied at level 600 Ciurug vein, Pongkor underground gold mine, PT Aneka Tambang Tbk. In *Mine Planning and Equipment Selections*; Taylor & Francis Group: London, UK, 2004.
116. Syafrizal, I.A.; Watanabe, K. Origin of Ore-forming Fluids Responsible for Gold Mineralization of the Pongkor Au-Ag Deposit, West Java, Indonesia: Evidence from Mineralogic, Fluid Inclusion Microthermometry and Stable Isotope Study of the Ciurug–Cikoret Veins. *Resour. Geol.* **2007**, *57*, 136–148.
117. Syafrizal, I.A.; Motomura, Y.; Watanabe, K. Characteristics of gold mineralization at the Ciurug vein, Pongkor gold-silver deposit, West Java, Indonesia. *Resour. Geol.* **2005**, *55*, 225–238. [CrossRef]
118. Carlile, J.C.; Davey, G.R.; Kadir, I.; Langmead, R.; Rafferty, W.J. Discovery and exploration of the Gosowong epithermal gold deposit, Halmahera, Indonesia. *J. Geochem. Explor.* **1998**, *60*, 207–227. [CrossRef]
119. Dilles, J.H.; John, D.A. Porphyry and Epithermal Mineral Deposits. In *Encyclopedia of Geology*, 2nd ed.; Alderton, D., Elias, S.A., Eds.; Academic Press: Cambridge, MA, USA, 2021.
120. Hoschke, T.G. Geophysical Signatures of Copper-Gold Porphyry and Epithermal Gold Deposits, and Implications for Exploration. Ph.D. Thesis, University of Tasmania, Hobart, Australia, 2011.
121. Burrows, D.R.; Rennison, M.; Burt., D.; Davies, D. The Onto Cu-Au Discovery, Eastern Sumbawa, Indonesia: A Large, Middle. *Soc. Econ. Geol.* **2020**, *115*, 1385–1412. [CrossRef]
122. PT Bumi Resources Mineral. PT Gorontalo Minerals (Copper & Gold). PT Bumi Resources Minerals Tbk. 2020. Available online: https://www.bumiresourcesminerals.com/pt-gorontalo-minerals-copper-gold/ (accessed on 3 September 2022).
123. Merdeka Copper Gold. wetar Porhyry Project. 2019. Available online: https://merdekacoppergold.com/en/operations/tujuh-bukit-porphyry-project/ (accessed on 1 September 2022).
124. Sutarto, I.A.; Harjoko ASetijadji, L.D.; Meyer, F.M.; Sindern, S. Mineralization style of the Randu Kuning porphyry Cu-Au and intermediate sulphidation epithermal Au-base metals deposits at Selogiri area, Central Java Indonesia. *AIP Conf. Proc.* **2020**, *2245*, 080006.
125. Hu, P.; Cao, L.; Zhang, H.; Yang, Q.; Armin, T.; Cheng, X. Late miocene adakites with the Tangse porphyry Cu-Mo deposit within the Sunda arch, North Sumatera, Indonesia. *Ore Geol. Rev.* **2019**, *111*, 102983. [CrossRef]
126. Li, X.-Y.; Zhang, Z.-W.; Wu, C.-Q.; Xu, J.-H.; Jin, Z.-R. Geology and geochemistry of Gunung Subang gold deposit, Tanggeung, Cianjur, West Java, Indonesia. *Ore Geol. Rev.* **2019**, *113*, 103060. [CrossRef]
127. McCarrol, R.J.; Graham, I.T.; Fountain, R.; Privat, K.; Woodhead, J. The Ojolali region, Sumatera, Indonesia: Epithermal gold-silver mineralisation within the Sunda Arch. *Gondwana Res.* **2014**, *26*, 218–240. [CrossRef]
128. Mulja, T.; Heriawan, M.N.; Supomo, B.D.H. The Miwah high-sulphidation epithermal Au-Ag deposit, Aceh, Indonesia: Geology and spatial relationships of gold with associated metals and structures. *Ore Geol. Rev.* **2020**, *123*, 103564. [CrossRef]
129. PT Merdeka Copper Gold, Tbk. *Quarterly Report: March 2020*; Merdeka Copper Gold: Jakarta, Indonesia, 2020.
130. Walker, S. Block Caving: Mining Specialization. *Eng. Min. J.* **2014**, *215*, 32.
131. Mining Technology. Ridgeway Gold and Copper Mine, New South Wales. 2008. Available online: https://www.mining-technology.com/projects/cadiavalley/ (accessed on 23 March 2022).
132. Mining Weekly. Oz Minerals gives the go-ahead at Carrapateena. Mining Weekly, Creamer Media's. 2022. Available online: https://www.miningweekly.com/article/oz-minerals-gives-the-go-ahead-at-carrapateena-2021-01-29/rep_id:3650/ (accessed on 17 September 2022).

133. Chung, J.; Asad, M.W.A.; Topal, E. Timing of transition from open-pit to underground mining: A simultaneous optimisation model for open-pit and underground mine production schedules. *Resour. Policy* **2022**, *77*, 102632. [CrossRef]
134. Tegachouang, N.C.; Bowa, V.M.; Li, X.; Luo, Y.; Gong, W. Study of the Influence of Block Caving Underground Mining on the Stability of the Overlying Open Pit Mine. *Geotech. Geol. Eng.* **2022**, *40*, 165–173. [CrossRef]
135. Ding, K.; Li, H. Monitoring analysis and environmental assessment of mine subsidence based on surface displacement monitoring. *J. Comput. Methods Sci. Eng.* **2022**, *22*, 399–410. [CrossRef]
136. Xia, Z.Y.; Tan, Z.Y.; Zhang, L. Instability Mechanism of Extraction Structure in Whole Life Cycle in Block Caving Mine. *Geofluids* **2021**, *2021*, 9932932. [CrossRef]
137. Vallejos, J.; Basaure, K.; Palma, S.; Castro, R.L. Methodology for evaluation of mud rush risk in block caving mining. *J. S. Afr. Inst. Min. Metall.* **2017**, *117*, 491–497. [CrossRef]
138. Ajayi, K.; Tukkaraja, P.; Shahbazi, K.; Katzenstein, K.; Loring, D. Computational fluid dynamics study of radon gas migration in a block caving mine. In Proceedings of the 15th North American Mine Ventilation Symposium, Blacksburg, VA, USA, 21–25 June 2015; pp. 341–348.
139. Erb, M.; Mucek, A.E.; Robinson, K. Exploring a social geology approach in eastern Indonesia: What are mining territories? *Extr. Ind. Soc.* **2021**, *8*, 89–103. [CrossRef]
140. Mata-Perello, J.M.; Mata-Lleonart, R.; Vintro-Sanchez, C.; Restrepo-Martinez, C. Social Geology: A new perspective on geology. *Dyna* **2012**, *79*, 158–166.
141. Stewart, I.S.; Gill, J.C. Social geology—Integrating sustainability concepts into Earth sciences. *Proc. Geol. Assoc.* **2017**, *128*, 165–172. [CrossRef]
142. Donaldson, S.N.; Harney, W.; Cox, T. Operational and real-time LHD dispatch at Rio Tinto's Argyle Diamond Mine. In *MassMin 2020: Proceedings of the Eighth International Conference & Exhibition on Mass Mining, Santiago, Chile, 9–12 December 2020*; University of Chile: Santiago, Chile, 2020; pp. 693–702.
143. Fernandez, F.; Evans, P.; Gelosn, R. Design and implementation of a damage assessment system at Argyle Diamond's block cave project. In *Caving 2010*; Potvin, Y., Ed.; Australian Centre for Geomechanics: Perth, Australia, 2010.
144. Glesson, D. OZ Minerals eyes up block cave opportunities at Carrapateena underground mine. Available online: https://im-mining.com/2020/06/23/oz-minerals-eyes-block-cave-opportunities-carrapateena-underground-mine/ (accessed on 5 September 2022).
145. Pitcher, A.; Hocking, R.J. Assessing the risk of a sublevel cave–block cave transition using dynamic decision tree analysis overlain with Monte Carlo analysis. In *Caving 2022, Proceedings of the Fifth International Conference on Block and Sublevel Caving, Adelaide, Australia, 30 August–1 September 2022*; Australian Centre for Geomechanics: Perth, Australia, 2022; pp. 815–824.
146. Poulter, M.; Ormerod, T.; Balog, G.; Cox, D. Geotechnical monitoring of the Carrapateena cave. In *Caving 2022, Proceedings of the Fifth International Conference on Block and Sublevel Caving, Adelaide, Australia, 30 August–1 September 2022*; Potvin, Y., Ed.; Australian Centre for Geomechanics: Perth, Australia, 2022.
147. Tukker, H.; Holder, A.; Swarts, B.; van Strijp, T.; Grobler, E. The CCUT block cave design for Cullinan Diamond Mine. *J. S. Afr. Inst. Min. Metall.* **2016**, *116*, 715. [CrossRef]
148. Brzovic, A. *Characterisation of Primary Copper Ore for Block Caving at The El Teniente Mine, Chile*; Curtin University of Technology: Perth, Australia, 2010.
149. Sieber, M.J.; Brink, F.J.; Leys, C.; King, L.; Henley, R.W. Prograde and retrograde metasomatic reactions in mineralised magnesium-silicate skarn in the Cu-Au Ertsberg East Skarn System, Ertsberg, Papua Province, Indonesia. *Ore Geol. Rev.* **2020**, *125*, 103697. [CrossRef]
150. Mernagh, T.; Leys, C.; Henley, R.W. Fluid inclusion systematics in porphyry copper deposits: The super-giant Grasberg deposit, Indonesia, as a case study. *Ore Geol. Rev.* **2020**, *123*, 103570. [CrossRef]
151. Debswana. Jwaneng Mine Goes Underground. 2022. Available online: http://www.debswana.com/Our-Projects/Pages/Jwaneng_underground.aspx (accessed on 21 April 2022).
152. Dunn, M.J.; Harris, R.O.; Boitshepo, B.; Chiwaye, H.T. Assessing the stress field for the Jwaneng underground project. In *Caving 2022, Proceedings of the Fifth International Conference on Block and Sublevel Caving, Adelaide, Australia, 30 August–1 September 2022*; Potvin, Y., Ed.; Australian Centre for Geomechanics: Perth, Australia, 2022.
153. Stegman, C.L.; Togtokhbayar, O.; Herselman, S.; Altankhuu, B. Oyu Tolgoi: Engineering a Mongolian caving dynasty. In *Caving 2022, Proceedings of the Fifth International Conference on Block and Sublevel Caving, Adelaide, Australia, 30 August–1 September 2022*; Potvin, Y., Ed.; Australian Centre for Geomechanics: Perth, Australia, 2022.
154. Philex Mining Corporation n.d. Philex Mining. *A Brave New Day, 2020 Annual and Sustainability Report*. 2020. Available online: http://www.philexmining.com.ph/zimbabwe/ (accessed on 23 February 2022).
155. Matayev, A.; Kainazarova, A.; Arystan, I.; Abeuov, Y.; Kainazarov, A.; Baizbayev, M.; Demin, V.; Sultanov, M. Research into rock mass geomechanical situation in the zone of stope operations influence at the 10th Anniversary of Kazakhstan's Independence mine. *Min. Miner. Depos.* **2021**, *15*, 103–111. [CrossRef]
156. Ernowo, E.; Meyer, F.M.; Idrus, A. Hydrothermal alteration and gold mineralization of the Awak Mas metasedimentary rock-hosted gold deposit, Sulawesi, Indonesia. *Ore Geol.* **2019**, *113*, 103083. [CrossRef]
157. Meldrum, S.J.; Aquino, R.S.; Gonzales, R.I.; Burke, R.J.; Suyadi, A.; Irianto, B.; Clarke, D.S. The Batu Hijau porphyry copper-gold deposit, Sumbawa Island, Indonesia. *J. Geochem. Explor.* **1994**, *50*, 203–220. [CrossRef]

158. Idrus, A.; Kolb, J.; Meyer, F.M. Chemical Composition of Rock-Forming Minerals in Copper–Gold-Bearing Tonalite Porphyries at the Batu Hijau Deposit, Sumbawa Island, Indonesia: Implications for Crystallization Conditions and Fluorine–Chlorine Fugacity. *Resour. Geol.* **2007**, *57*, 102–113. [CrossRef]
159. Anjarwati, R.; Idrus, A.; Setijadji, L.D. Geology at Beruang Kanan, Central Kalimantan, Indonesia. In *IOP Conference Series: Earth and Environmental Science*; IOP Publishing: Bristol, UK, 2018; Volume 212.
160. Idrus AFadlin Prihatmoko, S.; Warmada, I.W.; Nur, I.; Meyer, F.M. The Metamorphic Rock-Hosted Gold Mineralization At Bombana, Southeast Ulawesi: A New Exploration Target In Indonesia (Mineralisasi Emas Pada Batuan Metamorf Di Bombana, Sulawesi Tenggara: Target Baru Eksplorasi Di Indonesia). *Proc. Sulawesi Miner. Resour. 2011 Semin. MGEI-IAGI* **2011**, *22*, 35–48.
161. Basri; Sakakibara, M.; Sera, K. Current Mercury Exposure from Artisanal and Small-Scale Gold Mining in Bombana, Southeast Sulawesi, Indonesia—Future Significant Health Risks. *Toxics* **2017**, *5*, 7. [CrossRef] [PubMed]
162. Lubis, H.; Prihatmoko, S.; James, L. Bulagidun prospect: A copper, gold and tourmaline bearing porphyry and breccia system in northern Sulawesi, Indonesia. *J. Geochem. Explor.* **1994**, *50*, 257–278. [CrossRef]
163. Angeles, C.A.; Prohatmoko, S.; Walker, J.S. Geology and alteration-mineralization characteristic of the cibaliung epithermal gold deposit, Banten, Indonesia. *Resour. Geol.* **2002**, *52*, 329–339. [CrossRef]
164. Ambarini, E.; Hirnawan, F.; Guntoro, D. Sistem stabilitas lubang bukaan pengembangan dengan menggunakan baut batuan (rockbolt) dan beton tembak (shortcrete) di blok Cikoneng PT Cibaliung Sumberdaya, Kab. Pandeglang, Prov. Banten. *Pros. Tek. Pertamb.* **2015**, *6*, 168–177.
165. Rosana, M.F.; Matsueda, H. Cikidang hydrothermal gold deposit in wetern java, Indonesia. *Resour. Geol.* **2002**, *52*, 341–352. [CrossRef]
166. Davies, A.G.; Cooke, D.R.; Gemmell, J.B.; van Leeuwen, T.; Cesare, P.; Hartshorn, G. Hydrothermal breccias and veins at the Kelian gold mine, Kalimantan, Indonesia: Genesis of a large epithermal gold deposit. *Econ. Geol.* **2008**, *103*, 717–757. [CrossRef]
167. Febrian, I.; Wahyudin, A.; Gunadi, C.; Peterson, D.; Mah, P.; Pakalnis, R. Mining within a weak rock mass-Kencana Underground Mine case study-PT Nusa Halmahera Minerals (Newcrest Mining Ltd.), Indonesia. In Proceedings of the 1st Canada-US Rock Mechanics Symposium, Vancouver, BC, Canada, 27–31 May 2007.
168. Van Leeuwen, T.M.; Taylor, R.; Coote, A.; Longstaffe, F.J. Porphyry molybdenum mineralization in a continental collision setting at Malala, northwest Sulawesi, Indonesia. *J. Geochem. Explor.* **1994**, *50*, 279–315. [CrossRef]
169. Umar, T.W.; Maryanto, Y. Evaluating Stope Design Based on Equivalent Linear Overbreak/Slough (Elos) on Damar Vein Unit Gold Mining Toguraci Underground Mining. In *Prosiding Teknik Pertambangan*; Universitas Islam Bandung: Bandung, Indonesia, 2018.
170. Harrison, R.L.; Maryono, A.; Norris, M.S.; Rohrlach, B.D.; Cooke, D.R.; Thompson, J.M.; Creaser, R.A.; Thiede, D.S. Geochronology of the tumpangpitu porphyry Au-Cu-Mo and high-sulfidation epithermal Au-Ag-Cu deposit: Evidence for pre- and postmineralization diatremes in the Tujuh Bukit District, Southeast Java, Indonesia. *Econ. Geol.* **2018**, *113*, 163–192. [CrossRef]
171. Scotney, M.; Roberts, S.; Herrington, R.J.; Boyce, A.J.; Burgess, R. The development of volcanic hosted massive sulfide and barite-gold orebodies on Wetar Island, Indonesia. In *Mineralium Deposita*; Springer: Berlin/Heidelberg, Germany, 2005; Volume 40, pp. 76–79.

Article

Effects of Aeolian Sand and Water−Cement Ratio on Performance of a Novel Mine Backfill Material

Guodong Li [1,2], Hongzhi Wang [1,2,*], Zhaoxuan Liu [1], Honglin Liu [1,2,*], Haitian Yan [1] and Zenwei Liu [1]

1 College of Geology and Mining Engineering, Xinjiang University, Urumqi 830046, China
2 Key Laboratory of Environmental Protection Mining for Mineral Resources at Universities of Education Department of Xinjiang Uygur Autonomous Region, Xinjiang University, Urumqi 830046, China
* Correspondence: wanghongzhi@xju.edu.cn (H.W.); liuhonglin@xju.edu.cn (H.L.); Tel.: +86-15152111176 (H.W.); Tel.: +86-13999822448 (H.L.)

Abstract: The gob-side entry retaining (GER) technique, as the family member of the pillarless coal mining system, is becoming popular, mainly attributed to its high resource recovery rate and significant environmental benefits. Seeking cost-effective backfill material to develop the roadside backfilling body (RBB) is generally a hot topic for coal operators and scholars. Except for its relatively high cost, the other shortcoming of the widely used high-water backfill material is also obvious when used in arid, semi-arid deserts or Gobi mining areas lacking water. The modified high-water backfill material (MBM) mixed with aeolian sand was recently developed as an alternative to conventional backfill materials. Some critical parameters affecting both the physical and mechanical properties of the MBM, including the amount of the aeolian sand and water-to-powder ratio of the high water-content material, have been experimentally investigated in the present research. Test results showed that the MBM featured high early strength and bearing capability after a large post-peak deformation. In particular, the adjustable setting time of the MBM through changing the amount of sand widens its application in practice. Unlike the high-water backfill material, the MBM is a typical elastoplastic material; the stress-strain curves consist of pore compression, elastic deformation, yielding, and total failure. Note that both the peak and residual strength of the MBM increased as the doping amount of aeolian sand increased, which is probably because of the impacted aeolian sand and the uniform reticular structure of the ettringite in the MBM. Compared with the high-water backfill material, only limited cementitious material and water resources are requested to cast the RBB, which provides more economical and environmental benefits for the application of the GER technique in the arid, semi-arid deserts or the Gobi mining areas.

Keywords: aeolian sand; water-cement ratio; backfill material; compressive strength; micromorphology; gob-side entry retaining

Citation: Li, G.; Wang, H.; Liu, Z.; Liu, H.; Yan, H.; Liu, Z. Effects of Aeolian Sand and Water−Cement Ratio on Performance of a Novel Mine Backfill Material. *Sustainability* **2023**, *15*, 569. https://doi.org/10.3390/su15010569

Academic Editors: Longjun Dong, Yanlin Zhao and Wenxue Chen

Received: 3 November 2022
Revised: 16 December 2022
Accepted: 25 December 2022
Published: 29 December 2022

1. Introduction

Concern about many serious eco-environmental problems caused by coal exploitation is a global problem [1,2]. As a result, much research in recent years has focused on the development of mining techniques that are environmentally friendly [3]. Under the rapid development of the mining industry in recent decades, the continuous and irreparable impacts of contaminated air, degraded soil, polluted water, damaged biodiversity, and geological disaster of mining operations may increase the risks of mine development [4]. The coal reserves in Northwestern China account for about 70% of China's total coal resources [5,6]. However, most coal mines are distributed among these arid and semi-arid regions, where water resources are in short supply, and there is sparse vegetation, a barren ground surface, and ecological vulnerability [7]. As is well-known, aridity and water shortage are the major natural disasters in Northwestern China and the primary causes of ecological vulnerability in this region [8,9]. In recent years, however, large-scale coal mining activities in Western China have further aggravated ecological deterioration [10]. Environmental problems caused by general mining processes mainly included: (1) the influence on people, biota, biodiversity, the atmospheric environment, and surface ecosystems [11–13]; (2) the influence on the soil environment [14], the hydrological system [15], and severe geological disasters. These effects were combined with each other and caused more complicated consequences [11]. Therefore, maximally adapting the coal mining activities to the local ecological environment was an inevitable road towards "green coal mine construction and green mining in Western China".

Much research in recent years has been widely applied to control eco-environmental problems caused by coal exploitation, such as grout injection [16], room mining [17], strip mining [18], backfill mining [19], and pillarless coal mining [20]. The increased favor for the backfill mining method was due to several advantages: high extraction rate, effective control of ground pressure, reduction of ground surface subsidence, and eco-environment improvement of the mining region [21,22]. The backfill mining methods were divided into several types based on the backfill material used, such as hydraulic backfilling, gangue backfilling, and cemented past backfilling [23]. These backfilling methods offered viable pathways to liberate coal resources under the buildings. However, some drawbacks still existed, such as a complex backfilling process, slow backfilling speed, high backfilling cost, and unavailability of backfill materials. Recently, some new backfill materials have been developed, such as flexible formwork concrete, high-water backfill material, and cement mortar. Such novel backfill materials somehow reduce the backfilling cost and expand the application scope of the backfilling technology [24–28], such as roadway backfill mining [3] and pillarless coal mining [20].

The gob-side entry retaining (GER) technique enables pillarless coal mining by using a roadside backfilling body (RBB), which can effectively reduce the roadway-driven ratio, improve the resource recovery rate, and has significant economic and environmental benefits (Figure 1) [25,27]. Flexible formwork concrete and cement mortar have higher strengths but lower vertical deformation capacity. Therefore, such backfill materials hardly meet the requirement for a large deformation capacity of RBB for GER. On the other hand, high-water backfill materials have the benefits of having high early strength, high flowability of single slurry, fast setting of mixed slurry, and high bearing capability after a large post-peak deformation. Besides, such backfill materials were incompressible under triaxial compression and better adapted to the underground goaf environment, which was usually cold, damp, and enclosed [29]. Given the above discussion, high-water backfill materials are believed to be ideal for the RBB.

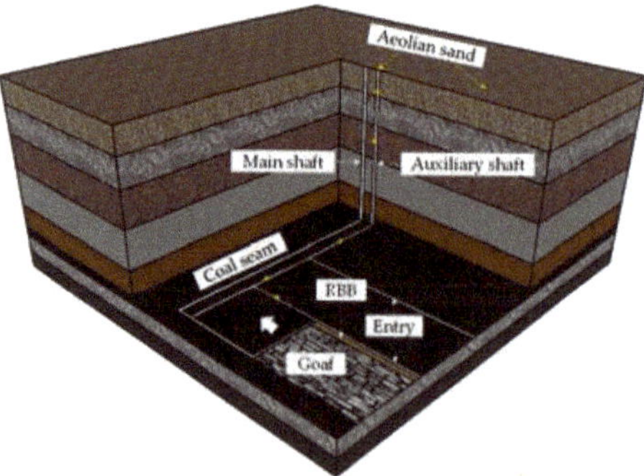

Figure 1. Three-dimensional flowchart of the GER technique.

However, high-water backfill materials have the following defects if used in mining regions in arid and semi-arid deserts or Gobi mining regions: (1) it consumes a large amount of water to prepare high-water backfill materials, which aggravates water shortage [30]; (2) despite the high content of water, the high-water backfill material is necessary in large quantities [31]; (3) due to the low strength of the backfill body, doping massive cementitious materials enhances the strength [32]. As is well known, aeolian sand extensively occurs on the ground surface in Northwestern China. Therefore, it is easily found in nature, is inert and harmless, and causes no pollution to water bodies and the surrounding environment. Moreover, aeolian sand provides a sufficient raw material source to prepare backfill materials for coal mines [33]. Therefore, it is necessary to study an alternative to conventional backfill materials with aeolian sand, which not only liberates coal resources but also limits high-water backfill materials and water resources in the arid, semi-arid deserts or Gobi mining areas.

In the present study, we developed a modified high-water backfill material (MBM) that met the coal exploitation requirements of coal mining in Northwestern China. In the proposed novel MBM, aeolian sand was the primary aggregate and a high-water backfill material was the calcium sulphoaluminate (CSA) based cementitious grout material. Some research has been conducted in past years to understand better the physical and mechanical characteristics of backfill material and their influencing factors, which showed that the physical and mechanical properties were affected by external and internal (e.g., binder type, content, water chemical properties and content, composition, and content of aggregate) factors. Although the influence of these variables has been extensively documented in the literature, the reported results were inconsistent due to large differences in the physical, chemical, and mineralogical properties of cementitious materials and aggregates [23,34–41]. In this study, in order to expand the application scope of the MBM in coal mines located in the arid and semi-arid deserts or Gobi region, we prepared 16 test samples with a diameter of 50 mm and a height of 100 mm. Then we studied the effect of the doping amount of aeolian sand and water−cement ratio on the physical and mechanical characteristics of MBM (e.g., initial setting time, unconfined compressive strength, and its microstructure). The main outcomes of this research may contribute to the safe and green exploitation of underground coal mines in the arid and semi-arid deserts or Gobi.

2. Experimental Scheme

2.1. Test Sample Design

The sample preparation and testing were conducted at Xinjiang University. Firstly, we prepared 16 test samples with a diameter of 50 mm and a height of 100 mm. Next, the water−cement ratio and the doping amount of aeolian sand were changed as the main parameters to discuss the influence of doping aeolian sand on the mechanical properties of the MBM. Then, the samples constituted three series and ten groups. In each group, there were two nominally identical samples.

As shown in Table 1, series 1 consisted of six standard short cylinders with three different water−cement ratios (1.0, 1.5, and 2.0) and the doping amount of aeolian sand being 0%. Samples in series 2 were differentiated from each other by the cementitious grout material's water-cement ratio (1.0, 1.5, and 2.0). The doping amount of aeolian sand was 60% in each sample. To study the influence of the doping amount of aeolian sand, we prepared eight test samples for series 3, each having a water−cement ratio of 1.0, but 4 different doping amounts of aeolian sand (0%, 20%, 40%, and 60%).

We named each test sample following a specific convention. The first number represented the water−cement ratio of the cementitious grout material, the second the doping amount of aeolian sand, and the last was a Roman numeral that differentiated two nominally identical samples. Take U-1.5-00-II as an example. U-1.5-00-II represented the second test sample, with a water−cement ratio of 1.5 in the cementitious grout slurry and the doping amount of aeolian sand being 0%.

Table 1. Details of specimens.

Series	Group	Specimen	Water−Cement Ratio (w/c)	Sand Content (a_s)
1	U-1.0-00	U-1.0-00-I,II	1.0	0%
	U-1.5-00	U-1.5-00-I,II	1.0	0%
	U-2.0-00	U-2.0-00-I,II	1.0	0%
2	U-1.0-60	U-1.0-60-I,II	1.0	60%
	U-1.5-60	U-1.5-60-I,II	1.5	60%
	U-2.0-60	U-2.0-60-I,II	2.0	60%
3	U-1.0-00	U-1.0-00-I,II	1.0	0%
	U-1.0-20	U-1.0-20-I,II	1.0	20%
	U-1.0-40	U-1.0-40-I,II	1.0	40%
	U-1.0-60	U-1.0-60-I,II	1.0	60%

2.2. *Raw Materials*

The CSA cementitious grout material was provided by Yangzhou Zhongkuang Construction New Material Technology Co., Ltd. (Yangzhou City, China). The high-water backfill material was composed of two main ingredients, A and B. It mainly consisted of sulfate aluminum cement, a suspending agent, and a set retarder. B was a mixture of lime, gypsum, a suspending agent, and an early strengthener [42]. The initial setting time was short after combining slurries A and B. Besides, the test sample developed sufficient strength within a short period. Therefore, such materials have the benefits of not causing pipeline blockage and have easy pumpability, high early strength, and environmental friendliness, with extensive application in mine backfilling, leak stoppage, and fire retardancy. The aeolian sand used for experiments came from the Kumtag Desert, one of the main deserts of northwestern China. We conducted grain size distribution on aeolian sand according to ASTMC136/C136M [43]. The cumulative grain size distribution curve (Figure 2) shows that the aeolian sand grain distributions were mainly between 0.075 mm and 0.25 mm. These grain sizes accounted for about 90% of the total mass, while the remaining were 0.25 to 1.0 mm.

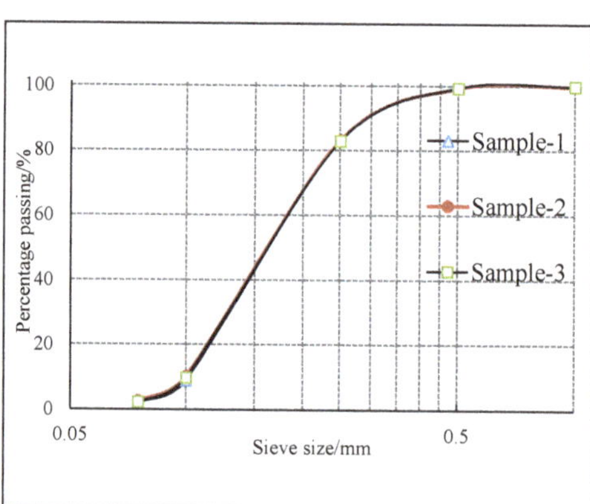

Figure 2. Grain size distribution curves of aeolian sand.

2.3. *Test Sample Preparation*

Next, we prepared the MBM samples (Figure 3). Material slurry A had aeolian sand mixed at a specific ratio. Material B slurry had added water adequately mixed. Finally, slurry A was doped with aeolian sand combined with the slurry B doped with aeolian sand, mixed for 5 min, and was left to stand still. The mixed slurry was poured into a mold and cured for 24 h before demolding. The test sample was covered with a preservative film and placed in an enclosed container. The subsequent tests began after 7 d of maintenance.

(a)

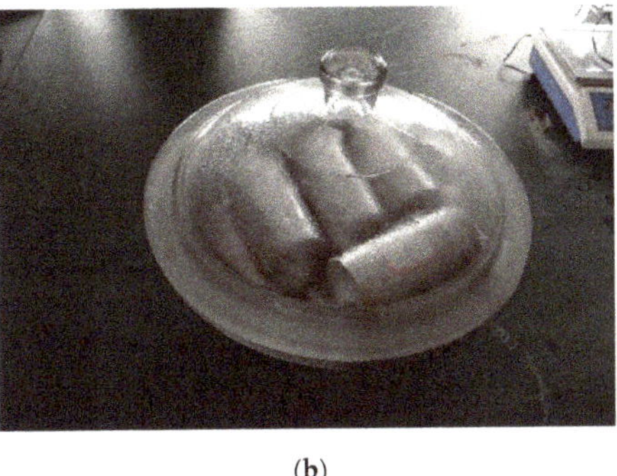

(b)

Figure 3. Preparation of MBM specimens: (**a**) casting MBM; (**b**) curing MBM.

2.4. Testing Equipment and Instruments

Our main discussions were on the influence of doping aeolian sand and the water−cement ratio on the physical and mechanical characteristics (e.g., initial setting time, unconfined compressive strength, and microstructure). The setting time of MBM should meet the requirements for the filling process for GER. If the setting time is too short, pipeline obstruction quickly happens, which will not be conducive to material transport. If the setting time is too long, MBM might not be easily formed in the goaf, thus failing to achieve the backfilling purpose. Therefore, the control of the setting time is of high importance. The Standard Vicat needle test for all test samples was as per GB/T1346-2001 [44]. We took the average initial setting time from two repeat tests on the same sample. When used to backfill the space underground, MBM was usually squeezed between the roof and floor of the coal seam. Therefore, the backfill body was subjected to uniaxial loading. The uniaxial compression test was crucial for the backfill material. We conducted the classical loading tests on MBM using the WAW-600D Hydraulic Universal Testing Machine. As shown in Figure 4, four strain gauges (SGs) were installed at the mid-height of each sample to measure axial and circumferential strains and the overall axial deformations of the samples. According to ASTMD7012-2010 [45], we used a loading rate of 1.5 mm/min for the compression test in all samples under the displacement loading mode. In order to understand the microscopic characteristics, we observed the morphology of MBM using the ZEISS LEO-1430 VP scanning electron microscope (Carl Zeiss AG, Jena, Germany).

(a)

(b)

Figure 4. Test set-up: (**a**) hydraulic universal testing machine; (**b**) layout of strain gauges.

3. Experimental Results and Discussion

3.1. Results of Vicat Needle Test

The standard Vicat needle test results are shown in Figure 5. As shown in Figure 5, doping aeolian sand would affect the initial setting time of MBM. With the doping amount of aeolian sand fixed at 60%, the initial setting times of the samples with the water−cement ratio being 1.0, 1.5, and 2.0 were shortened by 50%, 26%, and 30%, respectively, compared with those not doped with aeolian sand. When the water−cement ratio was fixed at 1.0, the initial setting time of the samples was 43 min, 53 min, 33 min, and 21 min when the doping amount of aeolian sand was 0%, 20%, 40%, and 60%, respectively. It is noteworthy that the initial setting time of the MBM decreased with the increased doping amount of aeolian sand and decreased with the water−cement ratio. It is logical that adjustment of the initial setting time by changing the doping amount of aeolian sand met the technical requirements of the construction of RBB.

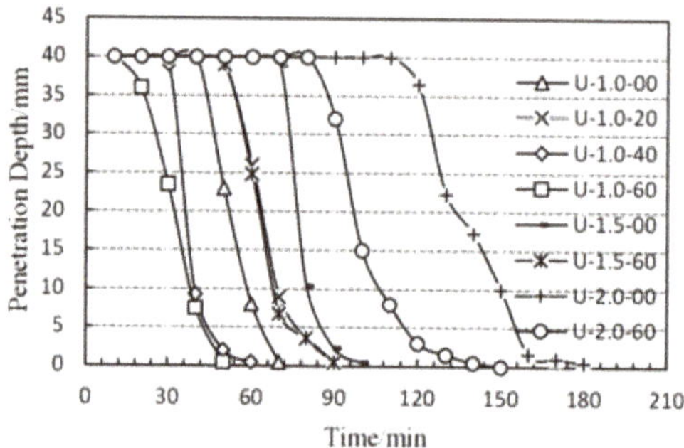

Figure 5. Evaluation of initial setting time and final setting time.

3.2. Failure Modes

The failure modes from the various tests of MBM are in Figure 6. As shown in the figure (Figure 6a,c), for the pure high-water materials (e.g., U-1.0-00, U-1.5-00 and U-2.0-00) and at a small doping amount of aeolian sand (e.g., U-1.0-20), failure first occurred at one end of the sample. Then, the cracks gradually propagated from the middle to the other end of the sample. Thus, a sharp tensile crack was formed, causing the failure of the entire sample. As the doping amount of aeolian sand continued to increase (Figure 6b,c), the sample underwent axial compression and radial expansion due to pressure from above. The friction at the end of the testing machine resulted in a three-way stress zone in an inverted cone shape at the top of the testing machine, where an X-shaped crack zone developed. With sustained axial pressure, a shear fissure intersecting with the axial line appeared in the lower part of the X-shaped crack zone. This fissure became a macroscopic crack, further leading to the sample's shear failure. Interestingly, as the doping amount of aeolian sand increased, the modified samples gradually transitioned from tensile failure to X-shaped shear failure. It seems likely that the failure modes of MBM can be affected by sand content.

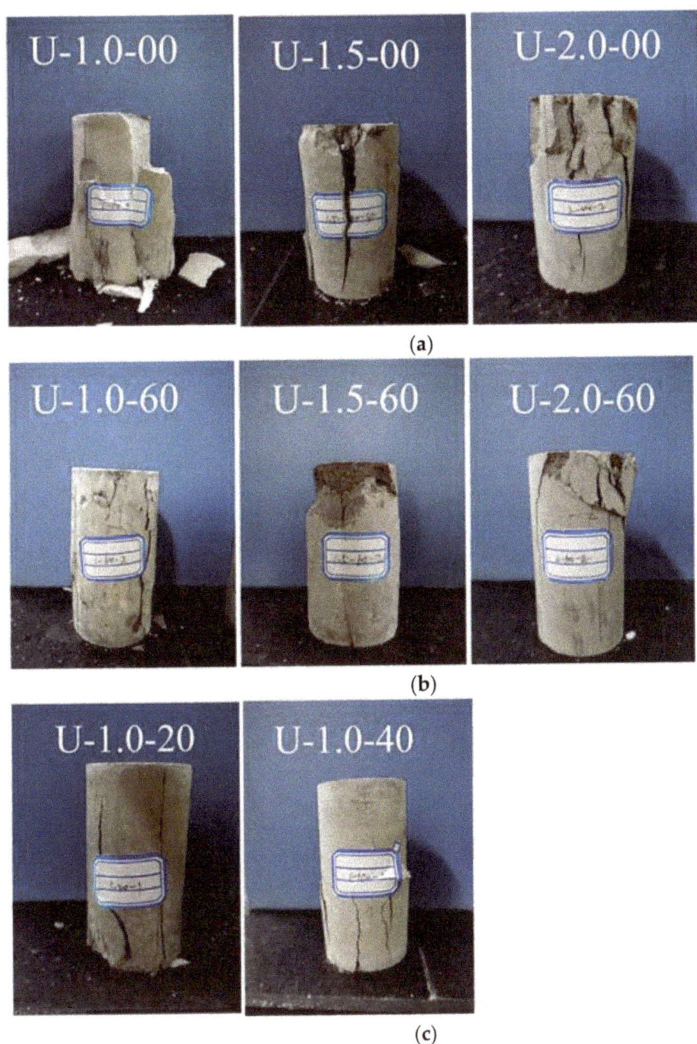

Figure 6. Typical failure modes of MBM specimens: (**a**) a_s = 0% and three different *w/c* (1.0, 1.5, and 2.0); (**b**) a_s = 60% and three different *w/c* (1.0, 1.5, and 2.0); (**c**) *w/c* = 1.0 and different a_s (20% and 40%). Herein, a_s is the sand content, *w/c* is the water-cement ratio.

3.3. Stress-Strain Curves

Using the WAW-600D Hydraulic Universal Testing Machine, we obtained the stress-strain curves of every sample. Figure 7 shows the axial stress–axial strain curves of MBM. Note that MBM is classic elastoplastic. Such material was not only compressible to adapt to surrounding rock deformation but also had residual strength even after failure. These two features are highly desirable for the construction of RBB [46].

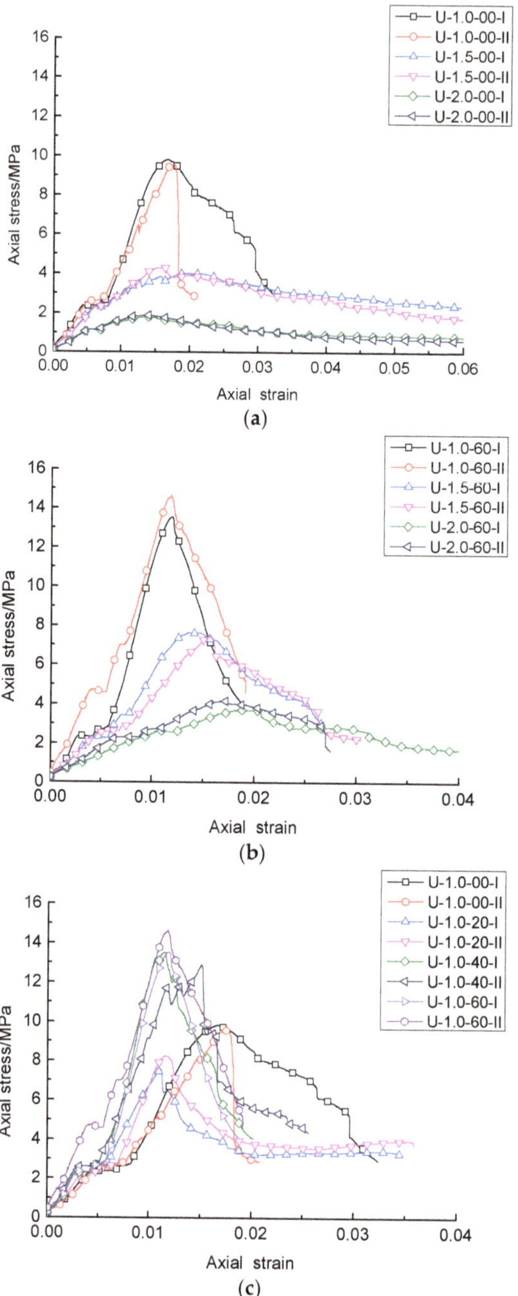

Figure 7. *Cont.*

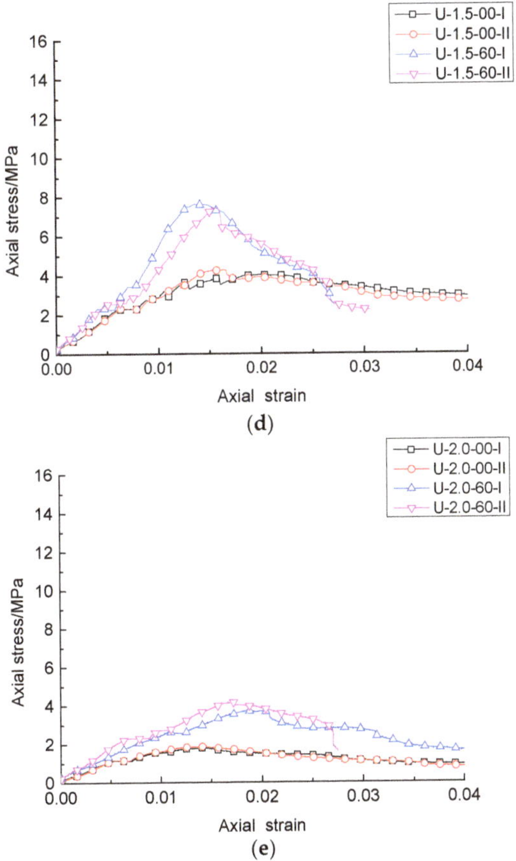

Figure 7. Stress–strain curve of MBM specimens: (**a**) a_s = 0% and three different w/c (1.0, 1.5, and 2.0); (**b**) a_s = 60% and three different w/c (1.0, 1.5, and 2.0); (**c**) w/c = 1.0 and different a_s (20% and 40%); (**d**) w/c = 1.5 and different a_s (0% and 60%); (**e**) w/c = 2.0 and different a_s (0% and 60%). Herein, a_s is the sand content, w/c is the water-cement ratio.

As shown in Figure 7c–e, the doping of aeolian sand dramatically altered the characteristics of the stress−strain curve of MBM. The stress−strain curve was divisible into four stages: pore compression, elastic deformation, yield, and failure. But compared with pure high-water materials (e.g., U-1.0-00, U-1.5-00 and U-2.0-00), MBM specimens display distinctive features at these four stages.

(1) Pore compression stage: This stage was more clearly distinct in the pure high-water materials (e.g., U-1.0-00, U-1.5-00, and U-2.0-00); as the water−cement ratio increased, the stage of pore compression became even more distinct (Figure 7a,b). But for MBM specimens, the scene became less precise as the doping amount of aeolian sand increased (Figure 7c). It would appear that the specimens with a larger water−cement ratio had a more remarkable pore compression stage. At the same time, the higher sand content that led to this stage was not significant. It may be because the doped aeolian sand particles filled the pores between the originally pure high-water material, reducing the number and space of pores.

(2) Elastic deformation stage: This stage became less distinct and had a shorter duration for pure high-water materials (Figure 7a) and a low doping amount of aeolian sand (Figure 7c). However, as the water−cement ratio increased, this stage became less

precise and had a shorter period (Figure 7a,b). In addition, as the doping amount of aeolian sand increased (Figure 7c–e), this stage became more distinct and had a longer duration. This phenomenon mainly dominated in the modified material samples U-1.0-60. It may be assumed that the more considerable amount of aeolian sand doped into the high-water material, the smaller the pores between aggregate particles. The friction between the aggregate particles could further increase due to the cementing effect of the high-water materials. Therefore, the inter-particle dislocation was more unlikely to happen.

(3) Yield stage: An apparent fracture plane appeared in the samples at this stage, and the fracture propagated constantly. At a higher doping amount of aeolian sand (Figure 7b–e), the failure occurred rapidly, resulting in a higher peak and more significant compressive strength on the curve at this stage. However, for pure high-water materials and samples doped with a small amount of aeolian sand (Figure 7a,c), the peak and the compressive strength were smaller on the curve. The above might be because the friction between the aggregate particles was lower due to the larger pores between them.

(4) Failure stage: The MBM displayed significantly different features at this stage. The stress−strain curve showed a more gentle decreasing trend in the pure high-water materials (Figure 7c–e). However, the stress was still high even at the maximum strain, indicating a high residual strength (Figure 7a). Apparently, the decrease was steeper on the stress−strain curve for the MBM (Figure 7b–e). The stress corresponding to the maximum strain was lower in MBM than in pure high-water materials. Besides, this stress decreased as the doping amount of aeolian sand increased (Figure 7c). It would seem that although the post-peak strength decreased slightly, the MBM had high bearing capability after a large post-peak deformation.

3.4. Consumption of Raw Materials

Figure 8 shows the compressive strengths and material consumption for different test samples. As shown in Figure 8a, the strength of the samples negatively correlated with the water−cement ratio when the doping amount of aeolian sand was 0% and 60%. When the doping amount of aeolian sand was above 40%, the strength of the sample positively correlated with the doping amount of aeolian sand, compared with the pure high-water material (e.g., U-1.0-00). Besides, the amounts of cementitious material and water negatively correlated with the doping amount of aeolian sand (Figure 8b). Compared with the pure high-water materials (e.g., U-1.0-00, U-1.5-00, and U-2.0-00), the samples with the water-cement ratio being 1.0, 1.5, and 2.0 and the doping amount of aeolian sand being 60% had an increase in strength of 45%, 83%, and 111% (Figure 8a), respectively; the amounts of cementitious material and water consumed decreased by 60% in all these doped samples. As shown above, the aeolian sand-modified high-water backfill materials had higher compressive strength. Besides, it consumed smaller amounts of cementitious materials and water in preparing these materials. It should be possible, therefore, to reduce cementitious material and water resources requested to cast the RBB, which provided new insight for the application of the GER technique in the arid and semi-arid deserts or Gobi mining areas.

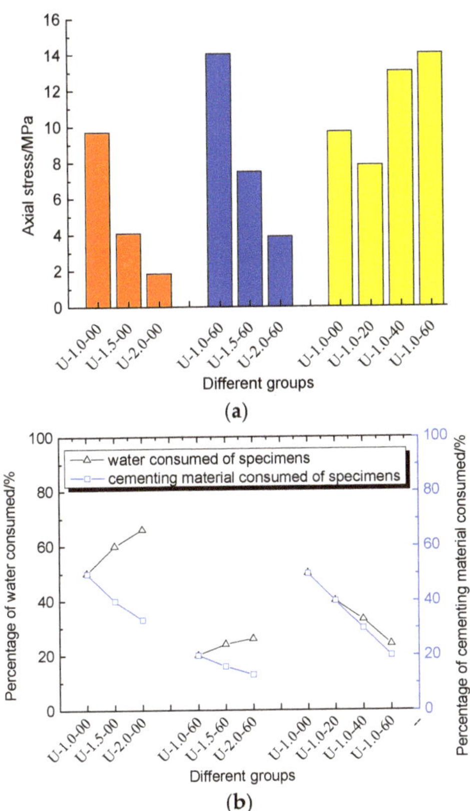

Figure 8. Consumption of raw materials: (**a**) Uniaxial compressive strength of specimens of different groups; (**b**) The percentage of water and cementitious material consumed by specimens of different groups.

4. Micromorphological Analysis

In order to understand the microscopic characteristics, we observed the morphology of MBM using the ZEISS LEO-1430 VP scanning electron microscope (SEM). Figure 9 shows 6000× SEM images of MBM. As shown in Figure 9, when the doping amount of aeolian sand was small (Figure 9a,c), the sample particles were slender and elongated, with a fibrous reticular structure. The fibrous substances were interwoven and overlapped with each other. Moreover, tight connections resulted, though some small cavities still existed. When the doping amount of aeolian sand was 20% (Figure 9c), the sample particles were short and prismatic. This may be the reason for the lower strength of the sample (e.g., U-1.0-20) compared to the pure high-water material (e.g., U-1.0-00). As shown in Figure 9b,c, when the doping amount of aeolian sand increased, the sample particles had a denser fibrous reticulate structure. This was unlikely to form large penetrating cavities. By contrast, the structure with the lower doping amount of aeolian sand was loose and more likely to have large penetrating cavities. Therefore, it seems plausible that the materials with the denser fibrous reticulate structure were stronger than those with the loose structure.

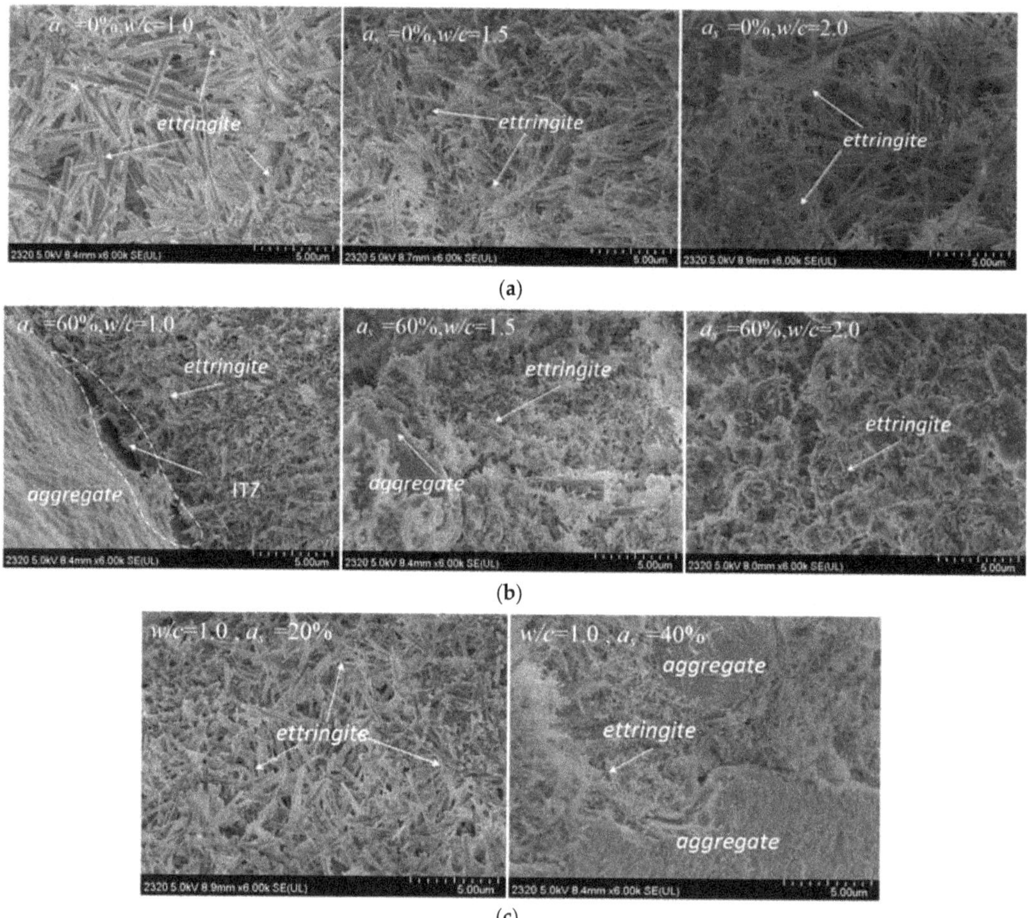

Figure 9. SEM images of MBM specimens: (**a**) a_s = 0% and three different *w/c* (1.0, 1.5, and 2.0); (**b**) a_s = 60% and three different *w/c* (1.0, 1.5, and 2.0) [31]; (**c**) *w/c* = 1.0 and different a_s (20% and 40%). Herein, a_s is the sand content, *w/c* is the water-cement ratio.

5. Conclusions

In this study, we introduced a modified high-water backfill material (MBM) mixed with aeolian sand. To fully understand both physical and mechanical properties of the MBM, a total of 16 samples with a diameter of 50 mm and a height of 100 mm were experimentally tested. Some critical parameters, such as the doping amount of aeolian sand and the water−cement ratio, have been well concerned. The initial setting time, unconfined compressive strength as well as the microstructure of the MBM were investigated in-depth. The main conclusions based on the results and discussions are listed below:

(1) The initial setting time of the MBM decreased with the increased doping amount of aeolian sand and decreased with the water−cement ratio of the high water backfill material. It is thus possible to regulate the setting time of the MBM by changing the doping amount of aeolian sand to meet the technical requirements of the construction of RBB.

(2) The typical stress−strain curve of the MBM consists of four portions: pore compression, elastic deformation, yield, and failure, indicating that the MBM is a typical elastoplastic material.
(3) The MBM consistent with the high-water backfill material has the benefits of having high early strength, high flowability of single slurry, fast setting of mixed slurry, and high bearing capability after a large post-peak deformation.
(4) Both the peak and the residual strength of the MBM increased with the doping amount of aeolian sand within a specific scope, which may be because the existence of the aeolian sand impacted the integrity and uniformity of the reticular structure of ettringite in the MBM.
(5) Compared with the high-water backfill material, only limited cementitious material and water resources are requested to cast the RBB, which provides new insight for the application of the GER technique in arid, semi-arid deserts or Gobi mining areas.

Author Contributions: Conceptualization, G.L., H.W. and H.L.; methodology, G.L. and Z.L. (Zhaoxuan Liu); validation, G.L., H.W. and H.L.; investigation, G.L.; data curation, G.L., Z.L. (Zhaoxuan Liu), H.Y. and Z.L. (Zenwei Liu); writing—original draft, G.L., H.W., Z.L. (Zhaoxuan Liu), H.L., H.Y. and Z.L. (Zenwei Liu); funding acquisition, H.W. and H.L.; project administration, G.L., H.W. and H.L.; writing—review and editing, H.W. and H.L. All authors have read and agreed to the published version of the manuscript.

Funding: This research was funded by the Key Project of Joint Funds of the National Natural Science Foundation of China (U1903209), the Natural Science Foundation of Xinjiang Uyghur Autonomous Region (2022D01E31), the National College Student Innovation Project (202110755008), the Natural Science Foundation of Xinjiang Uyghur Autonomous Region (2022D01A116) and the National Natural Science Foundation (51964043).

Institutional Review Board Statement: Not Applicable.

Informed Consent Statement: Not Applicable.

Data Availability Statement: Not Applicable.

Acknowledgments: The authors are grateful for the technical support and donations provided by Yangzhou Zhongkuang Construction New Material Technology Co., Ltd. (Yangzhou City, China).

Conflicts of Interest: The authors declare no conflict of interest.

References

1. Bai, E.; Guo, W.; Tan, Y. Negative externalities of high-intensity mining and disaster prevention technology in China. *Bull. Eng. Geol. Environ.* **2019**, *78*, 5219–5235. [CrossRef]
2. Xu, Y.; Shen, S.; Cai, Z.; Zhou, G. The state of land subsidence and prediction approaches due to groundwater withdrawal in China. *Nat. Hazards* **2008**, *45*, 123–135. [CrossRef]
3. Bai, E.; Guo, W.; Tan, Y.; Huang, G.; Guo, M.; Ma, Z. Roadway Backfill Mining with Super-High-Water Material to Protect Surface Buildings: A Case Study. *Appl. Sci.* **2019**, *10*, 107. [CrossRef]
4. Dong, L.; Shu, W.; Li, X.; Zhang, J. Quantitative evaluation and case studies of cleaner mining with multiple indexes considering uncertainty factors for phosphorus mines. *J. Clean. Prod.* **2018**, *183*, 319–334. [CrossRef]
5. Dong, S.; Wang, H.; Guo, X.; Zhou, Z. Characteristics of Water Hazards in China's Coal Mines: A Review. *Mine Water Environ.* **2021**, *40*, 325–333. [CrossRef]
6. Wen, Q.; Li, J.; Mwenda, K.M.; Ervin, D.; Chatterjee, M.; Lopez-Carr, D. Coal exploitation and income inequality: Testing the resource curse with econometric analyses of household survey data from northwestern China. *Growth Chang.* **2022**, *53*, 452–469. [CrossRef]
7. Wang, Y.; Chen, M.; Yan, L.; Zhao, Y.; Deng, W. A new method for quantifying threshold water tables in a phreatic aquifer feeding an irrigation district in northwestern China. *Agric. Water Manag.* **2021**, *244*, 106595. [CrossRef]
8. Hu, S.; Ma, R.; Sun, Z.; Ge, M.; Zeng, L.; Huang, F.; Bu, J.; Wang, Z. Determination of the optimal ecological water conveyance volume for vegetation restoration in an arid inland river basin, northwestern China. *Sci. Total Environ.* **2021**, *788*, 147775. [CrossRef]
9. Wang, G.; Xu, Y.; Ren, H. Intelligent and ecological coal mining as well as clean utilization technology in China: Review and prospects. *Int. J. Min. Sci. Technol.* **2019**, *29*, 161–169. [CrossRef]

10. Huang, Y.; Wang, J.; Li, J.; Lu, M.; Guo, Y.; Wu, L.; Wang, Q. Ecological and environmental damage assessment of water resources protection mining in the mining area of Western China. *Ecol. Indic.* **2022**, *139*, 108938. [CrossRef]
11. Dong, L.; Tong, X.; Li, X.; Zhou, J.; Wang, S.; Liu, B. Some developments and new insights of environmental problems and deep mining strategy for cleaner production in mines. *J. Clean. Prod.* **2019**, *210*, 1562–1578. [CrossRef]
12. Dudka, S.; Adriano, D.C. Environmental impacts of metal ore mining and processing: A review. *J. Environ. Qual.* **1997**, *26*, 516–528. [CrossRef]
13. Hilson, G.; Murck, B. Progress toward pollution prevention and waste minimization in the North American gold mining industry. *J. Clean. Prod.* **2001**, *9*, 405–415. [CrossRef]
14. Song, S.; Liang, L.; Zhou, Y.; Wu, H.; Zhang, X. The situation and remedial measures of the cropland polluted by heavy metals from mining along the diaojiang river. *Bull. Mineral. Petrol. Geochem.* **2003**, *22*, 152–155.
15. Zeng, Q.; Shen, L.; Yang, J. Potential impacts of mining of super-thick coal seam on the local environment in arid Eastern Junggar coalfield, Xinjiang region, China. *Environ. Earth Sci.* **2022**, *79*, 88. [CrossRef]
16. Jiang, B.; Oh, K.; Kim, S. Technical evaluation method for physical property changes due to environmental degradation of grout-injection repair materials for water-leakage cracks. *Appl. Sci.* **2019**, *9*, 1740. [CrossRef]
17. Ghasemi, E.; Shahriar, K. A new coal pillars design method in order to enhance safety of the retreat mining in room and pillar mines. *Saf. Sci.* **2012**, *50*, 579–585. [CrossRef]
18. Esterhuizen, G.; Dolinar, D.; Ellenberger, J. Pillar strength in underground stone mines in the United States. *Int. J. Rock Mech. Min. Sci.* **2011**, *48*, 42–50. [CrossRef]
19. Yang, K.; Zhao, X.; Wei, Z.; Zhang, J. Development overview of paste backfill technology in China's coal mines: A review. *Environ. Sci. Pollut. Res.* **2021**, *28*, 67957–67969. [CrossRef]
20. Zhao, H. State-of-the-art of standing supports for gob-side entry retaining technology in China. *J. South. Afr. Inst. Min. Metall.* **2019**, *119*, 891–906. [CrossRef]
21. Chen, F.; Liu, J.; Zhang, X.; Wang, J.; Jiao, H.; Yu, J. Review on the Art of Roof Contacting in Cemented Waste Backfill Technology in a Metal Mine. *Minerals* **2022**, *12*, 721. [CrossRef]
22. Zhu, X.; Guo, G.; Liu, H.; Chen, T.; Yang, X. Experimental research on strata movement characteristics of backfill-strip mining using similar material modeling. *Bull. Eng. Geol. Environ.* **2019**, *78*, 2151–2167. [CrossRef]
23. Zhao, H.; Ren, T.; Remennikov, A. Standing support incorporating FRP and high water-content material for underground space. *Tunn. Undergr. Space Technol.* **2021**, *110*, 103809. [CrossRef]
24. Prum, S.; Jumnongpol, N.; Eamchotchawalit, C.; Kantiwattanakul, P.; Sooksatra, V.; Jarearnsiri, T.; Passananon, S. Guideline for Backfill Material Improvement for Water Supply Pipeline Construction on Bangkok Clay, Thailand. In Proceedings of the 4th World Congress on Civil, Structural, and Environmental Engineering (CSEE'19), Rome, Italy, 7–9 April 2019.
25. Xie, S.; Pan, H.; Chen, D.; Zeng, J.; Song, H.; Cheng, Q.; Xiao, H.; Yan, Z.; Li, Y. Stability analysis of integral load-bearing structure of surrounding rock of gob-side entry retention with flexible concrete formwork. *Tunn. Undergr. Space Technol.* **2020**, *103*, 103492. [CrossRef]
26. Zhao, H.; Ren, T.; Remennikov, A. A hybrid tubular standing support for underground mines: Compressive behaviour. *Int. J. Min. Sci. Technol.* **2021**, *31*, 215–224. [CrossRef]
27. Sun, Q.; Zhang, J.; Zhou, N. Study and discussion of short-strip coal pillar recovery with cemented paste backfill. *Int. J. Rock Mech. Min. Sci.* **2018**, *104*, 147–155. [CrossRef]
28. Yan, B.; Ren, F.; Cai, M.; Qiao, C. Influence of new hydrophobic agent on the mechanical properties of modified cemented paste backfill. *J. Mater. Res. Technol. -JmrT* **2019**, *8*, 5716–5727. [CrossRef]
29. Sun, H.; Liu, W.; Huang, Y.; Yang, B. The use of high-water rapid-solidifying material as backfill binder and its application in metal mines. In Proceedings of the 6th International Symposium on Mining with Backfill, Brisbane, Australia, 14–16 April 1998.
30. Li, H.; Guo, G.; Zhai, S. Mining scheme design for super-high water backfill strip mining under buildings: A Chinese case study. *Environ. Earth Sci.* **2016**, *75*, 1017. [CrossRef]
31. Li, G.; Liu, H.; Deng, W.; Wang, H.; Yan, H. Behavior of CFRP-Confined Sand-Based Material Columns under Axial Compression. *Polymers* **2021**, *13*, 3994. [CrossRef]
32. Zhang, Z.; Deng, M.; Bai, J.; Yan, S.; Yu, X. Stability control of gob-side entry retained under the gob with close distance coal seams. *Int. J. Min. Sci. Technol.* **2021**, *31*, 321–332. [CrossRef]
33. Wu, Y.; Gong, P.; Liu, Q.; Chappell, A. Retrieving photometric properties of desert surfaces in China using the Hapke model and MISR data. *Remote Sens. Environ.* **2009**, *113*, 213–223. [CrossRef]
34. Zhao, Y.; Taheri, A.; Karakus, M.; Chen, Z.; Deng, A. Effects of water content, water type and temperature on the rheological behaviour of slag-cement and fly ash-cement paste backfill. *Int. J. Min. Sci. Technol.* **2020**, *30*, 271–278. [CrossRef]
35. Kwak, M.; James, D.; Klein, K. Flow behaviour of tailings paste for surface disposal. *Int. J. Miner. Process.* **2005**, *77*, 139–153. [CrossRef]
36. Wu, D.; Fall, M.; Cai, S. Coupling temperature, cement hydration and rheological behaviour of fresh cemented paste backfill. *Miner. Eng.* **2013**, *42*, 76–87. [CrossRef]
37. Wu, D.; Cai, S.; Huang, G. Coupled effect of cement hydration and temperature on rheological properties of fresh cemented tailings backfill slurry. *Trans. Nonferrous Met. Soc. China* **2014**, *24*, 2954–2963. [CrossRef]

38. Paterson, A. Pipeline transport of high density slurries: A historical review of past mistakes, lessons learned and current technologies. *Min. Technol. Trans. Inst. Min. Metall.* **2012**, *12*, 37–45. [CrossRef]
39. Wu, A.; Wang, Y.; Wang, H.; Yin, S.; Miao, X. Coupled effects of cement type and water quality on the properties of cemented paste backfill. *Int. J. Miner. Process.* **2015**, *143*, 65–71. [CrossRef]
40. Huynh, L.; Beattie, D.; Fornasiero, D.; Ralston, J. Effect of polyphosphate and naphthalene sulfonate formaldehyde condensate on the rheological properties of dewatered tailings and cemented paste backfill. *Miner. Eng.* **2006**, *19*, 28–36. [CrossRef]
41. Qian, Y.; Kawashima, S. Distinguishing dynamic and static yield stress of fresh cement mortars through thixotropy. *Cem. Concr. Compos.* **2018**, *86*, 288–296. [CrossRef]
42. Song, C.; Chen, K.; Wang, H. The Mechanism of hydration and hardening reaction of high-water material. *Bull. Mineral. Petrol. Geochem.* **1999**, *18*, 261–263.
43. *ASTMC136/C136M*; Standard Test Method for Sieve Analysis of Fine and Coarse Aggregates. American Society for Testing and Materials: West Conshohocken, PA, USA, 2014.
44. *GB/T1346-2001*; Test Methods for Water Requirement of Normal Consistency, Setting Time and Soundness of the Portland Cement. The Standards Press of China: Beijing, China, 2001.
45. *ASTMD7012-2010*; Standard Test Method for Compressive Strength and Elastic Moduli of Intact Rock Core Specimens under Varying States of Stress and Temperatures. American Society for Testing and Materials: West Conshohocken, PA, USA, 2010.
46. Zhang, Z.; Bai, J.; Chen, Y.; Yan, S. An innovative approach for gob-side entry retaining in highly gassy fully-mechanized long wall top-coal caving. *Int. J. Rock Mech. Min. Sci.* **2015**, *8*, 1–11. [CrossRef]

MDPI
St. Alban-Anlage 66
4052 Basel
Switzerland
Tel. +41 61 683 77 34
Fax +41 61 302 89 18
www.mdpi.com

MDPI Books Editorial Office
E-mail: books@mdpi.com
www.mdpi.com/books

www.ingramcontent.com/pod-product-compliance
Lightning Source LLC
LaVergne TN
LVHW071618170726
843515LV00009B/2426

* 9 7 8 3 0 3 6 5 6 3 9 8 5 *